NATURAL GAS HYDRATE

IN OCEANIC AND PERMAFROST ENVIRONMENTS

Coastal Systems and Continental Margins

VOLUME 5

Natural Gas Hydrate

In Oceanic and Permafrost Environments

Edited by

Michael D. Max

MDS Research,
Suite 302, 1211 Connecticut Avenue NW,
Washington DC, U.S.A.

KLUWER ACADEMIC PUBLISHERS
DORDRECHT / BOSTON / LONDON

A C.I.P. Catalogue record for this book is available from the Library of Congress.

ISBN 0-7923-6606-9

Published by Kluwer Academic Publishers,
P.O. Box 17, 3300 AA Dordrecht, The Netherlands.

Sold and distributed in North, Central and South America
by Kluwer Academic Publishers,
101 Philip Drive, Norwell, MA 02061, U.S.A.

In all other countries, sold and distributed
by Kluwer Academic Publishers,
P.O. Box 322, 3300 AH Dordrecht, The Netherlands.

Printed on acid-free paper

Printed in the Netherlands.

This book is dedicated to **Rodney Malone**, a friend to hydrate enthusiasts and science, who established the first National Gas Hydrate Research Program.

TABLE OF CONTENTS

Preface

1. THE BEGINNINGS OF HYDRATE RESEARCH

Until very recently, our understanding of hydrate in the natural environment and its impact on seafloor stability, its importance as a sequester of methane, and its potential as an important mechanism in the Earth's climate change system, was masked by our lack of appreciation of the vastness of the hydrate resource. Only a few publications on naturally occurring hydrate existed prior to 1975. The first published reference to oceanic gas hydrate (Bryan and Markl, 1966) and the first publication in the scientific literature (Stoll, et al., 1971) show how recently it has been since the topic of naturally occurring hydrate has been raised.

Recently, however, the number of hydrate publications has increased substantially, reflecting increased research into hydrate topics and the initiation of funding to support the researchers. Awareness of the existence of naturally occurring gas hydrate now has spread beyond the few scientific enthusiasts who pursued knowledge about the elusive hydrate because of simple interest and lurking suspicions that hydrate would prove to be an important topic.

The first national conference on gas hydrate in the U.S. was held as recently as April, 1991 at the U.S. National Center of the U.S. Geological Survey in Reston Virginia (Max et al., 1991). The meeting was co-hosted by the U.S. Geological Survey, the Naval Research Laboratory, and the U.S. Department of Energy. Two subsequent U.S. national meetings on the topic of gas hydrate have been held and a number of international scientific meetings also have been held; principally a major meeting in Ghent, Belgium in June, 1996 (Henriet and Meinert, 1997).

The first hydrate initiative conceived of as a broad national hydrate research program that had as its main practical aim the economic extraction of methane from hydrate, was established and funded by the National Energy Technology Laboratory (Morgantown WV) of the U.S. Department of Energy in 1982. A broad range of very important hydrate data was produced under the direction of the Program Manager, Rodney Malone. For the first time, the interrelated nature of various hydrate issues yielded data and new modeling which have proved to be the starting point of other national hydrate research programs.

In 1995, the government of Japan (Chapter 18) established a five-year hydrate research program. This program was well funded and involved assessment of Japanese hydrate resources and engineering and experimental work dealing with production issues (Chapter 5). A second five year hydrate research program is now underway (Chapter 18). This successor program is organized differently, with greater weight being given to gas distribution infrastructure implications and to industrial concerns associated with an anticipated indigenous supply of methane from hydrates in the Japanese sea area.

In 1996, the government of India (Chapter 17) also established a national gas hydrate research program under the direction of the Gas Authority of India Ltd (GAIL). The initial assessment included a very large sea area and sought out international consultants to review and reprocess existing data. Following recent changes in the Indian government, the Indian hydrate program was reorganized in 1999 .

As more information became available, the early estimates of very large volumes of hydrate have become generally confirmed (Chapters 2, 10), interest has also grown in the U.S. On September 30, 1997, the President's Committee of Advisors on Science and Technology recommended that hydrate may constitute a significant part of the U.S. national energy base and that a U.S. gas hydrate research program be established. (PCAST, 1997). On July 17, 1998, Senate bill S.1418, "To promote the research, identification, assessment, exploration, and development of methane hydrate resources and for other purposes", was passed in the U.S. Senate and completed hearings on September 15, 1998 before being sent to the House of Representatives. The final version of the 'Methane Hydrate Research and Development Act of 2000' has now (*April, 2000) been passed by the U.S. House of Representatives and after its passing by the U.S. Senate will be signed into law by the President. This act will empower the U.S. Department of Energy, working closely with the Department of Defense through the Secretary of the Navy, and the U.S. Geological Survey, and NOAA, to establish and implement a new national hydrate research program.

Recent interest in hydrate also has developed in the European Union, and individually some of its member countries, the Republic of South Korea, China (Ministry of Geology and Mineral Resources) and Taiwan, amongst others. The primary interest is the potential of hydrate to supply large quantities of methane to commercial energy markets in this century and beyond. However, other aspects of hydrate (Part 5, Chapters 10, 11 & 12, etc.) are also important.

2. ORGANIZATION OF THIS BOOK

Methane hydrate is the dominant species of naturally occurring gas hydrate on Earth. It appears to occur in huge quantities in the seafloor and in permafrost regions, but to date the methane resource in hydrate has only begun to be quantified and assessed. We are only beginning to understand how hydrate functions in the methane-carbon dioxide system of the biosphere and how it interacts with natural changes of pressure and temperature to shape the seafloor. The hydrate system is an important boundary mechanism between geological and ocean-atmosphere components of the Earth's biosphere.

The physical chemistry of methane and other gas hydrates is well dealt with in other books, particularly Dendy Sloan's Clathrate Hydrates of Natural Gases (Sloan, 1998). A number of areas in which hydrate has been identified, such as the Black Sea, the eastern Mediterranean Sea, some areas in the central Pacific, the Barents Sea, and the Bering Sea are not considered individually. This book does not intend to be a compendium of all hydrate localities but rather to introduce hydrate as a topic and to provide a broad introduction to naturally occurring methane hydrate.

This book mainly summarizes the present state of methane hydrate science rather than trying to present new results, as would be the case in a special publication of scientific papers. Nonetheless, contributors have included a number of new observations and insights, and in some cases, new data.

The form of the book is a series of chapters from contributors who are experts in hydrate science. Each contributor or group of contributors has been active in hydrate research in their own right. The contributors to this book have authored a significant proportion of existing publications in the field of gas hydrates.

Nine parts or topic areas relating to hydrate have been identified in an attempt to systematize description and explanation of the still imperfectly known science of gas hydrate:

Part 1. Hydrate as a material and its discovery. Methane hydrate is described as part of a the larger chemical system of container compounds. The salient features of methane hydrate as a naturally occurring material are outlined as is the history its history of discovery.

Part 2. Physical character of Natural Gas Hydrate. The physical chemistry and thermal state of methane hydrate in the natural environment are summarized.

Part 3. Oceanic and Permafrost-Related natural Gas Hydrate. Methane hydrate occurs in two general environments; in near surface sediments in permafrost regions (including now-flooded continental shelves containing permafrost hydrate), and in a deep water marine environment where it has been recognized mainly on upper and middle water depth continental slopes. This oceanic hydrate contains up to 95% of all naturally occurring hydrate. This part describes and contrasts hydrate in these different environments.

Part 4. Source of methane and its migration. The sources of methane, its migration and mechanisms of concentration within the seafloor are described. The functioning of the hydrate system, which alternatively sequesters and releases methane in response to environmental changes may affect a wide range of geobiological, chemical, oceanographic, and climatological issues.

Part 5. Major Hydrate-related Issues Hydrate appears to be important to three main issue-areas, 1. Their potential as a source of combustible energy, 2. Their impact upon global climate, and 3. Their impact upon seafloor stability and continental margin geological processes. Each of these is sufficiently important to justify close study of hydrate and the methane hydrate system.

Part 6. Some Examples of Natural Gas Hydrate Localities. The distribution of hydrate is as yet imperfectly known. The best known oceanic hydrate locations are described here. These locations may not prove to be the most important or the best known in relatively few years.

Mechanisms for concentrating methane and hydrate may differ substantially on active (collisional) and passive continental margins. In collisional margins sediment is thickened in accretionary complexes in which concentration of methane in hydrate may be facilitated not only by the prevalence of gas and fluid migration pathways caused by stresses and strain mechanisms, but by tectonic elevation of hydrate. In contrast, passive margins in which hydrate is found are characterized by tensional structures associated with establishment of a new plate margins and by successively younger being deposited primarily in sediment drapes without tectonism.

A number of examples of both active and passive margin hydrate deposits are described.

Part 7. How we see hydrate. Reflection and refraction seismic analysis is the principal tool for identifying the presence of hydrate (Chapter 15 for a brief description of electrical methods). The only sure inspection and quantification tool is drilling, direct sampling, and sample analysis. Interpretation of wide area survey methods such as seismics, must be controlled by direct observation.

Part 8. Laboratory studies of gas hydrates. Sea-going marine research is expensive. Considerable information can be derived from artificial hydrate formed in laboratory apparatus. Testing of various hydrate properties should result in data that will allow remote survey data, such as multi-channel seismics to be calibrated so that the volume and manner of hydrate formation can be deduced without extensive drilling. This part deals with two hydrate fabrication laboratories and their experimental methods and results.

Part 9. The Promise of Hydrate. Two aspects of the future of hydrate are presented here. The basis for carrying out hydrate research and learning to extract methane from hydrate is contrasted with a commercial analysis of the likelihood and timing of hydrate exploitation.

Acknowledgements. I would like to thank Rodney Malone and Hugh Guthrie of the Federal Energy Technology Center (FETC), Morgantown WV (U.S. Department of Energy), colleagues at the Naval Research Laboratory, and hydrate enthusiasts elsewhere. I especially thank the contributors to this book.

Michael D. Max
MDS Research, Suite 302, 1211 Connecticut Ave. NW
Washington DC 20036

Chapter 1

Introduction, Physical Properties, and Natural Occurrences of Hydrate

Robert E. Pellenbarg
Chemistry Division
Naval Research Laboratory
Washington DC 20375

Michael D. Max
MDS Research
Suite 302, 1211 Connecticut Ave. NW
Washington DC 20036

1. INTRODUCTION

In the early 1820's, John Faraday, working in England, was investigating the newly discovered gas, chlorine. He easily repeated the earlier experiments of Humphrey Davy (Davy, 1811) in which gaseous chlorine and water formed solid chlorine hydrate upon cooling in the "- late cold weather -". Faraday's lab curiosity chlorine hydrate has water as the host molecule, and chlorine molecules as the guest. These pioneering syntheses experiments are the first reported reference to a class of associative compounds now known as gas hydrates (Faraday, 1823, wvusd. 2000). Chlorine hydrate has persisted as a laboratory curiosity (Pauling et al., 1994) in part because its ease of formation lends it to laboratory demonstration. A variety of other molecules which can form hydrates specifically and clathrates in general. The non-bonding uniqueness of clathrates as "chemicals" has interested scientists for almost two centuries.

A clathrate is a compound formed by the inclusion of molecules of one kind within cavities in the crystal lattice of another (Webster, 1994). The generic name, clathrate, is taken from the Latin word 'clathratus', which means, 'enclosed by bars or grating' (Barrer and Stuart, 1957; Brown, 1962). Clathrates display no chemical bonding between the host and guest molecules, a condition which is a key characteristic of clathrates. Some clathrates can form without water being present where non-water molecules form the molecular structural array. There are many examples of clathrates (Table 1), which are also known as container compounds (Cram, 1992).

M.D. Max (ed.), Natural Gas Hydrate in Oceanic and Permafrost Environments, 1–8.

Conventional chemical wisdom holds that a chemical compound consists of atoms bonded to one another in a fixed ratio, yielding an atomic structure for the molecule. Thus, salt, or sodium chloride is NaCl, hexane is C_6H_{14}, or a chain of 6 carbon atoms with 14 hydrogen atoms attached. Either ionic (as in sodium chloride) or covalent (as in hexane) bonds serve to hold these and other molecular entities together, either as gases, liquids, or solids. Ordering on the molecular level is manifested by a crystal form, a material characteristic in all compounds save certain supercooled fluids such as glass. Some crystals of simple compounds, such as salt, are dense solids, and consist essentially of a series of spheres laid adjacent to one another in a three dimensional array. However, a variety of molecules will crystallize under certain conditions of temperature and pressure to give a rather open structure. The molecules of the clathrate-forming substances in question form an open lattice in the form of a rigid three-dimensional structure, rather like balled-up chicken wire, with open space (voids) regularly distributed within the lattice.

Host	Guest
Urea	Straight chain hydrocarbons
Thiourea	Branched chain and cyclic hydrocarbons
Dinitrodiphenyl	Derivatives of diphenyl
Phenol	Hydrogen chloride, sulfur dioxide, acetylene
Water (ice)	Halogens, noble gases, sulfur hexaflouride, low molecular weight hydrocarbons, CO_2, SO_3, N_2, etc.
Nickel dicyanobenzene	Benzene, chloroform
Clay minerals (molecular sieves)	Hydrophilic substances
Zeolites	Wide range of adsorbed substances
Graphite	Oxygen, hydrocarbons, alkali metals (in sheet-like cavities and buckyballs)
Cellulose	Water, hydrocarbons, dyes, iodine

Table 1. Common clathrates: Various hosts and guests.

Often the open space in the crystal lattice remains just that: open space. For example, many silicate minerals (especially zeolites) crystallize into solids with linear channels, or planar sheets of open volume in the crystal. However, under certain conditions, these crystalline voids can be occupied by foreign (guest) molecules of such size and configuration that the guest molecule fits into the crystalline voids formed by the host molecule. Since a host lattice has a well defined structure with a fixed void volume in the lattice, a clathrate can exhibit a definite formula, if the voids are completely occupied by guest molecules, quite analogous to that of a true chemical (bonded) compound. This state of affairs, that is, a combination of atoms with a fixed ratio in the mix implies a strong chemical bonding, a state which commonly does not exist in the case of clathrates.

Clathrates can form spontaneously under certain pressure temperature conditions. A host material, which can crystallize into an open lattice structure, is first needed, then a guest molecule of suitable size and molecular conformation to fit into the lattice voids is required to complete the clathrate crystalline structure. As an example, methane hydrates are dependant on pressures that are usually in excess of one atmosphere (STP), which forces a re-ordering of the lattice water molecules into their 3-dimensional array and inserts the guest methane molecule into the structure.

2. METHANE HYDRATE, A CLATHRATE SUBSPECIES

Hydrates are a subgroup of clathrates. The term "hydrate" is applied to clathrates in which the structural molecules are water (H_2O) and guest sites are occupied by gas molecules. Water, of course, is ubiquitous on Earth, and gas hydrate can form easily wherever conditions are suitable. Only methane hydrates are discussed here and elsewhere in this book. For information on other hydrates, see Sloan (1998).

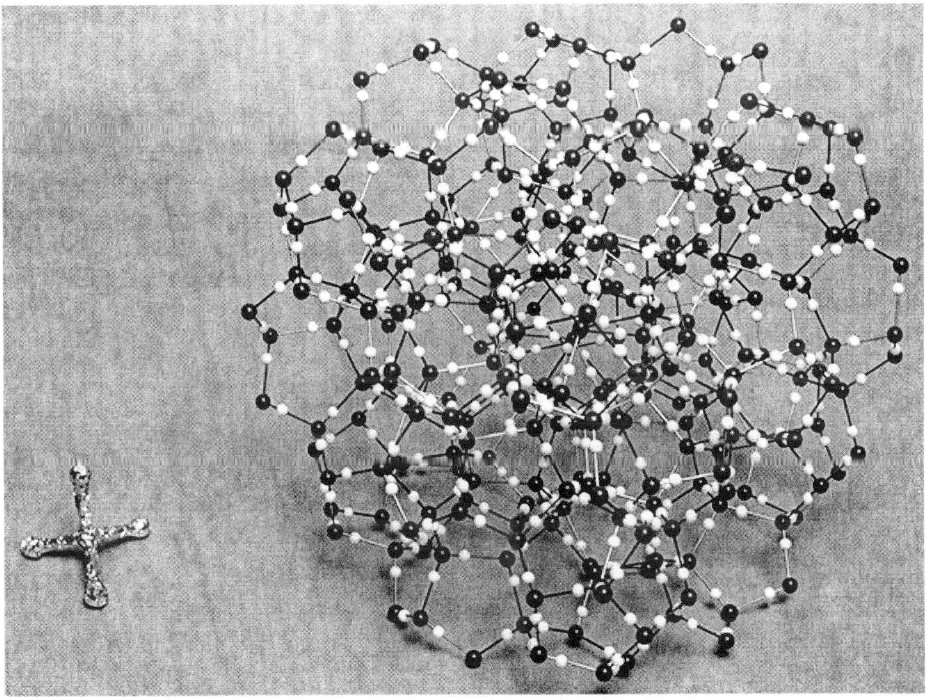

Figure 1. Photo of model of Type 1 hydrate lattice (atomic models, NRL Chemistry Division) showing water cage and adjacent methane molecule (atomic radii not to scale but size of lattice voids and guest methane molecule are to proportional scale.).

Crystallization forces methane molecules into tightly packed lattice sites, compressing the methane. Methane hydrate has the highest energy density of any naturally occurring form of methane (184,000 btu/ft^3 for the hydrate, and 1,150 btu/ft^3 for methane gas; in contrast LNG, which is an cryogenic industrial liquid form of methane, is about 430,000 btu/ft^3). Clearly, methane hydrate is an attractive economic target as a source of methane (i.e., energy), especially when it occurs relatively close to the Earth's or seabed surface.

The density of methane hydrate is about 0.9 g/cm^3; density may vary minutely according to the degree of methane saturation of the hydrate lattice and the local incorporation of other molecules (e.g., H_2S) taking the place of methane in the lattice. Note, however, that the molecular pressure of gas in the hydrate lattice can reach several kilobar with increasing saturation. The heat of hydrate formation and the heat of hydrate dissociation are equal in absolute magnitude but are of opposite sign. When hydrate forms heat is released from the system (exothermic) and when hydrates dissociate, heat is taken into the system (endothermic).

Property	Ice	Hydrate
Dielectric constant at 273 °K	94	~58
NMR rigid lattice 2nd moment of H2O protons(G2)	32	33 ± 2
Water molecule reorientation time at 273 °K (μsec)	21	~10
Diffusional jump time of water molecules at 273 °K (μsec)	2.7	>200
Isothermal Young's modulus at 268 °K (109Pa)	9.5	~8.4
Speed of longitudinal sound at 273 °K		
Velocity (km/sec)	3.8	80
Transit time (μsec/ft)	3.3	92
Velocity ratio Vp/Vs at 272 °K	1.88	1.95
Poisson's ratio	0.33	~0.33
Bulk modulus (272 °K)	8.8	5.6
Shear modulus (272 °K)	3.9	2.4
Bulk density (gm/cm3)	0.916	0.912
Adiabatic bulk compressibility at 273 °K 10-11Pa	12	=14
Thermal conductivity at 263 °K (W/m-K)	2.25	0.49±0.02
Heat of fusion (kJ/mol)	6	54 (meas), 57 (calc)

Table 2. Physical properties of water ice and methane hydrate (after Davidson, 1983; Prensky, 1995; Sloan, 1997 & Franks, 1973)

A nominal value for methane hydrate formation enthalpy at 273 °K is 54 kJ/mol (measured, Sloan, 1990; 1997). Hydrates have a constant pressure heat capacity of 257 kJ/mol (Handa, 1986). The heat of solution (absorption) of methane gas is 13.26 kJ/mol (Franks and Reid, 1973). The thermal conductivity of a hydrate-sediment mixture is 2.2-2.8 W/m-°K (Watt per meter-degree Kelvin) range (Sloan, 1997). For comparison the conductivity of a water-ice sediment mixture is 4.7-5.8 W/m-°K (Table 2).

Conductivity for a gas-sediment mixture, such as exists in natural gas pools is very low, in the 0.05-0.4 W/m-°K . Because hydrate is rarely developed uniformly and in a solid mass in nature, this high conductivity will rarely be attained. Taking into account physical heat transport by geothermal circulation cells, some free gas and a rapidly varying porosity and permeability, 1.5 W/m-°K, about 50% higher than for tight sediment with no hydrate, is a reasonable working estimate for modeling thermal conductivity where hydrate is widely developed in the pore space of sediments.

Methane hydrate occurs both in permafrost areas and in the marine sediment in the oceans and deep lakes where pressure-temperature conditions are suitable (Fig. 2), and where sufficient methane is delivered to the zone of hydrate stability in the uppermost sediments. Hydrate, especially in marine sediments, is composed of methane that has been produced largely by biogenic activity at relatively low temperatures and pressures (Chapters 7 & 8) and not through the same processes that produced most conventional gas and oil deposits (Max and Lowrie, 1993). This renders the methane in oceanic hydrate deposits virtually free from liquid petroleum and condensates and allows them to be considered as having a very low potential for pollution hazard.

Hydrate occurs in a relatively narrow zone termed the hydrate stability zone (HSZ or GHSZ) that lies about parallel to the terrestrial or seabed surface both in permafrost regions and in the oceans. In permafrost terrane, the extraordinary cold of the surface layers hydrate stabilizes relatively close to the surface (Chapters 2 & 5). In the oceans (Chapters 2 & 6), the pressure exerted by water stabilizes the hydrate from the surface downward to some depth determined by increasing temperature related to the geothermal gradient (temperature usually increases downward in the range of about 3 to 4 degrees C per 100 meters in continental slope sediments).

Hydrate in the HSZ bears a striking resemblance to well-known strata-bound mineral deposits such as lead-zinc deposits that are found in Carboniferous limestone host rocks (Max, 1997; Max and Chandra, 1998). Hydrate appears to be deposited slowly from groundwater. Hydrate forms in primary porosity sites and may cement sediment grains. Hydrate also forms in secondary porosity regimes created both by faulting and gas-charged fluid movement in finer grained sediments. Hydrate can occur as massive deposits, in fracture fill, and is often recovered in cores in apparent nodules and disseminated both randomly and along specific horizons where its development was apparently controlled by bedding porosity.

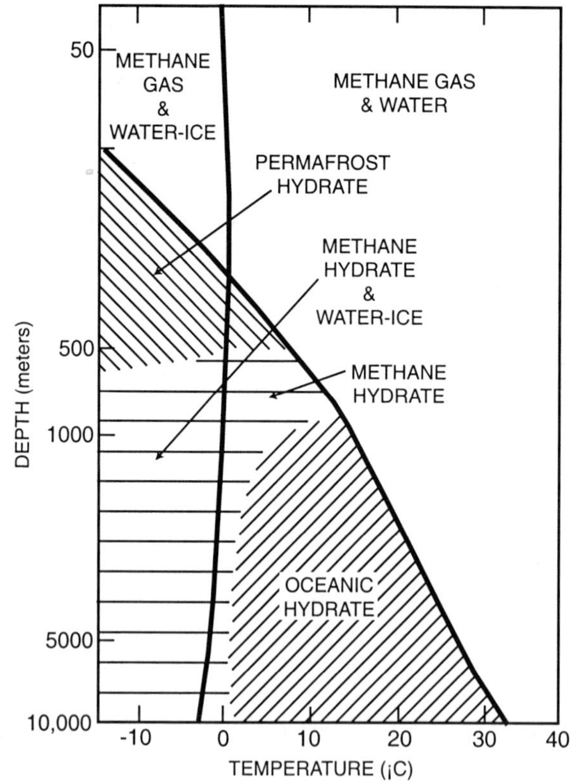

Figure 2. Stability fields of methane hydrate. Log plot carried to 10 km. Gas hydrate phase diagram, showing the stability fields of the water ice-methane-hydrate system with respect to temperature and total pressure and no effect from geothermal heat (dashed line). The presence of CO_2, H_2S, ethane and/or propane with methane in the hydrate will have the effect of shifting the hydrate phase boundary line to the right, thus increasing the P-T field in which methane hydrate is stable (Kvenvolden, 1993). Permafrost and oceanic hydrate fields of stability are the main potential economic regions in P-T space.

Hydrate is found in the seafloor because of a coincidence of rising pressure and diminishing temperature with increasing water depth. Opern oceans are characterized by having warmer water at the surface (at least during summers) than at the seafloor. The change of temperature with depth is called the hydrothermal gradient, which varies considerably depending on latitude and local heating attributes of bodies of water, such as the Gulf of Mexico. Generally, the rate of change of temperature with depth is more rapid near the surface. The rapid temperature decrase with depth often changes from a high rate of change to a lower rate of change at the 'thermocline' (Chapter 6, Fig. 2). Temperatures and pressure depths conducive to the formation of hydrate in the oceans is usually found below this thermocline. In the Polar regions and in restricted bodies of water such as the Mediterranean Sea, however, there is often no thermocline, especially in the winter. In the Arcitic hydrate is stable at shallower depth than in the more temperate open oceans because of the very cold water while in the Mediterranean-type water masses, hydrate is only stable at pressure-depths greater than the open oceans because of abnormally warm water at depth.

Pressure-depths and temperatures conducive to the formation of hydrate in open oceans are usually found just below the slope break between continental shelves and slopes. The cold sea water cools the sediments and rocks in the upper seafloor regime, and provides a lower temperature limit, which varies from place to place, from which temperature increases downward into the sediment along the local geothermal gradient (Chapter 6, Fig. 2). The base of the HSZ is located where the effect of increasing temperature renders the hydrate unstable, even though pressure is increased. The base of the HSZ occurs where the geothermal gradient, whose uppermost position is coincident with the deepest water temperature, intersects the local projected hydrate phase boundary. Thus, the base of the HSZ is itself a phase boundary. The thermal structure and exact position of the actual base of the hydrate is is subject to many local controls (Chapters 3 &4). Because the pressure at the seafloor, which is the top of the HSZ, increases with water depth (Miles, 1999), where similar geothermal gradients are found along a continental margin, the HSZ thickness increases with increasing water depth (Chapter 6).

3. INTEREST IN HYDRATE: ENERGY AND INDUSTRY

In the 1930s and 1940s, the natural gas industry responded to clogged gas and petroleum pipelines caused by spontaneously formed crystalline, wax-like incrustations, especially in colder regions of the country. The pipeline material often remained stable and was an economic nuisance even at ambient winter temperatures and pressures (Hammerschmidt, 1934). Research identified this material as hydrates of mixed hydrocarbon gases (e.g., mostly methane, with ethane, propane and/or butane) which have a stability field much broader than hydrate of pure methane. The problem was solved by drying the gas prior to transport. Without water, natural gas hydrates will not form.

Thus, hydrate was considered by scientists as a curiosity and by the hydrocarbon industry as a nuisance until its very recent recognition in the natural environment (Chapter 2, 28, etc.). Within twenty years of its discovery on the Earth, however, both industry and governments had begun to explore the potential of hydrate as an energy source (Preface).

The transition from a liquid petroleum to gas-based economy has already begun. Environmental and other considerations have mandated cleaner and less exhaust, especially from large power plants. We are seeing evidence now that the price of oil need not rise to intolerable levels before an alternate fuel such as methane becomes the fuel of choice. The stone age did not end because the people ran out of stones; they discovered something better - metal.

Methane must now be considered as a primary fuel of the future (see Chapter 28).

Clearly, methane derived from hydrate offers an unparalleled promise as an energy resource. The methane could be used directly as a fuel, or be converted to methanol, or higher molecular weight synthetic fluid fuels. Methane can be burned to produce energy in a variety of ways including direct use in methane fuel cells. In addition, methane could be transported to a point of use and transformed to hydrogen (with inert carbon byproduct) and used directly in virtually pollutionless hydrogen fuel cells.

3.1. Industrial Applications and Other Considerations

Hydrate science is only beginning in areas where other clathrates have already been identified as suitable for industrial applications such as purification of mixed materials. There clearly remains much to be done. Illustrations of some of the different aspects of industrial hydrate processes are:

1. Desalination and purification of water. Naval Research Laboratory scientists have defined the concept of using methane-based mixed gas hydrate as the basis of a new technology to desalinate seawater (Max and Pellenbarg, 1999). Further research and development of this promising new technology is now underway.
2. Specially fabricated methane clathrates which will compress methane safely and allow the material to be used as a methane transport and fuel storage media (Max et al., in press).
3. In-situ energy resource. Investigations are underway to determine if methane held in hydrate at or near the seabed can be obtained by submersible vehicles for use as an on-board fuel.
4. Basic fuel and industrial feedstock as part of interplanetary travel. Large amounts of methane hydrate may exist on Mars which would provide the basic elements for human habitation of the plànet (Max and Clifford, 2000).

Acknowledgments: Thanks to the Office of Naval Research, Program Element No. 0601153N, and U.S. Department of Energy, Grant DocumentOEA12697 FT 34238.

Chapter 2

Natural Gas Hydrate: Introduction and History of Discovery

Keith A. Kvenvolden
U.S. Geological Survey, Menlo Park, CA 94025

1. INTRODUCTION

Interest in natural gas hydrate is increasing rapidly as the multiple implications of its presence in the shallow geosphere are being recognized. The large amount of hydrate methane that is sequestered in shallow terrestrial and marine sediments makes this methane an attractive target for those concerned about future energy requirements and resources. The fact that natural gas hydrate is metastable and affected by changes in pressure and temperature makes any released methane an attractive agent that could globally affect oceanic and atmospheric chemistry and ultimately global climate. And finally this characteristic of metastability could explain major seafloor instabilities resulting in submarine slides and slope failures. Thus these ramifications of natural gas-hydrate occurrence all have potential effects on future human welfare, and hence explain the increasing worldwide interest. This chapter introduces natural gas hydrate, provides a background for understanding its occurrence, relates the early history of discovery, and describes the hydrate gas compositions that have been found. Following chapters will deal with the important aspects of natural gas hydrate as a potential (*a*) energy resource, (*b*) factor in global change, and (*c*) submarine geohazard.

2. BACKGROUND

2..1. Definition

In defining natural gas hydrate, it is useful to clarify the term "natural" which has two meanings. First, the term "natural" is used to indicate gas hydrate occurring naturally on Earth rather than being synthesized in the laboratory or inadvertently created during industrial transportation of petroleum gas. Second, the term "natural" indicates hydrate containing natural gas, defined in the oil industry as the gaseous phase of petroleum (Hunt, 1996), but which really includes all gases derived from naturally occurring chemical and biochemical processes. Typically, natural gas is composed of methane often accompanied by higher molecular weight hydrocarbon gases and by non-hydrocarbon gases. The

M.D. Max (ed.), Natural Gas Hydrate in Oceanic and Permafrost Environments, 9–16.

proportions of gases can vary from pure hydrocarbons to pure non-hydrocarbons, such as carbon dioxide, nitrogen, and hydrogen sulfide, all of which can form hydrates. As far as is known at present, the most widespread natural gas hydrates on Earth are those incorporating mainly methane, and these are the natural gas hydrates that are the main focus of this chapter.

Thus gas hydrate, also called gas clathrate, is a naturally occurring solid comprised of water molecules forming a rigid lattice of cages, with most cages containing a molecule of natural gas, mainly methane. Gas hydrate is essentially a water clathrate of natural gas. Three crystalline structures, I, II, and H, have been recognized in nature with structure I being the most common. In structure I, the cages are arranged in body-centered packing of the cubic crystallographic system, and the cages are large enough to include methane, ethane, and other gas molecules of similar molecular diameter, such as carbon dioxide and hydrogen sulfide. In structure II, diamond packing in the cubic system is present resulting in some cages being large enough to include not only methane and ethane, but also gas molecules as large as propane and isobutane. Structure H is least common in nature; some cages larger than in structure II are present in the hexagonal crystallographic system (Sloan, 1998). Structure I gas hydrate is emphasized in this chapter and is often referred to as methane hydrate. This gas hydrate can contain very large amounts of methane. The maximum amount is fixed by the clathrate geometry which translates to 164 volumes of methane at standard conditions to one volume of methane hydrate (Davidson et al., 1978).

2.2. Controls

The occurrence of natural gas hydrate is controlled by an interrelation among the factors of temperature, pressure, gas composition, and ionic strength of water (Kvenvolden, 1993). Presently the upper depth limit for methane hydrate is about 150 m below the surface in continental polar regions, where surface temperatures are below 0° C. In oceanic sediment gas hydrate occurs where bottom-water temperatures approach 0° C and the water depths exceed about 300 m (Chapter 1, Fig. 2). The lower limit of methane hydrate occurrence is determined by the geothermal gradient; the maximum lower limit is about 2000 m below the solid surface, although the lower depth limit is typically much less (<1000 m), depending on local conditions. The occurrence of gas hydrate is restricted to the shallow geosphere where the amount of methane exceeds its solubility in water and thus is present in large enough quantities to form methane hydrate.

2.3. Location

Gas hydrate occurs worldwide, but because of the pressure-temperature and gas volume requirements, it is restricted to two regions: polar and oceanic outer continental margins. In polar regions, gas hydrate is normally found where there is permafrost both onshore in continental sediments and offshore in sediment of the polar continental shelves. In oceanic outer continental margins, gas hydrate is found in sediment of the continental slope and rise where cold bottom water is present. The worldwide occurrence of known and inferred gas hydrate is shown in Figure 1.

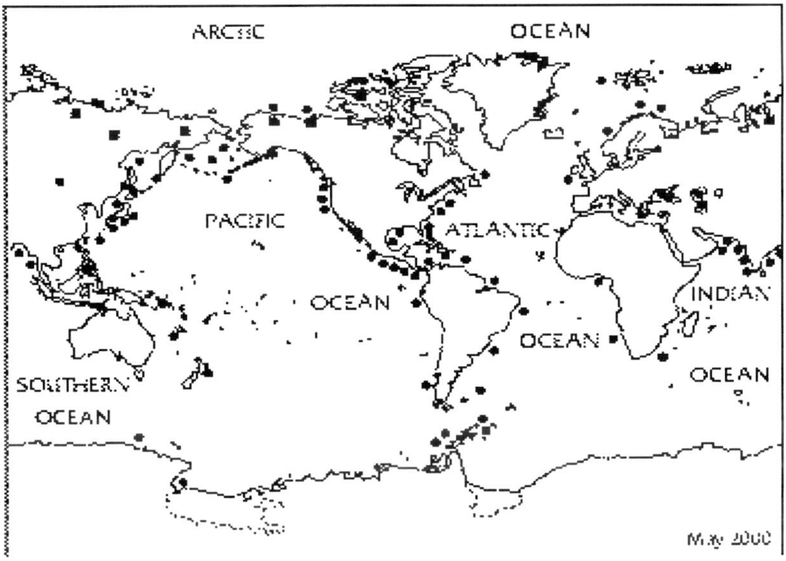

Figure 1. Map showing worldwide locations of known and inferred gas hydrate in oceanic (aquatic) sediments (solid circles) and in continental (permafrost) regions (solid squares). Updated from Kvenvolden (1988).

Samples of gas hydrate have been recovered on land in the western Prudhoe Bay oil field in Alaska [reviewed by Collett (1993)] and in the Mackenzie Delta of Canada (Dallimore et al., 1999). Thus far, gas hydrate has been found at about 30 oceanic locations (Kvenvolden et al., 1993; Booth et al., 1996; Ginsburg and Soloviev, 1998), providing irrefutable evidence for natural gas-hydrate occurrence. Deep-ocean drilling has recovered gas hydrate at locations offshore from Peru, Costa Rica, Guatemala, Mexico, western and eastern United States, Japan, and in the Gulf of Mexico. Gas hydrate at or near the seafloor has been recovered at locations in the Black Sea, Caspian Sea, Sea of Okhotsk, offshore from northern California, Oregon, Nigeria, in the northern Gulf of Mexico, offshore Norway (Ginsburg et al., 1999) and offshore Oregon (Suess et al., 1999).

2.4. Amount of Methane

Estimates of the methane content of natural gas hydrate all suggest that the methane quantities are very large (Kvenvolden, 1999). Cherskiy and Makogon (1970) proposed that the amount of methane in naturally occurring gas hydrate is potentially "enormous," but their estimates were highly speculative because not much was known about natural gas hydrate at that time. The Potential Gas Committee (1981) summarized early global estimates: methane in gas-hydrate deposits ranging from 3.1 x 10^{15} to 7600 x 10^{15} m^3 (at standard conditions) for oceanic sediments and from 0.014 x 10^{15} to 34 x 10^{15} m^3 for polar regions. Because oceanic gas hydrate apparently contains significantly more methane than polar region gas hydrate, the oceanic gas hydrate is emphasized in global estimations of methane content of gas hydrate.

During the period from 1980 to 1990, an better understanding of gas-hydrate occurrence has generally resulted in estimates within the lower ranges of the previous estimates. Kvenvolden (1988) and MacDonald (1990), working independently, estimated the methane content of the gas hydrate worldwide to be about 21 x 10^{15} m^3. That these estimates are equal is a coincidence, but the convergence of ideas has made this value the consensus estimate.

In the 1990s Gornitz and Fung (1994) and Harvey and Huang (1995) used plausible ranges of relevant variables to provide estimates of the methane content of gas hydrate based on considerations of various General Circulation Models. The estimates of Gornitz and Fung (1994) ranged between 26 x 10^{15} m^3 and 139 x 10^{15} m^3, with the most likely estimate of 26 x 10^{15} m^3 for their *in situ* microbial gas generation model and 115 x 10^{15} m^3 for their pore-fluid model. These estimates lie within the range of early values but are higher that the consensus estimate. Harvey and Huang (1995) calculated the total methane content of oceanic gas hydrate to be 23 x 10^{15}, 46 x 10^{15}, and 91 x 10^{15} m^3, depending on the assumptions regarding the pore space occupied by gas hydrate. They selected the intermediate value of 46 x 10^{15} m^3 as their best estimate.

Other global estimates of methane in gas hydrate were made during this same time period and resulted in values smaller that the consensus estimate of 21 x 10^{15} m^3. For example, Holbrook et al. (1996) estimated about 7 x 10^{15} m^3, whereas Dickens et al. (1997) concluded that global estimates of methane in gas hydrate in the range from about 2 x 10^{15} to 20 x 10^{15} m^3 are acceptable, based on gas-hydrate concentrations determined at the Blake Ridge. Other low estimates include a revised estimate by Makogon (1997) of 15 x 10^{15} and 1 x 10^{15} m^3 by Ginsburg and Soloviev (1995), who challenge all larger estimates. It is apparent that by 1990 the range of estimates has been greatly constrained from estimates available in 1980. The consensus value of 21 x 10^{15} m^3 remains about midway between the extremes. It is quite likely that the global amount of methane in gas hydrate is considerable less than 10^{17} m^3 but probably is greater than 10^{15} m^3, with the actual value in the lower or intermediate part of the range.

3. EARLY HISTORY

Interest in natural gas hydrate has increased steadily since about 1969 when the property of natural gas to form gas-hydrate deposits was first discovered in the Soviet Union (Vasil'ev et al., 1970). Russian scientists postulated that pressure-temperature conditions favorable for gas-hydrate development are present in sediment in polar regions as well as in sediment that covers about 90% of the ocean bottom (Makogon et al., 1971, 1972, 1973; Trofimuk et al., 1973, 1975, 1979). It is now known that gas hydrate is not present in 90% of the ocean bottom, but is restricted to continental slopes and rises of outer continental margins (summarized in Kvenvolden, 1993).

3.1. Gas Hydrate Associated with Permafrost

In 1970, well-logging and formation tests in the Messoyakha gas field in western Siberia indicated that methane and gas hydrate were present (Makogon et al., 1971, 1972). Methane was released from gas hydrate by injecting methanol into test wells that perforated the hydrate zone. The injection of methanol, which is a hydrate inhibitor, resulted in a large increase in gas production in the test wells.

Two oil and gas exploratory wells drilled in permafrost of the Mackenzie Delta penetrated shallow sand reservoirs containing gas hydrate at depths of 820 to 1100 m (Bily and Dick, 1974), but no gas hydrate samples were recovered. The amount of formation gas in the drilling mud increased significantly during penetration of the sands, which were porous but had low permeability. These characteristics suggested the presence of gas hydrate. Well-log responses, such as high resistivity and acoustic velocity, also indicated gas hydrate. The presence of gas hydrate in the Mackenzie Delta region has now been proven by the recovery of gas-hydrate samples during coring at the Mallik 2L-38 well (Dallimore et al., 1999).

The first actual recovery of gas hydrates from sediment associated with permafrost was obtained by means of a pressure-core-barrel at the Arco-Exxon N.W. Eileen State No. 2 wildcat well on the north slope of Alaska in the West Prudhoe Bay Oil Field (reviewed by Collett, 1993). Pressure reduction tests showed that gas hydrate had been recovered. Well logs from this well also indicated a gas-hydrate zone. The combination of resistivity, sonic, and mud logs all pointed to the presence of gas hydrate. In addition to the western Siberian basin, the Mackenzie Delta, and the North Slope of Alaska, the Arctic Archipelago of Canada (Hitchon, 1974) and the Vilyuy Basin of Russia (Makogon et al., 1972) also show evidence for the presence of gas hydrate.

3..2. Gas Hydrate in Oceanic Sediment

<u>Bottom Simulating Reflector</u> Most of the oceanic occurrences of gas hydrate shown on Figure 1 as filled circles are inferred, based mainly on the appearance on marine seismic reflection profiles of an anomalous bottom-simulating <u>reflection</u> (BSR). This reflection coincides with the depth predicted, based on pressure-temperature considerations, as the base of the gas hydrate stability zone (Shipley et al., 1979). The base of the gas hydrate stability zone is the bottom simulating <u>reflector</u>, also called the BSR, which produces the <u>reflection</u> on seismic records. BSRs mark the interface between higher sonic velocity, hydrate-cemented sediment above and lower sonic velocity, uncemented sediment with free gas below. The seismic reflection from the base of the gas-hydrate-stability zone is generally characterized by reflection polarity reversals and negative reflection coefficients. The reflection usually mimics the seafloor form and often increases in sediment depth with increasing water depth.

Inferred Gas Hydrate in Oceanic Sediment The bottom-simulating-reflector was a key factor in detecting gas hydrate in oceanic sediment, and BSRs were used to infer gas hydrate before any gas hydrate was ever recovered from oceanic sediment. Before the Deep Sea Drilling Program (DSDP) Leg 11 (prior to 1970) in the Blake Ridge region of the Atlantic Ocean offshore from southeastern United States, geophysicists were intrigued by BSRs on seismic records from the Blake Ridge (Markl et al., 1970; Stoll, et al., 1971). The observation that some of the seismic reflections intersected bedding reflections and paralleled the seafloor was unexplained. One objective of DSDP Leg 11 was to investigate the nature of these reflectors that caused the anomalous reflections. The strongest reflection mimicked the profile of the Blake Ridge at a depth of more than 500 m below the seafloor. Core samples across the region yielded mainly methane with traces of ethane (Claypool et al., 1973). No gas hydrate was observed, and it was not until DSDP Leg 76 that gas hydrate was observed in sediment of the Blake Ridge (Kvenvolden and Barnard, 1983). The prominent BSR was at the time correlated with a break in the drilling record and with a zone of carbonate minerals (Lancelot and Ewing, 1972). Another explanation was offered by Stoll et al (1971) and Ewing and Hollister (1972) that the BSR corresponded to the isotherm that separates the gas environment from the overlying gas hydrate environment. This latter explanation was confirmed by drilling on DSDP Leg 76 and is now the accepted interpretation.

Since then, other BSRs were reported from the western North Atlantic Ocean (Tucholke et al., 1977) and the Beaufort Sea (Grantz et al., 1976). In a now classic paper, Shipley et al. (1979) described BSRs in sediments off the western and eastern coasts of the United States, in the western Gulf of Mexico, off the northern coasts of Colombia and Panama, and along the Pacific coast of Central America from Panama to Mexico. Now more than 60 sites are known worldwide where gas hydrate occurs in oceanic sediment based on BSRs and/or sample recovery (Figure 1). A compilation of 47 of these sites can be found in Kvenvolden et al. (1993).

Early Direct Observations As far as can be ascertained, the first direct observation of gas hydrate in oceanic sediment was made by Yefremova and Zhizhchenko (1974). They described crystal gas hydrates in near-surface sediment recovered from the Black Sea: "These hydrates occurred 6.5 m below the seafloor, in large cavities, as micro-crystalline aggregates resembling hoarfrost and tended to disappear before one's eyes." This work was later amplified by Kremlev and Ginsburg (1989). Thus Russian scientists were not only the first to recognize the natural occurrence of gas hydrate in permafrost regions, they were also the first to recover gas hydrate in sediment from an oceanic setting (Black Sea).

During the course of DSDP, observations of deep-sea sediment cores that released large quantities of methane suggested that gas hydrate could exist beneath the seafloor in some areas, but gas hydrate samples were never seen. In reviews of gas data from DSDP, Claypool et al. (1973; Legs 10-19) and McIver

(1974; Legs 18-23) described instances where gas evolved from core samples taken on deck. Gas evolution sometimes continued for several hours, and the pressures generated were occasionally sufficient to extrude cores from the barrel and core liner. The expanding, cooling gas often resulted in ice formation on the exposed core. The quantity and rate of gas evolution suggested gas hydrate as one possibility. In most cases the gas was methane with traces of ethane.

In was not until DSDP Leg 66, offshore from Mexico, that gas hydrate was recognized in the recovered sediments (Shipley and Didyk, 1982). This important discovery provided quantitative evidence for naturally occurring, deep-sea gas hydrate. The gas hydrate was found at three different drill sites within unconsolidated sediments, generally associated with porous zones of volcanic ash or sand layers, interbedded with mud or mudstone. Gas hydrates were also recovered in DSDP Leg 67, offshore from Guatemala (Harrison and Curiale, 1982). The gas hydrate was invariably associated with stratigraphic sequences containing high-porosity sediments. Almost pure gas hydrate was recovered from a core catcher, and another core contained gas hydrate cementing coarse vitric sands. These kinds of observations were extended on Leg 84, also offshore from Guatemala, where more gas hydrate was recovered including a 1.05 m core of almost solid gas hydrate (Kvenvolden and McDonald, 1985). In addition to these early observations, gas hydrate has also been found on the following DSDP and Ocean Drilling Program (ODP) cruises: Leg 76 (Blake Ridge, Kvenvolden and Barnard, 1983); Leg 96 (Gulf of Mexico, Pflaum, et al., 1986); Leg 112 (Offshore Peru, Kvenvolden and Kastner, 1990); Leg 127 (Japan Sea, Shipboard Scientific Party, 1990); Leg 131 (Nankai Trough, Shipboard Scientific Party, 1991); Leg 146 (Cascadia Margin, Whiticar et al., 1995); Leg 164 (Blake Ridge, Paull et al., 1998); and Leg 170 (Offshore Costa Rica, Shipboard Scientific Party, 1997).

4. COMPOSITION OF GAS HYDRATE

One striking observation is the common occurrence worldwide of almost pure methane hydrate (Kvenvolden, 1995). That is, methane usually composes >99% of the hydrocarbon gas mixture. This composition suggests that these gas hydrates are likely structure I. Only in the Gulf of Mexico (Brooks et al., 1984) and the Caspian Sea (Yefremova and Gritchina, 1981; Ginsburg et al., 1992) has gas hydrate been found in which methane is accompanied by significant amounts of ethane and propane such that structure II hydrate is present. In the Gulf of Mexico, structure H gas hydrate has also been observed (Sassen and MacDonald, 1994). Gas hydrate containing about 90% methane and 10% H_2S has been recovered offshore from Oregon on ODP Leg 146 (Kastner et al., 1998), but because of the small molecular sizes of the two gases, this mixed gas hydrate is structure I.

Information on the molecular composition of hydrocarbon gases in gas-hydrate samples, coupled with measurements of carbon-isotopic compositions of methane, provide a basis for interpreting the origin of the methane in gas hydrate. In most of the methane hydrate samples, methane has a carbon-isotopic composition lighter than -60‰, suggesting that the methane is mainly microbial in origin. This microbial methane is believed to result from methanogenic processes, taking place in shallow sediment, in which CO_2, ultimately derived from organic matter, is reduced to methane. In contrast, the gas hydrate samples with possible structure II and H crystallography from the Gulf of Mexico and structure II crystallography from the Caspian Sea contain methane with carbon-isotopic compositions heavier than -60‰. This methane and accompanying heavier hydrocarbon gases are considered to be thermogenic, resulting from the thermal decomposition of organic matter at great depth.

Oceanic gas hydrate has been recovered in sediment ranging, thus far, from the seafloor to a depth of about 400 m. These gas-hydrate occurrences are generally formed in place from methane that is either generated microbially nearby or has migrated and recycled during gas hydrate dissociation and reformation (Paull et al., 1994). Microbial methane likely migrates only short distances to form gas hydrate occurring at or near the surface. Thermogenic methane, on the other hand, likely migrates long distances from deeply buried sediment in order to form gas hydrate at or near the seafloor in the Gulf of Mexico and the Caspian Sea.

5. CONCLUSIONS

Naturally occurring gas hydrate was first discovered in the 1960s by Russian scientists exploring for gas in permafrost regions of northern Russia. By the early 1980s, gas hydrate had been found in continental slope sediments of the Middle America Trench offshore from Mexico and Guatemala and off the U.S. SE coast in the Blake Ridge. Since then the rate of discovery of evidence for gas hydrate has accelerated. It is now known that gas hydrate occurs in sediments of continental and insular margins, both active (convergent) and passive (divergent), around the world. Most of the world's natural gas hydrate is composed mainly of methane and water and is in the structure Icrystallographic form. The amount of methane in natural gas hydrate is enormous, but the estimates are speculative, with a present range of 10^{15} to 10^{17} m^3, and a most likely value at the lower or intermediate part of this range. Determining the role that this methane plays in the global carbon cycle and perhaps will play in the global energy mix of the future is a major challenge.

Chapter 3

Practical Physical Chemistry and Empirical Predictions of Methane Hydrate Stability

Edward T. Peltzer and Peter G. Brewer
Monterey Bay Aquarium Research Institute
Moss Landing, CA 95039

1. INTRODUCTION

1.1. Background

Accurate and precise prediction of the temperature and pressure (P-T) conditions at the boundary of the methane hydrate stability field is an essential component of a variety of endeavors in the field of geochemistry. Kvenvolden (1988), Gornitz and Fung (1994) and others have used knowledge of the P-T stability conditions to define the geophysical limits of gas hydrates and thereby estimate the size of the global reservoir. As the thermal signature of global warming penetrates into the ocean (Levitus et al., 2000), precise knowledge of the stability of gas hydrates will be required to assess the risks of decomposition in this reservoir. Recently, Ruppel (1997) has suggested that a discrepancy exists between *in situ* temperature measurements on the Blake Ridge and the predicted base of the hydrate stability zone. This claim is based in part upon P-T predictions of gas hydrate stability. In our own research, we have conducted a series of *in situ* deep-sea gas hydrate synthesis experiments (Brewer, et al., 1998) and have begun using an ROV to prospect for gas hydrate out-crops and undersea gas vents, which potentially result from decomposing gas hydrate deposits. One of the goals of this field work is to explore for gas hydrates close to the limit of the stability zone and this creates the need for accurate and precise predictions. Given the small temperature gradients with depth in the deep-sea, an error of 0.5°C, could mean a depth error of more than 100 meters. With a shallow sloping bottom (1% grade), one could easily be ten kilometers or more off target if the wrong temperature is used.

A variety of empirical P-T relationships for predicting the stability of methane hydrates can be found in the literature. The purpose of this chapter is not to provide an exhaustive list of these relationships nor to resolve all the discrepancies among them. This would be an injustice to the varied purposes of the studies and the authors involved. Nor do we pretend to present here a

M.D. Max (ed.), Natural Gas Hydrate in Oceanic and Permafrost Environments, 17–28.

complete description of the thermodynamics of gas hydrates. That has already been done in an excellent textbook by Sloan (1990, 1998), and the reader is referred there if a more complete understanding is desired. Rather, our goal is to select several representative approaches, to examine their differences, strengths and weaknesses, and to present some practical tools for applying these methods to the study of methane hydrates in the natural environment.

1.2. Ground rules for comparisons

In order to compare the various approaches to estimating the P-T conditions of methane hydrate stability, we need to establish both a level and consistent playing field. The need for this is obvious. A variety of units have been used in the literature for both pressure and temperature. To use all of them here would lead to much confusion. Additionally, chemists working in the lab prefer units of pressure on an absolute scale; geologists and oceanographers prefer to use a vertical depth scale in meters. While a simple conversion of depth to pressure has often been used, we will see shortly that this shortcut introduces systematic and detectable errors. While these errors were initially thought to be too small to be important, they are now becoming significant as the limits of precision and accuracy of both the measurements and the predictions are constantly being improved. Therefore, in the discussions that follow we will restrict ourselves to absolute pressure and temperature. Where necessary, the data has been converted to these units.

There are other reasons beyond simple clarity for adopting this approach. Methane hydrates experience pressure not depth. Thus it makes sense to use pressure. Additionally, because gravity varies with latitude, phase diagrams prepared using depth will vary tens of meters from site to site. While this may seem a small error at depths of several thousands of meters, it is at present well within our abilities to measure and can lead to false interpretations. Similar errors are introduced when the compressibility of seawater is not accounted for. Pressure at depth in the ocean is often measured relative to the sea surface as zero. Failure to correct for atmospheric pressure when converting to absolute pressure will give a pressure error which is equivalent to 10 meters in depth. A similar error is introduced if lab data is reported as gauge pressure and not absolute pressure. These errors are cumulative and yield pressure errors of 2 to 3 atmospheres which is equivalent to depth errors of about 20 to 30 meters.

The equations to convert gauge and *in situ* hydrostatic measurements to absolute pressure from laboratory experiments, oceanographic research and in deep-sea sediments, respectively, are:

$$P(abs) = P(gauge) + P(atm). \qquad (1)$$

$$P(abs) = P(water\text{-}column) + P(atm). \qquad (2)$$

$$P(abs) = P(pore\text{-}water) + P(water\text{-}column) + P(atm). \qquad (3)$$

1.3. Converting depth to pressure

Typically, a constant density approximation has been made:

$$P(\text{at depth}) = \text{Depth(m)} \times \text{density} \times \text{acceleration of gravity}, \quad (4)$$

where, the density of the overlying water column is assumed to be 1.035 and the acceleration of gravity is 9.8 m/s^2. There are problems with this approach. Water is slightly compressible, so its density increases with depth. While small, the effect of this difference on pressure at depth is cumulative. The variation of gravity with latitude, makes this calculation site specific. These effects can be seen quite clearly in figure 1. Notice that below the seafloor, these effects work in reverse due to the geothermal gradient where pore-water temperature increases with depth. At higher temperatures, the density of the pore-water is greatly diminished and the cumulative pressure error can become negative.

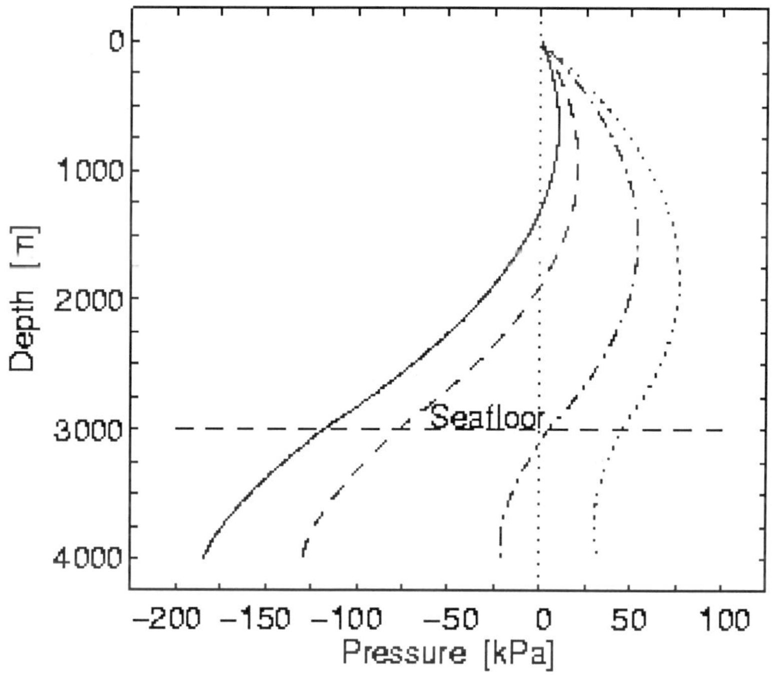

Figure 1. Calculated pressure differences *vs* depth between the simplified pressure relation and a calculation including the compressibility of seawater and the variation of gravity (Fofonoff and Millard, 1982) as a function of latitude (solid line, 90°; dashed line, 60°; dot-dash line, 30°; and dotted line, 0°). A standard ocean model was used: seawater salinity was 35 at a constant temperature of 0°C; porewater salinity was 32.5 and the geothermal gradient in the sediments was 35°C/km. Note that one atmosphere equals 101.3 kPA, which is equivalent to about 10m in depth.

2. P-T CONDITIONS OF METHANE HYDRATE STABILITY

2.1. Types of estimates

It would be a very large task to experimentally measure the P-T conditions for the boundary of the methane hydrate stability field at all of the possible combinations of temperature, pressure and salinity of interest to geochemists. Thus, various means to estimate these parameters have been developed and these methods fall into two types. The first type estimates the P-T relationship for the boundary by interpolating between experimentally determined dissociation data. Various equations have been fit. In some cases seawater data was available; in other cases freshwater data was used and the dissociation temperatures were adjusted for the effects of sea-salt. The second type of predictions are made based upon minimizing the Gibbs Free Energy of the system. These calculations have the advantage that both salinity and gas compositions can be freely varied.

2.2. Data Interpolations

Dickens and Quinby-Hunt (1994) were the first to make measurements of methane hydrate stability in seawater, and they fit a linear equation to their data:

$$1 / T = 3.79 \times 10^{-3} - 2.83 \times 10^{-4} \log_{10}P, \qquad (5)$$

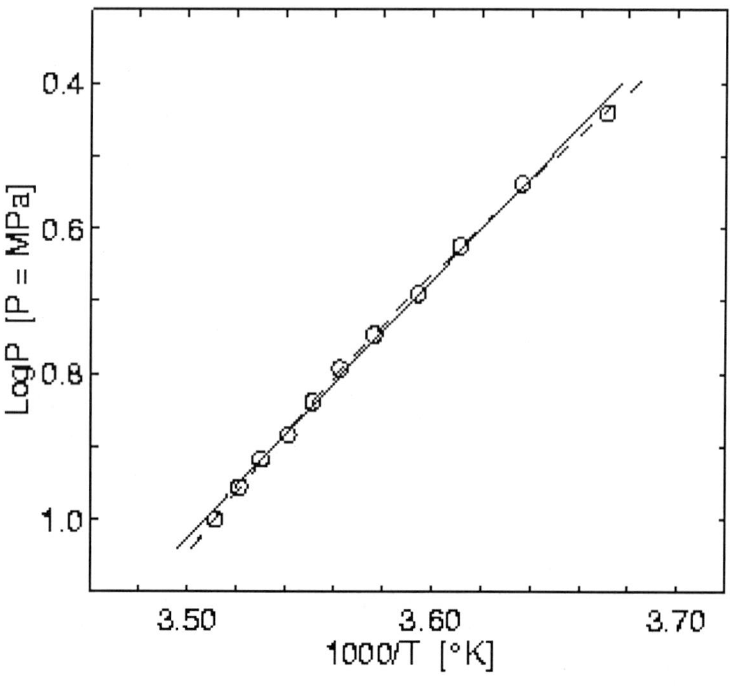

Figure 2. Plot of methane hydrate in seawater stability data (o) from Dickens and Quinby-Hunt (1994); their linear fit (—) to equation (5) and the second order polynomial fit (---) to equation (7). Curvature in the data is apparent for even this small range.

where, T is in degrees K and P is pressure in MPa (Fig. 2). This relationship is based upon the Clausius-Clapeyron equation:

$$d(\ln P) / d(1/T) = -\Delta H / zR, \tag{6}$$

where ΔH is the enthalpy of formation, z is the compressibility and R is the gas constant. Sloan (1990, 1998) has pointed out that this equation predicts straight lines over limited temperature ranges (assuming ΔH and z are constant). This statement raises the obvious questions, what is the extent of this limited temperature range, and how straight is straight? Fitting a simple second order polynomial to their data we obtain:

$$1 / T = 3.83 \times 10^{-3} - 4.09 \times 10^{-4} \log_{10}P + 8.64 \times 10^{-5} (\log_{10}P)^2. \tag{7}$$

This equation reveals significant curvature for even this limited temperature range (Fig. 2). When plotted in linear temperature *vs* pressure space, the offset between the data and equation (5) becomes more apparent (Fig. 3).

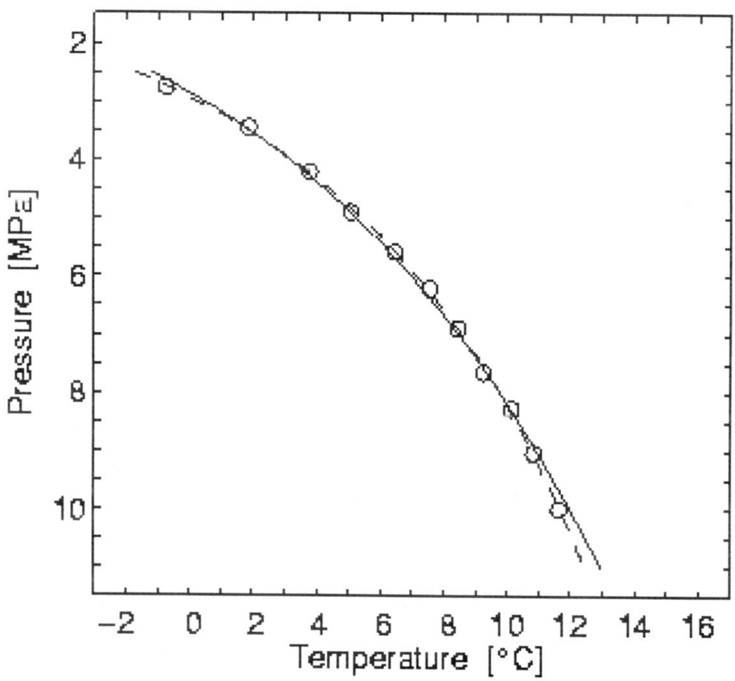

Figure 3. Plot of data from Dickens and Quinby-Hunt (1994) in linear temperature and pressure space; the linear fit (—) to equation (5) and the second order polynomial fit (---) to equation (7) are also shown. Although the offset between the linear fit and the data is small (<0.5°C), it is poorest at the ends of the line as well as at the mid-point of the data.

Extrapolation of these equations to higher P-T conditions reveals an ever increasing discrepancy between them (Fig. 4). After converting pressure to depth, and thereby introducing the error due to variations in gravity with latitude, Brown et al. (1996) fit the data of Dickens and Quinby-Hunt (1994) to a different second-order polynomial and obtained the following equation:

$$T = 11.726 + 20.5 \times \log_{10}z - 2.2 \times (\log_{10}z)^2, \qquad (8)$$

where, T is temperature in degrees C and z is depth in kilometers. This equation is also plotted in figure 4 after adjusting depth back to pressure, etc. Without data on the stability of methane hydrates in seawater at high temperature and pressures, it is impossible to tell which equation (5, 7 or 8) is correct. Dickens and Quinby-Hunt (1994, 1997) have found that there is a 1.1 – 1.2°C difference between the predictions for methane hydrate in pure water and seawater of 33.5 salinity at equivalent pressures. Handa (1990) fit the stability data for methane hydrates in pure water at 0.2 to 40 Mpa to obtain:

$$\ln (P/P_0) = -1205.907 + 44097.00 / T + 186.7594 \ln T, \qquad (9)$$

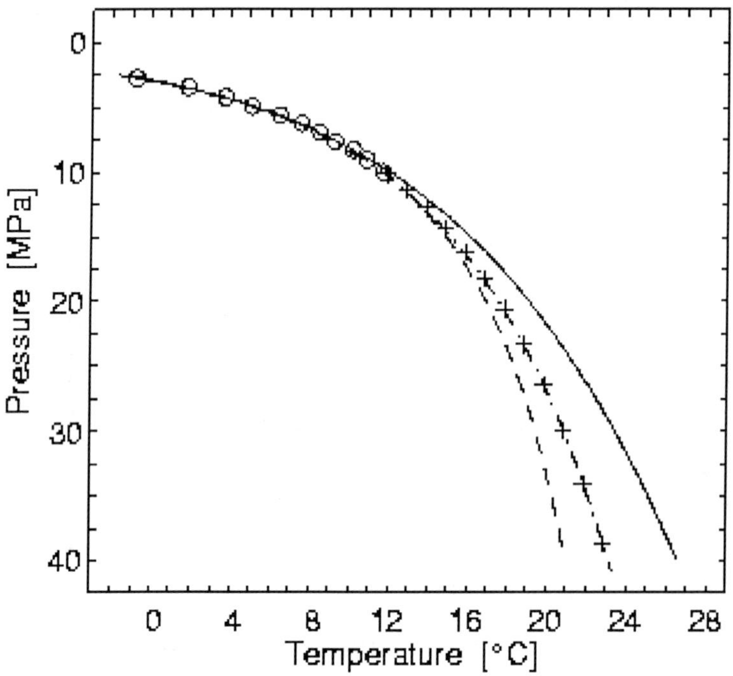

Figure 4. Plot of the linear fit (Eq. 5, —), the second-order polynomial fit (Eq. 7, ---) and equation 8 converted to absolute pressure (·-·-). Also shown is the stability data for methane hydrate in seawater (o) from Dickens and Quinby-Hunt (1994) and the freshwater stability data calculated at 1°C intervals with equation 9 offset by 1.15°C (+) to adjust for the destabilizing effects of sea-salt.

where T is in degrees K and P_0 = 1.01325 bar. Applying the mean offset to equation (9) for methane hydrate stability in pure water, allows us to test these three predictions at higher pressures (Fig. 4). Clearly, equation (8), when converted to pressure, is the closest extrapolation to the freshwater data from equation (9) adjusted to seawater.

2.3. Gibbs Free Energy Minimizations

Several methods to estimate the P-T conditions of methane hydrate stability by searching for the state which minimizes the Gibbs Free Energy of the system have been developed. These programs are computationally intensive and require sophisticated computer programming. However, computer programs to carry-out these calculations are now readily available. Sloan (1990, 1998) presents a detailed description of CSMHYD, a PC-DOS based computer program. His textbook includes a floppy disk with an executable version of the program. In addition to calculations of the stability temperature at a given pressure (or vice-versa) in pure water, the program also includes a variable composition salt component to allow seawater and pore-water predictions. We refer the reader to the textbook for the details of how the program works. Output from CSMHYD is shown in figure 5 for a pure methane hydrate in equilibrium

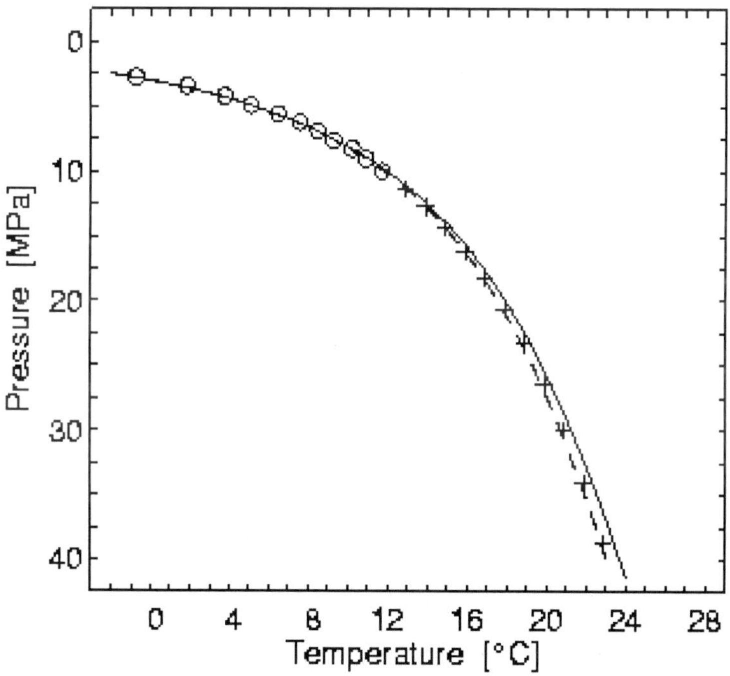

Figure 5. Output from CSMHYD (—) and Multiflash (---) plotted with the seawater data (o) from Dickens and Quinby-Hunt (1994) and the predictions from equation (9) offset for seawater (+) as in Figure 4.

with seawater. A salinity of 33.5 was chosen for the CSMHYD predictions so that a direct comparison could be made with the seawater data and the stability conditions calculated using equation (9) and adjusted to seawater, as above in figure 4. Clearly, the CSMHYD predictions better fit with the seawater data than the freshwater predictions adjusted to seawater at P > 12MPa.

Zatsepina and Buffet (1997, 1998) present an alternate Gibbs Free Energy minimization routine based, in part, on a very fast simulated annealing algorithm (Ingber, 1989). Their results are quite similar to the CSMHYD program and compare favorably to a prediction from Handa (1990). Recently, they have begun exploring whether the equilibrium calculations are sufficient or whether additional complexities lay in meta-stable phases that persist in nature outside their stability fields as decomposition of the gas hydrate is impeded by the free energy required to create small bubbles (Buffet and Zatsepina, 1999).

A commercially available program, Multiflash (Infochem Computer Services Ltd., London), also calculates the P-T conditions for methane hydrate stability using the Gibbs Free Energy minimization approach. It is more sophisticated that CSMHYD, running in the Windows® operating system, and it performs a wider variety of calculations. The Multiflash P-T predictions for methane hydrate stability in seawater of salinity 33.5 are also shown in figure 5. They are slightly different from the CSMHYD program and compare quite favorably with both the seawater data and the higher P-T predictions from equation (9) adjusted to seawater as before.

Given the close correspondence between these computer programs and the predictions based upon equations (8) and (9), it is logical to ask why incur the additional expense of the programs if the equations work so well? For pure methane hydrates in freshwater or seawater of salinity 33.5, these equations are the simplest approach. However, if one is dealing with substantially different salinity, or different salt compositions, as may be found in pore-water, or with mixed gas compositions, then computer programs offer the ability to deal with these situations and to extend the predictions beyond the range of the data. This can be seen quite clearly in figure 6, where the differences between the predicted equilibrium P-T conditions for mixed gas hydrate stability in salinity 33.5 seawater, where the gas is a composite of methane and other natural gases, and pure methane hydrate in the same salinity seawater are plotted as a function of pressure. For this example, we progressively increased the complexity of the gas mixture, starting with methane + ethane, then adding carbon dioxide and hydrogen sulfide (see the figure caption for the percent composition). These differences (0.4-2.0°C) are larger than the effects described earlier and they vary with pressure. Also shown is the destabilizing effect of high salinity (40.0) on pure methane hydrate. These lines represent complex functions for which it will be difficult to derive simple mathematical expressions. Thus, it is in situations that address the complex real-world problems of mixed gas compositions and varied salinity that the computer programs will find their greatest use.

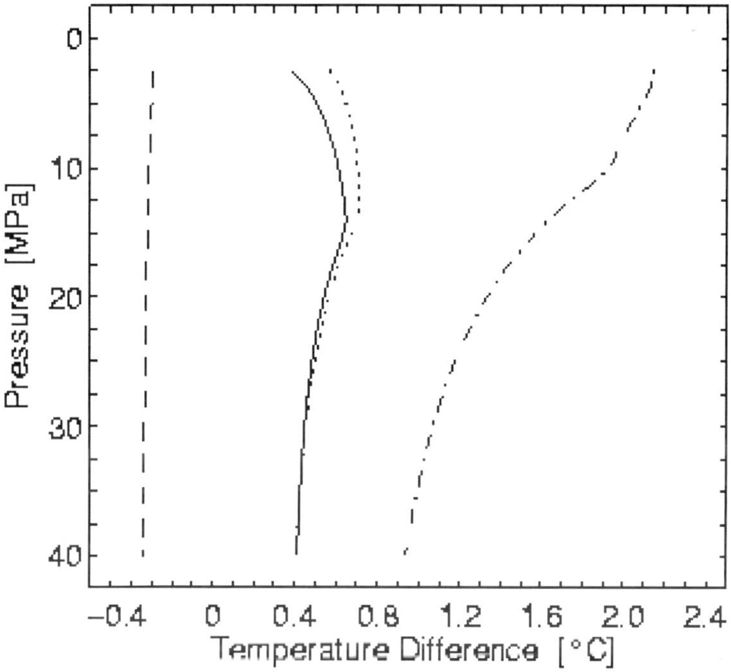

Figure 6. Plot of the temperature differences as a function of pressure between the Multiflash predictions for methane hydrate stability in high salinity (40.0) seawater (---); and methane + 2% ethane (—), methane + 2% ethane + 2% CO2 (····), and methane + 2% ethane + 2% CO2 + 2% H2S (·-·-) in seawater of salinity 33.5 relative to pure methane hydrate in seawater of salinity 33.5, respectively.

3. TWO REAL WORLD EXAMPLES

3.1. Eel River Basin

Brooks et al. (1991) reported the presence of gas hydrates (>99% CH_4) in sediment cores collected near the crest of a large sedimentary ridge in the Eel River basin off the coast of northern California. When this site was revisited with the MBARI ROV *Ventana* in 1997 and in 1999, no gas hydrates were found in the shallow sediment cores collected but several active gas vents were observed on both occasions. A temperature increase of 0.4-0.5°C was observed for the bottom water between the time when Brooks et al. (1991) collected their samples and our investigations. In order to determine whether this small change in bottom water temperatures was the reason that no gas hydrates were found, despite the presence of abundant free methane gas, we plotted the relevant hydrographic data on the methane hydrate phase diagram (Fig. 7). Given sufficient methane concentrations, hydrates would be stable in the region to the left and below the stability line. The bottom-water temperature observed by Brooks et al. (1991) when plotted at the appropriate temperature clearly falls within this region; whereas the temperature we observed does not. Thus, for a

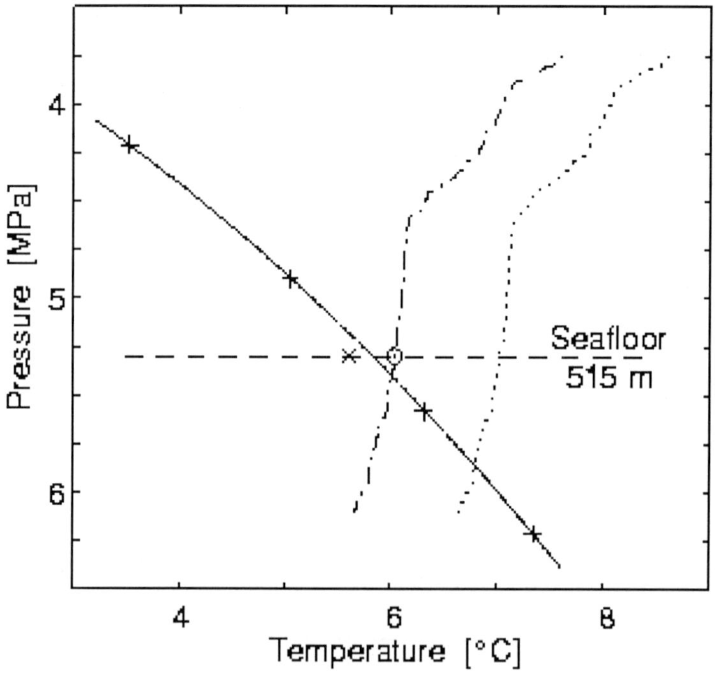

Figure 7. P-T stability curve for methane hydrate in seawater from equation 7 (—) and selected prediction using Multiflash (+). Bottom water temperatures from by Brooks et al. (1991) (x) and the authors (o). The *in situ* temperature profile nearby the dive site (-·-·) and offset 1°C (···).

first cut at an explanation, it would appear that the simple thermodynamic stability of methane hydrates explains why no gas hydrates were collected. It is not possible to say at this early stage whether the warming we observed at this site since Brooks et al. (1991) collected their samples is due to global warming or whether a more localized explanation is possible. Even so, if we shift the observed temperature profile 1°C, the potential impact of even a modest amount of warming on the stability of gas hydrate deposits is evident. At present, the temperature profile crosses the stability curve for methane hydrate at 5.41 MPa, equivalent to 526 m depth – ten meters into the sediments near the crest of the ridge. When shifted, the temperature profile crosses the stability curve at 5.86 MPa, equivalent to 571 m. This is an increase of 45 m. If such a shift in seawater temperature were to occur instantaneously, there would be a lag time before the thermal effect penetrated deep into the sediments. However, as it does, any existing gas hydrates within this zone would eventually decompose, releasing methane gas and water. Whether this is sufficient to destabilize the sediments and cause seafloor collapse remains to be seen.

3.2. Blake Ridge

Recently, Ruppel (1997) has pointed-out that when the *in situ* temperature gradient and the stability boundary for methane hydrate are plotted together, the geotherm and the stability boundary do not intersect at the depth of the bottom-simulating reflector (BSR) for several sites on the Blake Ridge. Since the BSR is considered to be at or near the top of the free gas zone, it is often considered to represent the base of the hydrate stability zone (Markl et al., 1970). The fact that it does not presents an interesting geochemical conundrum. This problem persists even after all the data are corrected to absolute pressure and the appropriate P-T stability conditions are calculated (Fig. 8). At first glance, it is easy to suggest that the problem lies in the geothermal gradient chosen by Ruppel (1997) since the geothermal gradient reported in *The Initial Reports of the ODP, Leg 164* nicely intersects the stability line near the depth of the BSR. However, this conclusion is inappropriate. The geothermal gradient used by Ruppel (1997) is anchored by a measurement only 50 m above the BSR, whereas the ODP geothermal gradient was anchored at the sea-floor. Given that both are extrapolations, it would be best to measure the *in situ* temperature at the

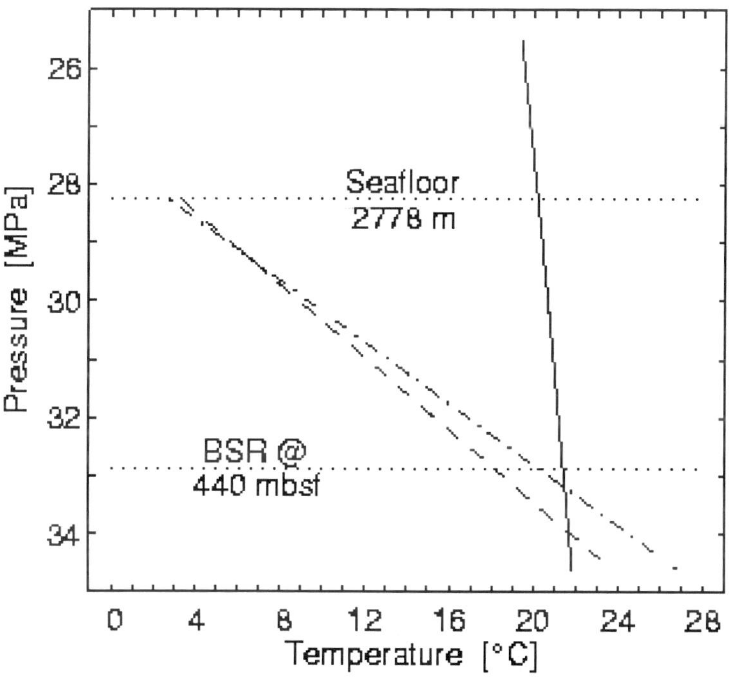

Figure 8. Temperature and pressure conditions for DSDP Site 995 on the Blake Ridge. Equilibrium stability conditions were calculated with Multiflash (—); the thermal gradients of 32.7°C/km (----)and 38.6°C/km (·--·-) are from Ruppel (1997) and the Initial Reports from ODP Leg 164.

BSR since there are no guarantees that the geothermal gradients must remain constant. Indeed, given that the BSR may coincide with a phase change, we should expect to find, at the very least, a change in the geothermal gradient in the region of the BSR. At the same time, given the effects that we saw earlier (Fig. 6), the importance of *in situ* measurements of the pore-water salinity, gas composition and pressure regarding predictions of the P-T stability conditions at this site can not be over-stated. Only then can we determine whether sufficient methane is present in the pore-spaces to achieve saturation relative to methane hydrate solubility (Xu and Ruppel, 1999). At present, the conundrum exists because of conflicts among extrapolations, not *in situ* observations; and it is only through additional observations that the solution to the riddle will be found.

4. SUMMARY AND CONCLUSIONS

Accurate and precise predictions of the temperature and pressure (P-T) conditions for the methane hydrate stability field have been shown to be an essential component of a variety of geochemical efforts. Careful correction of all measurements to absolute pressure is essential if a precision equivalent to a few meters depth is required. While linear interpolation of gas hydrate stability data works well in limited cases, second-order equations do a better job with regards to interpolating between existing data points. Our current inventory of geochemically relevant determinations of the dissociation temperatures and pressures of gas hydrates is very limited. Only one data set for seawater exists, and this is limited to pure methane gas measurements. Depending upon the form of the equation chosen to fit this data set, extrapolations to higher P-T conditions can be quite risky and lead to false interpretations. Adjustment of high pressure freshwater measurements can help extend the database, but these measurements are limited to pure methane gases as well.

The use of computer routines based upon various Gibbs Free Energy minimization algorithms to estimate the P-T stability conditions of gas hydrates has helped to expand the range of the predictions, but data to verify these estimates is quite scarce. Agreement between the computer predictions and the existing data is quite good, suggesting that the programs are accurate within reasonable limits. Their greatest utility will be with problems involving mixed gas compositions or at different salinities and ionic strengths not found in the data sets.

5. ACKNOWLEDGEMENTS

The authors would like to thank Jerry Dickens, Keith Kvenvolden, Charlie Paull, Carolyn Ruppel and Bill Ussler for helpful discussions and critical comments. This work was supported by a grant to the Monterey Bay Aquarium Research Institute by the David and Lucile Packard Foundation.

Chapter 4

Thermal State of the Gas Hydrate Reservoir

Carolyn Ruppel
School of Earth and Atmospheric Sciences
Georgia Institute of Technology
Atlanta, GA 30332-0340
cdr@piedmont.eas.gatech.edu

1. INTRODUCTION

The stability of gas hydrate is dependent on pressure (*P*), temperature (*T*), and the solubility of gas (e.g., Handa, 1990; Zatsepina and Buffett, 1997) as a function of pressure and temperature in the system. As illustrated in Chapter 1, the stability of hydrate is more susceptible to changes in temperature than pressure. Measurements that constrain thermal regimes in hydrate reservoirs therefore provide fundamental information about one of the most basic parameters controlling the stability of the deposits.

In addition to temperature, physical properties that control the transfer of heat through hydrate reservoirs are also critical for understanding thermal regimes. For example, thermal conductivity and thermal diffusivity are parameters needed for the analysis of thermal gradients measured by standard geophysical techniques, for accurate estimation of temperatures at the depth of the bottom simulating reflector (BSR), and for calibration of empirical relationships with other hydrate physical properties, such as seismic wavespeed. Perhaps most importantly, combining thermal properties measurements with predictive models for the distribution and concentration of hydrate (Rempel and Buffett, 1997; Xu and Ruppel, 1999) constrains the migration of hydrate dissociation fronts in marine sediments that experience disturbances due to exploration-related activities (e.g., pumping of hot drilling fluids through the hydrate reservoir), global climate change, sedimentation, erosion, submarine slide formation, or other processes.

This chapter focuses first on thermal property measurement techniques in both laboratory and in situ settings and on the physical and chemical factors controlling the temperature of hydrate dissociation. State-of-the-art downhole temperature measurements in a hydrate reservoir are discussed in terms of processes that may explain anomalies in the predicted temperature at the depth of the BSR. The chapter also reviews the prediction of hydrate reservoir thermal

M.D. Max (ed.), Natural Gas Hydrate in Oceanic and Permafrost Environments, 29–42.

regimes using seismic data that constrain the depth of the BSR and discusses the impact of sedimentation, erosion, catastrophic events (e.g., submarine slumps), and climate change on the stability of the hydrate reservoir.

2. THERMAL PROPERTIES

The fundamental properties important for understanding the thermal state of the marine hydrate reservoir are temperature, thermal conductivity, and thermal diffusivity. This section reviews how these quantities are measured in the laboratory or in situ during operations at sea.

2.1. TEMPERATURE

To date, most thermal studies in marine hydrate reservoirs have relied on traditional heat flow surveys in which temperatures are measured to depths of 2-5 m below the seafloor using one of several designs of heat flow probes. Although the details of the probes vary, each type is equipped with thermistors, which are sensors whose electrical resistance varies in a regular manner with very small changes in temperature. The thermistors are arrayed along the probe, which is dropped vertically into the seafloor (Figure 1a). The equilibrium temperatures T (units of °C or K) measured by these thermistors yield a direct constraint on the thermal gradient as a function of depth (dT/dz). Because these data are collected in the shallowest part of the sedimentary column, the measurements typically do not directly constrain temperatures within the gas hydrate reservoir itself, which is often located at depths far greater than the depth of maximum probe penetration.

The most direct measurements of the thermal state of the gas hydrate reservoir are obtained in situ, through the use of special probes deployed in undisturbed sediments at the bottom of boreholes (Fisher and Becker, 1993; Davis et al., 1997). This technique is known as downhole temperature measurement (Figure 1b). In the marine environment, such measurements have typically been acquired as part of Ocean Drilling Program (ODP) studies using one of three types of probes. These probes remain by far the best means for characterizing the thermal state of the hydrate reservoir. However, none of the probes can measure thermal conductivity, only one can determine a true thermal gradient during a single deployment, and one of the probes can only be deployed at depths at which sediments are not indurated.

Logging of fluid temperatures in open boreholes can sometimes be useful for constraining equilibrium temperatures above, below, and within the gas hydrate reservoir. Unfortunately, borehole fluid temperature logs do not usually provide an accurate characterization of equilibrium temperatures in the sediments. One difficulty is that borehole temperature logging is typically completed immediately following drilling, meaning that the fluid filling the borehole is not yet in equilibrium with the surrounding formation, where gas hydrate may or may not be present. A second problem is that fluids in open channels move more freely than pore fluids in the adjacent marine sediments and that boreholes may intersect fractures or other asperities that enhance migration of fluid into the borehole. Such processes can have a dramatic impact on fluid temperatures measured in boreholes.

By the time hydrate-bearing sediments reach the deck of a ship during marine coring operations, small pieces of hydrate disseminated throughout the sediment matrix have usually begun to break down into constituent components (gas and water). Because dissociation of gas hydrate is an endothermic process, temperature probes inserted in the core detect relatively colder regions where even very small amounts of hydrate were present in situ. Mapping of the thermal anomalies in cores immediately following recovery is therefore one of the best means for developing a qualitative constraint on the distribution of gas hydrate. Such temperature measurements, which are sometimes referred to as catwalk temperature measurements to denote their completion on the catwalk of the drilling ship, proved critical for locating gas hydrate in the cores during ODP Leg 164 (Paull et al., 1996).

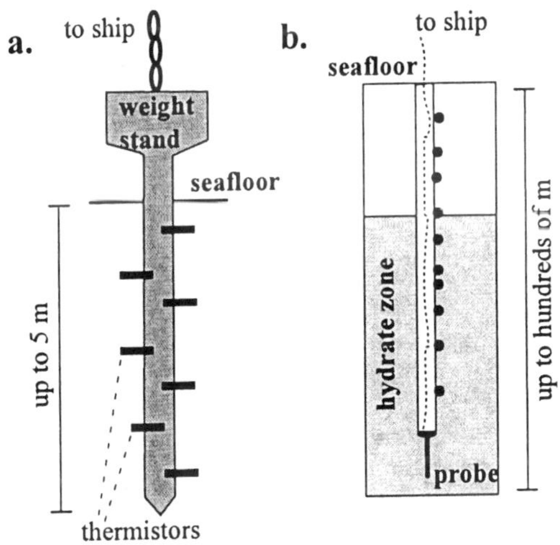

Figure 1. Common methods used to characterize the thermal state of the hydrate reservoir. (a) Traditional marine heat flow measurements are conducted in the shallowest part of the sedimentary section, often far shallower than the top of the hydrate reservoir. (b) Downhole temperature measurements in ODP boreholes determine equilibrium sediment temperature at depths as large as hundreds of meters below the seafloor. Circles schematically represent measurement depths.

2.2. THERMAL CONDUCTIVITY

Thermal conductivity K (units of W m^{-1} K^{-1}) describes a material's ability to transfer heat. Materials like salt ($K\approx 6$ W m^{-1} K^{-1}) transfer heat efficiently (conductors), while air ($K\approx 0.025$ W m^{-1} K^{-1}) and other gases are insulators. Most modern heat flow probes measure the thermal conductivity of the sediments in situ immediately following determination of equilibrium temperatures. Combining these conductivity values with thermal gradient measurements through the application of Fourier's law of heat conduction yields heat flux $q=-K\bullet dT/dz$ (units of mW m^{-2}). The negative sign indicates flux of heat out of the seafloor, in the direction opposite the increase in thermal gradient with increasing depth. If q is constant within a vertical column of the seafloor, the low thermal conductivity of hydrate (~0.5 W m^{-1} K^{-1}) relative to that of typical saturated marine sediments (0.7 to 1.3 W m^{-1} K^{-1}) implies that thermal gradients within the hydrate-bearing zone must be larger than those in the zone lacking hydrate, as shown in Figure 2a.

Two types of laboratory measurements are relevant to the determination of the thermal conductivity of hydrate and hydrate-bearing sediments. The first kind of measurement is conducted on sediments recovered during conventional piston coring or ocean drilling and uses needle probes (Von Herzen and Maxwell, 1959) inserted into the sediment through the core liner. After testing to ensure that the sediments are in thermal equilibrium, a known amount of heat is introduced and the temperature response of the sediments is measured as a function of time. Interpretation of temperature records as a function of time yields an estimate of thermal conductivity (Jaeger, 1956, 1958). Because these measurements can only be completed after the core has thermally equilibrated to ambient conditions, the thermal conductivity value applies only to the sediment matrix and associated pore fluids after hydrate dissociation. To estimate the conductivity of the original three component (hydrate, pore fluid, sediment) mixture requires additional information about the concentration of hydrate in situ, the thermal conductivity of pure hydrate, and the effective media models that govern the thermal conductivity of multicomponent mixtures.

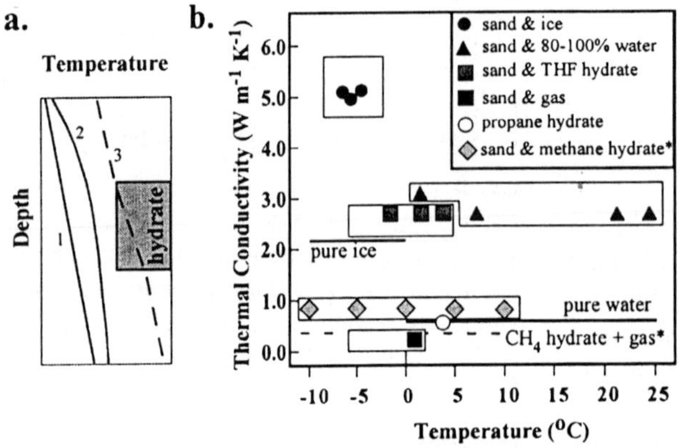

Figure 2. (a) Examples of a (1) a conductive thermal gradient, (2) a thermal gradient affected by upward advection of fluids, (3) a thermal gradient that increases within a zone of hydrate (low thermal conductivity) to maintain constant heat flux q. (b) Updated compilation of thermal conductivity results for hydrate, hydrate-sediment mixtures, water ice, and other substances after Sloan (1998). Measurements denoted by an asterisk were completed by deMartin et al. (1999) or provided by deMartin (pers. comm.).

The second type of laboratory thermal conductivity measurement is conducted on synthesized hydrate either in pure form or as part of hydrate-sediment mixtures. Until recently, most laboratory thermal conductivity measurements have used Structure II propane (Stoll and Bryan, 1979) or tetrahydrofuran (Ross and Anderson, 1982) hydrates or Structure I methane hydrates maintained at P-T conditions not characteristic of typical marine sediments (e.g., Stoll and Bryan, 1979). These experiments and others on hydrates containing various guest gas molecules (Sloan, 1998) confirm that hydrate has low thermal conductivity, with an estimated value of ~0.5 W m^{-1} K^{-1}. In contrast, water ice, a substance to which hydrate is often compared, has thermal conductivity of 2.23 W m^{-1} K^{-1}. Recently, deMartin et al. (1999) have reported on the first modern experiments to constrain the thermal conductivity of methane hydrate and hydrate-sand mixtures at P-T conditions more characteristic of those on continental margins, using hydrate synthesized with the Stern et al. (1996) technique. Experiments on pure hydrate and calculations to understand how thermal conductivity depends on the proportion of gas, sediment, and hydrate in the system remain in progress. Figure 2b summarizes the existing thermal conductivity data for hydrate, hydrate-sediment mixtures, and ice.

2.3 THERMAL DIFFUSIVITY

Thermal diffusivity α (units of m^2 s^{-1}) is related to thermal conductivity K through the relationship $\alpha=K/(\rho C_p)$, where ρ denotes density and C_p represents heat capacity. The thermal diffusivity of hydrate can be measured during the determination of thermal conductivity by using two needle probes separated by a known distance (Drury, 1988) or in specialized experiments using the Angstrom method (e.g., Durham et al., 1987). Due to the interdependence of thermal conductivity, thermal diffusivity, and heat capacity, combining laboratory thermal diffusivity determinations with thermal conductivity results should yield an estimate of heat capacity, a fundamental thermodynamic parameter that has only been the subject of a few studies to date (e.g., Ross and Anderson, 1982). There are no published direct measurements of the thermal diffusivity of hydrate although efforts are currently underway to determine this quantity for compacted methane hydrate under pressure conditions characteristic of those in marine sediments on continental margins.

3. STABILITY CONDITIONS

As discussed above, the stability of hydrate depends on pressure and temperature and on gas solubility as a function of these two variables. This section focuses on some of the physical and chemical factors that promote or inhibit hydrate stability. These factors, whose effects on the stability field are summarized in Figure 3, include: (a) gas properties, (b) pore fluid properties, (c) sediment composition, and (d) sediment physical properties. Although the effects of these factors on hydrate formation may to first order be kinematic, thermodynamic, or purely physical or chemical, these effects are cast here only in terms of their impact on the temperature of hydrate stability. It should also be noted that, at least anecdotally, many hydrate reservoirs appear to be associated with overpressured sediments. Compared to the tens of megapascals of ambient hydrostatic pressure in most marine systems though, the degree of overpressure in shallow hydrate reservoirs is typically not large enough to significantly displace the stability curve in *P-T* space. The effect is therefore not considered here.

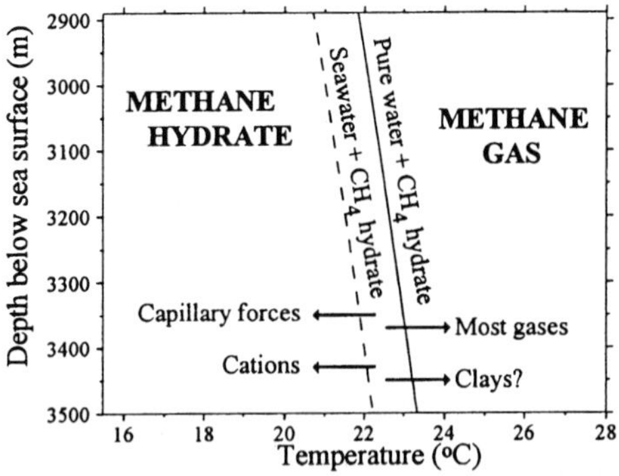

Figure 3. Stability of methane hydrate for pure water (solid line) and seawater (dashed line) systems. Arrows schematically show how various physical and chemical factors affect the stability of hydrate. After Ruppel (1997).

The phase equilibria shown in Figure 3 apply to pure methane systems and Structure I gas hydrate in the presence of fresh or saline water. The presence of even a fraction of a percent of a Structure II hydrate-former (e.g., ethane, propane) will produce Structure II hydrate, which is stable to higher temperatures than Structure I at a given pressure. Even when the concentration of a gas other than methane is not high enough to affect the structure that forms, addition of even very minor amounts of CO_2, ethane (C_2H_6), H_2S, and high-order thermogenic hydrocarbons may shift the stability curve toward higher temperatures, increasing the stability of methane hydrate.

In contrast, the presence of dissolved ions (Na^{2+}, K^+, Mg^{2+}, Ca^{2+}) in pore fluids inhibits the stability of gas hydrate. Laboratory and theoretical thermodynamical calculations (de Roo et al., 1983; Dickens and Quinby-Hunt, 1994; Dickens and Quinby-Hunt, 1997) confirm a -1.1°C offset between methane hydrate dissociation temperatures in 33.5‰ NaCl seawater vs. pure water. For the entire range of naturally-occurring seawaters (33-37‰ NaCl), the dissociation temperature does not vary by more than 0.14°C. The inhibitory effect that ionic compounds exercise on the stability of hydrate has critical implications for the evolution and long-term stability of hydrate deposits in areas characterized by salt tectonism, evolution of seafloor brine basins, and circulation of briny fluids (e.g., Gulf of Mexico; Sassen and Macdonald, 1997).

The composition of the sediment matrix may also exercise a strong influence on the temperature of hydrate dissociation in some settings. As hydrophilic substances with unbalanced surface charge and large surface area, clays are the most likely minerals to influence hydrate formation by sorbing water and providing nucleation surfaces for hydrate crystals. Laboratory experiments on mixtures of water + bentonite or montmorillonite ± other compounds have yielded contradictory results: Clays did not affect hydrate stability (Kotkoskie et al., 1990; Englezos and Hall, 1994) in some experiments, but appeared to promote hydrate formation both thermodynamically and kinetically in other experiments (Cha et al., 1988).

Physical properties of the sediment may play a critical role in the stability of hydrate as well. In fine-grained materials, large capillary forces will arise and may inhibit the entry of fluids into the interstices between grains. Theoretical calculations (Clennell et al., 1999) and laboratory measurements using fine-grained (clay-sized) particles (Handa and Stupin, 1992; Zakrzewski and Handa, 1993; Melnikov and Nesterov, 1996) indicate that capillary forces may significantly inhibit hydrate stability, depressing hydrate dissociation temperatures.

4. IN-SITU THERMAL REGIMES IN HYDRATE RESERVOIRS

Direct measurements of in situ equilibrium temperatures using sensors placed in the sediments at the bottom of Ocean Drilling Program boreholes is the best method for characterizing the thermal state of the hydrate reservoir. Such measurements have now been conducted in hydrate-bearing sediments in a number of settings, including both active (Cascadia, Chile, and Costa Rica; Hyndman et al., 1992; Brown et al., 1996; Ruppel and Kinoshita, 2000) and passive (Blake Ridge; Paull et al., 1996) margins.

4.1 IN SITU TEMPERATURES IN THE BLAKE RIDGE RESERVOIR

To date, only the Blake Ridge data set was acquired expressly for the purpose of understanding the thermal state of gas hydrate reservoirs and was complemented by a full suite of hydrate-specific measurements. Equilibrium sediment temperatures measured to a depth of ~415 mbsf on the crest of the Blake Ridge (~2700 m water depth) using downhole temperature methods are shown in Figure 4a. Several key observations emerge from an analysis of these data:

1. The measurements, which were obtained with three different types of probes and which include conservatively estimated uncertainties, show a nearly linear increase in temperature as a function of depth, implying heat transfer by largely conductive, not advective, processes (Figure 2a). The nearly linear gradient also indicates either that gas hydrate concentrations are nowhere high enough to appreciably affect the bulk thermal conductivity of the sediments or that the measurements are too coarsely spaced to detect variations in thermal gradients associated with conductivity changes as a function of depth.

Because the hydrate lattice excludes salts upon formation, dissociation of hydrate leads to local freshening of pore waters, as noted at discrete depths in the hydrate zone in Figure 4a. The temperatures measured within the depth ranges at which chloride anomalies are most pronounced (where hydrate is likely present in situ; Hesse and Harrison, 1981) do not deviate significantly from the linear gradient.

2. The Blake Ridge temperature measurements show a high degree of internal consistency and reproducibility. At several locations on the Blake Ridge, two different types of probes were deployed at the same depth in the same borehole and yielded similar equilibrium temperature values.
3. The best-fit thermal gradient does not intersect the measured bottom water temperature (BWT), indicating that the sediments are not in thermal equilibrium with the overlying seawater. Increasingly, marine hydrogeologists note that BWT perturbations related to oceanographic processes (periods of days to months; e.g., Ruppel et al., 1995) and longer-term climate change (Fisher et al., 1999) have an important impact on the thermal structure of the sedimentary column. Thus, it is not surprising that sediments at depths > 50 m are not in equilibrium with overlying bottom water. Changing deposition and compaction patterns may also play a role in producing different thermal regimes at various depths within the sedimentary section.
4. Extrapolation of the thermal gradient ~50 m downward from the deepest in situ measurement to the depth of the BSR yields a temperature lower than that of methane hydrate dissociation at these pressures, regardless of whether the seawater or pure water curve is adopted as the most appropriate (Figure 4b). The seismic anomaly associated with the BSR is widely believed to mark the P-T boundary at the base of the stability zone for hydrate. Thus, the temperature anomaly observed at the BSR on the Blake Ridge implies that our understanding of the physics and chemistry of the hydrate province must be incomplete.

4.2. UNDERSTANDING BLAKE RIDGE THERMAL REGIMES

To date, the Blake Ridge downhole temperature data set is the most complete available within the sediments of a marine hydrate reservoir. The relatively close spacing of the measurements with depth, the high precision of temperature determinations, and the fact that the data define a nearly linear gradient even when weighted by large, subjectively assigned uncertainties make it likely that the predicted temperature at the BSR is not simply incorrect. Several possible explanations for the temperature disparity at the BSR have been advanced, including: (1) Ongoing thermal equilibration of the sediments in response to Holocene climate change; (2) Lack of coincidence between the base of the HSZ and the BSR, meaning that the temperature at the BSR need not be equivalent to that predicted for the base of the HSZ; (3) Inhibition of hydrate stability owing to the presence of salts or gases other than methane; (4) Inhibition due to capillary forces; (5) Poor knowledge of the stability curves for methane hydrate at these pressures.

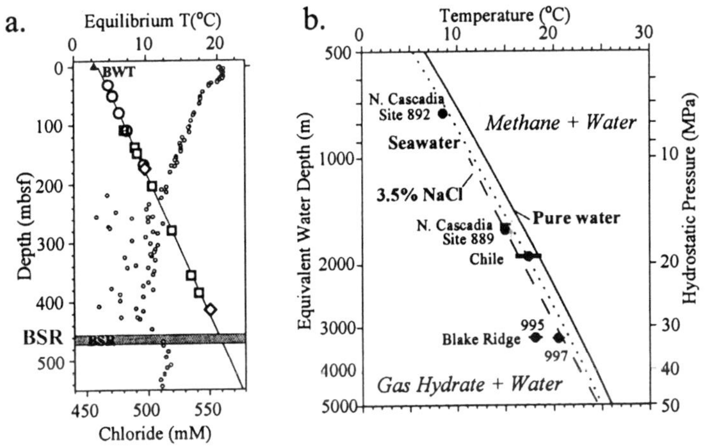

Figure 4. (a) In situ equilibrium temperature measurements at Site 997 on the Blake Ridge (Ruppel, 1997). Open symbols represent temperature data collected with different probes. The small circles denote pore water chloride anomalies (bottom scale). (b) Compilation of stability curves and estimated temperatures (solid circles) at the BSR in various hydrate provinces. The 3.5% NaCl curve is from theoretical calculations by Tohidi et al. (1995). Numbers denote sites that were part of ODP legs.

The simple analysis of Ruppel (1997) demonstrated that it was relatively unlikely that ongoing adjustments of the hydrate reservoir to Holocene climate change could explain the observed temperature disparity at the BSR on the Blake Ridge. Considering the remaining hypotheses in order, we first examine the possibility that the BSR is, in fact, not coincident with the base of the HSZ. Observational evidence (Paull et al., 1996), theoretical modeling (Xu and Ruppel, 1999), and synthetic studies (Wood and Ruppel, 2000) indicate that gas hydrate may indeed be present in situ even when an underlying BSR is absent. Thus, a BSR is not a necessary condition for the presence of hydrate. In fact, a BSR is probably only present when the supply of gas exceeds a threshhold value (Figure 5) that brings the base of the HZ into coincidence with the top of the free gas zone (Xu and Ruppel, 1999). A BSR, which is defined on the basis of the negative seismic impedance contrast created by the transition from sediment ± hydrate above to free gas below, is therefore not inherently the base of the HSZ. However, if a BSR only forms when the base of the HZ is coincident with the top of free gas, then, for practical purposes, the BSR does in fact mark the stability boundary (base of the HSZ) and should be associated with predicted temperatures equivalent to the dissociation temperature.

The section on stability constraints included a detailed examination of the different factors that can inhibit or promote the stability of hydrates. On the Blake Ridge, geochemical analyses of pore waters (Paull et al., 1996) show that the low temperatures predicted at the BSR cannot be explained by anomalous concentrations of ionic compounds in pore waters. Although small amounts of gases other than methane were detected, the presence of these gases actually implies that temperatures at the dissociation boundary should be higher, not lower, for a given pressure. Strong capillary forces arising between the clay-sized particles of the Blake Ridge sediments may inhibit stability of hydrate by several degrees (Clennell et al., 1999), which is alone sufficient to explain the observed temperature disparity at the BSR on the Blake Ridge.

A final, and somewhat troubling, explanation for the temperature disparity is that good constraints on the stability field of hydrate at the pressures characteristic of marine sediments on continental margins may still be lacking. For example, the seawater stability curve shown in Figure 4b represents an extrapolation of the results of Dickens and Quinby-Hunt (1994) to higher pressures using the formulation described by Brown et al. (1996). Comparison of the predicted BSR temperature on the Blake Ridge to a stability curve calculated using a statistical thermodynamics approach (Tohidi et al., 1995) yields a much smaller estimated temperature disparity at the BSR. As experimental procedures improve for working at the pressures (~10-30 MPa) required for methane hydrate stability at realistic temperatures (0-25°C), we will gain better knowledge of the dissociation temperatures and a clearer understanding of the significance of BSR temperature estimates.

4.3. BSR TEMPERATURES IN OTHER HYDRATE RESERVOIRS

A compilation of estimated BSR temperatures for a variety of settings in which downhole measurements have been collected as part of ODP research is shown in Figure 4b. Note that most of the estimated BSR temperatures are lower than the theoretically- or experimentally-constrained dissociation temperatures for methane hydrate. The disparity may imply either imprecision in the P-T curves for in situ conditions or a systematic problem with the downhole temperature data from which the BSR temperatures are estimated through extrapolation.

5. PREDICTING HEAT FLOW FROM BSR DEPTHS

Because the BSR coincides with the base of the HSZ, the seismically-constrained depth of the BSR can be used in combination with the stability curve and water depth to estimate first the temperature at the BSR and thence the thermal gradient and heat flux in the overlying sediment column. This type of interpretation, first detailed by Yamano et al. (1982), has been successfully applied in the Cascadia hydrate province, where the heat flux values predicted based on BSR depths are similar to true surface heat flux values measured using traditional marine heat flow probes (Davis et al., 1990). In such settings, seismic surveys, which are far less time consuming than heat flow studies, provide an important proxy (BSR depth) that constrains the thermal state of the reservoir.

In the Blake Ridge hydrate province, which has also been well-characterized thermally through the acquisition of coincident marine heat flow data (Ruppel et al., 1995), seismic data (e.g., Paull et al., 1996), and downhole temperature measurements (Ruppel, 1997), prediction of heat flux from the BSR depth is more problematic. Using the BSR depth in combination with a range of acceptable stability curves yields a prediction of surface heat flux significantly lower than the direct measurements of surface heat flux. In addition, thermal calculations based on the BSR depth predict a thermal gradient at least 11% higher than that determined directly from downhole temperature measurements. For this data set, the obvious corollary is that there is a significant disparity (~30% in some places) between thermal gradients measured by traditional heat flow methods and those determined from downhole temperature measurements.

6. TEMPERATURE PERTURBATIONS: SEDIMENTARY PROCESSES AND CLIMATE CHANGE

Under static (steady-state) conditions, the thickness of the zone in which hydrate forms in marine sediments can be predicted by combining the gas solubility curve, the geothermal gradient, and three fluxes: energy (heat) flux, fluid flux, and methane flux. Figure 5 summarizes how changing the flux rates will alter the relative thicknesses of the gas hydrate zone, the free gas zone, and the dissolved gas zones. Higher heat flux causes the base of the stability zone to migrate to shallower depths and reduces the thickness of the hydrate-bearing sediments. Higher fluid flux causes the top of the hydrate zone to lie closer to the seafloor. For a constant value of fluid flux, higher methane flux will increase the depth to the base of the gas hydrate zone until it becomes coincident with the bottom of the stability zone.

The sensitivity of the thickness of the gas hydrate zone to variations in flux rates underscores the dynamic nature of the reservoir, which will respond to changes that affect thermal regimes, fluid flux rates, or the rate of methane supply. While the effects of these fluxes on the reservoir have been here presented as largely separable, there are actually complex feedback loops. For example, superposing an upward advective fluid flux on a pre-existing conductive geotherm will lead to increased thermal gradients near the surface and lower gradients at depth (Figure 2a). In this section, we simplify the approach and focus only on the response of the gas hydrate reservoir to the temperature changes that may accompany climate change events, sedimentation, erosion, or catastrophic slumping.

The qualitative impact of temperature perturbations on the gas hydrate reservoir can be deduced by considering the effects of various processes in redistributing heat in the system (Figure 6). During climate change events, perturbations to oceanic circulation patterns or the temperature of deep currents cause long-term changes in the average BWT. Such BWT fluctuations

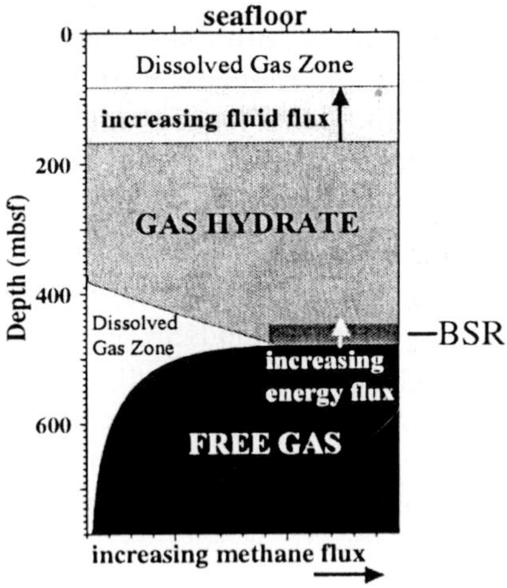

Figure 5. Effects of changes in energy (heat) flux, fluid flux, and methane flux on the thickness and positions of the HZ and free gas zone. Increasing heat flux causes the base of the HZ to lie closer to the seafloor. After Ruppel and Kinoshita (2000), based on Xu and Ruppel (1999).

propagate into the sediments on a time scale controlled by the sediment's thermal diffusivity. An increase in BWT (Figure 6a) will eventually cause temperature changes that may destabilize hydrate near the base of the HSZ and lead to partial degassing of the reservoir into the overlying sediments. If the entire reservoir is dissociated or if significant gas is emitted from the seafloor, measurable isotopic anomalies may result (Dickens et al., 1997; Ruppel and Dickens, 1999).

Erosional and depositional events may also have an important impact on the thermal state of the hydrate reservoir if re-equilibration of thermal gradients cannot keep pace with the changes caused by these processes. For typical sediment thermal diffusivity values and marine thermal gradients, even sedimentation or erosion rates as rapid as 50 mm yr^{-1} will not produce an appreciable change in the depth of the BSR if the initial geotherm is conductive and if addition or removal of sediment is perfectly isostatically compensated (no net change in water depth). On the other hand, slope failure events, which remove a substantial thickness of sediment from one area and deposit it in another, may result in a sudden change in the BSR depth. However, after thermal equilibration, the final BSR depth should be equivalent to the original BSR depth (Figures 6b and 6c), unless some other factor (e.g., BWT or background heat flux) changes. Following slumping events, individual sediment particles may cross from the hydrate zone to the free gas zone (sedimentation) or vice versa (erosion).

Dissociation of hydrate is endothermic ($\Delta H \approx$-407 kJ kg^{-1}), meaning that heat is consumed during the breakdown of the hydrate lattice to its constituent gas and water components. The endothermic nature of dissociation reduces the impact of increasing temperatures within the reservoir, since some of the heat introduced must necessarily be consumed in the dissociation process. Conversely, hydrate formation is exothermic ($\Delta H \approx 674$ kJ kg^{-1}), and cooling of an arbitrary depth within the reservoir by an externally imposed perturbation will result in net cooling by an amount smaller than the perturbation (Tzirita, 1992). Such thermodynamic effects have often been ignored by researchers in determining the impact of thermal perturbations on the hydrate reservoir and are clearly important in modulating the outcome of perturbations such as those associated with global climate change.

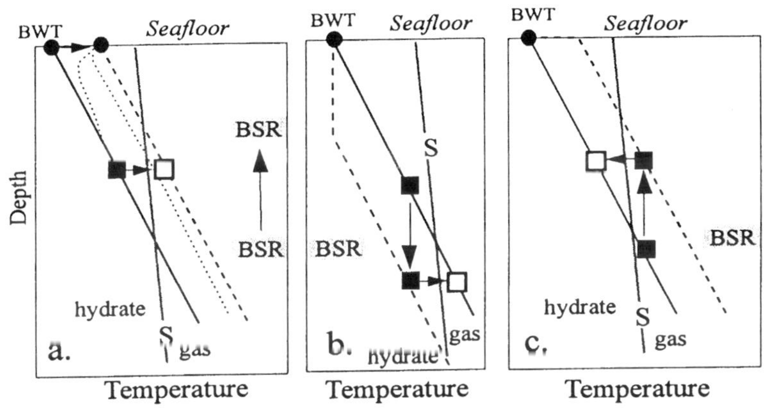

Figure 6. Qualitative effects of BWT variations and slumping (massive removal and re-deposition of sediments). In each panel, the gray line *S* represents the phase equilibrium for hydrate, and the BWT is shown by the circle. The path followed by a sediment parcel is shown by the squares, with the final position of the parcel after thermal equilibration denoted by the open square. (a) An increase in BWT with constant background heat flux causes the initial geotherm (solid line) to move to the position of the dashed line. Intermediate temperatures as a function of depth are shown by the dotted curves. The increase in BWT will lead to shoaling of the base of the HSZ. At some depths, the hydrate contained in a sediment parcel may dissociate during thermal equilibration. (b) Instantaneous, catastrophic deposition of a large thickness of sediment (e.g., during slumping) with no change in water depth (perfect isostatic compensation) will produce the perturbed geotherm indicated by the dashed line. The geotherm eventually equilibrates to its initial position (solid line), resulting in no net change in the BSR depth. (c) The effects of instantaneous erosion (during slope failure) are the opposite. Panels (b) and (c) show catastrophic events. For rates of normal sedimentation or erosion as high as several centimeters per year, thermal equilibration can easily keep pace with the slight thermal perturbations associated with the addition or removal of sediment. In such cases, there is little or no change in the BSR depth throughout the entire deposition or removal event.

This section has primarily focused on perturbations to the thermal state of the hydrate reservoir from above. However, in some settings, the thermal perturbation may be derived from below or within the reservoir. The most likely mechanism for such thermal perturbations in marine settings is likely to be disruption of the flow of energy or fluids (including gas) and establishment of new pathways for channeling fluid and energy in response to tectonic activity or even pressure fluctuations associated with oceanographic phenomena. Rapid changes in fluid flux rates and patterns (Tryon et al., 1999) and near-bottom temperatures (Macdonald et al., in review) have been observed in hydrate settings characterized by a high degree of spatial and temporal variability (e.g., Gulf of Mexico and the Hydrate Ridge feature on the Cascadia margin). Such hydrate reservoirs must clearly be characterized by a highly three-dimensional regime of energy (heat), fluid, and gas flux.

5. CONCLUSIONS

Temperature is one of the most fundamental parameters controlling the stability of the hydrate reservoir and is easily affected by a variety of processes (e.g., BWT variations, erosion, sedimentation, slumping, subsidence, and uplift) and physical parameters (e.g., bulk thermal conductivity of hydrate-bearing sediments). The best hope for constraining the thermal state of the hydrate reservoir probably remains direct measurement of equilibrium temperatures in the bottom of seafloor boreholes during drilling through hydrate-bearing sediments. In some settings, though, seismic constraints on the depth to the BSR can be combined with knowledge of hydrate stability curves to roughly constrain average heat flux and thermal gradients in the overlying hydrate-bearing sediments (Yamano et al., 1982). Observed disparities between the predicted hydrate dissociation temperature and the temperature inferred at the BSR from direct measurements in a variety of marine settings remain difficult to fully explain. If the measurements themselves are correct, then factors such as high capillary pressures (Clennell et al., 1999) or poor knowledge of the hydrate stability curves at elevated pressures may account for the disparities. Future research should provide not only higher quality in situ temperature measurements, but also better constraints on hydrate stability curves and on such important parameters as the thermal conductivity and diffusivity of hydrate.

Acknowledgements Research related to this article has been supported by the National Science Foundation (OCE-9730846), the Petroleum Research Fund of the American Chemical Society (AC8-31351), the Joint Oceanographic Institutions (F000319 and F000555), and the Ocean Drilling Program. I am grateful to R. Von Herzen for introducing me to thermal measurements, C. Paull and R. Matsumoto for supporting the in situ temperature program on ODP Leg 164, G. Dickens, W.S. Holbrook, I. Pecher, and B. Clennell for many fruitful discussions, and S. Kirby and W. Durham for their generous collaboration on the laboratory thermal measurements program. B. deMartin and J. Nimblett provided comments that improved this chapter, and B. deMartin permitted use of unpublished results obtained with W. Waite, S. Kirby, and others.

Chapter 5

Permafrost-Associated Gas Hydrate

Timothy S. Collett
U.S. Geological Survey
Denver, Colorado, USA

Scott R. Dallimore
Geological Survey of Canada
Ottawa, Ontario, Canada

1. INTRODUCTION

Gas hydrate in onshore arctic environments is typically closely associated with permafrost. It is generally believed that thermal conditions conducive to the formation of permafrost and gas hydrate have persisted in the Arctic since the end of the Pliocene (about 1.88 Ma). Maps of present day permafrost reveal that about 20 percent of the land area of the northern hemisphere is underlain by permafrost (Fig. 1). Geologic studies (MacKay, 1972; Lewellen, 1973; Molochushkin, 1978) and thermal modeling of subsea conditions (Osterkamp and Fei, 1993) also indicate that permafrost and gas hydrate may exist within the continental shelf of the Arctic Ocean. Subaerial emergence of portions of the Arctic continental shelf to current water depths of 120 m (Bard and Fairbanks, 1990) during repeated Pleistocene glaciations, subjected the exposed shelf to temperature conditions favorable to the formation of permafrost and gas hydrate. Thus, it is speculated that "relic" permafrost and gas hydrate may exist on the continental shelf of the Arctic Ocean to present water depths of 120 m. In practical terms, onshore and nearshore gas hydrate can only exist in close association with permafrost, therefore, the map in Figure 1 that depicts the distribution of onshore continuous permafrost and the potential extent of "relic" sub-sea permafrost also depicts the potential limit of onshore and nearshore gas hydrate.

The primary objective here is to assess the occurrence and distribution of permafrost-associated gas hydrate accumulations within the circumarctic of the northern hemisphere. Regions examined (Fig. 2) include northern Alaska, the Mackenzie Delta-Beaufort Sea region and Sverdrup Basin of Canada; and four physiographic provinces of Russia: West Siberian Basin, Lena-Tunguska, Timan-Pechora and several sedimentary basins in northeastern Siberia and the

M.D. Max (ed.), Natural Gas Hydrate in Oceanic and Permafrost Environments, 43–60.

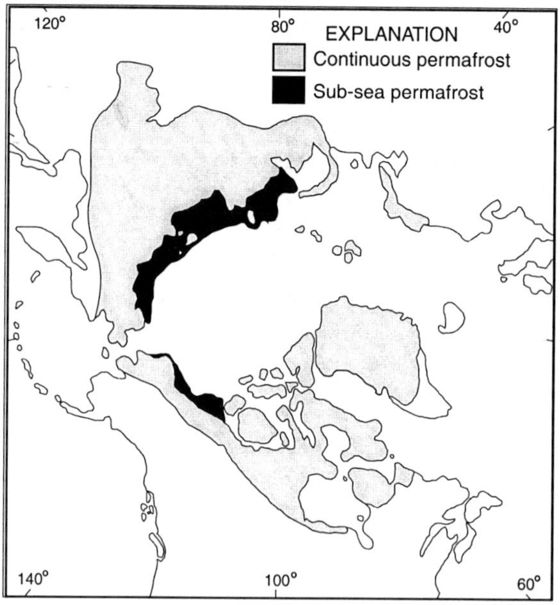

Figure 1. Distribution of permafrost in the Northern Hemisphere (modified from Pewe, 1983).

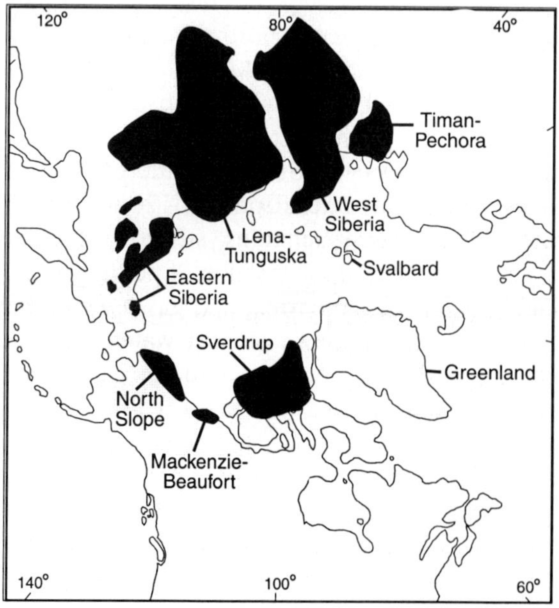

Figure 2. Location of sedimentary basins in the Northern Hemisphere that may contain gas hydrate.

Kamchatka area. The potential occurrence of gas hydrate on Svalbard and under the Greenland ice cap is also considered. Each regional discussion contains a brief description of the regional geology and a comprehensive review of the geologic parameters controlling the stability (formation temperature, pore-pressure, gas chemistry, and pore-water salinity) of gas hydrate accumulations. When available, each regional review contains descriptions of confirmed and/or inferred gas hydrate occurrences.

2. NORTHERN ALASKA, UNITED STATES

The North Slope of Alaska, encompasses all of the land north of the Brooks Range drainage divide and is generally subdivided into three physiographic provinces, from south to north: the Brooks Range, the Foothills, and the Coastal Plain. The three main structural elements that compose the North Slope are the Brooks Range orogen, the Colville trough, and the Barrow arch, all of which correspond generally to respective physiographic provinces. The geology and petroleum geochemistry for rocks on the North Slope of Alaska are described in considerable detail in a number of publications (reviewed by Bird and Magoon, 1987).

2.1. Gas Hydrate Stability Conditions

On the North Slope, subsurface temperature data come from high-resolution, equilibrated well-bore surveys in 46 wells and from well log estimates of the base of ice-bearing permafrost in 102 other wells (Collett et al., 1993). A comparison of geothermal gradients calculated from the high-resolution temperature surveys and projected from known ice-bearing permafrost depths are similar over most of the North Slope, with gradient values in the ice-bearing sequence ranging from about 1.5°C/100m in the Prudhoe Bay area to about 4.5°C/100m in the National Petroleum Reserve in Alaska (NPRA). The calculated and projected geothermal gradients from below the ice-bearing sequence range from about 1.6°C/100m to about 5.2°C/100m.

On the North Slope, pressure data from petroleum drill-stem testing in 17 wells, and log evaluation of discontinuities in overburden compaction profiles in 22 wells have been used to evaluate pore-pressures within the near-surface sediments (0-1,500 m). In general, pore-pressures in the wells we have examined in northern Alaska are a product of hydrostatic pore-pressure gradients (9.795 kPa/m; 0.433 psi/ft).

The analysis of mud-log gas-chromatographic data from 320 wells indicates that methane is the dominant hydrocarbon gas in the near-surface sedimentary rocks of the North Slope (Collett, 1993). Analysis of gas evolved from recovered gas hydrate samples in the Prudhoe Bay area suggest that the in-situ gas hydrates are composed mostly of methane (87 to 99 %). Therefore, methane gas chemistry is generally assumed for most assessments of gas hydrate stability conditions in northern Alaska.

Analyses of 55 water samples collected during petroleum formation testing in rock units from below the permafrost sequence in northern Alaska indicate that pore-water salinities range from 0.5 to 19.0 ppt. Analysis of cores from within the ice-bearing sequence of the BP 12-10-14A well (located in the Prudhoe Bay oil field) indicate that salinities within the sands of the ice-bearing permafrost sequence are also low, ranging from 0.15 to 0.50 ppt (Howitt, 1971). Gas-hydrate stability calculations in northern Alaska generally assume a low pore-water salinity of about 0 to 20 ppt.

The methane-hydrate stability zone, as mapped in northern Alaska (Fig. 3), is thickest (>1,000 m) in the area east of Prudhoe Bay. The offshore extent of the gas-hydrate stability zone is not well established; however, "relic" permafrost is known to exist on the Beaufort Sea continental shelf to a present water depth of 90 m.

2.2. Gas Hydrate Occurrences

The only direct confirmation of gas hydrate on the North Slope was obtained in 1972 when a core containing gas hydrate was recovered from a well in the Prudhoe Bay area (Collett, 1993). Well-log data from an additional 445 North Slope wells were examined for possible gas-hydrate occurrences. This review of all available data revealed that gas hydrate occurs in 50 of the surveyed wells. Many of these wells have multiple gas-hydrate-bearing units, with the thicknesses of individual occurrences ranging from 3 to 31 m. The well-log inferred gas hydrate occurs in six laterally continuous sandstone and conglomerate units geographically restricted to the east end of the Kuparuk River production unit and the west end of the Prudhoe Bay production unit. Open-hole logs from wells in the west end of the Prudhoe Bay field also indicate the presence of a large free-gas accumulation trapped stratigraphically downdip below four of the log-inferred gas-hydrate-bearing units. The potential volume of gas within the identified gas hydrates of the Prudhoe Bay-Kuparuk River area is approximately 1.0×10^{12} to 1.2×10^{12} m^3 of gas (Collett, 1993).

3. MACKENZIE DELTA-BEAUFORT SEA REGION, CANADA

The Mackenzie Delta-Beaufort Sea region, as described by Procter et al. (1984), is composed in part of modern deltaic sediments and older fluvial deposits of Richards Island, the Tuktoyaktuk Peninsula and offshore areas extending out onto the continental shelf to a water depth of about 200 m. The post Paleozoic sedimentary rocks of the Beaufort Sea continental shelf are subdivided into two major sections: pre-Upper Cretaceous and Upper Cretaceous to Quaternary strata. A major regional unconformity marks the boundary between the Upper Cretaceous and older strata. Above this regional unconformity, sedimentation was dominated by deltaic processes, resulting in a series of thick, generally northward prograding delta complexes (reviewed by Dixon and Dietrich, 1990).

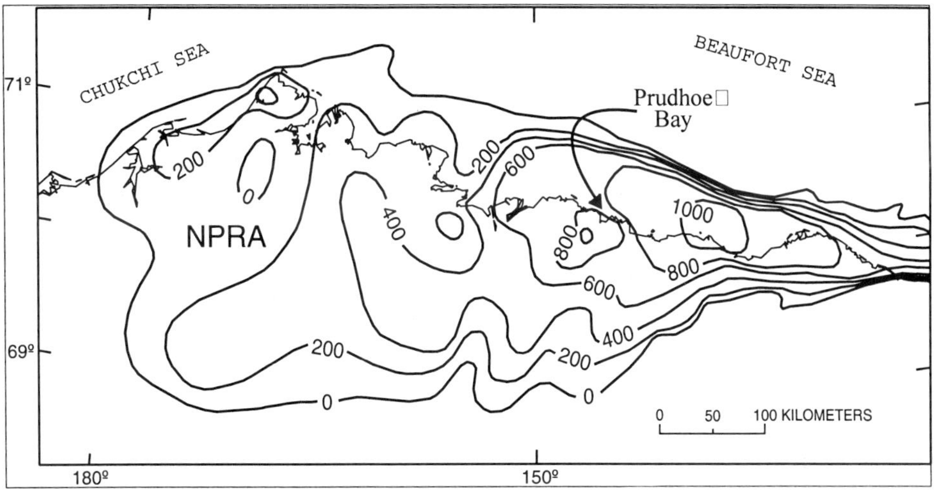

Figure 3. Map showing the thickness (in meters) of the methane hydrate stability zone in northern Alaska (modified from Collett, 1993).

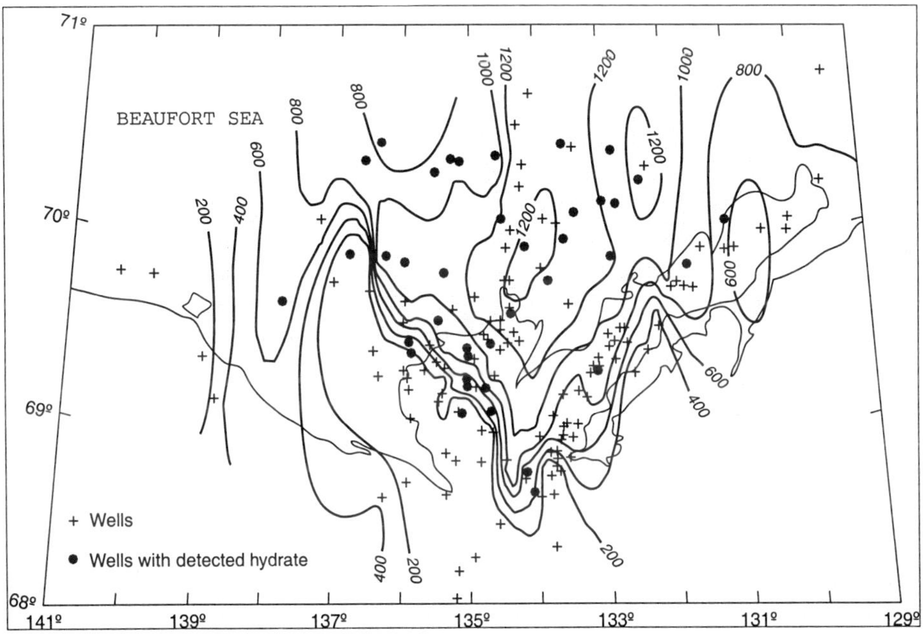

Figure 4. Map showing the depth (in meters) to the base of the methane hydrate stability zone in the Mackenzie Delta-Beaufort Sea region of northern Canada (modified from Judge and Majorowicz, 1992).

3.1. Gas Hydrate Stability Conditions

In the Mackenzie Delta area, subsurface temperature data come from industry acquired production drill stem tests, bottom hole well log surveys, and long term precise temperature studies undertaken in approximately 50 instrumented exploration wells (Judge et al., 1981; Taylor et al., 1982). The thickness of ice-bearing permafrost and related permafrost temperatures are known to vary considerably over relatively short distances in the Mackenzie Delta (Taylor et al., 1996). Beneath the permafrost interval the geothermal gradients in the Mackenzie Delta-Beaufort Sea region are relatively uniform, ranging from about 3.0°C/100m to 4.0°C/100m (Majorowicz et al., 1990; Majorowicz et al., 1995).

Pore-pressure information from beneath the permafrost in the Mackenzie Delta region suggests a variable stress regime. Data from four wells drilled offshore on the continental shelf indicate that pore-pressures are abnormally high immediately beneath the base of ice-bearing permafrost, possibly as a result of gas hydrate dissociation (Weaver and Stewart, 1982). Limited pore-pressure data from onshore wells suggest near hydrostatic pore-pressures (9.795 kPa/m; 0.433 psi/ft) immediately below the base of permafrost (Hawkings and Hatelid, 1975).

Analyses of gas samples and mud log gas chromatography data from industry wells reveal that the formation gases within the upper 2,000 m of sediment in the Beaufort Sea region consists almost entirely of methane (99.5 %) (Weaver and Stewart, 1982). Four drill stem production tests of suspected gas hydrate occurrences in two wells drilled on Richards Island in the Mackenzie Delta yielded gas composed principally of methane (99.19 to 99.53 %) (Bily and Dick, 1974). These data confirm that Structure-I methane hydrate should be expected as the primary gas hydrate form in the Mackenzie Delta-Beaufort Sea region.

The pore-water salinity of formation waters in the Mackenzie Delta-Beaufort Sea region, within the depth interval from 200 to 2,000 m are low ranging from values near 5 to 35 ppt (Weaver and Stewart, 1982; Hitchon et al., 1990; Dallimore and Matthews, 1997), which would have little effect on gas hydrate stability.

In a review of the geothermal conditions controlling gas hydrate stability, Judge and Majorowicz (1992) mapped the depth to the base of the methane-hydrate stability zone in the Mackenzie Delta-Beaufort Sea region. As shown in Figure 4, the zone in which methane hydrate can occur extends to depths greater than 1,200 m on Richards Island and is extensive beneath most of the continental shelf area of the Mackenzie Delta-Beaufort Sea region.

3.2. Gas Hydrate Occurrences

Assessment of gas hydrate occurrences in the Mackenzie Delta-Beaufort Sea area have been made mainly on the basis of data obtained during the course of hydrocarbon exploration conducted over the past three decades (reviewed by Judge et al., 1994). In addition, two dedicated scientific drilling programs (Dallimore and Collett, 1995; Dallimore et al., 1999) have included the collection of gas-hydrate-bearing core samples. A database presented by Smith and Judge (1993) summarizes a series of unpublished consultant studies that investigated well log data from 146 exploration wells in the Mackenzie Delta area. In total, 25 wells (17%) were identified as containing possible or probable gas hydrate (Fig. 4). The frequency of gas hydrate occurrence in offshore wells was greater, with possible or probable gas hydrate identified in 36 out of 55 wells (65%).

Prior to a recently completed gas hydrate research drilling program, the most extensively studied gas hydrate occurrences in the Mackenzie Delta-Beaufort Sea region were those drilled in the onshore Mallik L-38 and Ivik J-26 wells (Bily and Dick, 1974) and those in the offshore Nerlerk M-98, Koakoak O-22, Ukalerk C-50, and Kopanoar M-13 wells (Weaver and Stewart, 1982). On the bases of open-hole well log evaluation, it is estimated that Mallik L-38 encountered about 100 m of gas-hydrate-bearing sandstone, and Ivik J-26 penetrated about 25 m of gas hydrate. The well-log inferred gas-hydrate-bearing sandstone units in the Mallik L-38 well occur within the depth interval from 820 to 1,103 m, while in Ivik J-26, gas hydrate occupies a series of fine-grained sandstone and conglomeratic rock units within the depth interval from 980 to 1,020 meters. Analyses (Weaver and Stewart, 1982) of open-hole well logs and mud-gas logs, indicate that the offshore Nerlerk M-98 well penetrated about 170 m of gas-hydrate-bearing sediments, while the Koakoak O-22, Ukalerk C-50, and Kopanoar M-13 wells drilled approximately 40 m, 100 m, and 250 m of gas hydrate respectively. In all four cases, the well-log inferred gas hydrate occurs in fine-grained sandstone rock units.

The JAPEX/JNOC/GSC Mallik 2L-38 gas hydrate research well, drilled in 1998 near the site of the Mallik L-38 well, included extensive scientific studies designed to investigate the occurrence of in-situ natural gas hydrate in the Mallik field area (Dallimore et al., 1999). Approximately 37 m of core was recovered from the gas hydrate interval (878-944 m) in the Mallik 2L-38 well. Pore-space gas hydrate and several forms of visible gas hydrate were observed in a variety unconsolidated sands and gravels interbedded with non-hydrate bearing silts. The cored and downhole logged gas hydrate occurrences in the Mallik 2L-38 well exhibit both high electrical resistivities and rapid acoustic velocities. In total, the gas hydrate-bearing strata was approximately 150 m thick within the depth interval from 889 to 1,101 m.

During a permafrost-coring program in the Taglu area on Richards Island in the outer Mackenzie Delta, ice-bearing cores containing visible gas hydrate and possible pore-space gas hydrate were recovered (Dallimore and Collett,

1995). The visible gas hydrate occurred at a depth of about 330 to 335 m and appeared as thin ice-like layers that released methane upon recovery. Gas yield calculations suggest that other ice-bearing cores from a corehole in the Niglintgak field area on Richards Island also contained non-visible pore-space gas hydrate.

Estimates of the amount of gas in the gas hydrate accumulations of the Mackenzie Delta-Beaufort Sea region vary from $9.3x10^{12}$ to $2.7x10^{13}$ m^3 (Smith and Judge, 1995; Majorowicz and Osadetz, 1999); however, these estimates are generally poorly constraint. In a recent study by Collett et al. (1999), industry acquired reflection seismic data and available open-hole well logs were used to identify and map the distribution of four distinct gas hydrate accumulations on Richards Island. The total volume of gas trapped as hydrate in the four gas hydrate accumulations on Richards Island is estimated at $187{,}178x10^6$ m^3 (Collett et al., 1999).

4. SVERDRUP BASIN, CANADA

The Sverdrup Basin is a structural depression near the northern margin of the North American Craton (Figs. 2 & 5), it is about 1,300 km long and as wide as 400 km in the north-central portion of the basin. The Sverdrup Basin is bordered to the northwest by the Sverdrup Rim and to the south and east by the Franklin Foldbelt; it contains up to 13 km of Lower Carboniferous to upper Tertiary marine and nonmarine terrigenous clastics, carbonates, evaporates, basalt flows, and gabbro dikes and sills. The petroleum geology of the Sverdrup Basin has been described in numerous publications (Smith and Wennekers, 1977; Balkwill, 1978; Nassichuk, 1983, 1987; Procter et al., 1984; and Haimila et al., 1990).

4.1. Gas Hydrate Stability Conditions

Precise temperature surveys have been obtained from 32 petroleum wells drilled in and around the Sverdrup Basin (Taylor, 1988). Temperature logs from the Cape Allison C-47 well drilled in 244 m of water off the southern coast of Ellef Ringnes Island indicates that thick permafrost does not occur beneath the deeper parts of the inter-island channels. In coastal regions, however, permafrost is present and further inland beyond the marine limit, permafrost has been measured to depths as great as 700 m (Taylor et al., 1982). Thus, gas hydrate may exist on or near the subaireraly exposed islands in the Sverdrup Basin.

Temperature data from five onshore wells drilled on Ellef Ringnes Island show substantial variations due to effects of permafrost dynamics, recent marine regressions, and variable paleoclimatic histories. Geothermal gradients calculated from the temperature surveys in the offshore Cape Allsion C-47 well average about 1.3°C/100m in the Lower Cretaceous Isachesen Formation and about 2.5°C/100m in the Middle to Upper Jurassic Deer Bay Formation (Taylor et al., 1988). Geothermal gradients calculated from the temperature profiles for the five onshore wells drilled on Ellef Ringnes Island range from about

4°C/100m to 8°C/100m within the permafrost sequence and from 3°C/100m to 6°C/100m below permafrost (Taylor et al., 1988).

The review of all known technical sources have yielded no evidence of abnormal formation pore-pressure conditions within the Sverdrup Basin. Due to the lack of data, it is generally assumed that the near-surface sedimentary section in the Sverdrup Basin is characterized by hydrostatic pore-pressure conditions.

All of the known conventional gas fields in the Sverdrup Basin contain dry gas composed almost exclusively of methane (Smith and Wennekers, 1977), suggesting the potential occurrence of only Structure-I methane hydrate.

A review of available data sources uncovered no information on pore-water salinities within the Sverdrup Basin. Again, due to the lack of data, the gas hydrate stability calculations for the Sverdrup Basin have been made in the past assuming no affect from dissolved pore-water salts.

The computer program described in Collett et al. (1993) has been used to calculate the limit of the gas-hydrate stability zone in 30 onshore wells in the Sverdrup Basin (Fig. 5). The gas-hydrate stability program, as described in Collett et al. (1993), requires the following input: (1) Mean annual surface temperature which is assumed to be -20°C in the Sverdrup Basin (Taylor et al., 1988), (2) depth to base of ice-bearing permafrost (modified from Taylor, 1988), (3) temperature at the base of ice-bearing permafrost which in this case is assumed to be 0°C, and (4) the ratio between the geothermal gradient from above to below the base of the ice-bearing permafrost which is assumed to be 1.0 in this study. When present, the thickness of the methane-hydrate stability zone in the Sverdrup Basin (Fig. 5), extrapolated from available permafrost data, ranges from about 36 to 1,138 m. Due to the highly variable nature of the gas hydrate stability zone in the wells assessed, no attempt has been made to contour the stability data in Figure 5.

4.2. Gas Hydrate Occurrences

A study of downhole logs from 138 onshore exploratory wells in the Sverdrup Basin indicate that about 71% of the surveyed wells may have encountered gas hydrate, while about 17 of 30 offshore wells may have penetrated gas hydrate (Smith and Judge, 1993). Most studies dealing with the occurrence of gas hydrate in the Sverdrup Basin have been concerned with gas hydrate induced drilling hazards (Franklin, 1980, 1981). The review of limited information obtained from reports on drilling in the Sverdrup Basin infer the possible occurrence of gas hydrate on or near King Christian, Ellef Ringnes, and Mellville Islands. In 1971, during drilling of the King Christian Island N-06 well, gas leaked into the rig cellar around the outside of the surface casing,

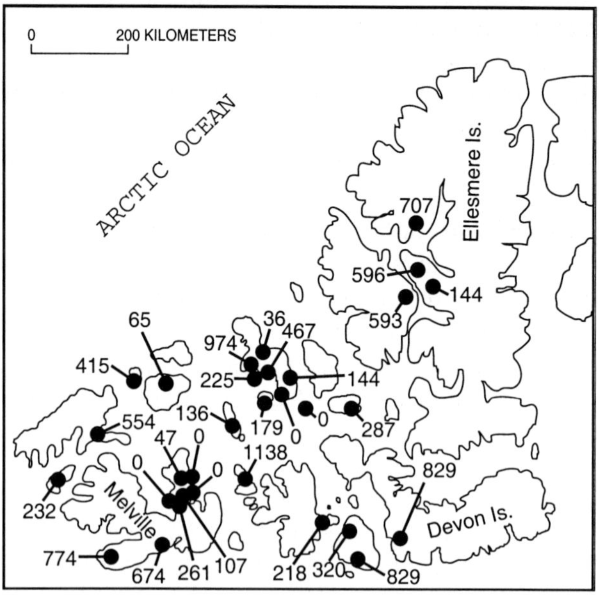

Figure 5. Map of the Sverdrup Basin (Canada) showing the location of 30 onshore petroleum wells. Also shown is the thickness of the methane-hydrate stability zone (in meters) in each well.

which was set at 160 m. A well drilled on Ellef Ringnes Island also experienced significant gas flow from behind casing (405 m) while drilling at 2,560 m. Similar gas leaks have been reported throughout the basin, which may be the result of drilling activity thermally disturbing by-passed gas hydrate occurrences. Drilling on Mellville Island has also revealed the possible occurrence of gas hydrate. For example, while drilling a well at Hearne Point several significant gas flows where encountered, one at a depth of 356 m and a second at 895 m. Hydrocarbon production test of both zones yielded classical gas hydrate test results, with low gas flow rates and shut-in pressures that slowly increased beyond hydrostatic during testing. While drilling the Jackson G-16 well off the southwestern coast of Ellef Ringnes Island (approximately 60 m water depth) gas was detected at a depth of 453 m and again at 567 m, indicating possible gas hydrate occurrences. Based on drilling reports, it appears that gas hydrate likely exists in the Sverdrup Basin; however, no direct evidence of gas hydrate has been obtained.

5. WEST SIBERIAN BASIN, RUSSIA

The geology and petroleum geochemistry of the West Siberian Basin is described in considerable detail in many English language publications (reviewed by Grace and Hart (1986). Gas production in the northern part of the West Siberian Basin is principally from the Neocomian reservoirs of the Vartov and Megion "Suites" (average depth of 2,800 m) and the Cenomanian reservoirs of the Pokur "Suite" (average depth 1,100 m). The Pokur "Suite" is overlain by the shale sequence of the Kuznetsov "Suite," which forms a regional seal for most of the underlying sandstone reservoirs.

5.1. Gas Hydrate Stability Conditions

In the West Siberian Basin, permafrost thickness increases gradually from areas of discontinuous permafrost in the south to 580 m thick in the northern part of the basin. Measured geothermal gradients range from 4.0°C/100m to 5.0°C/100m in the central and southwest portion of the basin and geothermal gradients as low as 2.0°C/100m to 3.0°C/100m are reported from the northern part of the basin (Cherskiy et al., 1985).

A review of available data uncovered no evidence of significant pore-pressure anomalies in the near-surface (0-1,500 m) sedimentary section of the West Siberian Basin. Therefore, a hydrostatic pore-pressure gradient (9.795 kPa/m; 0.433 psi/ft) can be assumed when considering gas-hydrate stability calculations in the West Siberian Basin.

The Cenomanian reservoirs of the Poker "Suite" in northern West Siberia contain mostly methane (92.5 to 99.0 %) (reviewed by Grace and Hart, 1986). Because methane appears to be the dominant hydrocarbon gas within the Cenomanian reservoirs of the basin, a pure methane gas chemistry can be assumed for gas-hydrate stability calculations in the West Siberian Basin.

Analyses of water samples collected during petroleum formation testing in Cenomanian reservoirs from below the permafrost sequence indicate that the (bulk) pore-water salinities are low (5 to 14 ppt) and would have little effect on gas hydrate stability.

Cherskiy et al. (1985) have calculated the depth to the top and base of the methane-hydrate stability zone at 230 locations in the West Siberian Basin. They determined that the depth to the base of the methane-hydrate stability zone in the West Siberian Basin ranges from zero along the Oba River to the south and reaches a maximum depth of about 1,000 m along the northeastern margin of the basin (Fig. 6).

5.2. Gas Hydrate Occurrences

Production data and other pertinent geologic information have been used to document the presence of gas hydrate in the Messoyakha field, located in the northeastern corner of the West Siberian Basin (Makogon et al., 1972; Makogon, 1981, 1988; Cherskiy et al., 1985; Krason and Ciesnik, 1985). The Messoyakha gas accumulation is confined to the Dolgan Formation of the Pokur "suite," and production has been established from the depth interval between 720 and 820 m. The upper part (about 40 m) of the Messoyakha field lies within the zone of predicted methane-hydrate stability; thus, separating the Messoyakha field into an upper gas-hydrate

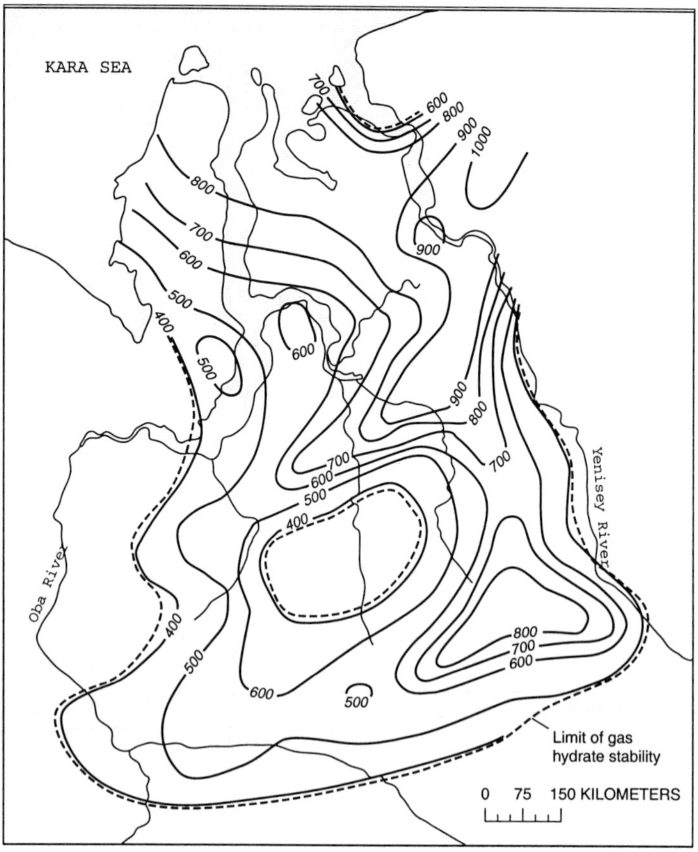

Figure 6. Map showing the depth (in meters) to the base of the methane hydrate stability zone in the West Siberian Basin, Russia (modified from Cherskiy et al., 1985).

accumulation and a lower free-gas accumulation. Prior to production, calculated total gas reserves within the gas-hydrate and free-gas parts of the Messoyakha accumulation were estimated to be about 80 x 10^9 m^3, with about one-third of the reserves within the gas hydrate (Krason and Ciesnik, 1985).

Many Russian researchers believe that long-term production from the gas-hydrate part of the Messoyakha field has been achieved by a simple depressurization scheme (reviewed by Collett and Ginsburg, 1998). When production began from the Messoyakha field in 1969, the reservoir pressure decline curve followed the predicted path; however, in 1971 measured reservoir-pressures began to deviate from predicted values. This deviation has been attributed to the liberation of free-gas from dissociating gas hydrate. Throughout the production history of the Messoyakha field it is estimated that about 36 percent (about 5.17 x 10^9 m^3) of the gas withdrawn from the field has come from gas hydrate (Makogon, 1988). Recently, however, several studies suggest that gas hydrate may not be contributing to gas production in the Messoyakha field and that the potential resource significance of gas hydrate may have been overestimated (reviewed by Collett and Ginsburg, 1998).

6. LENA-TUNGUSKA, RUSSIA

In this paper, the Vilyuy and Anabar-Khatanga basins are included in the Lena-Tunguska province of the eastern Siberia Craton. The geologic setting of the northern oil and gas provinces of Russia indicate that the Vilyuy Basin is the most promising region for the occurrence of gas hydrate. The Vilyuy Basin covers an area of about 250,000 km^2 and it is superimposed on the margin of the early Paleozoic Siberian Platform. The Vilyuy Basin opens to the east into the Pre-Verkhoiansk marginal trough, which together with the Vilyuy Basin forms the Lena-Vilyuy Basin.

6.1. Gas Hydrate Stability Conditions

Most of the Lena-Tunguska province is underlain by continuous permafrost, with thicknesses greater than 1,400 m in the north-central portion of the province (Cherskiy et al., 1985). In general, the permafrost thins toward the margins of the province and is absent to the southwest along the Yenisey River. Locally within the Vilyuy Basin permafrost is about 300 to 750 m thick and the geothermal gradient below permafrost averages approximately 2°C/100m.

Formation under-pressuring has been observed within the Lena-Tunguska province, with calculated pore-pressures being 1.5 to 3.0 MPa lower than normal hydrostatic pore-pressures. The origin of the abnormally low formation pore-pressures is unknown. Gas hydrate stability calculations in the Lena-Tunguska province must take into account the apparent affect of low pore-pressure conditions.

Relatively few gas samples have been collected from the Lena-Tunguska province due mainly to the lack of drilling. Analysis of mud-log data from wells drilled in the sandstone units overlying the Lower Jurassic Suntar shale seal indicate that methane is the dominant hydrocarbon gas in the near-surface (0-1,000 m) sedimentary rocks of the Lena-Tunguska province.

Formation pore-waters within the Middle Jurassic-Cretaceous sedimentary section of the Lena-Tunguska province have low dissolved salt content; ranging from 1 to 10 ppt. Therefore, gas hydrate stability is not likely affected by pore-water salts in the Lena-Tunguska province.

Assuming low pore-pressure gradients, methane gas chemistry, and no dissolved pore-water salts; Cherskiy et al. (1985) determined that the base of methane hydrate stability is about 2,000 m deep within the west-central portion of the Lena-Tunguska province and in the Vilyuy Basin about 800 to 1,000 meters deep (Fig. 7).

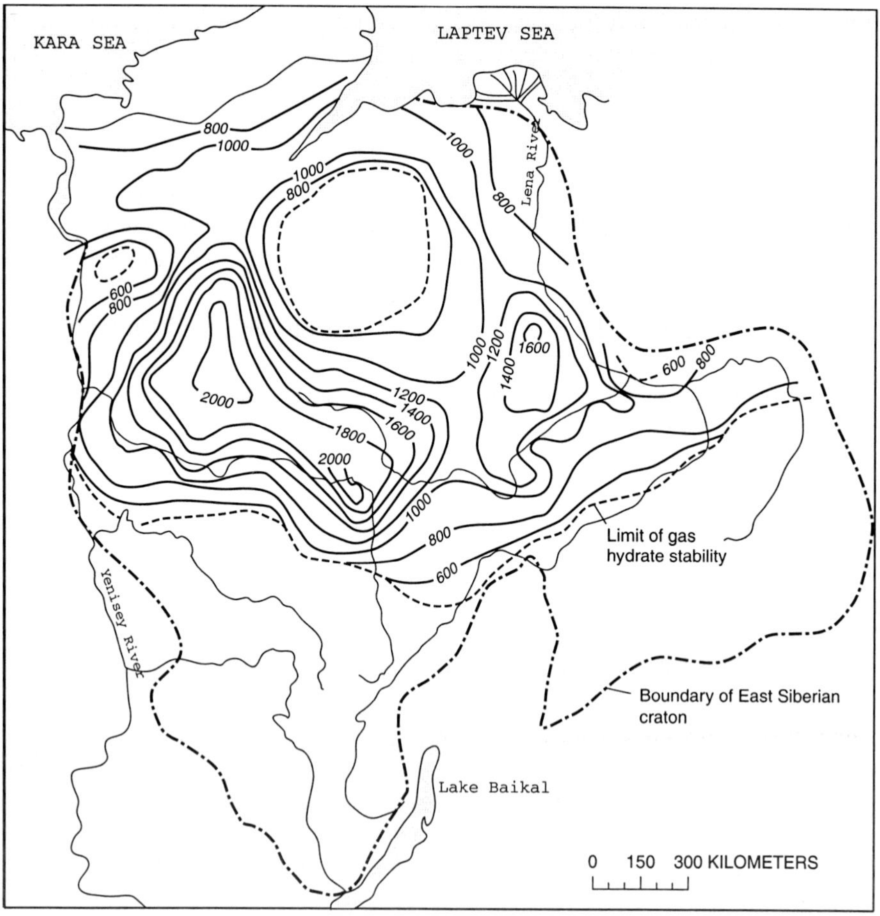

Figure 7. Map showing the depth (in meters) to the base of the methane hydrate stability zone in the Lena-Tunguska province, Russia (modified from Cherskiy et al., 1985).

6.2. Gas Hydrate Occurrences

Well data from the first 1,000 to 1,200 m of the Vilyuy Basin often show evidence of significant gas flows in the zone of predicted gas-hydrate stability. For example, while drilling at a depth of approximately 700 m in the Badaran field (Badaran Well Number 7) a gas flow of 120,000 m^3 per day was encountered, a similar flow of gas (2,000 to 3,000 m^3 per day) and water was encountered in the Bogoronts region at a depth of 500 m. Near-surface (0-1,000 m) gas accumulations were also reported from the Mastakh area. The near-surface sedimentary sequence within the Vilyuy Basin is virtually barren of conventional reservoir seals, however, permafrost may be an effective seal which may contribute to the formation of in-situ gas hydrate accumulations. Based on the occurrence of near-surface gas accumulations, it is likely that gas hydrate exists in the Vilyuy Basin, however, no direct evidence of gas hydrate has been obtained.

7. TIMAN-PECHORA BASIN, RUSSIA

The Timan-Pechora Basin occupies an area of about 322,000 km^2 in the northwestern portion of Russia (Figs. 2 & 8). The basin is bounded by the Ural Mountains on the east, by the Pay-Khoy Ridge on the northeast, and by the Timan Ridge on the northwest. To the north, the Timan-Pechora Basin opens into the Barents Sea. In the Timan-Pechora Basin the upper Proterozoic basement is overlain by a thick (3 to 4 km) sedimentary sequence of Ordovician through Lower Devonian rocks. The next sedimentary cycle, Late Devonian through Triassic, was characterized by the deposition of deep-water organic rich shales, limestones, and cherts. After a long break in sedimentation, clastic deposition in the Timan-Pechora Basin was renewed in Middle Jurassic time and continued to the end of Early Cretaceous time. Younger sediments in the basin are represented by Late Pliocene and Quaternary marine clastics and glacial deposits.

7.1. Gas Hydrate Stability Conditions

About 40% of the Timan-Pechora Basin is underlain by permafrost, with thicknesses greater than 600 m along the northeastern margin of the basin. For the most part, permafrost does not extend south of the Pechora River. The geothermal gradient in the near-surface (0-1,000 m) stratigraphic section of the Timan-Pechora Basin range from 1.0°C/100m to 3.0°C/100m.

Formation under-pressuring has been observed in the Timan-Pechora Basin (Sergiyenko and Maydak, 1982). Hydrodynamic studies show that the calculated pore-pressures are as much as 1.8 MPa lower than normal hydrostatic pore-pressures, which may significantly affect gas hydrate stability conditions.

Most of the natural gas within the near-surface stratigraphic section of the Timan-Pechora Basin is associated with coals, which yield mostly methane. Therefore, a pure methane gas chemistry can be assumed when considering gas-hydrate stability conditions in the Timan-Pechora Basin.

Permafrost appears to be absent beneath the Pechora River, which allows the meteoric recharge of low salinity waters into the Timan-Pechora Basin. Thus, the formation pore-waters within the near-surface stratigraphic section of the basin have low dissolved salt contents and would have no affect on gas hydrate stability.

Chersky et al. (1985) have calculated the depth to the top and base of the methane-hydrate stability zone at 114 locations in the Timan-Pechora Basin.

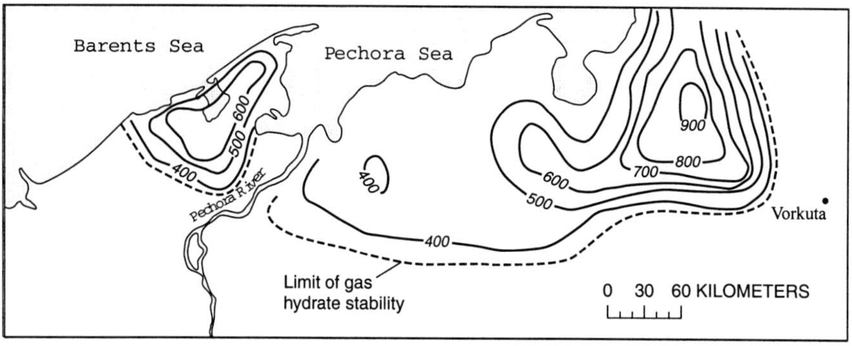

Figure 8. Map showing the depth (in meters) to the base of the methane hydrate stability zone in the Timan-Pechora Basin, Russia (modified from Cherskiy et al., 1985).

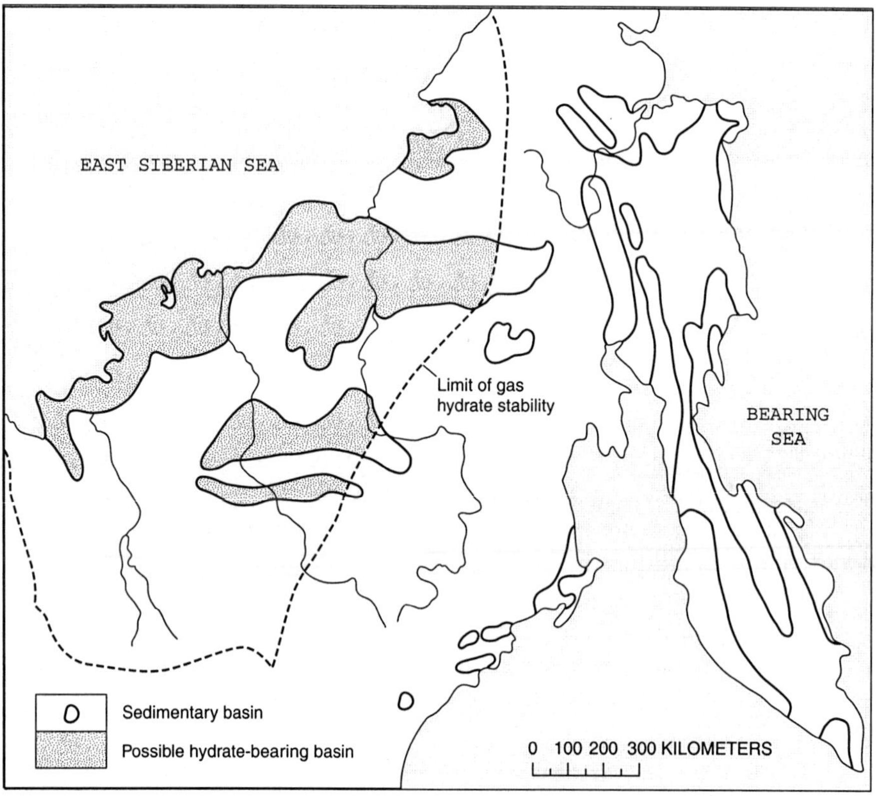

Figure 9. Map showing the location of sedimentary basins in the Northeastern Siberia and Kamchatka region of Russia that may contain conditions favorable for the occurrence of gas hydrate (modified from Cherskiy et al., 1985).

Their stability calculations assume a low pore-pressure gradient, methane gas chemistry, and no affect from pore-water salts. The map in Figure 8, of the methane-hydrate stability zone in the Timan-Pechora Basin, reveals two areas in which methane hydrate may occur. In the area east of the Pechora River the methane hydrate stability zone reaches a maximum depth of about 800 m, while to the northwest of the Pechora River a maximum depth of 600 m was calculated.

7.2. Gas Hydrate Occurrences

Little to no hydrocarbons have been discovered within the post-Permian stratigraphic section of the Timan-Pechora Basin. Small amounts of gas have been discovered with coals in the near-surface Cretaceous section, which may indicate the presence of gas hydrate. However, there is no other evidence of gas hydrate in the Timan-Pechora Basin.

8. NORTHEASTERN SIBERIA AND KAMCHATKA, RUSSIA

This region of northeastern Russia (Fig. 9), extends from the Lena and Aldan Rivers on the west to the Pacific Ocean on the east. More than 70 large to small intermountain basins have been mapped in eastern Russia. Most of these basins appear to be filled with thick (5 to 10 km) sections of upper Mesozoic and Cenozoic clastic sediments. In general, the geology and hydrocarbon potential of the basins (Fig. 9) are poorly known.

Literature reviews yield almost no information on the geologic parameters that control gas hydrate stability within the unexplored sedimentary basins of eastern Russia. Cherskiy et al. (1985) report that the temperature and pressure conditions conducive to gas hydrate formation are present in only the northwestern portion of the study area (Fig. 9); elsewhere subsurface temperatures appear to be high. Cherskiy et al. (1985) indicate that within basins with subsurface data, the base of the predicted gas hydrate stability zone ranges from a depth of about 500 to 1,000 m. In these frontier basins, however, there is no data available to assess the actual occurrence of gas hydrate.

9. SVALBARD, NORWAY

The Svalbard archipelago is located in the Norwegian Arctic between the cold Barents Sea and the relatively warm Atlantic Ocean (Figs. 1 & 2). The geology of the Svalbard archipelago is dominated by the Spitsbergen Basin, a very pronounced synclinal feature that covers most of central Svalbard. A 5-km-thick late Paleozoic through Tertiary sedimentary section has been preserved in the Spitsbergen Basin (Nottvedt et al., 1992).

Approximately 60% of the land area of Svalbard is covered by glaciers. Information from scientific and industry exploratory boreholes indicate that permafrost may cover the entire land area Svalbard with known depths ranging between 100 and 460 m (Landvik et al., 1988). From studies in Alaska and the Sverdrup Basin, it is known that in areas with permafrost depths greater than about 200 m, in-situ thermal conditions may be favorable for the occurrence of gas hydrate. Therefore, pressure and temperature conditions conducive to the formation of gas hydrate does exist in at least some portion of Svalbard.

The only direct evidence for gas hydrate on Svalbard also comes from scientific and industry drilling projects. Government and industry operators have reported significant shallow gas flows while drilling the permafrost and sub-permafrost section on Svalbard. Gas shows during drilling are often the first and only evidence for gas hydrate in many frontier regions. However, there is no data that confirms the occurrence of gas hydrates on Svalbard.

10. GREENLAND, DENMARK

A vast ice cap covers most of Greenland, and about a third of this area is underlain by sedimentary basins. Data from climate research coreholes suggest that temperatures near the base of the ice cap on Greenland are very low and these low in-situ temperatures likely extend into the underlying sedimentary basins where gas hydrate may exists. Geologic studies of regions covered by glaciers during the Pleistocene, such as the Mackenzie Delta area (Dallimore and Matthews, 1997), suggests that thick ice masses elevate pore-pressures within the underlying sedimentary basins. Thus, it is likely that the pressure and temperature conditions conducive to the formation of gas hydrate are prevalent beneath most of Greenland. However, there is no evidence of gas hydrate beneath the Greenland ice cap.

11. CONCLUSIONS

The primary objectives of this paper were to document the potential distribution of permafrost-associated gas hydrates within the circumarctic of the northern hemisphere and to assess the geologic parameters that control the stability of in-situ natural gas hydrate accumulations. Two primary factors affect the distribution of the gas-hydrate stability zone--geothermal gradient and gas composition. Other factors, which are difficult to quantify and often have little affect, are pore-fluid salinity and formation pore-pressures. Geologic studies and thermal modeling indicate that permafrost and gas hydrate may exist in all of the sedimentary basins examined in this study. However, gas hydrate has only been conclusively identified in the Mackenzie Delta-Beaufort Sea region and on the North Slope of Alaska.

Chapter 6

Oceanic Gas Hydrate

William P. Dillon
U.S. Geological Survey
Woods Hole, MA, 02543, USA

Michael D. Max
Marine Desalination Systems
Suite 302, 1211 Connecticut Ave. NW
Washington DC

1. INTRODUCTION

Many gas hydrates are stable in the deep ocean conditions, but methane hydrate is by far the dominant type, making up >99% of hydrate in the ocean floor (Chapter 2). The methane is almost entirely derived from bacterial methanogenesis, predominantly through the process of carbon dioxide reduction. In some areas, such as the Gulf of Mexico, gas hydrates are created by other thermogenically-formed hydrocarbon gases, and other clathrate-forming gases such as hydrogen sulfide and carbon dioxide. Such gases escape from sediments at depth, rise along faults, and form gas hydrate at or just below the seafloor, but on a worldwide basis these are of minor volumetric importance compared to methane. Methane hydrate exists in several forms in marine sediments. In coarse grained sediments it often forms as disseminated grains and pore fillings, whereas in finer silt/clay deposits it commonly appears as nodules and veins). Gas hydrate also is observed as surface crusts on the sea floor. Methane hydrate samples have been obtained by drilling (Fig. 1) .

2. THE GAS HYDRATE STABILITY ZONE IN OCEANIC SEDIMENT

Gas hydrate forms wherever appropriate physical conditions exist - moderately low temperature and moderately high pressure - and the materials are present - gas near saturation and water. These conditions are found in the deep sea commonly at depths greater than about 500 m (shallower in the Arctic, where water temperature is colder). The physical conditions that control the presence of methane hydrate are usually diagrammed in terms of the temperature/depth field (Fig. 2). The phase boundary (heavy line) separates colder, higher pressure conditions where methane hydrate is stable to the left of the curve from

M.D. Max (ed.), Natural Gas Hydrate in Oceanic and Permafrost Environments, 61–76.

conditions to the right where it is not. The dashed line shows how temperature conditions typically vary with depth in the deep ocean and underlying sediments.

Figure 1: Photograph of a gas hydrate sample drilled in the Atlantic Ocean on the Blake Ridge, 500 km east of Savannah, GA (Ocean Drilling Project Leg 164, hole 997A, water depth 2770 m, core depth 327-337 m below the sea floor). Photograph courtesy of William Winters, U.S. Geological Survey.

We chose typical western North Atlantic Ocean thermal conditions and imagine a sea floor at 2 km water depth (Fig. 2). Near the ocean surface, temperatures are too warm and pressures too low for methane hydrate to be stable. Moving down through the water column, temperature drops and an inflection in the temperature curve is reached, known as the main thermocline, which separates the warm surface waters, in which "geostrophic" currents are driven by winds, from the deeper cold waters, in which "thermohaline" currents are driven by density variations that are caused by temperature and salinity differences. At about 500 m, the temperature and phase boundary curves cross; from there downward temperatures are cold enough and pressures high enough for methane hydrate to be stable in the ocean.

If methane is sufficiently concentrated (near saturation), gas hydrate will form. However, like ice, the density of crystalline methane hydrate is less than that of water (about 0.9), so if such hydrate formed in the water (e.g. at methane seeps) it would float upward and would dissociate when it crossed the depth where the curves intersect. However, if the gas hydrate forms within sediments, it will be bound in place. Minimum temperature occurs at the sea floor (Fig. 2).

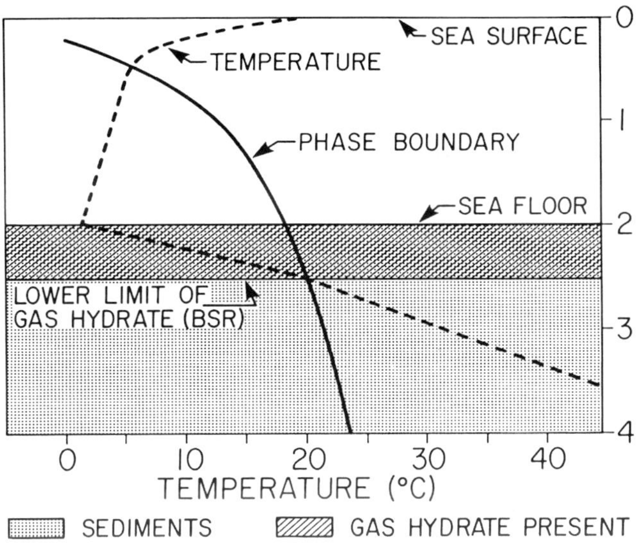

Figure 2: Stability region of methane hydrate in sea water as defined by temperature (T) and pressure (P, indicated as water depth). The heavy line defines the limit in P/T of the stability field of gas hydrate, known as the "phase boundary". We indicate the effects of having a sea floor at 2 km water depth in an area with typical temperature distribution. The variation in T with depth is indicated by the dashed line. Gas hydrate can exist where the P/T conditions are to the left of the phase boundary, thus gas hydrate cannot exist in the shallow water, nor at depths in the sediment below 2.5 Km. The gas hydrate stability zone in the sediments, in this example, will extend from the sea floor to about 0.5 km below it.

Downward through the sediments, the temperature rises along the geothermal gradient toward the hot center of the earth. At the point where the curve of conditions in the sediments (dashed line) crosses the phase boundary, we reach the bottom of the zone where methane hydrate is stable.

The precise location of the base of the gas hydrate stability zone (GHSZ) under known pressure/temperature conditions varies somewhat depending on several factors, most important of which is gas chemistry. In places where the gas is not pure methane, for example the Gulf of Mexico, at a pressure equivalent to 2.5 km, the base of GHSZ will occur at about 21° C for pure methane, but at 23° C for a typical mixture of approximately 93% methane 4% ethane, 1% propane and some smaller amounts of higher hydrocarbons. At the same pressure (2.5 km water depth) but for a possible mixture of about 62% methane, 9% ethane, 23% propane. plus some higher hydrocarbons, the phase limit will be at 28°C. These differences will cause major shifts in depth to the base of the GHSZ as you can see from Figure 2. Below the base of the GHSZ (500 m in our example in Fig. 2) methane and water will be stable and methane hydrate will not be found.

The thermal gradient tends to be quite uniform across broad regions where sediments do not vary, so, for a given water depth, the sub bottom depth to the base of the GHSZ will be quite constant. However, because a change in water depth causes change in pressure, we anticipate that the base of GHSZ will extend further below the sea floor as water depth increases (Fig. 3).

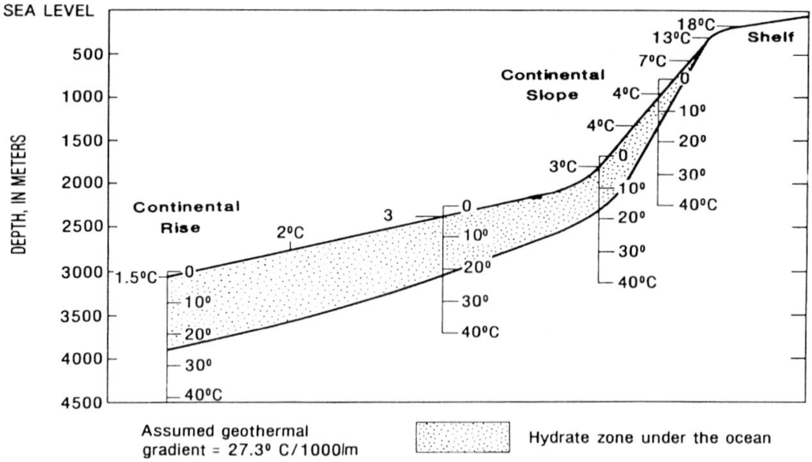

Figure 3: Inferred thickness of the GHSZ (dot pattern) in sediments of a continental margin assuming a typical geothermal gradient (from Kvenvolden and Barnard, 1982)

Fortunately, the base of the the GHSZ is often easy to detect by remote sensing acoustic methods. That will be discussed further in Chapter 21 by Peter Miles, but briefly, free gas bubbles commonly accumulate just beneath the base of GHSZ, where free gas is stable and gas hydrate will not exist. Presence of bubbles in intergranular spaces reduces the acoustic velocity of the sediment markedly. Conversely, in the GHSZ the velocity is increased slightly by the presence of gas hydrate, which in the pure state has twice the velocity of typical deep sea sediments. A large velocity contrast generates a strong echo when an acoustic pulse impinges on it. Thus we can create an image of the base of the GHSZ by measuring the return time of echoes along a profile. This approach of using seismic reflection profiles shows that the base of GHSZ generally acts as predicted and that it roughly parallels the sea floor; hence the reflection in seismic profiles has become known as the "Bottom Simulating Reflection" (BSR). The BSR is a sure sign that gas exists trapped beneath the base of GHSZ, and strongly implies that gas hydrate exists, since free gas, which has a tendency to rise, exists just below and in contact with the zone where gas would be converted to gas hydrate.

The discussion so far has implied that the zone where gas hydrate exists forms a more-or-less uniform layer below the sea floor, thickening toward greater depths. This commonly is true, but exceptions exist, generally because the thermal structure in the sediments has been disturbed so that it is not uniform. The thermal structure can be disrupted in several ways. Seafloor landslides remove the cooler near-surface sediments, leaving warmer-than-normal materials near the sea floor; that causes local shallowing of the base of the GHSZ. A second common cause of thermal disruption is the presence of salt diapirs, which produce warm spots because the salt has greater thermal conductivity than other sediments; this forces the base of the GHSZ shallower. A secondary affect associated with salt diapirs is that ions dissolved out of the salt act as inhibitors (anti-freeze) to gas hydrate formation, just as salt does to ice. A third way that thermal structure can be disrupted is by circulation of warm fluids up to shallow subbottom regions, using faults as channelways. The region around faults can be warmed enough so that a conduit is created in which temperatures are so high that gas hydrate cannot form. Thus methane and other gases moving with the fluids can reach the sea floor, often in high concentrations. Of course, on reaching the ocean water the fluids are abruptly chilled, and gas hydrate is commonly formed directly on the ocean floor. Such seafloor deposits of gas hydrate often co-exist with the escape of free methane and form distinctive biological environments that are characterized by unique organisms

3. WHERE IS OCEANIC GAS HYDRATE FOUND?

The amount of gas hydrate in the sediments of the world ocean is clearly immense, as discussed in Chapter 2. It has been identified almost anywhere that anyone has looked intensively around the edges of the continents. Methane accumulates in continental margin sediments probably for two reasons. 1. The margins of the oceans are where the flux of organic carbon to the sea floor is greatest. That is because oceanic biological productivity is highest there and organic detritus from the continents also collects to some extent. 2. The continental margins are where sedimentation rates are fastest. The rapid accumulation of sediment covers and seals the organic material before it is oxidized, allowing the bacteria in the sediments to use it as food and form the methane that becomes incorporated into gas hydrate.

4. WHERE IS OCEANIC GAS HYDRATE CONCENTRATED?

Most of the reports of gas hydrate in marine sediments, such as those mapped by Kvenvolden (Chapter 2, Fig. 1) are only indications that gas hydrate exists at some place. Almost every natural resource that we extract for human use, including petroleum, is taken from the unusual sites where there are natural high concentrations. Much of the large volume of gas hydrate may be dispersed material, and therefore may have little significance for the extraction of methane from hydrate as an energy resource. However, even if only a small fraction of the estimated gas hydrate exists in extractable concentrations, the resource could be extremely important. Little mapping of the variations in concentration of gas hydrate has been done in the world (see Dillon, Chapter 13 for the Atlantic continental margin of the United States). The important goal for the future use of methane hydrate as an energy resource is to be able to predict where methane hydrate concentrations exist.

Research still is in an early stage, but field studies in marine settings (seismic profiling, velocity studies, and drilling) suggest that the highest concentrations of hydrate commonly exist near the base of the GHSZ. This suggests that methane is being introduced below the base of the GHSZ by some process, and that the gas is probably being trapped there before entering the lower part of the GHSZ. Of course the presence of trapped gas just beneath the base of the GHSZ is demonstrated by the common occurrence of BSR's. Some of the methane may come from bacterial activity at considerable depth below the base of the GHSZ, but most is probably recycled from above. Many continental margin settings have ongoing sediment deposition. As the sea floor builds up, the thermal gradient tends to remain constant, so the isothermal surfaces must rise with the sea floor. A sediment grain or bit of gas hydrate in the shallow sediments effectively sees the GHSZ migrate upward past it as the sea floor builds up. Eventually it ends up sufficiently far below the sea floor that it is beneath the base of the GHSZ, outside the range of gas hydrate stability. Then the hydrate breaks down (dissociates) and releases its methane, which will tend to rise through the sediments because of the low density of gas and ultimately accumulate at the base of the GHSZ (Fig. 4).

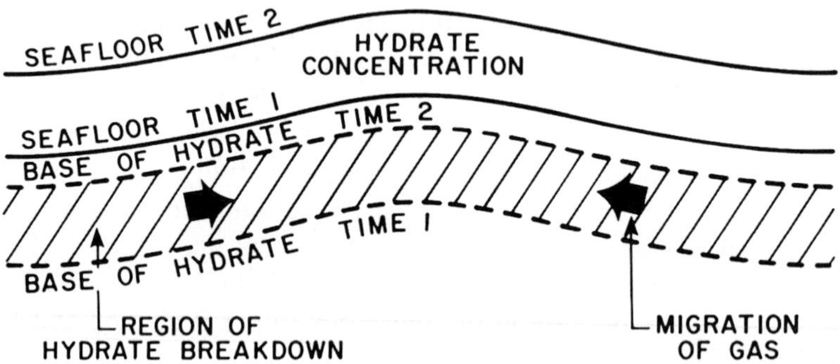

Figure 4: Diagram showing the effect of sedimentary accretion of the sea floor, which causes the base of the gas hydrate to migrate upward over time in order to maintain a constant thickness of the GHSZ. This causes continued dissociation of gas hydrate and release of gas at the base of hydrate. The gas becomes available to be recycled back up into the gas hydrate zone.

Another cause for dissociation of gas hydrate at the base of the GHSZ is present in active continental margins where sediments are thrust into a subduction zone. In such situations a tectonic accretionary wedge is built in which new sediment that is brought into the wedge is thrust in along a fault at the base. This has the effect of folding and lifting the previously accreted

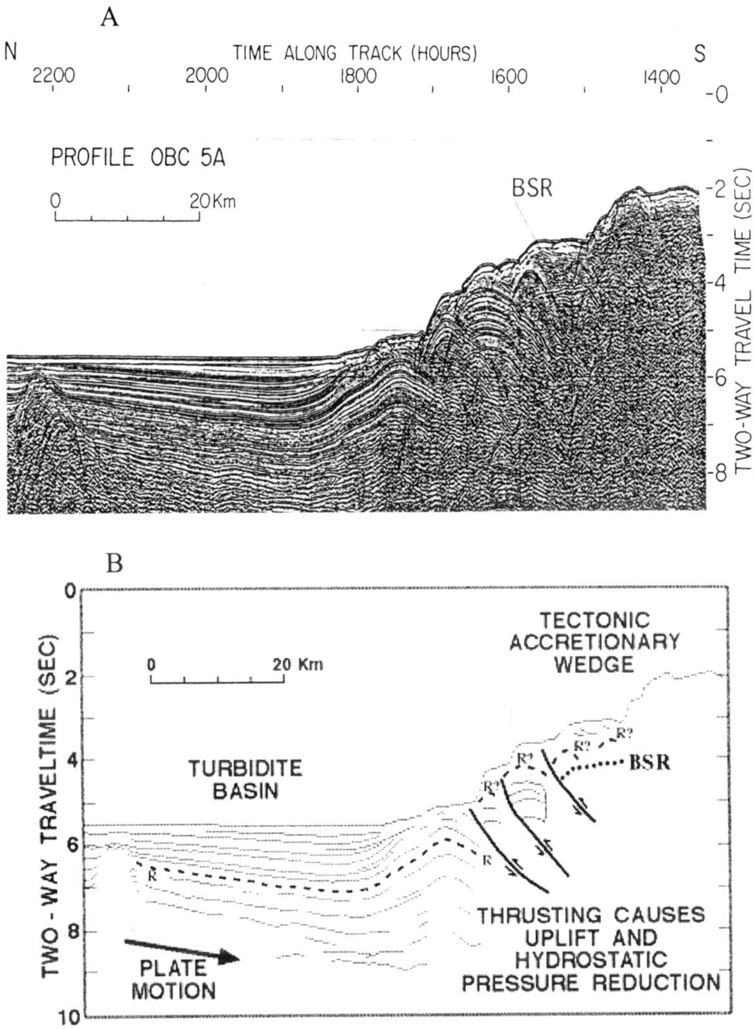

Figure 5: Seismic reflection profile (5A) and interpretation (5B)across the northern Haiti margin, where plate motions cause underthrusting of sediment that results in folding and uplift. The uplift reduces pressure, causing dissociation of gas hydrate at the base of the GHSZ and release of gas. The released gas rises, returns to the GHSZ and forms gas hydrate (from Dillon et al., 1992).

sediments. The upward movement transports the thrusted sediments into shallower water, thus reducing the pressure and causing the gas hydrate to dissociate at the base of the GHSZ, where it is at its phase limit. A seismic profile (Fig. 5) indicates the result of this mechanism and shows a BSR that transects reflectors from strata.

The gas that is released can migrate laterally, of course. When it does so, the gas will often reach a site where it becomes trapped at a culmination (shallow spot) on the sealing surface formed by the base of the GHSZ. Gas trapped at a place will tend to continuously nourish the gas hydrate above that place and produce high concentrations of gas hydrate. Gas traps at the base of the GHSZ can take several forms. The simplest is formed at a hill on the sea floor, where the base of the GHSZ parallels the sea floor and forms a broad arch or dome that acts as a seal to form a gas trap (Fig. 6). Such a hill can be a

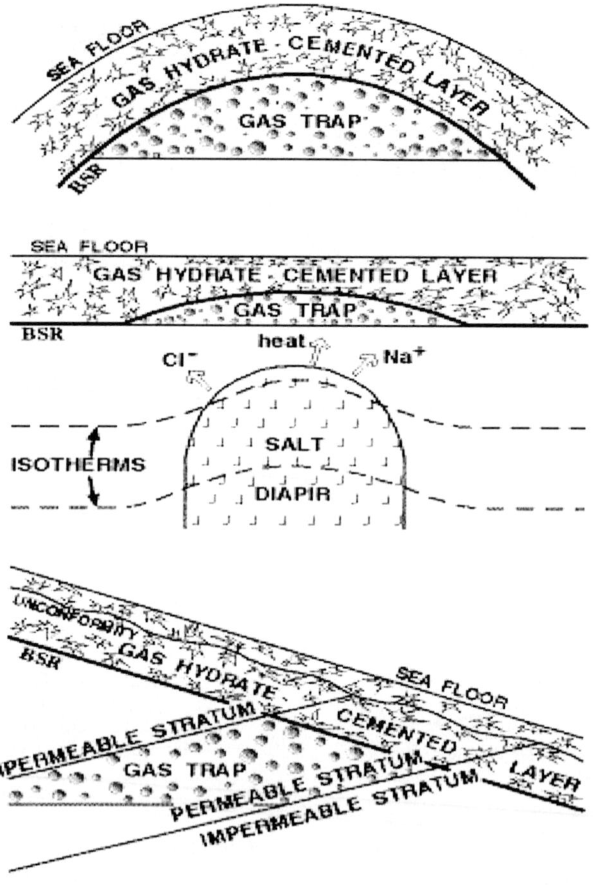

Figure 6: Diagrams of situations that can act as gas traps in which the gas hydrate layer serves as a seal.

sedimentary buildup, such as at the Blake Ridge off South Carolina, a well known site of gas hydrate concentration, or it can be a fold in a tectonic setting.

In some cases, a culmination that traps gas at the base of the GHSZ can form independent of the morphology of the sea floor. This happens over salt domes as a result of control by two parameters. First, salt has a higher heat conductivity than sediment, so a warm spot will exist above a salt dome. Secondly, the ions that are dissolved out of the salt act as inhibitors (antifreeze) to gas hydrate, just as salt lowers the freezing temperature of water ice. This double effect of chemical inhibition and disturbance of the thermal structure causes the base of the GHSZ to be warped upward above a salt dome, creating a gas trap (second panel in Fig. 6).

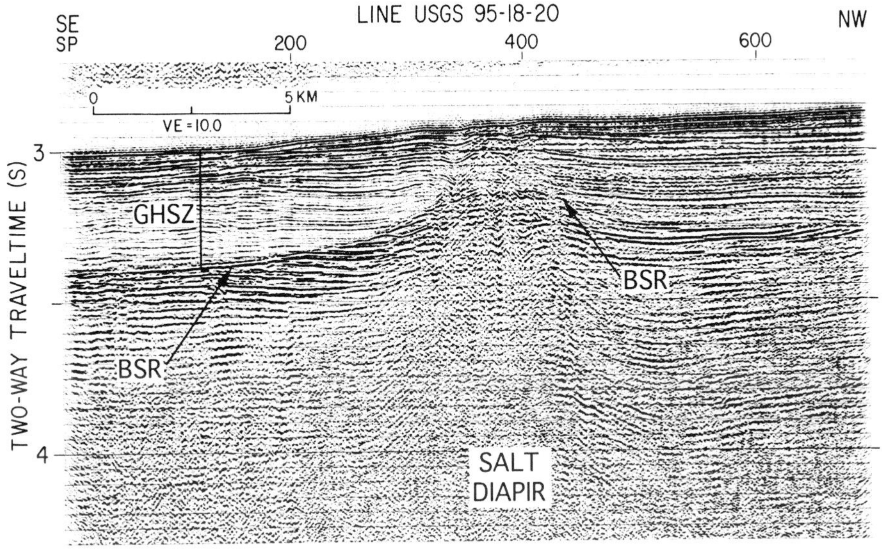

Figure 7: Seismic profile across a salt dome on the Carolina continental margin.

Figure 7 is a seismic reflection profile at a site off South Carolina (southeastern United States), where the gas hydrate layer is strongly affected by a rising salt dome (diapir). In a sense any seismic profile represents a cross section through the seabed, but keep in mind that the image is formed by reflections of sound from reflectors formed by density and acoustic velocity changes that are inherent to sedimentary layers and gas hydrate phenomena. The vertical axis is imaged in travel time of sound. The image has been computer processed so that it is essentially stretched such that vertical distances are proportionately about ten times horizontal ones (in a true section, features would appear much flatter). Water depth is about 2200 m. Notice that the deep strata appear to be bent up sharply by the rising salt. The base of the GHSZ, as indicated by the BSR, also rises over the dome and the BSR cuts through reflections that represent sedimentary layers. This does not actually represent a physical bending of the base of the GHSZ, but rather a thermal/chemical inhibition that prevents gas hydrate from forming as deeply as it would if the salt dome were not present. The extremely strong reflections just below the BSR to the left of the diapir indicate that considerable gas is trapped beneath the GHSZ. Notice that the reflections above this trapped gas, in the GHSZ are weak. We infer that the weakening of reflections, known as "blanking", may be caused by the preferential accumulation of (fast) gas hydrate in the more porous (slower) strata, thus reducing velocity contrasts that are required to create strong reflections. The blanking suggests that gas hydrate is concentrated there.

To review, we have considered three situations where a dome-like trap would form at the base of the GHSZ. In the first two (formed at a hill on the sea floor or a fold) the trap is formed because the base of the GHSZ parallels the sea floor. In the third case (at a salt diapir) the base of the GHSZ does not follow the sea floor, but rather the shape that traps gas is formed by thermal and chemical control on gas hydrate phase stability. The three cases also demonstrate the three influences that cause dissociation of gas hydrate and release of methane to the traps. In the first case, where sediments are accumulating, the controlling factor is thermal, in the second case, where the trap is being raised to shallower water depths by tectonic forces, the controlling factor is pressure, and in the third case, at a salt dome, the controlling factors are both thermal and chemical.

Of course, there are innumerable ways in which gas can be trapped beneath the gas hydrate-bearing zone and geologists need to be imaginative in searching for such situations. A simple trap (Fig. 6) in which dipping strata of alternating permeability could be sealed at their updip ends by the gas hydrate-bearing layer, forming traps in the more permeable layers. Such a system might develop in a turbidite fan, for example.

The analysis of gas hydrate concentrations as energy resources may employ an analogy to strata-bound mineral deposits, such as the Upper Mississippi Valley Pb-Zn deposits. Gas hydrate deposits are analogous in a number of important respects..

1. Oceanic hydrates are frequently strata-bound because the GHSZ is parallel not only to the seafloor, but commonly to the bedding of recent marine sediments. Where older sediments have been affected by gas hydrate formation or tectonically disturbed, the seafloor-following base of the GHSZ may pass across strata. The zone of gas hydrate formation is the important feature even though it is diagenetic or secondary to the sediment within which it occurs.
2. Oceanic hydrate may generally be horizontally distributed, rather than vertically distributed along fault wall or fault stockwork. Small-scale faulting may be important to fine-scale gas-hydrate distribution within the GHSZ, though.
3. Oceanic hydrate appears to be deposited slowly from low-temperature groundwater fluids carrying the economic material in small quantities to the GHSZ where hydrate forms.

Following this model, the first part of the process of extracting methane from gas hydrrate may utilize techniques that are adaptive analogs of mining methods that are designed to preserve mine integrity.

5. TRANSFER OF METHANE FROM THE GAS HYDRATE RESERVOIR TO THE ATMOSPHERE; CLIMATE IMPLICATIONS

The gas hydrate reservoir in the ocean sediments has significant implications for climate because of the vast amount of methane situated there and the strong greenhouse warming potential of methane in the atmosphere. Methane absorbs energy at wavelengths that are different from other greenhouse gases, so a small addition of methane can have important effects. The climate issue will be considered more extensively by Haq (Chapter 11), but briefly, if a mass of methane is released into the atmosphere it will have an immediate greenhouse impact that will slowly decrease as the methane is oxidized to carbon dioxide in the air. The global warming potential (GWP) of methane is calculated to be 56 times by weight greater than carbon dioxide over a 20 year period. That is, a unit mass of methane introduced into the atmosphere would have 56 times the warming effect of an identical mass of carbon dioxide, over that time period. Because of chemical reactions in the atmosphere, this factor decreases over time; for example, the GWP factor is 21 for a 100 year time period (Houghton et al., 1995, p. 22). Reasonably conservative estimates (Kvenvolden,1988) suggest that there is roughly 3,000 times as much methane in the gas hydrate reservoir as there is in the present atmosphere.

The methane that reaches the atmosphere can be gas released by dissociation of gas hydrate and/or gas that escapes from traps beneath the gas hydrate seal, but even in the latter case the gas will escape most easily when the seal is disturbed by dissociation of the hydrate. Warming or pressure reduction can accomplish the dissociation.

Obviously, warming will occur if ocean bottom waters warm up. However, gas hydrate will only dissociate at its phase boundary. In Figure 2, for example, the gas hydrate at the sea floor exists well within its zone of stability and a few degrees of warming will not cause it to dissociate. In Figure 2 the phase boundary is at about 500 meters below the sea floor. If bottom water became abruptly warmer the warming front would have to propagate downward through the sediment to the depth where the gas hydrate is at the phase limit, which might take hundreds or thousands of years. The change would have to occur as a conductive heat flow, as downward flow of water that could transfer (advect) heat is extremely limited in ocean sediments. Obviously, the place where warming of bottom water will have a rapid influence is where the base of the GHSZ is very close to the sea floor (see Figure 3). Atmospheric warming is presently occurring. Global surface air temperature has probably increased by roughly 0.8° C over the last century. This warming is probably being transferred to the ocean in a manner comparable to chemical tracers that have been observed, which means that warmed surface water can be expected to circulate down to depths of the shallower gas hydrates in several tens of years. In specific cases in the Gulf of Mexico, where warm bottom currents sometimes sweep through the region, and off northern California, active dissociation of gas hydrate at the seafloor has been observed (evidence is the absence of previously observed gas hydrate and release of methane bubbles). Some of this activity has been related to identifiable water temperature changes and present atmospheric warming may be leading to hydrate dissociation that would reinforce the warming trend.

Another process of gas-hydrate warming has been proposed for Arctic areas, where the ongoing sea level rise is causing ocean waters to spread across the coastal plains. In the Arctic, ground temperatures are cold enough that gas hydrate exists in association with permafrost. (see Chapter 5). Ocean water is warm compared to Arctic ground temperatures, so heat transferred from the ocean would be expected to cause dissociation of gas hydrate (Fig. 8). Attempts to verify this process in the field have not been successful to now, but transfer of methane seems likely by this process.

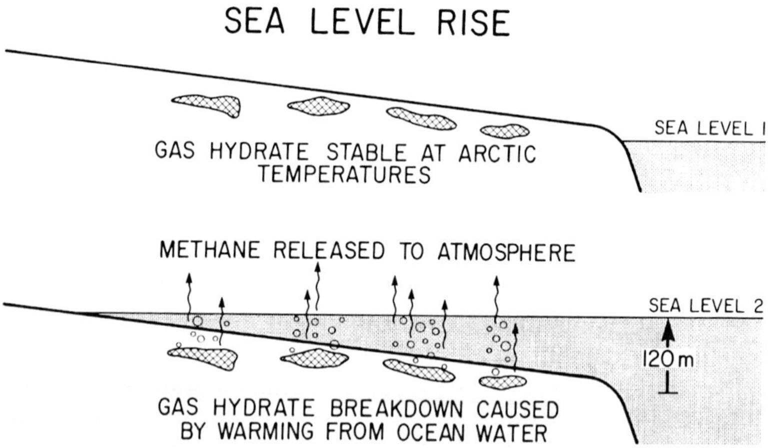

Figure 8: Methane release caused by relatively warm sea water that covers cold, hydrate-bearing coastal plain deposits in the Arctic.

Evidence for dissociation of oceanic gas hydrate by pressure reduction seems clearer at present than for the effect of warming. When sea level drops, the pressure at the seafloor and down into the sediments decreases instantaneously. Pressure is dependent on the weight of a column of water and sediment above a spot, and if that changes there is no delay in changing pressure at all depths, as there is in changing temperature. Thus a lowering of sea level, as occurred during buildup of continental icecaps that removed water from the ocean, will immediately reduce the pressure at the base of the GHSZ and cause gas hydrate to dissociate.

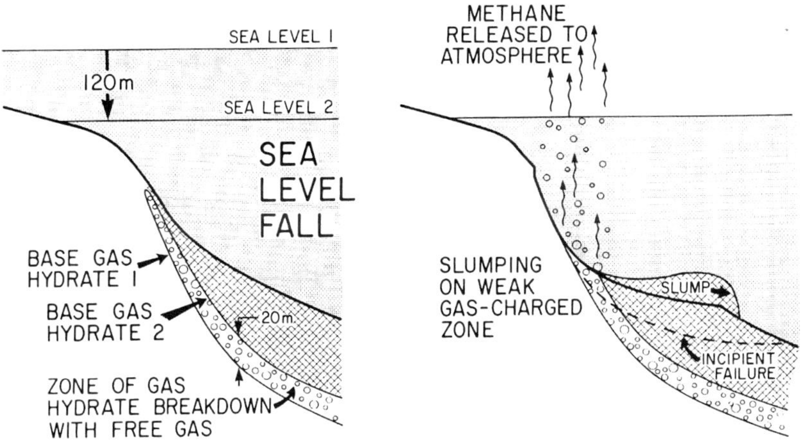

Figure 9: A drop in sea level can reduce pressure on the gas hydrate, which will cause dissociation. That can introduce water and gas into sediment pore space previously occupied by solid gas hydrate and generate overpressures. The presence of fluids and excess pressures will weaken the sediments and cause sediment slides.

The most recent glacial sea level lowering of about 120 m ended about 15,000 years ago, and must have caused significant dissociation of gas hydrate. The process would release of gas and water into a sediment that previously contained solid gas hydrate, which to some extent may have acted as a cement. The products of gas hydrate breakdown, gas + water, commonly occupy greater volume than the hydrate they were derived from, and thus the dissociation will increase the internal pressure in the pore space. Such pressures are called "overpressures", pressures greater than the column of water plus sediment above the spot.

These changes, conversion of solid hydrate to gas + water and generation of overpressures, weaken the sediment significantly and are likely to initiate sediment slides on continental slopes and rises (Figure 9). When the slide takes place, the removal of sediment reduces the load on the sediment that was below it, and thus creates another pressure reduction that may cause further gas hydrate dissociation, resulting in cascading slides. The methane released at the base of the GHSZ can escape to the ocean/atmosphere system when the capping sediment is slid away. Even on relatively flat slopes where slides are not triggered, evidence for buildup of pressure, mobilization of gas + water-rich sediments, and escape of methane, water, and sediment has been interpreted in apparent blowout structures in the Gulf of Mexico and U.S. Atlantic margin.

A seismic profile that shows the structure of such an inferred blowout off South Carolina is shown in Figure 10. Notice that the only strata disturbed are those within or just below the base of the GHSZ; its base is shown by the BSR. The subsidence of the seafloor is significant - more than 100 m over a large area.- and the loss of volume has been calculated to exceed 13 cubic kilometers. Such a volume released might have increased the methane content of the atmosphere by 4% on the basis of gas concentrations sampled by drilling in the area. This is only a single event of course. During a sea level drop, we might expect comparable events all over the world, most of which would have occurred on continental slopes and rises where the event would have triggered a slide. Further evidence that gas hydrate processes and landslides are related is provided by a map of the shallow limit of gas hydrate stability compared to the tops of known landslide scars on the U.S. Atlantic margin that is shown in Figure 11.

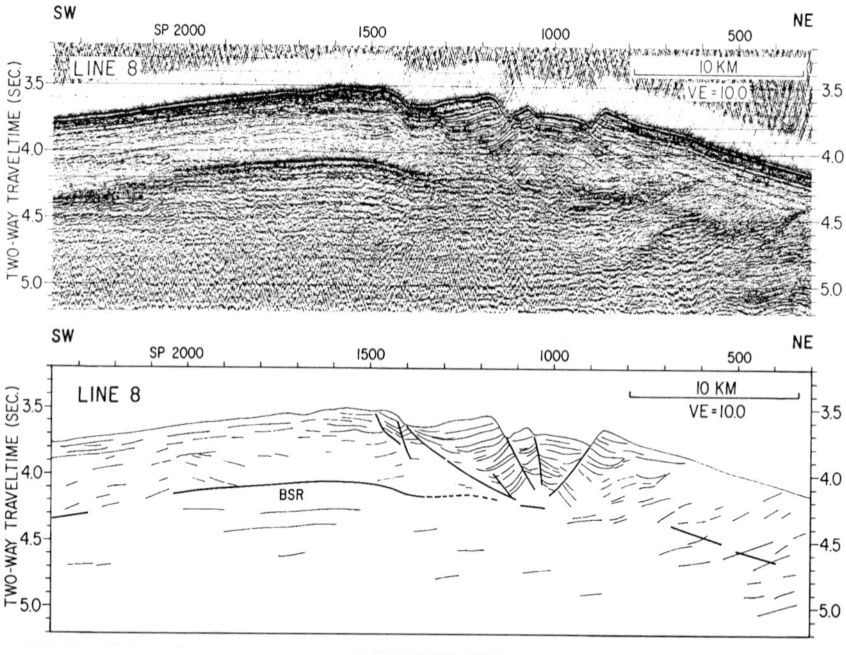

Figure 10: Seismic profile and interpretation across a blowout and collapse caused by gas hydrate processes at the Blake Ridge, a sedimentary accumulation off South Carolina.

In the long term, methane from the gas hydrate reservoir might have had a stabilizing influence on global climate. When the Earth cools at the beginning of an ice age, expansion of continental glaciers binds ocean water in vast continental ice sheets and thus causes sea level to drop. Lowering of sea level would reduce pressure on sea-floor hydrates, which would cause hydrate dissociation and gas release, possibly in association with seafloor slides and collapse. Such release of methane could increase the greenhouse effect, and cause global warming. Through such a mechanism, global cooling might cause release of a greenhouse gas and result in warming. Thus we speculate that gas hydrate might be part of a great negative feedback mechanism leading to stabilization of Earth temperatures.

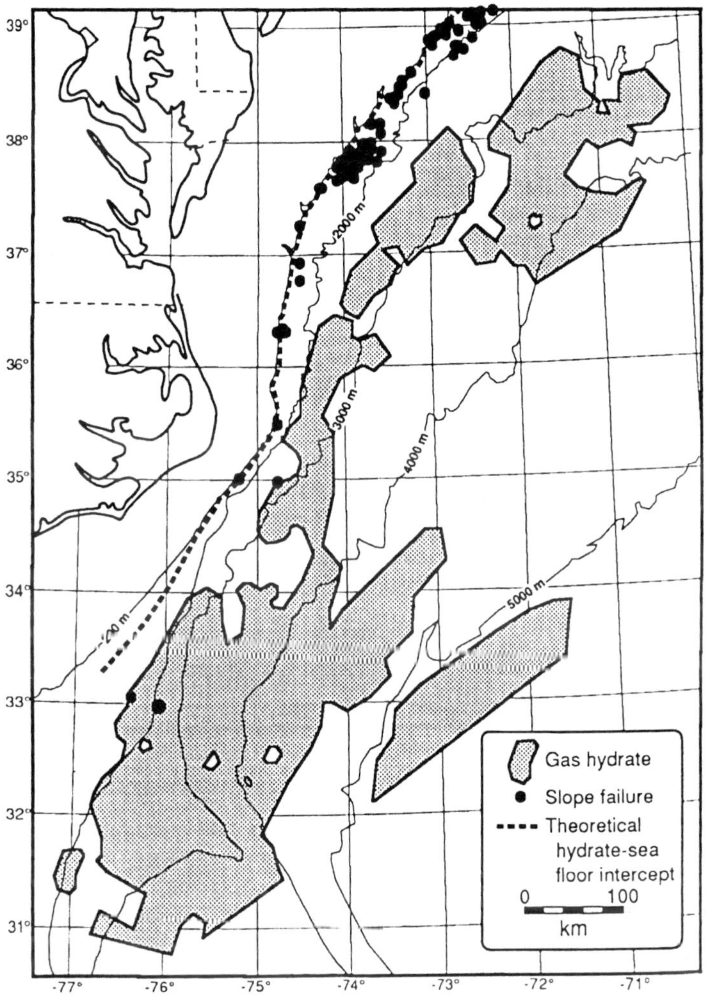

Figure 11: Map showing locations of the tops of slide scars on the U.S. Atlantic continental margin compared to the depth of the theoretical upper (shallow) limit of gas hydrate stability (Booth et al. 1994). Clustering of slope failures within the zone of gas hydrate stability, just below its upper limit, is considered to be circumstantial evidence for a relation between slumps and gas hydrate processes comparable to those diagramed in Figure 9.

At present we are just beginning to analyze the potential effects of this huge reservoir of methane on the global environment and much study is needed to understand the processes and determine which hypotheses are correct.

6. CONCLUSIONS

Gas hydrate is stable at ocean water depths exceeding about 500 m (shallower in the Arctic). Its presence is controlled by temperature, pressure and availability of appropriate gases. Almost all gas hydrate in the world is formed from bacterially generated methane, although other gases can and do form hydrate when they are available. The methane is most easily formed and sequestered as hydrate in the organic-rich, rapidly depositing sediments of the continental slopes and rises. Hydrate has also been identified in abyssal rise sediments well away from terrestrial sediment sources (Auzende et al., 2000).

Gas hydrate may form and be stable in sediments just beneath the sea floor, but temperature invariably increases downward in the sediments, so conditions will change out of the gas hydrate stability range at some depth below the sea floor. Because increased pressure makes gas hydrate more stable, the gas hydrate stability range tends to extend to greater depths below the sea floor at progressively increased water depths. Often the depth to the base of the gas hydrate stability zone (GHSZ) in a restricted region will approximately parallel the sea floor, but that relationship may break down as a result of variability in temperature, chemistry, and pressure, parameters that control gas hydrate stability. Temperature anomalies may result from such phenomena as circulation of warm fluids upward through faults, presence of salt domes (salt has a high thermal conductivity), or landslides that remove shallow, cooler sediments. Chemical changes that effect hydrate stability include variations in gas composition and salinity changes in pore water. Pressure changes can result from sea-level movement, sediment slides, or tectonic forces that cause vertical movements of the sea floor.

High concentrations of gas hydrate (possible places where methane might be extracted for energy) seem to occur where gas becomes trapped beneath a seal formed at the base of the GHSZ. We infer that the trapped gas nourishes the region above and creates zones of rich gas hydrate concentrations.

Methane is a potent greenhouse gas and methane from the gas hydrate reservoir may have escaped to the atmosphere and affected climate in the past. Some evidence suggests that the methane probably escaped to the atmosphere as a result of seafloor collapses and sediment slides. These mass sediment movements most likely were generated by dissociation of gas hydrate that weakened the sediment by generating overpressures and introduction of additional water and gas where a solid gas hydrate previously existed.

The potential significance of gas hydrate in ocean sediments to energy resources, climate, and seafloor stability are compelling reasons to continue our study of gas hydrate in the marine setting.

Chapter 7

The Role of Methane Hydrate in Ocean Carbon Chemistry and Biogeochemical Cycling

Richard B. Coffin and Kenneth S. Grabowski
Naval Research Laboratory
Washington DC 20375

Jeffrey P. Chanton
Department of Oceanography
Florida State University
Tallahassee FL 32306

1. INTRODUCTION

The microbiological role in the production of methane and the formation and stability of methane hydrates is critical to our understanding of ocean carbon cycling and global warming, and has important ramifications for sources of alternative energy, and the global economy. The methane hydrate reservoir vastly exceeds other carbon energy reservoirs (Kvenvolden, 1988). The amount of methane that is present in the ocean floor depends on the distribution of hydrates and the methane content. The estimated range of ocean gas hydrates is 26.4 to 139.1 x 10^{15} m^3 (Gornitz and Fung, 1994). The maximum content in 1 m^3 of hydrate is calculated to be 164 m^3 methane and 0.8 m^3 water (Kvenvolden et al., 1993; Max and Lowrie, 1997). Variability in hydrate methane content is controlled by geothermal gradients, methane and other hydrocarbon gas contents and by the rate of biological formation. Currently only the sketchiest details of the ocean carbon cycle in the sediment and water column are understood. This is a primary emerging research topic in ocean science.

Methane hydrates have been identified in marine sediments deposited along continental slopes. Analysis of bottom seismic reflectors indicates a large presence of hydrates from the ocean water column-sediment interface down to 1100 m (Hyndman and Davis, 1992). National and international research, driven by the possibility of commercial exploitation, has begun to obtain information about the oceanic hydrate system (gas plus hydrate). If the methane in these deposits can be obtained commercially, they may provide an extremely large energy source. The potential gas hydrate province is vast (Collett, this book), although the total magnitude of these deposits is still uncertain. A coastal region with a large and active

M.D. Max (ed.), Natural Gas Hydrate in Oceanic and Permafrost Environments, 77–90.

investigation by US scientists is along the Texas-Louisiana Shelf in the Gulf of Mexico (Fig. 1). This region has a large concentration of outcropping methane hydrates at the sediment-water column interface. This region provides the possibility to examine methane hydrate formation and stability.

Gas hydrate crystallizes, under conditions of high pressure and low temperature, when a regular lattice of water molecules includes molecules of gas, often methane, but also higher hydrocarbons--ethane through pentane (Sloan, 1990). The resulting solid is stable at temperatures below about 7°C and pressures greater than 50 bar. Recognition of the significance of hydrates for marine geology began when deeply buried layers with contours which exactly followed the seafloor were detected by seismic and deep-sea drilling studies (Markl et al., 1970; Shipley et al., 1979).

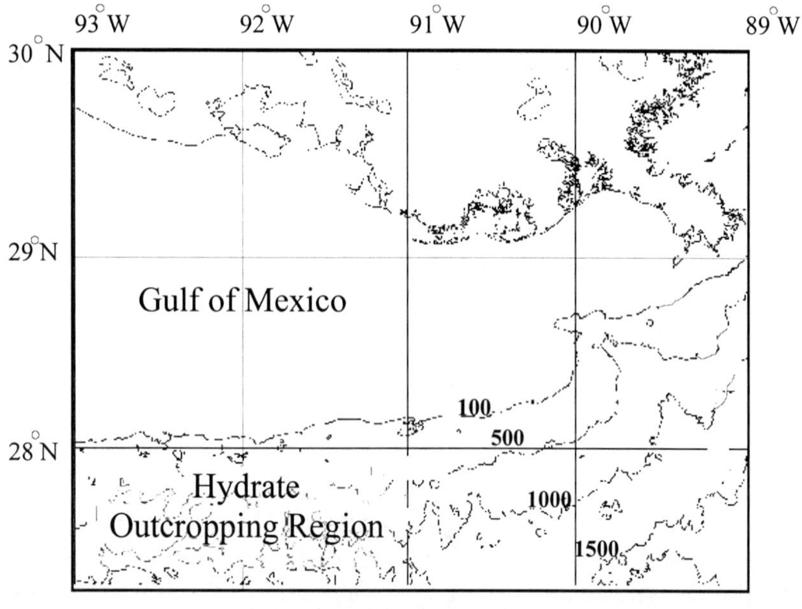

Figure 1: Region of coastal region along the Texas-Louisiana Shelf in the Gulf of Mexico that has a large abundance of surface outcropping of methane hydrate.

These BSR (bottom simulating reflectors) features are prevalent at the base of continental margins at sub-bottom water depths of about 300 m. The problem with estimating the quantity of methane stored in these reservoirs is in determining how much methane--free or in hydrates--will produce the detected change in sound velocity. In other words, some sort of calibration needs to be developed between gas quantity and seismic signal.

The primary focus of global ocean modeling is surface ocean primary production. Recently the models have been enhanced with the addition of coastal inputs. However, while surface ocean carbon cycles are their primary focus there has been realization over the last 15 years that there are numerous active regions on the ocean floor, which are more biologically active than the surface ocean. The enhanced biological activity at these sites is driven by a number of processes including geothermal energy at vents, and reduced compounds seeping from cold seeps and mud volcanoes. Methane hydrates are commonly observed at these sites.

Modeling and predicting global climate is, in part, a function of modeling carbon cycling in the ocean. It may be that the contribution of the significant biological activity on the ocean floor should be incorporated into such models, allowing for greater heterotrophic activity in the water column fueled by substrates from the seafloor (Kelley et al., 1998). Certainly, the abundance of methane hydrates in coastal oceans at relatively shallow depths could significantly impact atmospheric composition and global warming. Warming of ocean waters by 0.5 to 1.0°C has the potential to result in an enormous flux of methane to the atmosphere, which would result in further global warming (Macdonald, 1990; Paull et al., 1991).

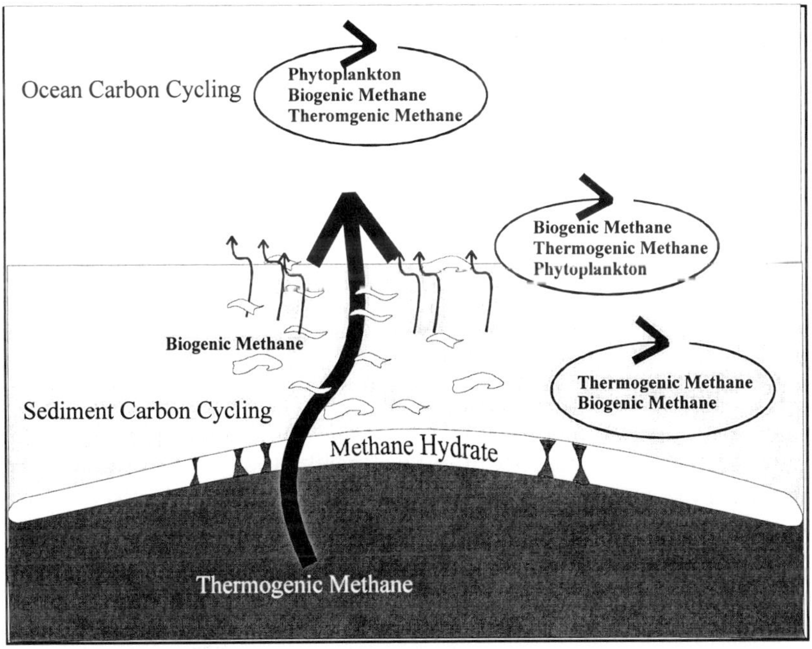

Figure 2: This manuscript outlines the contribution of biogenic and thermogenic methane to methane hydrates, and the influence of this energy on biological cycles at the sediment-water column interface and the ocean water column.

The content of this chapter will provide an overview of sedimentary processes that influence the carbon cycle within the sediment, at the sediment-water interface, and in the water column (Fig. 2). The presentation will concentrate on the microbial role in the geochemical cycles and will describe the use of carbon isotope data to assist in interpreting the cycles. Isotopic data on carbonate hard-grounds and organic rich sediments provide a record of geochemical pathways and metabolic processes that are active in the formation, stability and fate of methane hydrates.

2. APPLICATION OF CARBON ISOTOPE CHEMISTRY

There has been a broad effort in the development of elemental isotopic analyses. Stable isotope ratios have been used to examine carbon, nitrogen, and sulfur cycling to determine the major sources of organic matter that support food webs in a variety of ecosystems (Peterson and Fry 1987; Fry, 1986; Fogel et al., 1989). Further development of this technology has provided the ability to pursue cycling of carbon at the molecular level (Coffin et al., 1990; Silfer et al., 1991; Meier-Augenstein, 1995; Hullar et al., 1996) allowing identification of specific microbial roles in the biogeochemical cycling of carbon and nitrogen. In most ecosystems, a complex mixture of physical, chemical and biological factors control the carbon sources that are available to support bacterial production. Isotope analysis provides delineation of bacterial assimilation of substrates in complex ecosystems, and has provided a more thorough understanding of microbial processing (Coffin et al., 1989; Coffin et al., 1994; Hullar et al., 1996). Such methods have been developed to identify the substrate sources that support bacterial production in aquatic environments (Coffin et al., 1989; 1990).

These techniques have been combined with _--C analysis of dissolved and particulate organic carbon (DOC and POC), dissolved inorganic carbon (DIC) and methane to examine the roles of bacterioplankton in aquatic carbon cycles (Peterson et al., 1994; Coffin et al., 1994, 1997; Coffin and Cifuentes, 1999). The development of these new approaches in isotope biogeochemistry has provided the ability to determine carbon sources that support bacterial production from a variety of anthropogenic, autochthonous and allochthonous sources.

There is a large data base available on carbon isotope ratios for the components of carbon cycles in the ocean. For example, there is a long history of _^{13}C measurements of oceanic plankton provide values that vary between -18‰ and -30‰ in the surface ocean (Gearing et al., 1977; Sackett, 1991). The variation of the stable carbon isotope signature of this pool has been found to be a function of the water temperature and CO_2(aq) concentration and the phytoplankton type (Rau et al., 1989; 1991). As a result, in polar regions algal biomass is substantially more depleted in ^{13}C with values ranging from –23.2‰ at 53.3° S and a decreasing to –30.3‰ at 62°S in a transect across Drakes Passage from the coast of Chile to the Northern reaches of Antarctica (Rau et al., 1991).

In addition, carbon isotope analysis is used to delineate ocean regions where carbon cycling is controlled by terrestrial input, phytoplankton production, autotrophic fixation of carbon during chemosynthetic oxidation of reduced compounds, and assimilation of thermogenic carbon sources (Peterson and Fry, 1987; Rau et al., 1989; Sassen and MacDonald, 1997). An example key to this chapter is methane that is produced from thermal alteration of organic matter is relatively ^{13}C enriched (varying for -39 to -45‰, while methane of biogenic origin is ^{13}C depleted, varying from -55 to -110‰ (Martens et al., 1992).

An additional development in this area has been the use of molecular assays of specific bacterial markers to determine the contribution to complex cycling. Compounds such as d-alanine provide the ability to trace carbon isotope signatures through the eubacterial component of the microbial web (Pelz et al. 1997; 1998). Further delineation of the role of specific bacterial groups in carbon cycling can be obtained with analysis of phosolipids, that select key components of the archeabacterial and eubacterial community (White et al., 1979; Zelles et al., 1992). In an additional step, with the analysis of specfic biomarkers such as saturated and unsaturated isoprenoids, bacterial and archaea roles in methane cycling is determined (Elvert et al., 1999; Thiel et al., 1999). This approach provides a thorough evaluation of bacterial roles in carbon cycling.

In addition to stable isotopes and molecular tracers, radiocarbon (_—C), is particularly useful in constraining the complex carbon cycle of the ocean. Radiocarbon is formed in the atmosphere when ^{14}N is altered by cosmic rays. The quantity of ^{14}C in organic material is determined from the ^{14}C content of substrates which have been input to the material and the time since the isolation of the material from the production cycle, since radiocarbon decays with half life of 5370 years. In the ocean, _—C ranges between approximately +100‰ and an undetectable value, near -1000‰ (Cherrier et al. 1999). The _—C of dissolved organic carbon in the ocean varies widely and is related to thermohaline ocean circulation, among other factors. Open ocean surface waters range between –150‰ to –258‰ (Williams and Druffel, 1987; Bauer et al. 1992, 1998; Druffel et al., 1992). Deeper ocean water, below 1000m characteristically have _—C values that range from –393‰ to –525‰ (Williams and Druffel, 1987). A somewhat different range occurs for dissolved CO_2, as reported in results of the World Ocean Circulation Experiment (WOCE) for the Pacific Ocean (von Reden et al 1997). In that work, the surface ocean waters were reported to have a _—C range from about –50 to +150‰, while deeper waters below 1000 m have a range from about –160 to –220‰.

Radiocarbon isotope analysis is especially useful in tracing carbon cycling at the interface of petroleum, geothermal or seep carbon sources which contain little to no radioactive carbon (Brooks et al., 1987, Paull et al., 1989, Bauer et al., 1990), and contemporary carbon sources fixed from atmospheric CO_2, which contain abundant radiocarbon. Analyses of _—C have also been employed to study the cycling of aged and recent carbon sources in river, estuarine and wetland aquatic systems (Schiff et al., 1990; Chanton et al., 1995, Chasar et al., in press). Recently, methods have been developed to use _—C to differentiate between bacterial assimilation of new relative to aged carbon (Bauer et al., 1990, 1992, 1995; Cherrier et al., 1999). Cherrier et al. (1999) extended methods developed to study _—C of bacteria (Coffin et al. 1990) to radiocarbon analysis. A result of this effort was the finding that approximately 20% the carbon assimilated by bacteria in the surface waters in the mid-Pacific Ocean was older, presumably recalcitrant material. This result initiates a new understanding of ocean carbon cycling. In addition to _—C of bacteria, current methods include analysis of _—C for DIC, POC and DOC (Bauer et al. 1990, 1992, 1995; Chasar et al., in press).

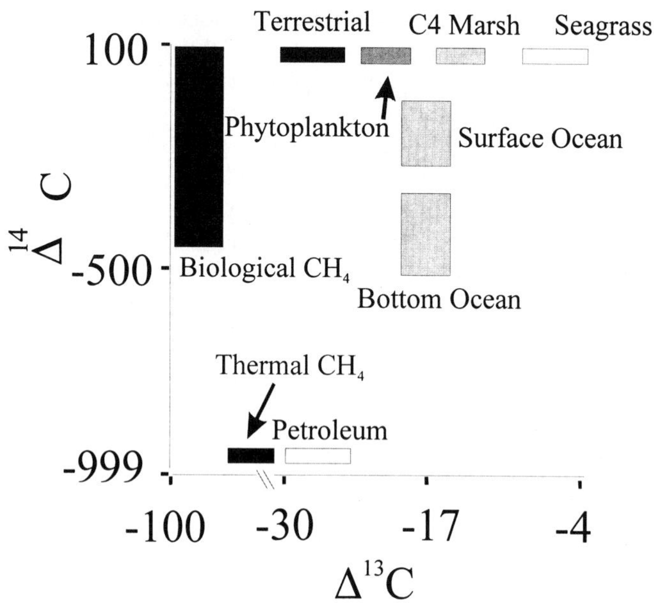

Figure 3: Two-dimensional carbon isotope analysis applied to define carbon sources in the ocean. This diagram is a composite of carbon isotope signatures referenced in Coffin et al. 1989, 1994; Bauer et al. 1992, 1998; Cherrier et al. 1999; and Peterson and Fry, 1985.

The combination of radio and stable carbon isotope analysis provides a useful tool to delineate carbon sources and cycles in complex environments (Fig. 3). Analysis of a series of carbon pools can be used to determine sources and fate of carbon. This approach is being used for a understanding the sources of methane in hydrates and the geochemical and biogeochemical cycles that control methane hydrate stability and formation.

3. METHANE CYCLING IN THE SEDIMENT

Ocean floor carbon cycling is controlled by a complex mixture of biological, chemical and physical processes. Physical factors include the motion of the continental plates resulting in expulsion of fluids charged with reduced volatile compounds, flows of geothermal energy toward the sediment-ocean interface, cold pressure mediated seeps of reduced compounds, and transport of terrestrial carbon to the sea floor. These processes control chemical speciation through key elemental pools that affect biological diversity. For example ocean floor thermal seeps have a high flux of reduced compounds, e.g. Fe^{+2}, CH_4, H_2S and NH_4 that supports the chemoautotrophic population. Enzymatic oxidation of reduced compounds results in fixation of CO_2 into cellular biomass which support the foodweb.

In contrast petroleum seepage along the Texas-Louisiana Shelf fuels some portion of heterotrophic activity on the ocean floor. In many of these systems methane hydrates are present. Comparisons of different sites have shown that there is a wide range of reduced compounds which support both heterotrophic and chemosynthetic bacterial activity. The relative portion of heterotrophic and autotrophic carbon fixation and energy transport to higher trophic levels need to be determined. Nor has the relative portion of the biogeochemical and thermal input to the hydrate formation and stability been widely determined. The following text provides an overview of the variation of biogeochemical cycles in the ocean floor and known abundance and formation of methane hydrates in these regions.

Methane hydrates are present in sediments with carbon contents greater than 0.5% and methane concentrations that are above 10 ml L^{-1} (Kennicutt et al., 1993). Methane hydrate formation is associated with geothermal and biological sources, however, at the majority of sites that contain methane hydrates there are geothermal source of reduced compounds induce methane production. In addition to the direct conversion of thermal methane into the hydrate compounds such as H_2S, NH_4^+, and Fe^{+2}, that supports the autotrophic bacterial assemblage at a rate where CO_2 becomes the dominant electron acceptor and biological methane is incorporated into the hydrate. This must occur at a rate where all of the energetically more efficient electron acceptors, e.g. nitrate, iron, manganese, and sulfate are consumed. The formation of the hydrate through the vertical profile in the sediment depends on the presence of an upward advective or diffusive flux of geothermal compounds. Blake Ridge, in the Atlantic Ocean, off the coast of the Carolinas and Georgia is an exception where the dominant input of organic matter originates from sedimentation derived from coastal origin and water column primary production (Paull et al., 1994). For biogenic methane to support significant hydrate formation, organic input must occur at a rate where all of the energetically more efficient electron acceptors, e.g. nitrate, iron, manganese, and sulfate are consumed. A process that detracts from the formation of hydrates is anaerobic oxidation of methane using sulfate as the electron acceptor (Borowski et al., 1997).

The Texas-Louisiana shelf of the Gulf of Mexico is a region rich in gas hydrates. This region is a salt basin that formed during the Late Triassic rifting of the Pangea super continent, and was subsequently flooded by a thick salt deposit during the middle of the Jurassic marine incursions (Salvador, 1987). Thermogenic petroleum and gas deposits produced from Miocene to Pleistocene reservoirs in the region originate from deep buried marine sources of Mesozoic age (Kennicutt et al 1993; Sassen et al. 1994.). In this region are over pressured fracture zones that surround moving salt diapers and sheets, as well as dynamic faults that provide active conduits for vertical migration from deep reservoirs to shallow pools and the surface (Brooks et al., 1984, Sassen et al. 1993a, in Sassen et al. 1994). This physical structure and the availability of reduced compounds in the site support a wide range of chemosynthetic communities (Sassen et al. 1994). Carbon isotope analysis of the sediment and structural analysis of the hydrates demonstrates a high input of thermogenic methane in this region. For example methane _^{13}C ranges between –36 to –40‰, indicating little or no biological recycling (Sassen and MacDonald, 1997). Structural analysis of the hydrates also supports this observation. In this region hydrate structure II is dominant as a result of the hydrate formation in the presence of a mixture of reduced thermogenic compounds (Sassen and MacDonald, 1997).

While thermal methane is present in the hydrates in the Gulf of Mexico this should not be assumed to be the only source of methane. Microbial productivity is known to be an important component of the carbon cycle at natural seeps where free-living bacteria consume hydrocarbons resulting in anoxic sediments and sulfate reduction (Sassen, 1980; Sassen et al., 1994; Sassen et al., 1993). Sediments in this region have been measured with mean concentrations of petroleum and C1-C5 gases at mean concentrations of 5650 ppm and 12,979 ppm, respectively (Sassen et al., 1994). In addition to the advection of methane in this system, petroleum migrates through the sediment and stimulates the heterotrophic biological activity which results in the reduction of electron acceptors and the production of biogenic methane. Evidence for biogenic methane at this site comes from carbon isotope ratios as light as -57‰ in tube worms and mussels (Brooks et al., 1987). This combination of methane cycles (Figure 4) is likely to result in multiple sources of methane in the hydrates.

Preliminary data suggests that the Håkkon Mosby Mud Volcano in the Norwegian-Greenland Sea is also a complex system with a combination of biological and geologic cycles that control the methane hydrate formation. This location is one of the more recent sites where methane hydrate samples have been recovered. A recent issue of *Geo-Marine Letters* (Volume 19, Number 1/2, 1999) describes various aspects of this site in a number of papers. Unlike many other hydrate-rich regions, their formation here is associated with venting caused by past glaciations. Mud flow in the volcano is thought to be caused by a combination of 1) the rise of lower density preglacial biogenic silica oozes buried beneath higher density glacial marine sediments, and 2) the regions history of massive submarine slides (Vogt et al 1999). Associated fluid flows carry methane to the surface of the sea floor, allowing hydrates to form.

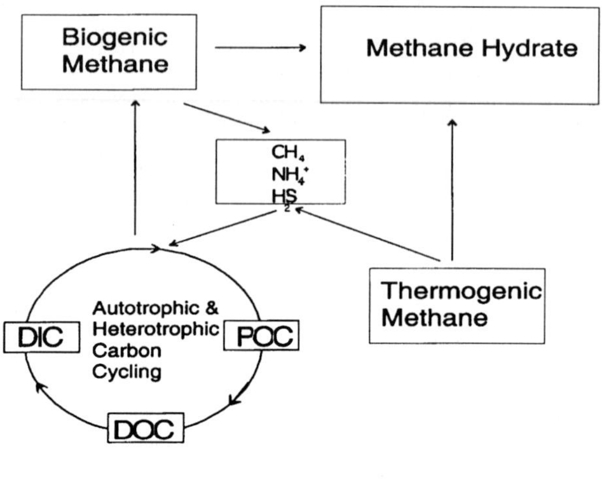

Figure 4: Contribution of thermogenic and biogenic methane to the formation of methane hydrates.

In earlier work on methane from Haakon-Mosby hydrates, $\delta^{13}C$ values around –60‰ were reported, suggesting methane in these hydrates is a mixture of thermogenic and biogenic sources (Lein et al 1999). Vertical profiles of the microbial assemblage document active sulfate reduction, methane generation and methane oxidation (Pimenov et al. 1999). At this location both aerobic and anaerobic methane oxidation is observed (Pimenov et al., 1999). As noted in the Gulf of Mexico a diverse series of geochemical cycles results in the methane production. In the Håkkon Mosby Mud Volcano the hydrates were survey in cores taken across the mud volcano. Hydrate content through the transect range from 10-20% to 0%, by weight, over a 750 m span of the sediment (Ginsburg et al., 1999). The average through this region is 1.2% hydrate. Given the high methane content of hydrates this is a substantial methane loading.

In contrast, Blake Ridge methane gas is microbial in origin and does not have a significant thermogenic input (Paull et al., 1994.). At this site isotope analysis of the methane and CO_2 pools provide a thorough understanding of the sources. Here the $\delta^{13}C$ indicates that methane is produced by biological reduction of CO_2 and that there are diverse mechanisms for the transport of methane through the sediment (Borowski et al., 1997). Ranges in the methane $\delta^{13}C$ are –85 to –103‰. The $\delta^{13}C$ that is more depleted represents diffusive methane transport through the vertical gradient in this region. With the slow vertical transport there is cycling of the methane and CO_2 that results in isotope depletion (Figure 1). This conclusion is supported with an observation of light $\delta^{13}C$, -30 to –51‰, in the DIC pool in the same region.

Further understanding of the diffusive methane transport in this system is obtained with the analysis of sulfate and $\delta^{34}S$ analysis in the vertical profiles (Borowski et al., 1996). In the sediment, sulfate profiles provide an understanding of the carbon sources that is being cycled (Chanton et al., 1993). If the concentration of sulfate declines from the sediment-water column interface to depth, the source of organic matter is sedimentation from the surface ocean. At Blake Ridge sulfate concentrations are linear through the upper region of the sediment indicating the methane originates from deeper deposits of organic matter. Sulfate concentrations decline in the deep sediments as a result of anaerobic methane oxidation where the sulfate is the terminal electron acceptor. In this region there is a strong gradient in the $\delta^{34}S$ profile of sulfate with values that are approximately 60‰ enriched. This gradient in the isotope signature results from the reduction of sulfate.

The role of bacterial sulfate metabolism in methane hydrate stability and formation has been investigated on the Cascadia Margin, in the northeastern Pacific Ocean. In this region bacterial biomass and activity are stimulated in specific regions of the hydrate zone; 74 and 225 mbsf (Cragg et al. 1996). Bacterial activity at these sites was analyzed in terms of total production and methane oxidation. Preliminary analysis of this region indicates that there is a fluid flux through the accretionary wedge that provides electron acceptors to these specific depths. Primary findings at these depths were that the production and activity of the bacterial population in vertical profiles was greater in the regions with methane hydrates. Also, there was lower bacterial production in regions with high H_2S concentrations. It was speculated that the lower production resulted from H_2S toxicity. Subsequent analysis of bacterial activity and community structure showed a large diversity through vertical profiles where arechaea are dominantly responsible for methane oxidation in the deep anoxic sediments and aerobic methanotrophs oxidized methane in the shallow sediments (Elvert et al. inpress).

4. AUTOTROPHIC ACTIVITY AT THE SEDIMENT WATER COLUMN INTERFACE

Methane hydrates have also been observed in shallow sediments (Sassen et al., 1993; Vogt et al., 1999). In the Gulf of Mexico these surface sediment hydrates are observed to be present in a region where the stability as a function of pressure and temperature is not predicted. There appears to be a continuous formation and dissociation of these surface hydrates (MacDonald et al. 1993). Carbon isotope analysis suggests that there is a wide range of carbon sources that control the presence of the hydrates in this region. Radio carbon isotope analysis indicates that the flux of thermal methane and hydrocarbons supports a large part of the biological activity (Brooks et al., 1987). For clams, mussels and tube worms _ ^{14}C ranged between –840 and –204‰.

While the more negative values indicate that the thermogenic carbon sources are supporting the surface organisms, the large range in values indicates a host of carbon sources and the more positive value demonstrates an alternate source. In these cases there is active biological cycling on the hydrate surface which may contribute to methane hydrate stability. In addition, the dissociation of the hydrates provides energy to the active benthic community. More recent carbon isotope analysis of organic sediments in hydrate rich regions supports the observation that there is a complex mixture of thermogenic and biogenic carbon sources that influence the methane hydrate formation and stability (Coffin and Grabowski, unpublished). These regions of the ocean floor-water column interface have been observed to have high biological activity.

Closer examination of the biology at these sites indicates that the higher trophic levels are supported with a diverse array of chemosynthetic carbon fixation cycles. A large number of the organisms obtain energy through symbiotic cycles with the chemosynthetic bacteria. Along the Texas-Louisiana Shelf there has been several years of analysis of the biota on the ocean floor. In this region active crude oil and gas seepage is widely distributed along the continental slope.

This energy supports a wide variety of organisms at the base of the food chain, the petroleum appears to drive the system to hypoxia which results in high availability of hydrogen sulfide and biogenic and thermogenic methane (Sassen et al. 1993). This activity in the reduced environment results in the formation of energy sources for the chemosynthetic bacterial community. The high microbiological activity at this site results in mounds of authigenic carbonate minerals. The formation of the carbonate base is a result of methane oxidation associated with sulfate reduction which increases alkalinity causing the carbonate formation (Paull et al., 1992). The activity by the bacterial consortium occurs independently in the sediments and in symbiotic association with a broad range of organisms. Sulfur oxidizing bacteria, *Beggiatoa*, are known to fix CO_2 in association with H_2S oxidation (Zobell, 1963). *Beggiatoa* are abundant in mats through the regions that contain high loading of methane and oil in the sediments (Sassen et al., 1994).

In addition, chemosynthetic activity supports a large amount of the food chain through symbiotic associations. Tube worm bushes are abundant through this region of the ocean floor, on the crest of gas hydrate mounds, with bacterial symbionts that oxidize H_2S (MacDonald et al., 1989). Symbiotic bacteria are also the basis of very high metazoan biomass at seeps (MacDonald et al., 1989) exhibited by methane-oxidizing mytilids (Childress et al., 1986; MacDonald et al., 1990) and sulfide-oxidizing vestimentiferans (Fisher, 1990; MacDonald et al., 1990). This symbiotic relationship is a complex balance of elemental cycling.

While the host receives carbon in the form of methane from the chemotrophic bacterial assemblage, the bacteria require a mix of elements to account for the nutritional balance. In studies of the bivalve, *Solemya reidi* Bernard, it was observed that ammonium uptake and assimilation by the chemotrophic bacteria depletes nitrogen from the host and promotes the passive membrane transport of exogenous ammonium (Lee et al. 1992a). However, further analysis of deep-sea bivalves (*Bathymodiolus* sp., undescribed) indicated that a broad combination of nitrogen sources may be used to support the nutrient requirements of the host and chemoautotrophic organisms (Lee et al., 1992b). Depending on the availability of nitrogen at the site sources could include ammonium and particulate and dissolved organic nitrogen.

5. OCEAN WATER COLUMN CARBON CYCLING

Current ocean modeling focuses on surface ocean primary production, transmission of sun light through the ocean and transport of land based carbon sources through the coastal margin. In most open ocean environments the dominant carbon cycle is thought to be a balance of primary production and heterotrophic cycling with subsequent key element mineralization and transition of energy through the food chain. Over the last twenty years refined analysis of chemical and biological cycles suggest that other factors may contribute to the ocean water column carbon cycling (Fig. 5). A wide variety of observations promote the need to reorganize modeling of ocean carbon cycling. While the early analysis of methane cycling in the ocean was believed to be dominated by activity in anoxic microniches the more recent work strongly suggests that the sediments also contributes to this cycle.

For example, pockmarks in sediments on the ocean floor along eastern Skageerak between the coast of Denmark and Norway suggesting that there is a high flux of methane from the ocean floor into the water column (Hovland, 1992). Methane fluxes have been measured from the sediment, into the water column in the Cariaco Trench with values ranging from 12.5 to 17.5 mmol cm^{-2} yr^{-1} (Scranton, 1998). High concentrations of methane have been observed in open ocean environments with concentrations that are 30-70% supersaturated relative the concentration at atmospheric equilibrium (Lamontagne et al., 1973). Studies demonstrate the presence of methane oxidation in the open ocean water column (Sieburth 1993), leading to speculation that the methane oxidation resulted from microscale anoxic cycles in floating organic rich marine seston. Bacterial species have been observed in the seston which are obligatory anaerobic. Such observations are supported by a wide range of other microbiological and biogeochemical surveys. Water column 16s rRNA surveys demonstrate that in deep ocean waters a significant fraction of the bacterial assemblage are archaea bacteria (Fuhrman et al., 1992; Massana et al., 1997). A large segment of this phylum is autotrophic.

Further support for autotrophic carbon cycling in the water column comes from _^{13}C analysis of bacteria. An examination of _^{13}C of particulate organic matter and bacterioplankton in transects from the Mississippi River outflow to the high salinity end member indicated that besides carbon from the Mississippi river and primary production a significant, alternate source of carbon support microbial production (Kelley et al., 1998). The range of _^{13}C found in bacteria and seston was from –25‰ to –32‰. Future investigation will determine if petroleum or methane are the primary carbon source. While the early analysis of methane cycling in the ocean was believed to be dominated by activity in anoxic microniches the more recent studies demonstrate that the sediments also contributes to this cycle.

In addition to methane, ammonium oxidation has been shown to be a substantial fraction of carbon cycling through the microbial assemblage in the Mississippi River Plume in the Gulf of Mexico (Pakulski et al., 1995). For this study 20-60 % of the total oxygen demand was attributed to ammonium oxidation, resulting in a substantial portion of the total bacterial production. This is an important observation because, when present in sufficient concentrations, methane will be cycled before ammonium (Jones and Morita, 1983).

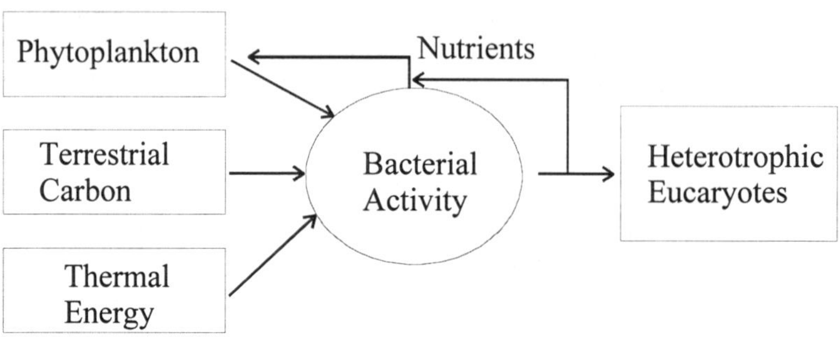

Figure 5: Carbon sources that contribute to the water column biological activity.

Comparisons of Nitrosomonas oceanus selecting an energy source demonstrated that the presence of methane inhibited ammonium oxidation (Ward, 1987). However, subsequent studies showed that both methane and ammonium oxidation could occur simultaneously (Ward and Kilpatrick, 1990). These studies initiate the need for further analysis of ocean carbon cycling. In addition to methane other geothermal energy sources need to be examined as an energy source in the water column. High active petroleum seeps have been observed in the Gulf of Mexico surface ocean (MacDonald et al., 1993 Mitchell et al., 1999). In addition, ethene to butene (C2-C4) have been reported to be in concentrations that range from 134 to 37 pM, respectively (Plass-Dülmer et al. 1995).

The turnover time of these carbon pools need to be measured to quantify the support to the bacterial assemblage relative to other forms. Understanding the role of methane hydrates in the transport of carbon to the water column is an interesting topic. Methane hydrate regions are highly populated with diverse biological species. There is likely to be high concentrations of carbon that seeps from these regions and into the water column. Recent estimates of the methane flux to the water column stated that 1-10% of the total methane hydrate was transported to the water column (Ginsburg et al., 1999).

These events can be difficult to survey. The flux of compounds from the sediment is not always constant. An earthquake along the Klamath River Delta in Northern California was observed to produce active venting, then reduced substantially after one year and totally subsided with in five years (Field and Jennings, 1987). The observation promotes the need for long term understanding of the flux of reduced compounds into the watercolumn for a more thorough understanding.

6. CONCLUSIONS

Methane hydrates, a potentially harvestable energy, form at low temperature and high pressure from methane which has resulted from the thermal or microbiological decomposition of organic material. Prior to the capture and utilization of this resource the biogeochemical factors which affect hydrate stability and formation must be determined. Methane is a high energy product of the result of a terminal electron accepting process; its migration towards aerobic waters can fuel chemoautotrophic production and growth. Natural abundance radiocarbon and stable isotopic composition serve as effective tools to trace the sum of the processes which have affected methane formation and consumption (Fig. 3).

The isotope analysis of bacterial biomarkers, dissolved inorganic carbon, organic carbon and methane will assist in analysis the relative contribution thermogenic and biogenic methane to hydrates (Fig. 4). A combination of analyses of microbial cycling rates and community diversity, coupled with the two-dimensional carbon isotope analysis provide the potential to access the rate of methane hydrate formation and stability and the biogeochemical processes that control these cycles. These isotopes also allow the tracing of sea floor autotrophically fixed carbon to higher trophic levels. Research following these tracers into the water column has led to the suggestion that high energy componds derived from the benthos may fuel heterotrophic processes in the water column. These data suggest future development of ocean modeling requires addition of geothermal energy that influences the carbon cycling.

Chapter 8

Deep Biosphere: Source of Methane for Oceanic Hydrate

Peter Wellsbury and R. John Parkes
Department of Earth Sciences
University of Bristol, UK

1. INTRODUCTION

Methane is an important product of anaerobic bacterial metabolism. Bacterial methane makes a substantial contribution to global methane reserves. Methanogenesis is the final step in the anaerobic degradation of organic matter, and can continue in deeply buried sediments. Methane can also be produced abiologically at elevated temperatures and pressures e.g., thermal breakdown of organic matter, crustal and hydrothermal processes. The boundary between biological and abiological processes is not always clear. Bacteria can be active at temperatures up to 113°C and pressures in excess of 1000 atm, and abiological processes can produce energy sources for bacterial methanogenesis. In addition, deep sourced thermogenic methane can diffuse to the surface, and under certain conditions, biogenic methane can have a chemical and stable isotope signature indicating an abiological origin.

Oceanic gas hydrates contain predominantly methane with a bacterial origin (Kvenvolden 1995). Recent studies have demonstrated the presence of bacteria to depths of over eight hundred metres below the sea floor (mbsf) in marine sediments (Parkes et al. 2000). Bacterial populations and their activity are stimulated in oceanic hydrate deposits to such an extent that some processes are more intense at depth than in near-surface sediment (Cragg et al. 1996, Wellsbury et al. 1997), demonstrating that gas hydrates are a unique deep subsurface habitat.

The role of bacteria in methane production, and their activity in sediments containing hydrate, is reviewed in detail below.

2. BACTERIAL METHANOGENESIS

Bacteria which produce methane are called methanogens. They produce methane to generate energy for survival and growth. Methanogens are obligate anaerobes, killed by traces of oxygen, and thrive only in anoxic conditions. Methanogens can only use a limited range of small molecules for metabolism. These small molecules are supplied as the end product of the metabolism of

M.D. Max (ed.), Natural Gas Hydrate in Oceanic and Permafrost Environments, 91–104.

other types of bacteria. Thus methanogenic bacteria form only a sub-set of the total microbial population in their environment. In these circumstances, one bacterium's metabolic waste product is another's food; thus it is important to understand the interactions of different types of bacteria in effecting the breakdown of organic matter to methane.

In oxic conditions, organic material can be degraded fully to carbon dioxide by a single bacterium. Conversely, under anoxic conditions, the degradation of organic matter to carbon dioxide or methane is carried out by a sequence of interacting bacterial types (Fig. 1).

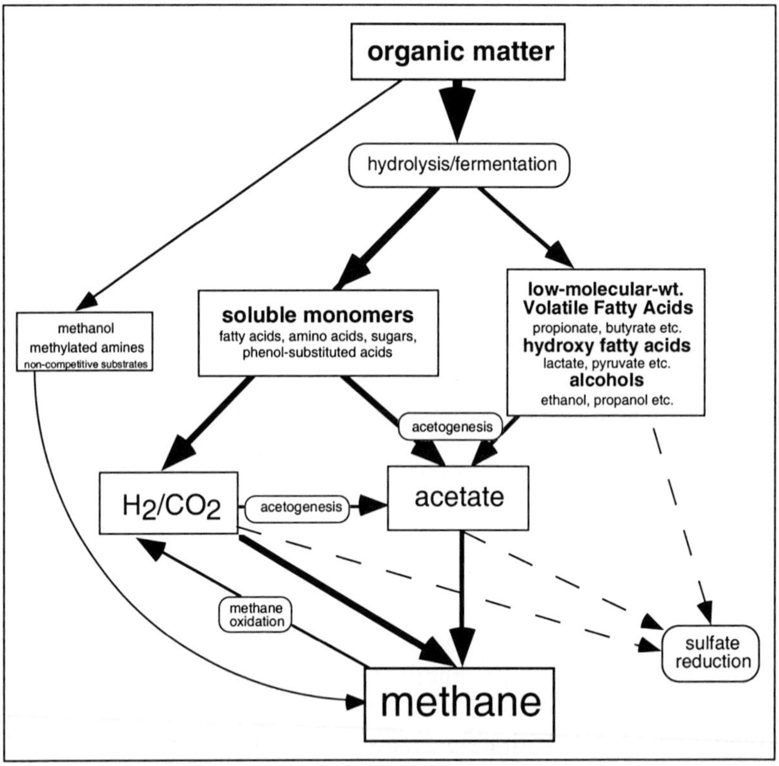

Figure 1. Pathways involved in methanogenesis from the bacterial decay of organic matter in marine sediments. Arrow thickness denotes the relative importance of each process. Solid arrows show pathways to methane; dashed arrows show substrates to sulfate reduction.

Large, complex organic polymers are first broken down by hydrolytic and fermentative bacteria to form a range of small molecules known as 'metabolic intermediates'. Some of these metabolic intermediates may be used by other types of bacteria in addition to methanogens. H_2 and acetate are 'competitive substrates', and are utilized by whichever organism in the system is the most efficient. For example, sulfate-reducing bacteria have a higher affinity for hydrogen and acetate (Schonheit et al. 1982), than methanogens (Lovley et al. 1994) and thus outcompete them for these substrates.

Thus, in environments where sulfate is present, such as near-surface marine sediments, sulfate reduction is the dominant terminal organic matter mineralization process. In some (rare) cases, when organic matter concentrations are very high, competitive substrates for methanogenesis may be present at high enough concentrations to enable both methanogenesis and sulfate reduction to proceed. Other metabolic intermediates, such as methanol and methylated amines, are used by methanogens alone and are thus 'non-competitive' substrates. The presence of non-competitive substrates means methanogenesis can occur in environments dominated by other processes, such as sulfate reduction.

The relative flow of organic carbon through the degradation pathways is suggested in Fig. 1 by the thickness of the arrows. The most important substrates for bacterial methanogenesis are acetate and H_2:CO_2. Methane formation from hydrogen and carbon dioxide, or 'carbonate reduction', is an almost universal processes in isolated cultures of methanogens. This has a secondary significance in that these species are autotrophic and do not require complex organic matter for their metabolism. Methane formation from acetate, or 'acetoclastic methanogenesis', is restricted to fewer species, but it is still a significant process in many environments, as discussed below.

Processes which remove hydrogen and thus prevent its accumulation are fundamentally important (Fig. 1). This 'interspecies hydrogen transfer' occurs when a hydrogen-utilizing organism, such as a methanogen (or acetogen), grows syntrophically with a fermentative bacterium. Syntrophic organisms are those which interact as partners to degrade a substance neither can degrade alone. In this case, the methanogen will consume the hydrogen produced by the fermenter and thus maintain a low hydrogen concentration. Hydrogen build-up would inhibit fermentation. If hydrogen levels are kept low by syntrophs, more energy will be produced by fermentation, and more oxidised compounds produced. One such compound is acetate, a key substrate for methanogenic activity. Acetogens can play a critical role as they remove hydrogen and form acetate, and thus increase the importance of acetoclastic methanogens.

Methane oxidation, which typically occurs aerobically (Madigan et al. 2000), can also occur anaerobically (Iversen and Jørgensen 1985). Both produce carbon dioxide from methane. Although anaerobic methane oxidation has been measured using radiotracers (eg. Iversen and Jørgensen 1985; Cragg et al. 1996; Wellsbury et al. 2000), no anaerobic methane-oxidizing bacterium has been isolated. The current perception is that a consortium of syntrophic organisms are involved in anaerobic methane oxidation. These organisms are possibly a methanogen carrying out "reverse methanogenesis" to produce hydrogen (Figure 1), coupled with a sulfate-reducing bacterium utilizing this hydrogen (Hoehler et al. 1994, Hinrichs et al. 1999).

3. PHYSIOLOGY OF METHANOGENS

Methanogens are a morphologically diverse group of strictly anaerobic Archaea, which obligately produce methane as an end-product of their metabolism. Archaea are distinct from other bacteria, and many are extremophiles, capable of living in extremes of temperature, pressure, salt, and pH.

3.1. Temperature and pressure

Methanogens can exist at a wide range of temperatures, usually from 4-55°C. Most methanogens are mesophiles, with optimum growth temperatures around 35°C. *Methanobacterium frigidum* lives in an Antarctic Lake in near-freezing temperatures (Franzmann et al. 1997). Viable methanogens (both H_2:CO_2 and acetate utilizers) have been found in permafrost (Rivkina et al. 1998). At the upper temperature limits, organisms such as *Methanopyrus* have been isolated from hydrothermal marine sediments. *Methanopyrus* grows up to 110°C (Stetter, 1996), and other hyperthermophilic methanogens may exist at temperatures up to 113°C, the upper temperature limit for bacteria. Growth at ~110°C is only possible under pressure to prevent water boiling, and hence all bacteria able to grow at these temperatures must also withstand elevated pressure. Other methanogens have been isolated from deep hot marine environments such as oil reservoirs, e. g. *M. thermoaggregans* (Blotevogel and Fischer, 1985).

3.2. Salt

Methanogens have been isolated from high salinity environments, including *Methanohalophilus*, which can grow at salinities up to 4.4M NaCl (Lai and Gunsalus 1992).

3.3. pH

Methanogens normally grow in the pH range 6-8, although they are found in acidic peat bogs and in alkaline lakes (to pH 9.7) (Oremland 1988). At lower pH values, acetate methanogenesis may become more significant than H_2:CO_2, because of more efficient competition for H_2 by acetogenic bacteria (Phelps and Zeikus 1984).

4. ECOLOGY OF METHANOGENIC ENVIRONMENTS

Methanogenic bacteria, being strict anaerobes, are only active in oxygen-free environments, such as sediments (freshwater and marine), wetlands, the digestive systems of ruminants and termites, and other locations where the respiration of aerobic organisms has removed all traces of oxygen. Methanogenesis is also important in a range of anthropogenic ecosystems such as landfills, anaerobic sewage treatment plants and rice paddies.

4.1. Freshwater environments

In freshwater sediments, rice paddies, swamps and other terrestrial wetland environments, the principal source of carbon for methanogens is acetate, which can account for >70% of the methane produced (e.g. Cappenberg and Prins 1974). Terrestrial environments are the chief source of methane to the atmosphere, and account for ~80% of global emissions each year (Crill et al. 1991). By comparison, the oceans and underlying sediments are responsible for only 0.2-3% of the total methane flux, despite their huge surface area.

4.2. Coastal, Marine and Estuarine Sediments

The high concentration of sulfate in seawater allows sulfate-reducing bacteria to outcompete methanogens (Fig. 1). Thus, in anaerobic near-surface sediments, sulfate reduction dominates the terminal oxidation of organic matter, accounting for up to 60% of all organic matter degradation (Jorgensen 1982). Overall, methanogenesis represents only ~0.8% of total carbon flow in near-surface estuarine sediments (Wellsbury et al. 1996). However, once the sulfate concentration decreases below ~3 mM (Capone & Kiene 1988), significant methanogenesis occurs, and thus methanogenic activity peaks below the zone of sulfate reduction. If non-competitive substrates are present, however, there will be methanogenesis even in the presence of significant quantities of sulfate.

5. DEEP ENVIRONMENTS

Although the bulk of organic carbon is degraded by bacteria in near-surface sediments, a small amount survives to become deeply buried. Over geological time this accumulates to form the largest global organic carbon reservoir.

Buried organic carbon is resistant to further degradation, enabling only slow methanogenesis. However, direct substrates for methanogenesis, such as H_2 and acetate, are also supplied thermogenically at depth, potentially stimulating CH_4 production. These processes include thermogenic breakdown of organic matter, deep-sourced fluid flow, or diffuse hydrothermal venting.

5.1. The Deep Marine Biosphere

As depth increases, the temperature and pressure within marine sediments increase, creating conditions more hostile for life. However, as ~70% of the Earth's surface is marine, and the sediments below the sea floor can be up to 10 km deep, this represents an enormous potential habitat for anaerobes.

Over the last 15 years, the presence of significant populations of bacteria at depth in marine sediments has been demonstrated (Parkes et al. 2000). Previously, deep marine sediments had been thought too hostile for life, despite the presence of biogenic methane and other indirect evidence for their presence, such as i) chemical changes in pore water; ii) modification of organic compounds; iii) chemical biomarker compounds suggesting the existence of living bacteria; and iv) changes in the stable isotope composition of compounds.

Bacteria have been detected at all sites studied in the Pacific and Atlantic Oceans, and the Mediterranean Sea. Bacterial abundances decrease with increasing depth from the sediment surface, but persist to great depths, generally conforming to the model published by Parkes et al. (1994) (Fig. 2). The current reported maximum depth is 842 mbsf in sediments from the Woodlark Basin, Papua New Guinea, where there were 320,000 cells ml^{-1} (Taylor et al. 1999). The average depth of sediments beneath the oceans is 500 m, and at this depth, the model predicts a population of 2.76×10^6 cells ml^{-1}. This is only ~3% of the typical near-surface abundance, but remains a considerable bacterial population. Indeed, the model indicates that bacteria in deep-sea sediments account for ~10% of living biomass on Earth.

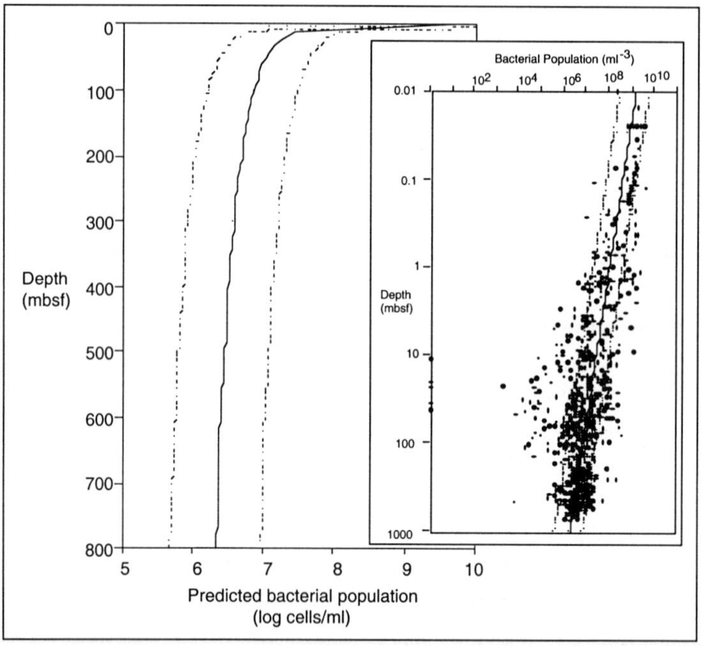

Figure 2. Depth distribution of bacterial populations in deep marine sediments. The solid line depicts the general regression line model of Parkes et al. (2000), where log cells = 7.98 – 0.57 log depth. The dashed lines show the 95% prediction limits of the model.

Bacterial depth distributions in marine sediments vary systematically from the model (Fig. 2) according to the oceanographic setting (Parkes et al. 2000). In high productivity zones such as the Peru Margin, high organic input to the sediments is reflected in relatively large bacterial populations. Conversely, in low productivity zones such as the Eastern Equatorial Pacific, bacterial populations are relatively small. Importantly, both the population size and its activity can actually increase in the deeper sediment layers in situations with fluid and gas flows, such as brine incursion in sediments (eg. the Peru Margin), and deep thermogenic methane (eg. the Japan Sea).

Methane is ubiquitous in deep marine sediments once sulfate reduction has removed sufficient sulfate to enable methanogenesis to dominate anaerobic organic matter degradation. Methanogens have been enriched from depths of 50 m in the Peru Margin (Cragg et al. 1990), although growth was slow, taking 18 months for detectable methane production. Methanogens are notoriously difficult to grow, and maintaining anaerobic conditions whilst sampling for methane over long periods just compounds this problem. Hence, culturing is likely to underestimate or fail to detect methanogens. ^{14}C-labelled substrates are a much more sensitive technique to measure methanogenic activity, and methanogenesis has been detected at a range of deep sediment sites (Parkes et al. 2000), including those with gas hydrates (Fig. 3). At the Blake Ridge hydrate site, low rates of H_2:CO_2 methanogenesis occurred at ca. 700 mbsf, whilst the rate of acetate methanogenesis was much higher (see section 6.2). Specific gene sequences for methanogens have also been detected in deep sediments.

5.2. Oil Reservoirs and Coalfields

Methanogenic bacteria have been isolated from oil reservoirs where significant methanogenesis occurs (Nazina et al. 1995). In many oil formation waters, acetate occurs in high concentrations (Lewan and Fisher 1994). Also, oil reservoir methanogens are able to use unsaturated and monoaromatic hydrocarbons as substrates when growing syntrophically with other anaerobic meso- and thermophiles. A number of potential syntrophic partners have recently been isolated from oil reservoirs, and include *Thermotoga elfii* (Ravot et al. 1995), *T. subterranea* (Jeanthon et al. 1995), *Desulfacinum infernum* (Rees et al. 1995), *Desulfotomaculum thermocisternum* (Nilsen et al. 1996) and *Thermodesulforhabdus norvegicus* (Beeder et al. 1995).

More recently, a methanogenic consortium has been shown to metabolize saturated hydrocarbons (e.g. hexadecane) to methane in the laboratory (Zengler et al. 1999). The key to the process seems to be the presence of acetogenic bacteria which metabolize hexadecane to acetate and hydrogen, both of which are then utilized by methanogens. The formation of methane from acetate, other volatile fatty acids and hydrocarbons may represent a significant source of methane for hydrates, which are often associated with petroleum deposits.

In some coalfields there is also bacterial methane formation from fossil fuels. Most notable is the San Juan Coalfield, the most prolific coalbed gas basin in the world (Scott et al 1994), where subsurface groundwater flow transports bacteria and probably nutrients through permeable rock strata to the coal beds. The bacteria metabolize hydrocarbons and other organic compounds in the coal to produce secondary biogenic gases, predominantly methane and carbon dioxide. Continuous biogenic gas production is essential to maintain the exceptional gas production of this field. Non-conventional gas resources, such as coalbeds and organic-rich shales, have largely been attributed to thermogenic processes, yet they may contain substantial quantities of biogenic gas.

5. 3 Hydrothermal systems

Hydrothermal systems at mid-ocean ridges, subduction zones, back-arc basins and mid-plate hot spots, are a major source of methane to the oceans (Karl, 1995). This methane is thought to be produced by abiological reactions, either degassing of the mantle or high temperature water-rock reactions. During these reactions, substrates for methanogens are also produced (McCollom and Shock 1997) and, therefore, some of the methane may be of bacterial origin (Baross et al. 1997). This suggestion is reinforced by the isolation of very high temperature H_2-utilizing methanogens from some hydrothermal systems, including *Methanopyrus* sp., which grows optimally at 98°C. In sediment-covered hydrothermal systems, such as Guaymas Basin, thermogenic breakdown of organic matter produces a range of low molecular weight organic compounds, including acetate, which may.stimulate methanogenesis (see also section 6.2).

5.4. The terrestrial subsurface

Considerable populations of anaerobic bacteria (10^6-10^7 cells/g, Sinclair and Ghiorse 1989), including methanogens, are present in the terrestrial subsurface. Viable methanogens were obtained throughout a depth profile of U. S. Atlantic Coastal Plain sediments down to 300 m (Jones et al. 1989). In addition, sediment slurries produced acetate and methane when incubated anaerobically. Thus deep terrestrial sediments of Cretaceous age still contained metabolizable organic matter. Acetogens can utilize this organic matter to produce volatile fatty acids, substrates for some methanogens (Chapelle and Bradley 1996). Methanogenesis from ancient organic matter also occurs in limestone rocks of the Florida Escarpment (Martens et al. 1991).

The deep crystalline rock aquifers of the Columbia River Basalt Group are 3-5 km thick, and contain only trace amounts of organic carbon. However, high concentrations of CH_4 (up to 160 mM) and H_2 (up to 60 μM) are present. Furthermore, the $\delta^{13}C$ of methane suggests it is biological, consistent with bacterial methanogenesis from H_2:CO_2. Active methanogens were present in the system, and formed the base of a unique deep bacterial ecosystem (Stevens and McKinley 1995). This community is entirely independent of photosynthesis and hence the surface biosphere. Experiments have shown that weathering of these basalt rocks produces H_2 for methanogens, although recently questions have been raised about the environmental applicability of these experiments (Anderson et al. 1998). However, this aquifer environment demonstrates the potential for a bacterial ecosystem supported by abiotic processes i.e. H_2 generation. If this process is widespread in the Earth's subsurface, it will have significant implications for geosphere processes fuelling the biosphere.

Autotrophic methanogens and homoacetogens, which produce methane and acetate respectively and utilize subterranean hydrogen and bicarbonate, have also been detected to 446 m depth in granitic aquifers in Sweden (Kotelnikova and Pedersen 1997).

5.5. Cold Terrestrial Environments

Methanogenesis in cold environments is an important component of the global methane budget. Wetlands are the largest natural source of methane and about 50% of all wetlands occur in cold high latitude environments. Cold-adapted psycrophilic methanogens have been isolated from an Antarctic Lake (Franzmann et al. 1997), glaciers (Sharp, personal communication) and permafrost sediments (Rivkina et al. 1998). In ice sheets, methane increases with depth, suggesting that methanogens may be utilizing organic matter ground from bedrock by the advancing glacier. A range of different bacteria exist in permanently cold environments, both to create anoxic conditions and to supply substrates for methanogens (Vorobyova et al. 1997, Sharp et al. 1999).

6. MICROBIOLOGY OF HYDRATE DEPOSITS

Oceanic hydrate deposits contain predominantly methane, usually ~99% of the gas present. Other gases comprising the remaining 1% are generally carbon dioxide, C_2-C_5 hydrocarbons and hydrogen sulphide. The high ratio of methane to higher hydrocarbons (C_1/C_2+ ratio), combined with the stable isotope signatures, denotes a bacterial origin for the methane (Kvenvolden 1995).

Two deep ocean sites containing gas hydrates have been the subject of comprehensive microbiological analysis. These sites were located at the Cascadia Margin (Ocean Drilling Program Leg 146) and at Blake Ridge (ODP Leg 164). Analyses included measurements of bacterial numbers and their activity, the presence of different culturable metabolic groups, and estimates of biodiversity using molecular genetic analysis (Marchesi et al. unpublished).

6.1. Microbiology of Hydrate at Cascadia Margin and Blake Ridge

Hydrate was not actually recovered during sampling at Cascadia Margin (Cragg et al. 1995), but its presence was inferred from geochemical and geophysical data, in a discrete zone between 215 and 225 mbsf at Site 889/890. Bacteria populations decreased with increasing depth similar to other sites (Fig. 2). However, the population increased dramatically (x 10) in the discrete hydrate zone. Culturable bacteria, and bacterial activity (methanogenesis from H_2:CO_2, sulfate reduction and methane oxidation) also increased in the same zone. These observations suggest that hydrate stimulated deep sediment bacteria. The high rates of anaerobic methane oxidation indicate that methane is an important substrate for bacteria in this deep habitat. Molecular genetic analysis confirmed the presence of diverse bacterial and methanogenic populations at the site (Marchesi et al. unpublished), and similar microbial diversity has been confirmed in near-surface hydrate deposits (Bidle et al. 1999).

Hydrate was present in all three Blake Ridge sites between ~200-450 mbsf, and increased with the strength of the BSR (Paull et al. 1996). Bacteria decreased in number with increasing depth (Wellsbury et al. 2000), similar to the model of Parkes et al. (1994, Fig. 2). However, bacterial populations were significantly stimulated at certain depths within the hydrate zone and directly below the hydrate (Fig. 3), reflecting high concentrations of free gas (Dickens et al. 1997). The number of cells involved in division also increased significantly at these depths at all three sites, suggesting a more active population. A solid sample of methane hydrate recovered from 331 mbsf at Site 997 contained only 2% of the predicted bacterial population for a sediment at that depth. Thus the active bacterial populations associated with hydrate are living in the sediment around the hydrate rather than within the hydrate lattice.

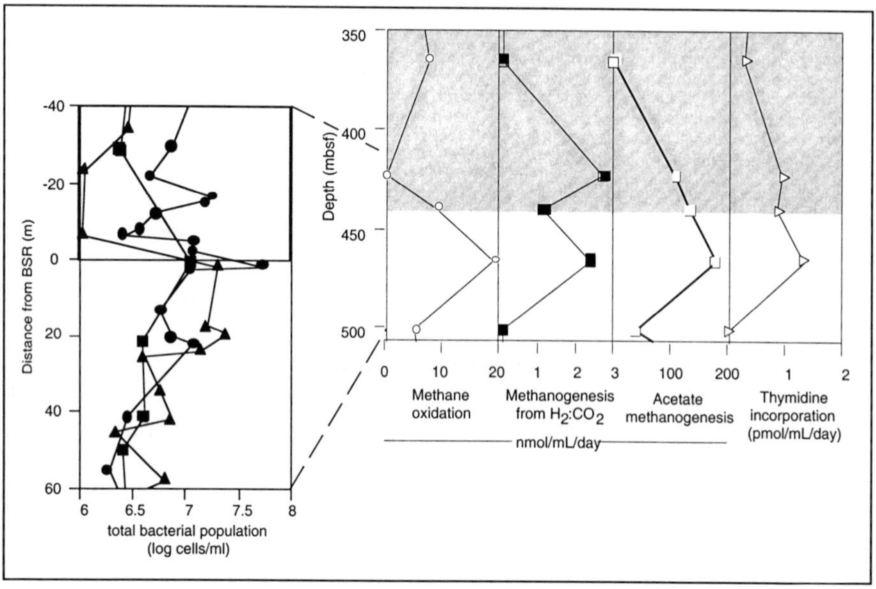

Figure 3. Stimulation of bacterial populations and activity around the BSR at Blake Ridge. The shaded area represents the hydrate zone.

Radiotracer techniques were used to determine bacterial activity at these sites (Parkes et al. 2000). Sulfate reduction was the most important process near the sediment surface, and rates decreased rapidly with depth. Sulfate was removed by ~20-30 mbsf. Methanogenesis, methane concentrations, and methane oxidation all increased below this depth. By ~100 mbsf, bacterial activity was low, but then increased dramatically around and below 450 mbsf (1.5 to 15 times), associated with the base of the hydrate zone and the free gas beneath (Fig. 3). Deep hydrate deposits are therefore biogeochemically active and this is reflected in increased bacterial growth (thymidine incorporation). Carbon cycling is occurring in this zone through methane, acetate, and carbon dioxide as rates of methanogenesis from H_2:CO_2 and acetate and methane oxidation are all stimulated (Fig. 3).

6.2. Sources of methane in deep marine sediments

Accurately determining the source of methane in hydrates is difficult. Biogenic gas deposits consist predominantly of methane, and are distinct from thermogenic gases, which contain considerable amounts of C_2 and higher hydrocarbons. Thus thermogenic gases have a lower C_1/C_2+ ratio than bacterially produced natural gas.

Stable carbon and hydrogen isotope signatures are also used to distinguish sources of methane. Thermogenic methane tends to be enriched in ^{13}C relative to bacterial methane. Thus, thermogenic methane has a distinctive $\delta^{13}C$ signature of ca. –20‰ to –50‰ (Whiticar, 1999). In contrast, bacterially derived methane has a $\delta^{13}C$ value of between –50‰ and –110‰, with methane derived from H_2:CO_2 being more ^{13}C-depleted (–60 to –110‰) than that derived from acetate (–50 to –65‰).

Various environmental factors can reduce the extent of carbon isotope discrimination during bacterial methanogenesis and thus obscure the $\delta^{13}C$ distinction between different methane sources, e.g., substrate limitation (reservoir effects) and increased temperature. Furthermore, methane oxidation causes an increase in the ^{13}C content of residual methane. All these effects yield methane with a $\delta^{13}C$ value which more closely resembles thermogenic gas. Despite these factors, the majority of methane in oceanic hydrate is recognized as biogenic, including the hydrate deposits at Blake Ridge and Cascadia Margin.

The hydrogen isotope signature of the methane can also be used to distinguish the particular substrate for bacterial methanogenesis. Methane from acetate has a δD value more negative than –250‰, whilst H_2:CO_2 methanogenesis generally results in a δD value of between –150‰ and –250‰. This difference is thought to arise because three-quarters of the hydrogen in methane derived from acetate comes from the deuterium-depleted methyl group, and the remaining quarter from water (Whiticar et al. 1986). In H_2:CO_2 methanogenesis, all the hydrogen in methane is ultimately derived from water.

Hydrogen isotope signatures must be used with caution, as bacterial acetate methanogenesis involves hydrogen isotope exchange (de Graaf et al. 1996). In de Graaf's experiments, labelled acetate (CD_3COO^-) was utilized for methanogenesis. As acetate was metabolized, other labelled acetate molecules were formed (CD_2HCOO^- and CDH_2COO^-) demonstrating that methyl hydrogen atoms exchanged with water, affecting the δD of methane produced. This could also result in the δD of methane from acetoclastic methanogenesis being more similar to thermogenic or H_2:CO_2 methane.

Methanogenesis from H_2:CO_2 and methane oxidation were measured in the hydrate zone at Cascadia Margin (Table 1). As subsurface hydrate increased, the rate of methane oxidation increased. However, H_2:CO_2 methanogenesis was around 5 orders of magnitude lower than methane oxidation, suggesting that an alternate source must exist at Cascadia Margin to maintain the methane supply. Cragg et al. (1996) suggested there may be gas and fluid flux into the sediment.

			Activity rates (nmol/ml/d)		
Leg	Site	Depth	Methane	Methanogenesis	
			Oxidation	H_2:CO_2	acetate
Cascadia *Increasing ←hydrate*	888	Near-seafloor	0.033	0.0071	nd
		Hydrate	0.71	0.0029	nd
	891	Near-seafloor	0.006	0.0048	nd
		Hydrate	76.91	0.00087	nd
	889/890	Near-seafloor	0.124	0.027	nd
		Hydrate	134.54	0.0077	nd
Blake Ridge	994	Near-seafloor	0.013	0.098	nd
		Hydrate	43.034	0.0045	nd
	995	Near-seafloor	0.046	0.164	14.291
		Hydrate	19.233	2.312	339.65

Table 1. Rates of methanogenesis from H_2:CO_2, acetate, and methane oxidation in sediments near the seafloor and at depths associated with hydrate in sediments from Cascadia Margin and Blake Ridge. Gas and fluid venting increases at Cascadia from 888<891<889/890; Hydrate abundance was similar at Blake Ridge at Sites 994 and 995. Methanogenesis from acetate was only determined at Site 995.

Acetate methanogenesis could be another source of methane for hydrate. The rate of acetate methanogenesis was determined at one of the Blake Ridge Sites (Site 995). Near-surface acetate concentrations were typically low (7μM at 50 cm depth), indicating that bacterial production and consumption of acetate were closely balanced. In deeper sediment, however, acetate concentrations began to increase and reached ~15 mM at 691 mbsf (Wellsbury et al. 1997). This increase was confirmed by three independent methods: isotachophoresis, ion chromatography, and a specific enzymatic technique. High concentrations of acetate at depth stimulated acetoclastic methanogenesis (Table 1; Fig. 3).

Rates of acetate metabolism were sufficiently high that they could not be sustained without a supply of organic carbon into the sediments. Again, fluid flow may be involved as there is evidence for upward migration of high concentrations of dissolved organic carbon (DOC) into the sediments beneath the hydrate zone at this site (Egeberg and Barth 1998)

Importantly, acetate methanogenesis was two orders of magnitude higher than H_2:CO_2 methanogenesis. This high rate of acetate methanogenesis was associated with high concentrations of free gas beneath the BSR. Acetate methanogenesis also exceeded methane oxidation at and below the BSR. Under these conditions, a local supply of bacterial methane to the hydrate is possible, although this is unexpectedly acetate rather than H_2:CO_2.

What supplies acetate to sediments beneath hydrate at Blake Ridge? There is DOC flux, but the mechanism for its generation is unclear. However, as the depth of burial in sediment increases, so does the temperature. Wellsbury et al. (1997) demonstrated that even a small rise in temperature will increase the biological availability of buried organic carbon, before further temperature increases result in thermogenic processes. This newly reactivated organic carbon will allow continued bacterial activity and thus provide substrates for methanogens, including acetate. At Blake Ridge there was a dramatic downcore increase in the Mass Accumulation Rate (Paull et al. 1996) which resulted in the rapid burial of relatively high amounts of organic carbon (up to 1.8%) below ~500 mbsf. Thus heating during burial (37°C/km) would have had a particularly marked effect at this site and contributed to the high acetate concentrations. At even higher temperature and greater depth, purely thermogenic processes may also produce low-molecular-weight compounds, including H_2. These substrates could migrate upwards to fuel deep methanogenesis.

6.3. Methane oxidation and hydrate

Defining the upper boundary of a gas hydrate stability zone (HSZ) is difficult as there is no clear geophysical feature. However, any methane degassing from the hydrate and subsequently migrating upwards is a potentially significant substrate for methane-oxidizing bacteria. This is demonstrated at both sites on Blake Ridge where methane oxidation was measured using ^{14}C-labelled methane (Wellsbury et al. 2000), as there is a clear maximum in bacterial methane oxidation at 83 mbsf (Fig. 4). The methane oxidation maximum occurs above the zone where hydrate is present, and may reflect the upper boundary of gas hydrate stability. The presence of anaerobic methane-oxidizing bacteria in sediment from above the hydrate zone at Sites 994 and 995 were confirmed in enrichment cultures in the laboratory (Wellsbury et al. 2000). Sterile culture medium was inoculated with sediment and incubated anaerobically under a methane headspace. After 1 year, a decrease in methane and a concomitant increase in carbon dioxide was measured.

Similarly, anaerobic methane oxidation was measured near the seafloor at the Cascadia Margin Site (Table 1), and thus appears to be characteristic of hydrate sites. Methane oxidation would result in highly ^{13}C-depleted carbonate, and the most isotopically depleted ^{13}C found in the marine environment was measured at Cascadia Margin (Elvert et al. 1999).

The capacity of deep sediment bacteria to oxidize methane that has migrated upwards from a gas hydrate deposit is of fundamental significance for global climate. Methane is ~21 times more effective as a greenhouse gas than carbon dioxide on a per molecule basis (Andrews et al. 1996). Furthermore, methane release from hydrates could create a positive feedback for global warming; temperature increases would result in hydrate instability and more methane release to the atmosphere. However, if bacterial methane oxidation 'mops up' the slow, continuous release of methane from the upper boundary of the hydrate, then significant quantities of methane will be released to the atmosphere only in catastrophic circumstances, such as sediment slumping events.

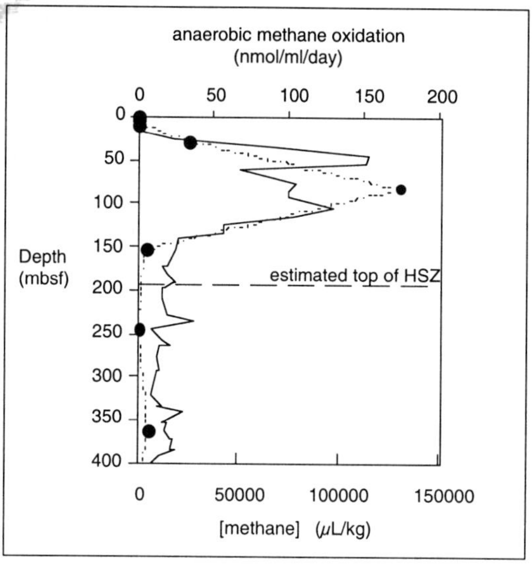

Figure 4. Methane oxidation above the Hydrate stability zone (HSZ) at Blake Ridge, site 995. Methane is represented by the solid line; the dashed line with solid circles shows rates of anaerobic methane oxidation measured from $^{14}CH_4$.

7. SIGNIFICANCE OF THE DEEP MARINE BIOSPHERE

Globally, methane hydrate is estimated to contain around 1600 Gigatonnes of carbon as CH_4 (Kvenvolden 1995), and the bulk of this methane is of bacterial origin. This reservoir contains more than twice the amount of carbon present in global fossil fuel reserves.

The results from Cascadia Margin and Blake Ridge demonstrate that hydrates constitute a unique deep bacterial habitat in marine sediments, as the abundance and activity of bacteria are elevated at depth. Deep fluid and gas flow from the geosphere may play a role in stimulating these bacterial populations. It has been estimated that about 60% of all bacteria on Earth live in sub-seafloor sediments (Whitman et al. 1998). As hydrates are so widespread, their stimulated bacterial populations may substantially increase this estimate and hence a significant proportion of subsurface marine bacteria may be associated with gas hydrates.

Acknowledgements

We thank ODP for allowing us to obtain samples, and participate in a number of cruises. This research was funded by grants from the Natural Environment Research Council, UK. We are grateful to Drs Ian Mather, Ed Hornibrook and Fiona Brock for their comments on a draft version of this chapter.

Chapter 9

Movement and Accumulation of Methane in Marine Sediments: Relation to Gas Hydrate Systems

M. Ben Clennell
Universidade Federal da Bahia, Brazil.

Alan Judd
University of Sunderland, United Kingdom

Martin Hovland
Statoil, Stavanger, Norway

1. INTRODUCTION

Hydrates may occur where thermodynamic conditions permit and where methane concentration in the water exceeds a threshold level, but they will only concentrate where gas flow is focused. Existing models of submarine gas hydrate occurrence encapsulate the system of transport and reactions into a one dimensional model (e.g. Rempel and Buffet 1998, Zatsepina and Buffett 1998, Xu and Ruppel 1999). With this simplification we can constrain key parameters, but it is difficult to capture the geological complexity of real systems. To predict the spatial distribution of hydrates we need to account for the range of mechanisms by which methane can move though the sediments.

Submarine (and lake bottom) gas hydrates form part of a system of methane production, transport, and reaction known as the shallow gas realm. The general characteristics of gases in the first kilometer or so of the sediments are known from geological studies (Kaplan 1974, Floodgate and Judd 1992, etc.) but important questions about the interactions of the biological, chemical and physical components of the system remain to be answered. Clathrate-forming gases, predominantly methane may be produced by bacteria within the hydrate stability zone itself, but generally there is not sufficient labile organic matter deposited with the sediment to produce an appreciable quantity of methane hydrate (Claypool and Kvenvolden 1983, refer to previous chapters). Recycling of biogenic gas at the base of the stability zone is therefore cited as an important process to concentrate hydrates (Paull et al. 1994; Chapter 6). Deeper in the sediments, gases come predominantly from temperature-dependent reactions. These thermogenic gases may migrate long distances vertically and laterally before reaching the hydrate stability zone.

M.D. Max (ed.), Natural Gas Hydrate in Oceanic and Permafrost Environments, 105–122.

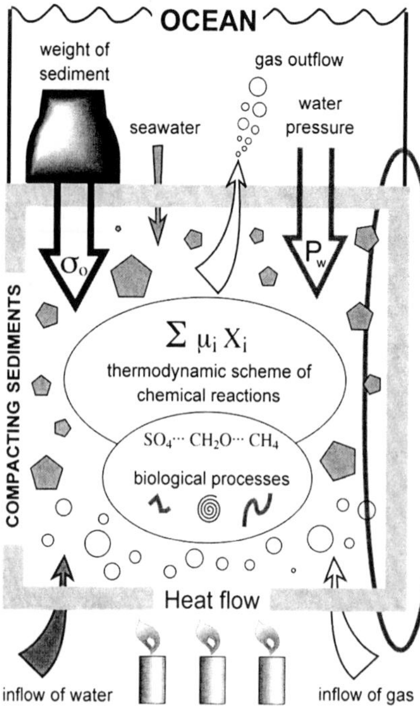

Figure 1. The gas hydrate system.

Exactly how hydrocarbons move through sediments, especially those of low-permeability and across structural barriers is a long-standing headache known as the "migration problem" (Schowalter 1979). However, this problem is most commonly studied in the context of petroleum reservoirs and seals. Whilst these generally occur in solid rocks, gas hydrates generally occur in unlithified sediments as the stability zone may be a shallow as the seabed.

In this paper we examine natural gas hydrates in the broader perspective of sediment gas realms (Fig. 1) (cf. Hovland et al. 1995). Porosity, pore size distribution and permeability of the sediments control the rate and modes of transport of the gas and other components. The reaction rates are coupled to the transports, and the proper fluxes must account for sediment compaction (Bonham 1980, Boudreau 1997). The macroscopic and microscopic state of stress in the sediments, and their textures determine the physical distribution of liquid, gas and hydrate phases (Clennell et al. 1999, Henry et al. 1999). To these ingredients we must add geological realism. That is, the inevitable presence of heterogeneity and features such as faults and fractures which disrupt the idealised continuum.

2. HABITAT OF GAS IN MARINE SEDIMENTS

Judd and Hovland (1992) describe major features of the shallow gas realm, with emphasis on seabed seepages (Hovland, & Judd 1988) and other manifestations of gas migration. Hovland et al. (1994) review geophysically-detectable features of shallow gas in marine sediments, and define the terminology for seismic surveys and sub-bottom profiles. Heggland (1997) and Cooper et al. (1998) highlight the role of high resolution 2D and 3D seismic suveys in identifying gas within the sedimentary section. The role that gas has in shaping the geomorphological evolution of the continental margins is emphasised by Yun et al. (1999) and references therein. From these studies it is evident that shallow gas systems are dynamic: literally in a state of flux, as demonstrated by the existance of active gas seeps.

In rigid rocks it is simple to envisage gas bubbles exisitng within the pore spaces. In soft sediments, however, gas may either expand into existing pore spaces, or it can create its own space by deforming the surrounding matrix; this amounts to an extra degree of freedom in the system (see Clennell et al. 1999). In the first case the gas pressure, P_g, must exceed the water pressure, P_w, in the space to be invaded. In the latter case the strength of the sediment must be exceeded by the deforming agent.

Free phase gas in sediments near the sea bed can exist wholly within the original pore space of coarse sediments (sands and gravels), but it is not stable as small bubbles within the very small pores of fine-grained sediments (clays and silts). Rather, in fine-grained sediments gas forms voids of up to a few millimetres across (Wheeler et al. 1990, Abegg et al. 1997) in a matrix of water-saturated sediment (Fig. 3). These gas voids have much lower capillary pressure than interstitial gas. The principles of gas invasion of pore spaces and the formation of gas voids are described in Chapter 2.0.

2.1. The pressure environment. In considering the pressure / stress environment of a marine sediment the sediment and the pore fluids must be considered separately. In normal circumstances, that is when there is no gas present and there is no over or under pressure, the pore water pressure approximates hydrostatic, i.e., the pressure imposed by a continuous column of water extending downwards, through the pore spaces of the sediment, from the sea surface. The overburden pressure equates to the vertical stress imposed by the bulk mass of the overlying sediment. This vertical stress and confining (horizontal) stresses tend to compact the sediment, giving it strength (although fine-grained sediments also posses cohesion as a result of electrochemical forces). However the pore fluid pressure resists compaction.

2.2. Gas saturation and the solubility of methane in pore water

Gas saturation refers to a property of the sediment and means the proportion of the total pore space filled with the free gas phase (bubbles - consisting mainly of methane in the shallow gas realm), the remainder being filled with water. This is unrelated to the amount of methane that is dissolved in the water. Thus any non-zero gas saturation in the pores implies that the coexisting pore water is holding the maximum amount of methane in solution at that P,T. Water with less dissolved methane is undersaturated with respect to CH_4. Pore water with more dissolved methane is supersaturated and is out of chemical equilibrium. In this sense, methane saturation is a property of the fluid relating to the relative amount of dissolved CH_4.

Solubility of methane is roughly 0.5 to 2.5 x 10^{-3} mole fraction in water and brines at depths from 300m to 5000 m (Hunt 1997). Pressure, temperature and salinity are sufficient to paramaterize an Equation of State (e.g, Duan et al. 1992) that can predict accurately the solubility of methane in water over a range of conditions. If pressure falls or temperature changes so that the methane saturation of the pore fluid is exceeded, methane gas will exsolve as bubbles (*effervescence*). A pressure rise can lead to bubbles being reabsorbed and eventually disappearing (evanescence). A certain level of supersaturation is necessary before gas bubbles will nucleate in a liquid, though in the presence of a solid matrix, surface sites for heterogeneous nucleation will be available.

To predict methane solubility in subsea sediments we also have to consider surface interactions and capillarity. As in case of imbibition or drainage, effervescence and evanescence in pores is likely to occur in jumps rather than continuously (Miller 1980). Supersaturation may, however arise in fine grained porous media because of the difficulty of nucleating bubbles of small size and thus high capillary pressure (Claypool 1996). Capillary supersaturation can be significant. Water charged up with oxygen at 10 MPa will after depressurization to atmospheric conditions only exsolve bubbles inside pores of a radius greater than 25 microns (Brereton 1998). This is a supersaturation factor of a few hundred times. Effective pore radii in fine sediments are on order 0.1 to 1 micron, so not only is the capillary entry pressure high, there is also a barrier to exsolution of gas bubbles from the pore fluid.

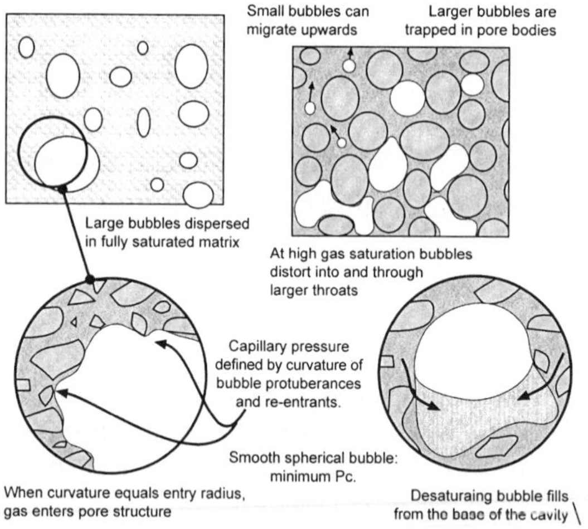

Figure 2. Gas voids from Wheeler et al. 1990.

2.3. Capillary Theory

Gas cannot enter into the porous structure of a sediment or rock until the capillary pressure exceeds a value comensurate with the radius of pore throats that provide access into the interior pore space (Corey 1994). This threshold pressure is also known as the capillary entry pressure. As the saturation of gas increases within a porous medium, the bubbles must distort to fit into the more confined interstices (Fig. 3). Thus we have a functional relationship between saturation and capillary pressure, defining a capillary characteristic curve. The capillary pressure P_c is the pressure difference between immiscible phases, is termed the capillary pressure P_c and this is proportional to the surface tension and to the tightness of curvature κ of the interface. When the fluid interfaces int-

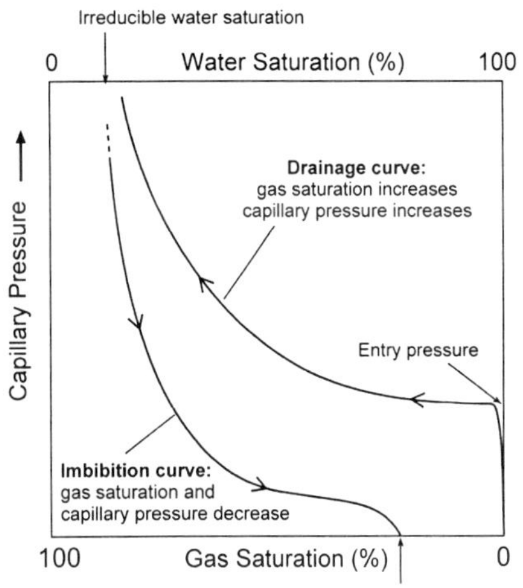

Figure 3. Capillary characteristic. Bubbles are trapped in pore bodies as gas saturation declines and water is imbibed. For a given gas saturation the imbibition curve has a lower capillary pressure than the gas invasion / water drainage curve. The *residual gas saturation* trapped in this way is typically between 10 and 25%. Some water remains in interstices and films even when the gas pressure is high enough to force out most of the mobile liquid. This *irreducible water saturation* is typically 8-15%, attaining higher values in clays.

-eract with the pore walls (i.e. the sediment grains), the pressure difference is controlled by the relative attraction between the solids and fluids; the more strongly attracted fluid is termed the wetting fluid, the other being non-wetting. The relative degree of wettability changes the angle of contact θ between the two fluid phases subtended at the pore wall. This factors into the Young-Laplace equation for the capillary pressure Corey (1994):

$$P_c = P_g - P_w = \gamma \kappa \cos\theta \tag{1}$$

In marine sediments containing gas and water the contact angle is close to zero and cos q is unity. The interfacial tension of water with methane gas is about 72 mN m^{-1} at STP, and decreases with temperature (Schowalter 1979, Sachs and Meyn 1995). For a slit-like pore where r is the radius of curvature for the short axis of the slit:

$$P_c = \frac{\gamma}{r} \quad \text{(2a)} \qquad \text{and for a cylindrical pore:} \quad P_c = \frac{2\gamma}{r} \quad \text{(2b)}$$

Even though real pore spaces are highly irregular, the uncertainties as to the sizes of pores and the bubbles of gas within them means that we can work with the assumption that pores are cylindrical or slit like.

The internal pressure of gas voids is constrained to lie between the capillary entry pressure of the sediment matrix and the water entry pressure, which is the capillary pressure in a smooth gas bubble inscribed in the cavity. In deformable sediment, the maximum gas pressure is further constrained to be less than the yield stress of the surrounding sediments and greater than the effective confining pressure (Wheeler et al. 1990, Sills et al. 1991). As depth increases matrix stresses force the bubbles to distort and eventually collapse. Little is known about how and over what depth range this occurs, in part because the microscopic stress field in sediments (< 2mm) is much more complicated than the macroscopic continuum scale (> 1 cm).

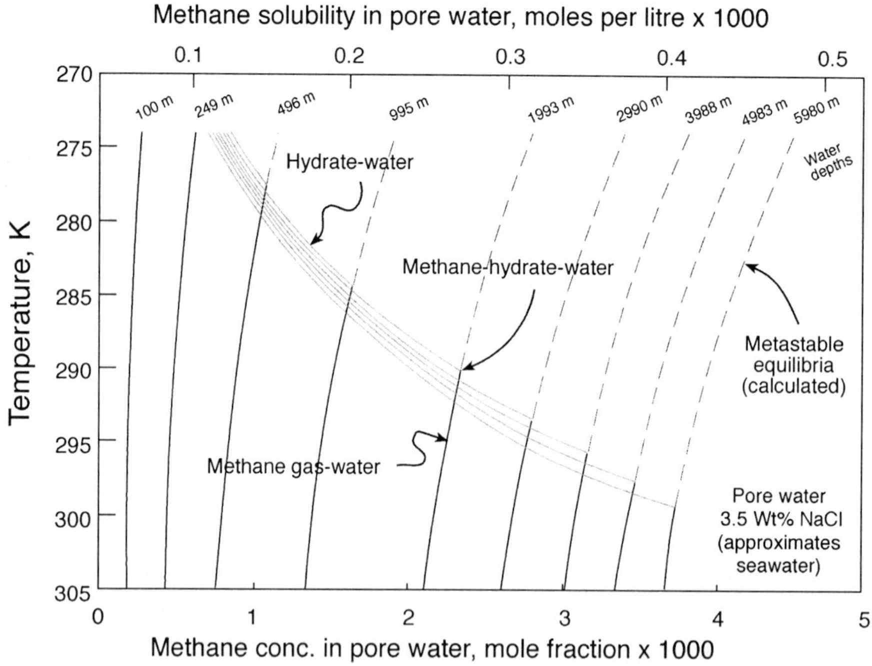

Figure 4. Gas solubility in the hydrate zone from Tohidi et al. (1997). Within the hydrate stability zone, methane is partitioned from the liquid phase into the hydrate phase (Handa 1990). Penetrating further upwards into the hydrate stability zone progressively less methane remains in solution. Thus there is a concentration gradient which exerts a dominant control on methane fluxes and distributions (Rempel and Buffet 1998, Xu and Ruppel 1999).

3. GAS MOVEMENT IN SEDIMENTS: MECHANISMS AND MODELS

3.1. Driving forces for flow.

The concentration of methane in pore waters generally increases with depth in the sediment due to a combination of *in situ* and deep thermogenic sources, methane recycling at the base of the hydrate zone and the thermodynamic equilibrium in the two phase region (Fig. 4). Free methane is strongly buoyant: even beneath 3-4 km of water its density is only 200-300 kg m^{-3} compared with 1024 kg m^{-3} for seawater. In most shallow gas systems these two factors contribute to a dominant flux of methane and associated gases towards the seabed. Water saturated with dissolved methane is also lighter than normal pore water (Park et al. 1990), and overpressure may exist at depth as a consequence of disequilibrium compaction or through the process of hydrocarbon generation itself (Hedberg 1974, Hunt 1997), so that additional mechanisms can drive methane-charged waters from depth.

3.2. Advection, Diffusion and Dispersion.

If the pressure field in the sediment deviates from a hydrostatic condition, then the pore water will flow *en masse* according to Darcy's Law:

$$\frac{Q}{A} = -\frac{k}{\eta}\nabla(P - \rho g h) \qquad (3)$$

Volumetric flow Q per unit sectional area A is proportional to the gradient in excess fluid pressure (total P_f - hydrostatic pressure), and inversely proportional to the viscosity (a property only of the fluid, for water this depends mainly on temperature and salinity; dissolved gas and pressure have little effect). The coefficient k in eqn. 3 is the *intrinsic permeability* (m^2), a function of the pore size and pore space geometry. Permeability of marine sediments range from 10^{-8} m^2 for sands to 10^{-19} m^2 for consolidated muds. Values for deep sea silty clays and biogenic oozes, typical sediments in deep continental margins where most hydrates are found, cluster around $3\text{x}10^{-16}$ to $3\text{x}10^{-18}$ m^2 (Schultheiss and Gunn 1985; Clennell et al. 1999). For flow in an open crack, it is possible to use the *cubic law* equation for the flow velocity. For a fractured medium with n parallel sided fractures per metre, each of aperture w the effective permeability (Roberts et al. 1996) is:

$$k_f = \frac{nw^3}{12} \qquad (4)$$

whereas flow along a pipe-like conduit we can calculate the volume flux from *Poiseuille's law* where l is the tube length and P_{xs} is the difference in excess pressure that drives flow along it:

$$Q = \frac{\pi r^4}{8\eta}\frac{\Delta P_{xs}}{l} \qquad (5)$$

The *advective flux* of methane is simply the product of the volumetric flow rate and the concentration of methane in solution. Since the solubility of methane is low, and seepage velocities though the bulk of the sediment using these values of permeability in eqn. 3 are small, then we can expect the advective flux through sediments to be low. In cracks or pipe-like features, or along highly permeable sands layers then the volume flow rates calculated using the above equations can be much higher providing there is a *source* of excess pressure at depth to drive the flow, and a *sink* (at hydrostatic potential, such as the ocean) to accept the fluid. Under these conditions, advection will dominate the methane transport. Where water flow is focused upwards, the entrained heat may perturb the geothermal structure and deflect the gas hydrate stability zone upwards (Zwart et al. 1996).

When the pore fluid itself is not in motion, the steady state *diffusive flux* of methane through water-saturated sediment can be computed from *Fick's First Law:*

$$\mathbf{j}^{D}_{CH_4} = -D^{*}\nabla c^{w}_{CH_4(aq)} \qquad (6)$$

The flux of methane, $\mathbf{j}_{CH4}$ is the flow rate per unit area, in moles per second per m^2 of sediment section, and is proportional to the *gradient* in the concentration of *methane in solution in the pore water*. The concentration must be expressed here in moles per cubic meter. The diffusion coefficient D^* is lower than the diffusion coefficient through bulk water because of the tortuosity of the flow paths through the sediment (Clennell 1997). Iversen and Jorgensen (1993) give values for the D^* for methane in unconsolidated marine sediments ranging from 10^{-8} to 10^{-9} m^2s^{-1}. Values for consolidated mudrocks can be as low as 10^{-13} m^2s^{-1} (Schlomer and Krooss 1997). According to Hunt, (1997):

"Diffusion acts to disperse rather than concentrate gas and is an exceedingly slow process."

Diffusion is very inefficient at transporting gas *over long distances*, because the concentration gradients are small. However, diffusion is omnipresent in systems that are out of chemical and thermal equilibrium.

When the pore water is moving it is appropriate to consider dispersion, which includes the effect of mesoscale to macroscale fluctuations in the velocity field as well as random thermal molecular motions that constitute molecular diffusion. Together these effects combine into an *effective dispersion coefficient*, with the same dimensions as D^*, i.e. m^2s^{-1} (cf. Xu and Ruppel 1999). If the fluid advection rate is very small, then the dispersion coefficient is of the same order as D*, however, on larger scale of observation the effective dispersivity can be much greater, particularly if flow makes use of fractures.

3.3. Multiphase flow phenomena

The low solubility of methane in water generally makes movement of methane via advection and diffusion very inefficient. Far more methane can be transported if free phase gas is in motion. In very coarse sediments, fractures and other conduits bubbles of methane may be entrained in the flow, adding to the advective flux. Microbubble transport along fractures and bedding planes is considered by some to be a major mechanism of gas migration (e.g. Neglia 1979, Saunders et al. 1999) but we note that in general small bubbles will evanesce or coalesce. Microbubbles are most important for short travel distances in the shallow subsurface (Klusman 1993). In all other cases, gas must overcome capillary resistance if it is to move through the sediment.

Less dense fluids can rise spontaneously through denser fluids because there is a net buoyant force as a consequence of a difference in hydrostatic pressures. The greater is the vertical extent of the lighter fluid column (however thin it may be; thin columns are called *stringers*), the stronger is this pressure difference to drive flow. Buoyancy gradients of gas in water or brine vary from about 4 to 11 kPa per meter (Schowalter 1979; Hunt 1997). Consider the case where free phase gas occurs in a sand layer, but its upwards motion under buoyancy is impeded by a finer layer of mud (Fig. 5).

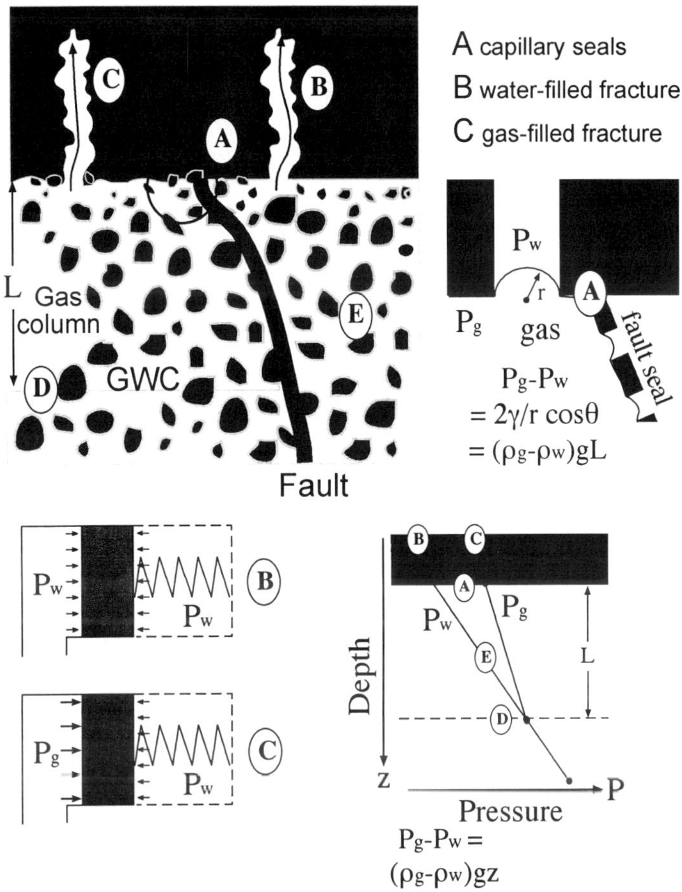

Figure 5. Buoyancy and capillary equilibrium. Pressure of the less dense gas phase P_g is higher than that of the water P_w when the phases are in hydrostatic and capillary equilibrium. The pressure difference generated by the buoyant gas column of height L relative the hyrostatic pressure gradient in the continuous water phase (E) is supported by the capillary resistance of small pores in the top seal and/or fault seal (A). A gas filled fracture penetrating the seal (C) will have a higher internal pressure than a water filled fracture (B) by an amount proportional to the difference in the densities of the phase and to the column height.

In order to penetrate further, the pressure difference across the gas/water interface at the top of the sand layer must exceed the capillary entry pressure of the mud. Say this mud has a characteristic pore entry radius of r, while the pore radius of the sand is R. By combining the equations for the hydrostatic pressure in gas and water columns, and the Young-Laplace equation, we can arrive at the classic criterion for the maximum height of the gas column; L_c that can be trapped:

$$L_c = \frac{2\gamma\cos\theta}{r(\rho_w - \rho_g)}\left(\frac{1}{r} - \frac{1}{R}\right) \quad (7)$$

(e.g., Schowalter 1979; Watts 1987). The curvature 1/R is so much smaller than 1/r that it can be neglected. Usually cos θ is unity for gas and water, and so we have an equation in which the driving force is a density contrast and the resistance is related to the inverse of the pore size of the mud layer:

$$L_c = \frac{2\gamma}{r(\rho_w - \rho_g)} \quad (8)$$

The maximum pressure that a seal can withstand ranges from hundreds of kPa for clayey silts to tens of MPa for mudrocks. With the fluid densities and surface tension given above, we can calculate from (8) that a mud with an entry radius of 100 nm can support a gas column nearly two kilometres high! The efficiency of capillary trapping explains the wide distribution of trapped shallow gas where methane is sourced biogenically or from greater depths.

3.4. Invasion Percolation

Free gas moving through a sediment must first overcome the entry pressure and thereafter find the path of least resistance. Generally capillary forces outweigh the viscous forces since the flow is very slow and so the favoured path is the one which connects the largest throats through the medium *however long and tortuous it may be*. The mathematical concept underlying such flow through a network of pores and throats is termed *percolation theory* (Sahimi 1994). Its relevance to fluid flow, and particularly oil and gas migration is obvious and has been exploited (Sahimi 1995; Wagner et al. 1996 and references therein). These studies mainly refer to sandstone carrier beds, but the principle for fine sediments is identical (Fig. 6).

If there is overpressure associated with the gas source at depth then a directed advective flux of water is generated that can assist or deflect the hydrocarbon movement (but see Bjorkum et al. 1998), particularly if the head drop is concentrated across a low permeability layer (Mandl 1981). Chapman (1983) discusses such situations from a petroleum geology perspective, and details for modelling of such flows can be found in Hindle (1997). At the high saturations which may arise if gas collects in coarser layers, pressure-driven flows of gas and water can be calculated according to a relative permeability curve (Klusman 1993). Permeability of unsaturated fine grained sediment to water is typically two or more orders of magnitude lower than the water-saturated case (Corey 1994). However, in fine sediments and especially close to the hydrate zone (Claypool and Matava 1999) the probability is that gas will not collect in such high concentrations that it is the mobile phase, and the concept of relative permeability for gas is meaningless.

When gas saturation drops below a critical level, it will not form continuous stringers capable of movement by buoyancy, and will form a mist of bubbles isolated in the larger pore bodies, and detached gas patches known as *residual ganglia* (Lowy and Miller 1995). These are unlikely to be detached by the modest water pressure gradient generated by overpressure. Donaldson et al.

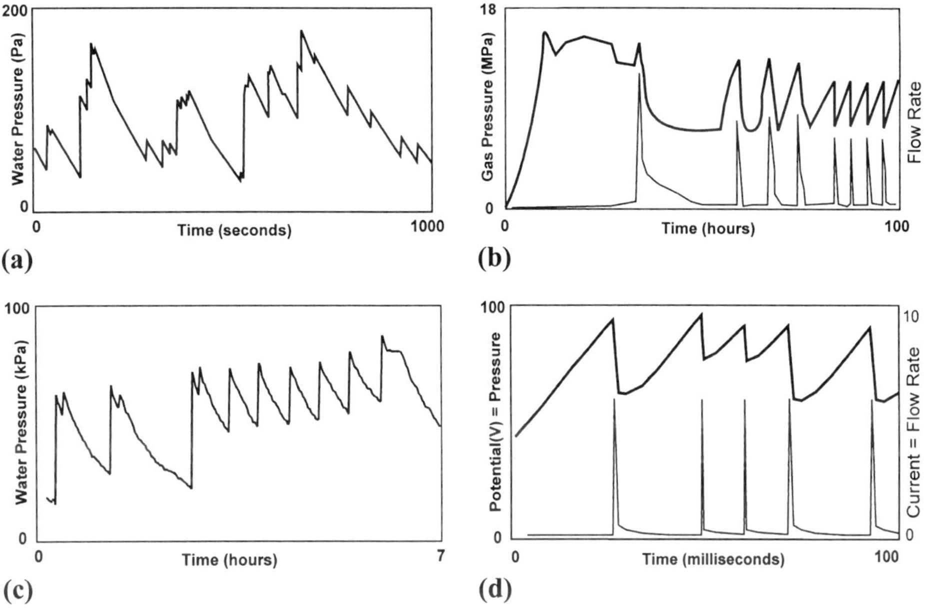

Figure 6. Critically-limited non-linear flow phenomena. *Invasion percolation:* (a) Drainage experiments conducted by Furuberg et al. (1996) on a bead array. (b) A dynamic invasion percolation model for gas migration though mudrocks (Impey et al. 1997). *Fracture flow.* (c) Injection of water into fractured unlithified mud (A.J. Bolton and M.B. Clennell, unpublished results). (d) Electrical circuit with a hysteric switch. Common features are evident: Pulsatile flow, regular cycling of flow episodes, and period doubling whereby the interval between pulses reduces by half during an experiment. Higher flux would drive the system into chaos (see Acheson 1997).

(1998) showed that in aquifers with a continuous water flow the dispersive transport of dissolved oxygen can be inhibited significantly by sequestration of the gas into these trapped bubbles. In compacting sediments the pressure of capillary gas seals could help maintain overpressures (Hovland and Judd 1988, Shosa and Cathles 1996, Kuo 1997, Revil et al. 1998).

The trapping ability of sediments clearly depends on the continuity of the layers that have the highest capillary entry pressure (see Ho and Webb 1998). A thin layer of mud is just as effective a seal to two phase flow as is a thick layer (Watts 1987), but thin muds do not stop diffusive methane transport. Gas can dissolve on one side of a capillary barrier, and without the entry pressure being exceeded, pass by diffusion through the layer of fine sediment, and reappear by evanescence of the other side (Miller 1980). If the bubbles on the upper side can move away though buoyancy, then the process can continue *ad infinitum*.

Further, gas bubbles can exsolve in positions along the flow pathway, and eventually coalesce into a stringer that spans across the capillary barrier, without the sequential bottom-to top process of invasion percolation necessarily occurring (Li and Yortsos 1995). Bubble strings do not even need to merge completely for transport to be greatly enhanced (Fig. 7).

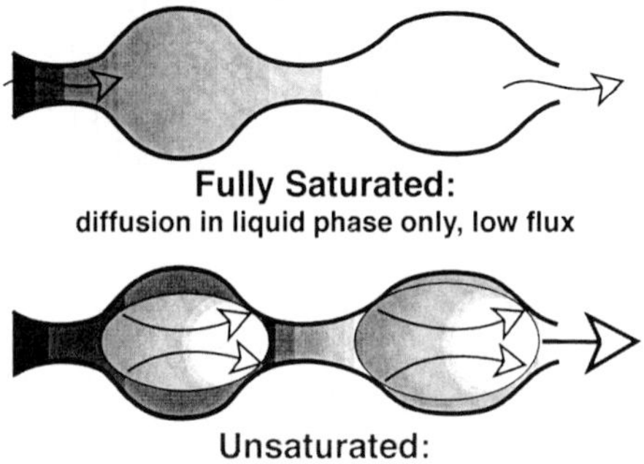

Figure 7. Catenary transport (Heard 1994). Methane can diffuse via a chain of gas filled bubbles separated by water filled pores in a process termed catenary transport. While the bubbles themselves are stationary, they permit a through fluxes of methane and water going from globally high to globally low chemical potentials.

The amount of methane that can be transported by advection, and the purely diffusive flux are both limited by the low level of methane saturation in the pore water. Mechanical and capillary effects on saturation discussed above are critically important when we try and predict these fluxes. Excess gas in supersaturated solution can move freely by advection though throats that would block the transport of the same gas in bubble form. If more gas is in solution in a fine layer because of capillary supersaturation, there will be a gradient in its chemical potential that has the capacity to drive a diffusive flux towards a region of normal saturation or undersaturation. On the other hand, if pore pressure drops due to mechanical disturbance, then gas can exsolve and become trapped in the new pores or fractures.

3.5. Flow in Faults and Fractures

Where present faults and fractures often dominate the flow pattern. Fractures not only increase the permeability of sediments and breach barriers but will also sequester gas from the matrix because of the lower capillary pressure that exists there (Bethke et al. 1991, fig. 17). Fluid can flow passively along fractures held open by asperties, or can actively generate the flow pathway through hydrofracture. If a clay or mud has previously experienced a greater effective stress it will be relatively brittle and can be fractured under tensile stress even though it may appear to be plastic and ductile when handled (Jones 1994). The classical criterion for generating an open tensile fracture in a water-saturated sediment is that the fluid pressure in the fracture P_f must exceed the sum minimum effective stress σ_3 and the tensile strength or cohesion of the sediment T (Secor 1965):

$$P_f = \sigma_3 + T \quad (9)$$

T is related to grain interlocking and cementation. To keep the fracture open once formed, P_f must be maintained at a value of at least σ_3. The cohesion in coarse uncemented sediments is often negligible but in muds it is related to electrochemical properties, so it is important to define whether the sediment is fractured open by water or by gas pressure (Harrington and Horseman 1999). If the maximum effective stress is vertical, then the pressure of fluid required to hydrofracture the mud works out at 75-90 % of the lithostatic pressure (Clayton and Hay 1994). As sediment is loaded by continued sedimentation, the lateral stress, σ_h in a basin without imposed tectonic stresses is a simple function of the vertical stress σ_v.

$$\sigma_h = K_0 \sigma_v \tag{10}$$

where K_0 is the coefficient of earth pressure at rest, (Jones 1994)). K_0 during monotonic loading takes on very conservative values, typically from 0.6 to 0.75. However, during unloading, or equivalently, overpressuring, the value of K_0 increases, even to values between 1 and 3 (Brown 1994). Not only does this inhibit the formation of fractures, but since σ_3 is now the vertical stress, it suggests that hydraulic fractures formed will be subhorizontal.

Given the existence of elongate flaws with favourable orientation, the criterion for propagation of the fracture in an elastic reduces to one of stress intensity at the tip (Murdoch 1993; Andersen et al. 1994) (Fig. 8).

The preceding analysis suggests that open fractures require an almost isotropic stress field, and almost lithostatic pore fluid pressures. Numerous studies show that fracture flow is important in sedimentary basins even when such extreme conditions are not met. There are several ways around this problem:

1. Fractures may flow under lower pore pressures if propped open by asperities or cements.

2. Secor (1965; cf Chapman 1983) note that the actual failure mode when pressure is increased in a basin with vertical maxium effective stress is usually by shear fracture. Mudrocks will dilate if sheared when overconsolidated, or pathways may open along fault jogs (Ingram and Urai 1999). However shear zones in mudrocks generally seal except at high pore fluid pressure (Dewhurst et al. 1999). Furthermore, P_f required to keep a shear fracture open is even higher than σ_3 is since it is not oriented normal to the minimum principle stress.

3. A tensile component of the tectonic stress field will reduce the value of σ_3. But this must propagate into the upper part of the sedimentary basin (Yassir and Bell 1994).

4. On a slope, the sediments will be under tension in the up-slope region, and compression at the base of the slope (Mandl 1981, Jones 1994, Bjorlykke and Hoeg 1998). Therefore seepage may be localised here at the head wall of slope failures (see Yun et al. 1999).

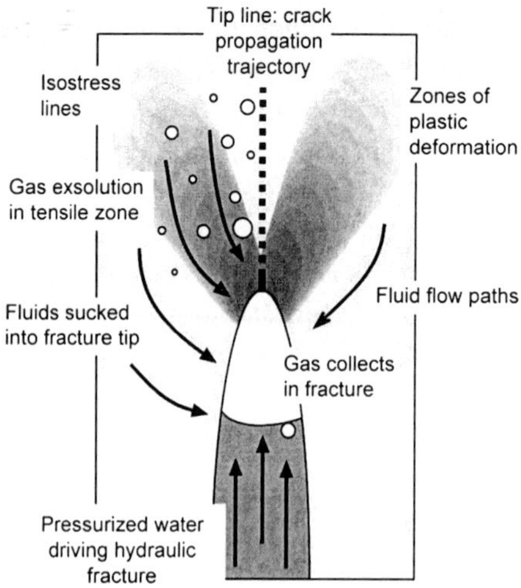

Figure 8. Fracture propagation (from Murdoch 1993). The mechanism proceeds initially by pressurizing the existing fracture until the elastic stresses in the wall region balance the internal fluid pressure excess. This is time-dependent because some injected water leaks off. Around the tip the sediments will be in tension, leading to a drop in pore fluid pressure. Water is sucked into the tip zone. Gas may exsolve in the tensile zone and pool in the fracture. When the pressure is sufficient, the rupture propagates as a mode I crack.

Watts (1987) stated that fractures can form in a seal when hydrocarbons build up beneath it because the buoyancy pressure in the non-wetting phase can exceed the fracture pressure if the hydrocarbon column height, L, becomes large (Fig. 5); i.e.,

$$P_g = P_w + (\rho_w - \rho_g)gL; \quad \text{and} \quad P_g > T + \sigma_3 \tag{11}$$

together lead to failure. Watts proposes that the gas pressure is the effective Pf in equation (9). These analyses presuppose that the fracture is completely filled with gas, and the walls are impermeable. When gas invades a fracture in porous sediment gas pressure increases will in part be balanced and dissipated by intrusion of gas into the pores, and the full gas pressure excess available from buoyancy will not be mobilised to do work on the surroundings. Bjorkum et al. (1998) note that gas buoyancy pressure cannot assist fracturing until the gas has already entered the cap rock, and that overpressure in the water will not assist upwards gas drive since the phases are in static capillary equilibrium. Nunn (1996) has modelled the propagation of single, fluid-filled vertical fractures in sediment. The fluid in the open fractures, which may be water, gas or a mixture is less dense than the surrounding saturated sediment, and so generates a buoyant upthrust. Existing flaws, or fractures generated by overpressure in the source bed can propagate upwards when this force is sufficient to overcome the fracture toughness of the sediment.

3.6. Other Mechanisms of Methane Transport

There are a number of other mechanisms by which gas may move closer to the surface. Diapirism (gravitational instability and sediment overturn) is common in fine grained sedimentary sequences, particularly when the permeability is low, and overpressure builds up through gravitational loading, or via fluid sources, including the generation of hydrocarbons. If fluid sources are strong enough, the sediment may become fluidised, and entrained in a localised upwards flow to form a diatreme (Brown 1990). The flow is driven by a combination of buoyancy of expanding gas and the overpressure. If it reaches the surface, a mud volcano will form. The relationship between methane and mud volcanoes has long been known and hydrate dissociation are implicated as a source of the gas in some cases (Hedberg 1974, Martin et al. 1996, Lance et al. 1998, Kopf et al. 1998).

During the tidal cycle gas bubbles expand and contract in response to the hydrostatic pressure change (Hovland and Judd 1988) and this couples with the poroelastic properties of the gassy sediment (Sills et al. 1991, Fredlund and Rahardjo 1993) to produce strong pressure gradients. If there is some form of valving mechanism, perhaps due to opening or closing of fractures, or movement of bubbles within the pores (Hovland and Curzi 1989) then a directed flux can result even when the driving forces are symmetrical (Wang et al. 1998). This "tidal pumping" action can drive a flow of water through sediments that present an impermeable capillary barrier to gas migration, and Hovland et al. (1999) suggest that such periodic force may contribute to the "pumping up" of fluid pressure in gassy sediments that are interlayered with stiffer sediments.

More sporadic, but potentially larger forcings can be generated by seismic activity. A stress drop in the sediment framework can, counter to intuition, produce significant transient overpressures, while localised dilational strain in the vicinity of faults can open up fracture pathways to a low pressure reservoir or to the oceanic/atmospheric sink (Brown 1994). Seismic pumping or valving has been implicated in the formation of pockmarks and other seabed seep features (Field and Jennings 1987, Hovland and Judd 1988, Hasiotis et al. 1996, Soter et al. 1999).

3.7. Discussion of Transport Mechanisms

In the first km or so of the sedimentary column the capillary entry pressure of fine sediments such as clays, silty clays and biogenic oozes (typically 10^5 to 10^6 kPa) is high with respect to the tensile strength and confining stress, so it is often assumed that in this zone any separate phase gas flow though fine sediments will occur along fractures (Clayton and Hay 1994; Judd and Sim 1998). The fact that chimney-like gas escape features exist on seismic lines across areas with a uniform, thick blanket of fine sediments suggest that gas "punches" its way through these capillary barriers, along some form of weakness (Hovland et al. 1994), or where gas accumulation in encouranged by the subtle geometry of coarse, gas-bearing sediments.

During gas migration and venting episodic flow patterns are the rule rather than the exception. This may reflect external forcings (e.g. earthquakes or tides) or may be a consequence of non-linearity in the system. The time series of potential and flux generated by invasion percolation experiments (Fig 6a) and numerical models (Fig 6b) match closely the results of some laboratory tests on fully saturated microfractured clays (Fig 6c) where there is no question of 2 phase flow occurring. The "success" of the invasion percolation model for gas flow is tempered by the fact that other mechanisms can generate the same oscillations. A simple electrical circuit model with a hysteric switch (Fig 6d) can reproduce the typical pattern. The hysteric switch valves the flow by opening only when a threshold potential is reached, but closing again only at a lower potential after permitting some current to flow. We can conclude that separate phase gas flow is critically-limited by the pressure and stress state and whether they be along structural features or percolation pathways, the migration routes into through and out of the hydrates zone will likely be highly localised and "one dimensional", exploiting the weakest links.

Different transport phenomena can act in series and in parallel with one another. That is to say, at different times and in different places the gas can move in all manner of ways in an overall process referred to as *migration.* A range of processes can breach or bypass capillary barriers, and the low permeability of a fine sediment layer is no guarantee of its seal capacity. As memorably summarised by Downey (1994):

> "Just as little comfort can be taken from a guarantee that your parachute will (on the average) open, explorationists are not really interested in the average properties of the enclosing, sealing surface."

4. IMPLICATIONS FOR HYDRATE HABITAT

4.1. Bottom Simulating Reflectors

One of the key questions relating to prediction of hydrate distribution and quantity is the nature of the interface at the base of the hydrate stability zone. Wherever a strong a bottom simulating reflector occurs, it is generally considered that free gas is present below, and that this marks the position of the three phase boundary (Pecher et al. 1998 and refs therein); though other models have been considered Brown et al. (1996).

The free gas at low saturations in the pore space will most likely be trapped as a mist of isolated bubbles, that will be buried with the sediment. Henry et al. (1999) suggest that a threshold saturation necessary for gas movment would typically be about 20%, which represents the percolation threshold for buoyant separate phase movement. This obviously impedes recycling of gas into the hydrate zone (Paull et al. 1994). Pecher et al. (1998) among others note that the free gas responsible for accentuating most geophysically mappable BSRs probably comes from melting of prexisting hydrate rather representing a trapped pool that is sourced from below. This inference is based on the finding that on the Central and South American margins BSRs are largely confined to geological units that are downwarping tectonically or subsiding due to continued sedimentation. Very importantly, in this scenario the amount of gas is tied to the amount of hydrate that has melted to source the gas (see Dickens et al., 1997; Henry et al. 1999).

In the (perhaps rare) instances that gas does come from below, it will form a mist rather than buoyant column of high saturation except where the flux is concentrated. This makes for an overall prevision of low saturations of gas and small, discontinous pockets rather than extensive BSRs in such settings (e.g. the

Niger Delta, Hovland et al. 1997). Indeed if the gas flux is small, the pore fluid may never reach supersaturation at any depth below the hydrate stability zone. Hydrate will only precipitate higher in the section as "solubility" drops due to increasing thermodynamic stability of the clathrate.

Rempel and Buffett (1998) and Zhu and Ruppel (1999), using computation reaction-transport models of different flavours showed how important was the flux of methane from below to the location and position of both the BSR and to the concentration profile. These works show that if there is a small gas flux then a gas- BSR will not form for the reason given above, and the greatest depth of hydrate occurrence will be encountered at a higher level in the sediments than the base of thermodynamic stability that is predicted from the phase diagram if methane is abundant.

4.2. Hydrates-Related Traps

Generating a large amount of gas by biogenic means is problematic; there simply is not that much labile organic carbon deposited and preserved for anaerobic consumption in most localities (Paull et al. 1994, Waseda 1998). However, those in search of commercial, hydrate-related hydrocarbon accumulations will probably consider thermogenic sources more rewarding.

Thermogenic and mixed thermogenic/biogenic gas hydrate plays will be associated with gas migration pathways, and therefore will be localised along permeable layers, faults or fractures, or above spillpoints in deeper (conventional reservoirs). Such focused flow will not give rise to a continuous BSR unless the gas should spread out for some reason (pervasive fracturing or rather uniform sediment properties). Rather the gas would be expected to pool and reach high saturations in coarser layers (sands) because the formation of hydrate at and immediately above the depth of three phase stability blocks the pore space in the updip direction. This hydrate-capillary seal concept is at least three decades old (Hunt 1997), but as yet is unproven commerically. However, as exploration and production frontiers advance into deep and superdeep waters (1000- 2000 m), there is a good chance that this situation will change (Grauls et al. 1999). To evaluate such prospects more work is needed to understand how hydrate grows within sediments, and the capillary seal capacity that can be produced by partial hydrate cementation.

4.3. Other Issues

In homogenous, unfractured fine sediments separate phase gas flow is not possible. Diffusion will dominate over short distances and when the upwards flow rate of water is less than about 0.5 mm per year. Advection will likely dominate if the water moves upwards at more then a few mm per year, and particularly along fractures, pipe-like conduits or sand layers.

Capillary supersaturation increases the efficiency of diffusion and advection of the water dissolved methane through fine sediments, but exsolved methane bubbles can be trapped in pore bodies and in coarse sediment layers. In the hydrate system, capillary supersaturation enhances hydrate stability (Henry et al. 1999) and can drive a methane flux from fine layers where hydrate will not form, to coarse layers where hydrate will concentrate (Clennell et al. 1999).

Dispersed methane flux to the sea bed is limited by the oxidation process in the sulfate reduction zone (Borowski et al. 1999). This zone can be bypassed if flow is along hydrate-lined fractures, or if overpressure and buoyancy generate a rapid bubble stream (Martens and Klump 1980) or combined flow of sediment, gas and water in a diatreme or mud volcano (Ginsburg and Soloviev 1994, 1998). The time series of pressure and flux in fracture driven flow and during invasion percolation in a capillary network are remarkably similar. This is because both are critically-limited phenomena that generate a pulse of flow when a threshold of driving pressure is reached. Seabed vents commonly exhibit episodic discharge, but it is not a simple matter to deduce the mechanism responsible. Hydrate formation may be irregular in time and place as flow switches around, and in fractured sediments a stockwork of hydrate-filled veins may form (Ginsburg and Soloviev 1997; Soloviev and Ginsburg 1997).

Gas charged sediment is less dense than sediment with no free gas, will exert a buoyant thrust. This could lead to diapirism (gravitational instability, and overturn), or simply may cause arching and fracturing. Hydrate is also lighter than water and thick layers may be gravitationally unstable. Permafrost-like sediment deformation features are a dostinct possibility.

5. CONCLUSION

The nature of the host sediment, particularly its texture, exerts a strong influence on the distribution and growth forms of hydrates. In the dynamic system, the habitat of gas and methane transport mechanisms will modulate this pattern, but ultimately the geological structures will control where the hydrate is concentrated.

Acknowledgements. MBC acknowledges United Kingdom NERC Fellowship GT5-97/IS/ODP held at the University of Leeds 1997-1999. At UFBA, MBC is supported by grants from the Brazilian Ministry of Science and Technology. This work was inspired by the late Gabriel Ginsburg, whose visit to the UK in 1998 was made possible by a generous grant from the Royal Society under the Ex-Quota Visits scheme.

Chapter 10

Natural Gas Hydrate as a Potential Energy Resource

Timothy S. Collett
U.S. Geological Survey
Denver, CO 80225

1. INTRODUCTION

The estimated amount of gas in the hydrate accumulations of the world greatly exceeds the volume of known conventional gas reserves. However, the role that gas hydrate will play in contributing to the world's energy requirements will depend ultimately on the availability of sufficient gas hydrate resources and the "cost" to extract them. Yet considerable uncertainty and disagreement prevails concerning the world's gas hydrate resources. Disagreements over fundamental issues such as volume of gas stored within delineated gas hydrate accumulations and the concentration of gas hydrate within hydrate-bearing reservoirs have demonstrated that we know very little about gas hydrate.

Despite the fact that relatively little is known about the ultimate resource potential of natural gas hydrate, it is certain that gas hydrate is a vast storehouse of natural gas and significant technical challenges need to be met before this enormous resource can be considered an economically producible reserve. It is proposed in this paper that the evolution of gas hydrate as a viable source of natural gas, like any other unconventional energy resource (e.g., deep gas, shale gas, tight gas sands, and coalbed methane), will follow a predictable path from research and discovery to implementation (Figure 1); however, insurmountable barriers may exist along this pathway. Today, most of the gas hydrate research community is focused on three fundamental issues: WHERE does gas hydrate occur, HOW does gas hydrate occur in nature, and WHY does gas hydrate occur in a particular setting. However, relatively little has been done to integrate these distinct research topics or evaluate how collectively they affect the ultimate resource potential of gas hydrate. Only after understanding the fundamental aspects of WHERE-HOW-WHY gas hydrate occur in nature will we be able to make accurate estimates of how much gas is trapped within the gas hydrate accumulations of the world. Even with the confirmation that gas

M.D. Max (ed.), Natural Gas Hydrate in Oceanic and Permafrost Environments, 123–136.

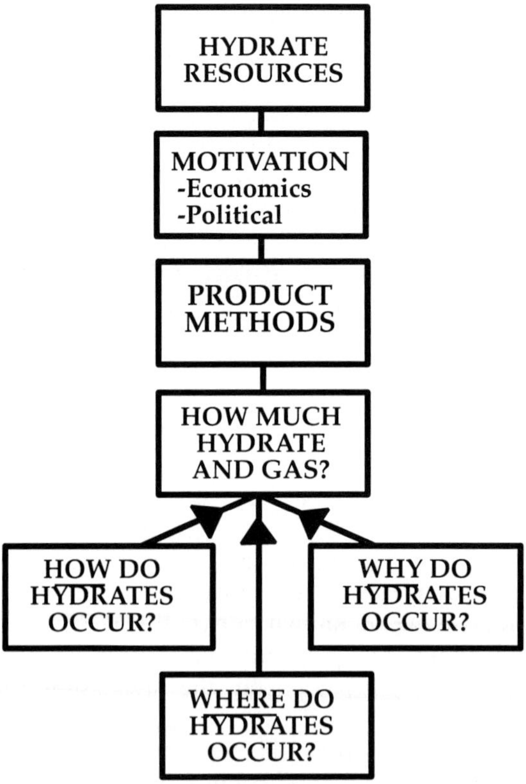

Figure 1. Flow chart depicting the evolution of gas hydrate understanding from a non-producible unconventional gas resource to a producible energy resource

hydrate may exist in considerable volumes, significant technical, economic, and political issues need to be resolved before gas hydrate can be considered a viable energy resource.

In this paper, I have attempted to review the status of gas hydrate as a future energy resource. The technical and non-technical factors controlling the ultimate resource potential of gas hydrate have been identified and assessed. The fundamental questions of WHERE does gas hydrate occur, HOW does gas hydrate occur in nature, and WHY does gas hydrate occur in a particular setting have been individually reviewed and discussed. In addition, published gas hydrate volume assessments have been summarized and the production technology needed to extract the world's gas hydrate resources are assessed. The paper concludes with a discussion of the economic and political motivations that may eventually lead to gas hydrate production.

2. WHERE DOES GAS HYDRATE OCCUR?

The geologic occurrence of gas hydrate has been known since the mid-1960s, when gas-hydrate accumulations were discovered in Russia (reviewed by Makogon, 1981). Gas hydrate is widespread in permafrost regions and beneath the sea in sediment of outer continental margins (reviewed by Kvenvolden, 1993). Cold surface temperatures at high latitudes on earth are conducive to the development of onshore permafrost and gas hydrate in the subsurface. Onshore gas hydrate are known to be present in the West Siberian Basin (Makogon et al., 1972) and are believed to occur in other permafrost areas of northern Russia, including the Timan-Pechora province, the eastern Siberian Craton, and the northeastern Siberia and Kamchatka areas (Cherskiy et al., 1985). Permafrost-associated gas hydrate is also present in the North American Arctic. Direct evidence for gas hydrate on the North Slope of Alaska comes from a core-test in gas hydrate exploration well (the Northwest Eileen State-2 well), and indirect evidence comes from drilling and open-hole industry well logs that suggest the presence of numerous gas hydrate layers in the area of the Prudhoe Bay and Kuparuk River oil fields (Collett, 1993). Well-log responses attributed to the presence of gas hydrate have been obtained in about one-fifth of the wells drilled in the Mackenzie Delta, and more than half of the wells in the Arctic Islands are inferred to contain gas hydrate (Judge and Majorowicz, 1992). The recently completed Mallik 2L-38 gas hydrate research well, confirmed the presence of a relatively thick, highly concentrated, gas hydrate accumulation on Richards Island in the outer portion of the Mackenzie River Delta (Dallimore et al., 1999).

The presence of gas hydrate in offshore continental margins has been inferred mainly from anomalous seismic reflectors (i.e., BSRs) that coincide with the predicted phase boundary at the base of the gas-hydrate stability zone. Gas hydrate have been recovered in gravity cores within 10 m of the sea floor in sediment of the Gulf of Mexico (Brooks et al., 1986), the offshore portion of the Eel River Basin of California (Brooks et al., 1991), the Black Sea (Yefremova and Zhizhchenko, 1974), the Caspian Sea (Ginsburg et al., 1992), and the Sea of Okhotsk (Ginsburg et al., 1993). Also, gas hydrate have been recovered at greater sub-bottom depths during research coring along the southeastern coast of the United States on the Blake Ridge (Kvenvolden and Barnard, 1983; Shipboard Scientific Party, 1996), in the Gulf of Mexico (Shipboard Scientific Party, 1986), in the Cascadia Basin near Oregon (Shipboard Scientific Party, 1994), the Middle America Trench (Kvenvolden and McDonald, 1985), offshore Peru (Kvenvolden and Kastner, 1990), and on both the eastern and western margins of Japan (Shipboard Scientific Party, 1990, 1991).

3. HOW DOES GAS HYDRATE OCCUR IN NATURE?

Little is known about the nature of gas hydrate reservoirs. For example, do hydrate occur as pore-filling constituents or are they only found in massive form. Information about the nature and texture of reservoired gas hydrate is needed to accurately determine the amount of gas hydrate and associated gas in a given gas hydrate accumulation. The textural nature of gas hydrate in the reservoir also controls the production potential and characteristics of a gas hydrate accumulation. The physical and chemical conditions that result in different forms (disseminated, nodular, layered, massive) and distributions (uniform or heterogeneous) of gas hydrate are not understood (reviewed by Sloan, 1998). It is necessary; therefore, to systematically review descriptions of known gas hydrate occurrences and evaluate existing gas hydrate reservoir models in order to assess the nature of gas-hydrate-bearing reservoirs.

For this review of the nature of gas hydrate occurrences, I have relied extensively on the offshore gas hydrate sample database published by Booth et al. (1996). In this database, Booth et al. (1996) systematically review and describe more than 90 marine gas hydrate samples recovered from 15 different geologic regions. In general, most of the recovered gas hydrate samples consist of individual grains or particles, which are often described as inclusions or disseminated in the sedimentary section. Gas hydrate also occur as, what has been described as, a cement, nodules, or as lamina and veins, which tend to be characterized by dimensions of a few centimeters or less. In several cases, thick, pure gas hydrate layers measuring as much as 3- to 4-m-thick have been sampled (DSDP Site 570; Shipboard Scientific Party, 1985). In both marine and terrestrial permafrost environments, the thickness of identified gas-hydrate-bearing sedimentary sections varies from a few centimeters to as much as 30 m (Collett, 1993; Booth et al., 1996; Dallimore et al., 1999). Most pure gas hydrate lamina and layers, however, are often characterized by thicknesses of millimeters to centimeters (Booth et al., 1996; Dallimore and Collett, 1995; Dallimore et al., 1999). Booth et al. (1996) conclude that gas-hydrate-bearing sedimentary sections tend to be tens of centimeters to tens of meters thick, but thick zones of pure hydrate are relatively rare and only represent a minor constituent of potential gas hydrate accumulations.

The Booth et al. (1996) review along with recently published gas hydrate sample descriptions from the Mackenzie Delta (Dallimore and Collett, 1995; Dallimore et al., 1999) and the Blake Ridge (ODP Leg 164, Shipboard Scientific Party 1996), confirm that gas hydrate are usually uniformly distributed within sediments as mostly pore-filling constituents.

4. WHY DOES GAS HYDRATE OCCUR IN A PARTICULAR SETTING?

A review of previous gas hydrate studies reveal that the formation and occurrence of gas hydrate is controlled by numerous factors (Collett, 1995); two of the most important appear to be formation temperatures and the composition of the gas molecule included in the gas hydrate clathrate. Other factors, including the availability of gas, also exhibit important controls on the occurrence of gas hydrate.

Gas hydrate exists under a limited range of temperature and pressure conditions such that the depth to the top of the zone and the thickness of the potential gas-hydrate stability zone can be calculated (Fig. 2) and shown

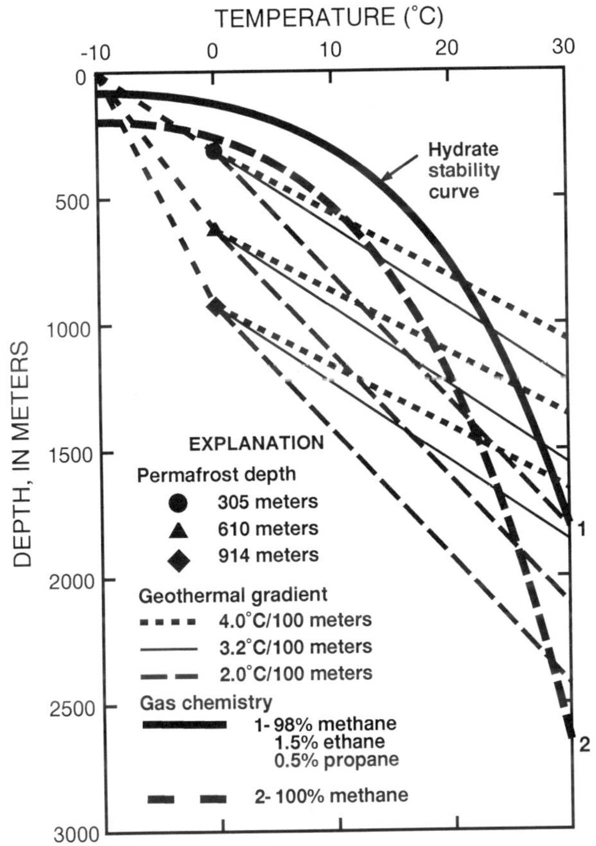

Figure 2. Graph showing the depth-temperature zone in which gas hydrate is stable in a permafrost region [assuming a 9.795 kPa/m pore-pressure gradient] (modified from Holder et al., 1987)

as a series of subsurface temperature profiles from an onshore permafrost area and two laboratory-derived gas-hydrate stability curves for different natural gases (modified from Holder et al., 1987). This gas-hydrate phase-diagram (Figure 2) illustrates how variations in formation-temperature and gas composition can affect the thickness of the gas-hydrate stability zone.

The zone of potential gas-hydrate stability in Figure 2 lies in the area between the intersections of the geothermal gradients and the gas-hydrate stability curves.

It is well documented that the formation of gas hydrate requires a large source of natural gas. In most gas hydrate resource assessment studies; the characterization of gas sources has been based on assessing a set of minimum source-rock criteria that includes organic richness (total organic carbon), sediment thickness, and thermal maturity. It has been shown that the availability of large quantities of hydrocarbon gas from both microbial and thermogenic sources are an important factor controlling the formation and distribution of natural gas hydrate (Kvenvolden, 1988; Collett, 1993). Carbon isotope analyses indicate that the methane in most oceanic hydrate is derived from microbial sources. However, molecular and isotopic analyses indicate a thermal origin for the methane in several offshore Gulf of Mexico and onshore Arctic gas-hydrate occurrences.

5. HOW MUCH GAS IS TRAPPED WITHIN GAS HYDRATE ACCUMULATIONS?

The amount of methane sequestered in gas hydrate is probably enormous, but estimates of the amounts are speculative and range over three orders-of-magnitude from about 3,114 to 7,634,000 trillion cubic meters (Table 1) (reviewed by Kvenvolden, 1993).

Terrestrial Gas Hydrates	
Cubic meters	Reference
1.4×10^{13}	Meyer (1981)
3.1×10^{13}	McIver (1981)
5.7×10^{13}	Trofimuk et al. (1977)
7.4×10^{14}	MacDonald (1990)
3.4×10^{16}	Dobrynin et al. (1981)
Oceanic Gas Hydrates	
Cubic meters	Reference
3.1×10^{15}	Meyer (1981)
5 to 25×10^{15}	Trofimuk et al. (1977)
2×10^{16}	Kvenvolden (1988)
2.1×10^{16}	MacDonald (1990)
4×10^{16}	Kvenvolden and Claypool (1988)
7.6×10^{18}	Dobrynin et al. (1981)

Table 1. World estimates of the amount of gas within hydrate.

It is likely, however, that the amount of gas in the hydrate reservoirs of the world greatly exceeds the volume of known conventional gas reserves. Before reviewing assessments of the world gas hydrate resources it is necessary to examine the quality and variability of gas hydrate assessments at the accumulation and reservoir scale.

5.1. Gas Hydrate at the Reservoir and Accumulation Scale

Estimates of the amount of gas hydrate and associated gas within a given gas hydrate accumulation can vary considerably (Table 2). For example, recent estimates of the volume of gas that may be contained in the gas hydrate and free-gas beneath gas hydrate on the Blake Ridge range from about 70 trillion cubic meters of gas over an area of 26,000 km^2 (Dickens et al., 1997) to about 80 trillion cubic meters of gas for an area of 100,000 km^2 (Holbrook et al., 1996). The difference between these two estimates has been attributed to the observation that the amount of free-gas directly measured within pressure-core samples (Dickens et al., 1997) from beneath the gas hydrate accumulation is significantly larger than that estimated from borehole vertical seismic profile data (Holbrook et al., 1996).

Cubic meters	References
Blake Ridge, USA	
18×10^{12}	Dillon and Paull (1983)
80×10^{12}	Holbrook et al. (1996)*
70×10^{12}	Dickens et al. (1997)*
38×10^{12}	Collett and Ladd (2000)
57×10^{12}	Collett and Ladd (2000)
Nankai Trough, Japan	
50×10^{12}	MITI/JNOC (1998)**
Andaman Sea, India	
6×10^{12}	GAIL/ONGC/DGH (1997)**
Prudhoe Bay Alaska, USA	
1.2×10^{12}	Collett (1993)
Mackenzie Delta (Richards Is.), Canada	
0.2×10^{12}	Collett et al. (1999)

Table 2. Gas Hydrate Accumulations In-Place Natural Gas Resources Estimates of the amount of gas within known and inferred gas hydrate accumulations. * Includes associated free-gas. ** Unpublished industry reports

Recent analysis of downhole electrical resistivity log data from ODP Leg 164 coreholes (Collett and Ladd, 2000) reveal that the gas hydrate and underlying free-gas accumulations on the Blake Ridge contains about 57 trillion cubic meters of gas (assuming a gas hydrate areal distribution of 26,000 km^2). Other published studies indicate that the gas hydrate at the crest of the Blake Ridge alone (area of about 3,000 km^2) may contain more than 18 trillion cubic meters of gas (Dillon and Paull, 1983). The broad range of these estimates demonstrates the need for high-resolution measurements of the gas hydrate and associated free-gas volumes within any gas hydrate accumulation of interest.

It has been suggested that the volume of gas that may be contained in a gas hydrate accumulation depends on five primary "reservoir" parameters (modified from Collett, 1993): (1) areal extent of the gas-hydrate occurrence, (2) "reservoir" thickness, (3) sediment porosity, (4) degree of gas-hydrate saturation, and (5) the hydrate gas yield volumetric parameter which defines how much free-gas (at STP) is stored within a gas hydrate (also known as the hydrate number). In the following section, the five "reservoir" parameters (Table 3) needed to calculate the volume of gas associated with the gas hydrate on the Blake Ridge (ODP Sites 994, 995, and 997; Shipboard Scientific Party, 1996), along the Cascadia continental margin (ODP Site 889; Shipboard Scientific Party, 1994), on the North Slope of Alaska (Northwest Eileen State-2 well; Collett, 1993), and in the Mackenzie River Delta of Canada (Mallik 2L-38 well; Dallimore et al., 1999) are reviewed.

The following "resource" assessment (modified from Collett, 1998) has been conducted on a site-by-site basis; that is, for each site examined the volume of gas hydrate and associated gas within a one square kilometer area surrounding each drill-site have been individually calculated (Table 3). For this "resource" assessment, the thickness of the gas-hydrate-bearing sedimentary section at both the marine and permafrost drill sites was defined to be the total thickness of the downhole log inferred gas-hydrate accumulation (Table 3). Average core derived sediment porosities for the gas-hydrate-bearing reservoirs at Sites 994, 995, 997 (Blake Ridge) and 889 (Cascadia continental margin) range from about 52 to 58 percent (Table 3). The "corrected" density log derived sedimentary porosities for three gas-hydrate-bearing units identified in the Northwest Eileen State-2 in northern Alaska (Units C, D, and E) range from an average value of about 36 to 39 percent (Table 3). The density log derived sediment porosities for the gas-hydrate-bearing interval in the Mallik 2L-38 well in northern Canada averages about 31% (Table 3). Gas-hydrate saturations at the marine drill-sites (Sites 994, 995, 997 and 889), calculated from available downhole logs, range from an average value of about 3 to 6 percent (Table 3). Gas-hydrate saturations in all three gas-hydrate-bearing units (Units C, D, and E) in the Northwest Eileen State-2 well, calculated from available downhole log

Site/Well ID	Log Depth of inferred gas hydrates	Hydrate zone thickness	Sediment Porosity (%)	Gas hydrate saturation	Gas Volume within hydrate per km^2 (m^3)
Ocean Drilling Program Drill Sites:					
Site 994	212.0-428.8	216.8	57.0	3.3	669,970,673
Site 995	193.0-450.0	257.0	58.0	5.2	1,267,941,673
Site 997	186.4-450.9	264.5	58.1	5.8	1,449,746,073
Site 889	127.6-228.4	100.8	51.8	5.4	466,635,705
Northwest Eileen State 2 Drill Site:					
Unit C	651.5-680.5	29.0	35.6	60.9	1,030,904,796
Unit D	602.7-609.4	6.7	35.8	33.9	133,382,462
Unit E	564.0-580.8	16.8	38.6	32.6	346,928,811
Total for Northwest Eileen State 2					1,511,216,069
Mallik 2L 38 Drill Site:					
HY Unit	888.8-1101.1	212.3	31.0	44.0	4,749,066,080

Table 3. Volume of gas within the downhole log inferred gas hydrate occurrences at ODP Sites 994, 995, 997 (Blake Ridge), and 889 (Cascadia contiental margin), and in the Northwest Eileen State-2 (northern Alaska) and Mallik 2L-38 (northern Canada) wells

data, range from an average value of about 33 to 61 percent (Table 3). The resistivity well log derived gas-hydrate saturations in the Mallik 2L-38 well average about 44% (Table 3). In this assessment, a hydrate number of 6.325 (90% gas filled clathrate) was assumed, which corresponds to a gas yield of 164 m^3 of methane (at STP) for each cubic meter of gas hydrate. The log inferred gas hydrate at Sites 994, 995, and 997 on the Blake Ridge contain between 670,000,000 and 1,450,000,000 cubic meters of gas per square kilometer (Table 3). The volume of gas within the log inferred gas hydrate at Site 889 on the Cascadia continental margin is about 467,000,000 cubic meters of gas per square kilometer (Table 3). Cumulatively, all three log inferred gas-hydrate-bearing stratigraphic units (Units C, D, and E) drilled and cored in the Northwest Eileen State-2 well may contain about 1,511,000,000 cubic meters of gas in the one square kilometer area surrounding this drill-site (Table 3). It was also determined that the log inferred gas-hydrate-bearing stratigraphic interval drilled in the Mallik 2L-38 well contains about 4,750,000,000 cubic meters of gas in the one square kilometer area surrounding the Mallik drill-site (Table 3).

A close examination of the gas hydrate saturations in Table 1 reveals a potential problem associated with production of gas from marine gas hydrate. Even though vast portions of the world's continental shelves appear to be underlain by gas hydrate, the concentration of hydrate within most marine accumulations appears to be very low. Low gas hydrate concentrations may significantly affect the economic production potential of marine gas hydrate (gas hydrate production is discussed in more detail later in this paper).

5.2. Gas Hydrate at the World and National Scale

World estimates for the amount of natural gas in gas hydrate deposits range from 14 to 34,000 trillion cubic meters for permafrost areas and from 3,100 to 7,600,000 trillion cubic meters for oceanic sediments (Table 1; modified from Kvenvolden, 1993). The estimates in Table 1 show considerable variation, but oceanic sediments seem to be a much greater resource of natural gas than continental sediments. Current estimates of the amount of methane in the world's gas hydrate accumulations are in rough accord at about 20,000 trillion cubic meters (reviewed by Kvenvolden, 1993). If these estimates are valid, the amount of methane in gas hydrate is almost two orders of magnitude larger than the estimated total remaining recoverable conventional methane resources, estimated to be about 250 trillion cubic meters (Masters et al., 1991).

The 1995 National Assessment of United States Oil and Gas Resources, conducted by the U.S. Geological Survey, focused on assessing the undiscovered conventional and unconventional resources of crude oil and natural gas in the United States (Gautier et al., 1995). This assessment included for the first time a systematic resource appraisal of the in-place natural gas hydrate resources of the United States onshore and offshore regions (Collett, 1995). In this assessment, 11 gas-hydrate plays were identified within four offshore and one onshore gas hydrate provinces. In-place gas resources within the gas hydrate of the United States are estimated to range from about 3,200 to 19,000 trillion cubic meters of gas, at the 0.95 and 0.05 probability levels, respectively. Although this wide range of values shows a high degree of uncertainty, it does indicate the potential for enormous quantities of gas stored as gas hydrate. The mean in-place value for the entire United States is calculated to be about 9,000 trillion cubic meters of gas.

6. GAS HYDRATE PRODUCTION TECHNOLOGY

Even though gas hydrate are known to occur in numerous marine and Arctic settings, little is known about the technology necessary to produce gas hydrate. Most of the existing gas hydrate "resource" assessments do not address the problem of gas hydrate recoverability. Proposed methods of gas recovery from hydrate usually deal with dissociating or "melting" in-situ gas hydrate by (1) heating the reservoir beyond hydrate formation temperatures, (2) decreasing the reservoir pressure below hydrate equilibrium, or (3) injecting an inhibitor, such as methanol or glycol, into the reservoir to decrease hydrate stability conditions. Gas recovery from hydrate is hindered because the gas is in a solid form and because hydrate is usually widely dispersed in hostile Arctic and deep marine environments. First order thermal stimulation computer models (incorporating heat and mass balance) have been developed to evaluate hydrate gas production from hot water and steam floods, which have shown that gas can be produced from hydrate at sufficient rates to make gas hydrate a technically recoverable resource (Sloan, 1998). However, the economic cost associated with these types of enhanced gas recovery techniques would be prohibitive. Similarly, the use of gas hydrate inhibitors in the production of gas from hydrate has been shown to be technically feasible (Sloan, 1998), however, the use of large volumes of chemicals such as methanol comes with a high economic and environmental cost. Among the various techniques for production of natural gas from in-situ gas hydrate, the most economically promising method is considered to be the depressurization technique. However, the extraction of gas from a gas hydrate accumulation by depressurization may be hampered by the formation of ice and/or the reformation of gas hydrate due to the endothermic nature of gas hydrate dissociation.

The Messoyakha gas field in the northern part of the West Siberian Basin is often used as an example of a hydrocarbon accumulation from which gas has been produced from in-situ natural gas hydrate. Production data and other pertinent geologic information have been used to document the presence of gas hydrate within the upper part of the Messoyakha field (Makogon, 1981). It has also been suggested that the production history of the Messoyakha field demonstrates that gas hydrate are an immediate producible source of natural gas, and that production can be started and maintained by conventional methods. Long-term production from the gas-hydrate part of the Messoyakha field is presumed to have been achieved by the simple depressurization scheme. As production began from the lower free-gas portion of the Messoyakha field in 1969, the measured reservoir-pressures followed predicted decline relations; however, by 1971 the reservoir pressures began to deviatc from expected values. This deviation has been attributed to the liberation of free-gas from dissociating gas hydrate. Throughout the production history of the Messoyakha field, it is estimated that about 36% (about 5 billion cubic meters) of the gas withdrawn from the field has come from thc gas hydrate (Makogon, 1981). Recently, however, several studies suggest that gas hydrate may not be significantly contributing to gas production in the Messoyakha field (reviewed by Collett and Ginsburg, 1998).

It should be noted, that our current assessment of proposed methods for gas hydrate production do not consider some of the more recently developed advanced oil and gas production schemes. For example, the usefulness of downhole heating methods such as in-situ combustion, electromagnetic heating, or downhole electrical heating have not been evaluated. In addition, advanced drilling techniques and complex downhole completions, including horizontal wells and multiple laterals, have not been considered in any comprehensive gas hydrate production scheme. Gas hydrate provinces with existing conventional oil and gas production may also provide us with the opportunity to test relatively more advanced gas hydrate production methods. For example, in northern Alaska existing "watered-out" production wells are being evaluated as potential sources for hot geopressured brines that will be used to thermally stimulate gas hydrate production.

As previously noted, the low concentration of hydrate in most of the world's marine gas hydrate occurrences raises a concern over the production technology required to produce gas from highly disseminated gas hydrate accumulations. In addition, the host-sediments also represent a significant technical challenge to potential gas hydrate production. In most cases, marine gas hydrate has been found in clay-rich unconsolidated sedimentary sections that exhibit little or no permeability. Most of the existing gas hydrate production models require the establishment of reliable flow paths within the formation to allow the movement of produced gas to the wellbore and injected fluids into the gas-hydrate-bearing sediments. It is unlikely, however, that most marine sediments possess the mechanical strength to allow the generation of significant flow paths. It is possible that in basins with significant input of coarse-grained clastic sediments, such as the Gulf of Mexico or along the eastern margin of India, gas hydrate may be reservoired at high concentrations in more conventional clastic reservoirs; which is more analogous to the nature of gas hydrate occurrences in onshore permafrost environments (Collett, 1993; Dallimore et al., 1999).

7. MOTIVATIONS LEADING TO GAS HYDRATE PRODUCTION

As previously discussed in this paper, significant if not insurmountable technical issues need to be resolved before gas hydrate can be counted as a viable option for future supplies of natural gas. In most cases, the viability of an energy resource is based almost solely on economics. It is important to note, however, that in some cases the viability of a particular hydrocarbon resource can be controlled by unique local economic and non-technical factors. For example, countries with little domestic energy production usually pay considerably more for their energy needs since they rely more on imported hydrocarbons, which often come with additional tariffs and transportation expenses. Energy security is often a concern to resource poor countries, which in comparison to energy rich countries will often invest more money in relatively expensive unconventional domestic energy resources. In some cases the uniqueness of a particular location, such as distance to a conventional energy resource, may lead to the development of otherwise non-economic unconventional resource. In the following section, the economic and non-economic motivations that may eventually lead to sustained production of gas from hydrate will be discussed.

7.1. Economic Motivations

Because of uncertainties about the geologic settings and feasible production technology, few economic studies have been published on gas hydrate. The National Petroleum Council (NPC), in its major 1992 study of gas (National Petroleum Council, 1992), published one of the few available economic assessments of gas hydrate production. This information, extracted from MacDonald (1990), assessed the relative economics of gas recovery from hydrate using thermal injection and depressurization. It also benchmarks the cost of gas hydrate production with the costs of conventional gas production on Alaska's North Slope. The NPC report concluded, that within countries with considerable production of cheaper conventional natural gas, hydrates appear not to be an economically viable energy resource in a competitive energy market.

Japan, India, and South Korea, like many other countries with little indigenous energy resources, pay a very high price for imported liquid natural gas (LNG) and oil. The high cost of imported hydrocarbon resources is one reason why in the last two years government agencies in Japan, India, and South Korea have begun to develop hydrate research programs to recover gas from oceanic hydrate. One of the most notable gas hydrate projects is underway in Japan, where the Japan National Oil Corporation (JNOC), with funding from the Ministry of International Trade and Industry (MITI), has launched a five year study to assess the domestic resource potential of natural gas hydrate. In numerous press releases, MITI has indicated that "methane hydrate could be the next generation's source of producible domestic energy". JNOC is scheduled to drill a gas hydrate test well in the Nankai Trough area, near Tokyo, in the later part of 1999. As much as 50 trillion cubic meters of gas may be stored within the gas hydrate of the Nankai Trough. In 1998, JNOC also drilled the Mallik 2L-38 gas hydrate research well with the Geological Survey of Canada in the Mackenzie Delta of northern Canada (Dallimore et al., 1999).

India (Chapter 17), like Japan (Chapter 18), has also initiated a very ambitious national gas hydrate research program. In March of 1997, the government of India announced new exploration licensing policies that included the release of several deep water (>400m) lease blocks along the east coast of India between Madras and Calcutta. Recently acquired seismic data have revealed possible evidence of widespread gas hydrate occurrences throughout the proposed lease blocks. Also announced was a large gas hydrate prospect in the Andaman Sea, between India and Myanmar, which is estimated to contain as much as six trillion cubic meters of gas. The government of India has indicated that gas hydrates are of "utmost importance to meet their growing domestic energy needs". The National Gas Hydrate Program of India calls for drilling as many as five gas hydrate test wells. Most recently the United States, through the U.S. Department of Energy, has launched a national level research program to assess the resource potential of both marine and permafrost-associated gas hydrate.

7.2. Political Motivations

The world will consume increasing volumes of natural gas well into the 21st century if reliable, low cost supplies can be discovered and exploited. In the near term, natural gas is expected to take on a greater role in power generation and transportation because of increasing pressure for cleaner fuels and reduced carbon dioxide emissions. Gas demand is also expected to grow throughout the first half of the next century because of the expanding role of gas as a competitive transportation fuel due to the commercial development of gas-to-liquids technology. The drive to increased reliance on natural gas will only be in part based on economics. Government regulatory and taxation policy may also dictate the viability of a particular energy commodity such as gas hydrate. In the recent past, government subsidies for unconventional gas resources such as coalbed methane contributed to their technical and economic viability. Similar forms of government support may have a significant impact on the resource viability of gas hydrate. Another non-economic factor that may affect the resource potential of gas hydrates in a particular country is the concerns dealing with national security and dependence on foreign energy resources. The governments of many countries, including the United States, often express concerns over reliance on imported energy resources. Most certainly the international gas hydrate research programs of Japan, India, and South Korea have been established in part to address concerns over their reliance on foreign energy resources.

8. CONCLUSION

Despite the fact that relatively little is known about the ultimate resource potential of natural gas hydrates, it is certain that gas hydrates are a vast storehouse of natural gas and the national gas hydrate research programs of Japan, India, and the United States will significantly contribute to our understanding of the technical challenges needed to turn this enormous resource into a economically producible reserve.

In conclusion, will gas hydrate become a significant energy resource? It is unlikely that we will see significant worldwide gas production from hydrate for the next 30 to 50 years (Chapter 26). However, in certain parts of the world characterized by unique economic and/or political motivations, gas hydrate may become a critical sustainable source of natural gas within the foreseeable future, possibly in the next five to ten years.

Chapter 11

Climatic Impact of Natural Gas Hydrate

Bilal U. Haq
National Science Foundation
Division of Ocean Sciences
Arlington VA 22230

1. Introduction

Gas hydrate occurrence in the sediments of the outer continental margins is sustained in place for relatively long periods by high hydrostatic pressure and low ambient temperature. Most naturally occurring hydrate is composed of molecules of methane trapped in an ice cage of water molecules. Thus, the breakdown of hydrate in response reduced hydrostatic pressure or increased bottom-water temperature can potentially introduce significant quantities of this potent greenhouse gas in the water column and atmosphere, encouraging accelerated warming. At higher latitudes hydrate also occurs in association with permafrost at depths ranging from 130 to 2000 m. Here methane is held captive in the clathrate enclosure by frigid temperatures. An increase in the mean temperature of the higher latitudes, therefore, also has the potential to dissociate the hydrate and emit methane directly into the atmosphere.

Hydrate methane is, in turn, oxidized into water and carbon dioxide, the latter being the dominant contributor to the atmospheric greenhouse forcing. Ice-core records of the recent geological past from Greenland and Antarctica reveal that climatic warming occurs in parallel with rapid increase in atmospheric methane and carbon dioxide (Fig. 1). It has been proposed that catastrophic release of methane from hydrate sources into the atmosphere during periods of lowered sea level (and, thus, reduced hydrostatic pressure) may have been a contributing factor for rapid climate change in the past. It is also feared that escalating methane emissions from the hydrates and the permafrost may soon lead to further strengthening of the on-going global warming trend and may cause unpredictable and abrupt changes in future climatic patterns. Thus, environmental consequences of the release of methane from hydrate sources have become an important issue of societal relevance.

M.D. Max (ed.), Natural Gas Hydrate in Oceanic and Permafrost Environments, 137–148.

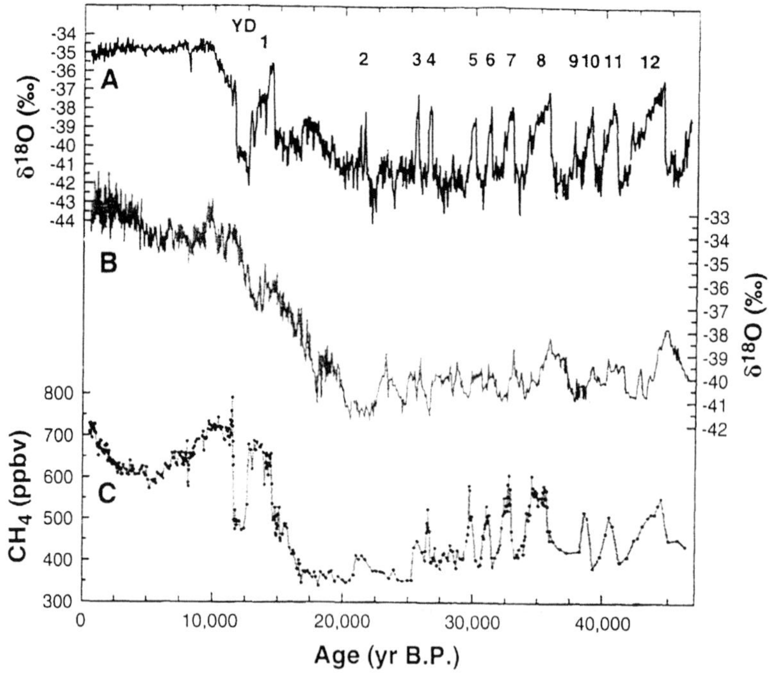

Figure 1. Record of $\delta^{18}O$ (paleotemperature) changes in air bubbles trapped in ice from Greenland ice core (A) and Antarctic ice core (B). Methane fluctuations (C) in the Greenland core parallel those in temperature. Numbers 1-12 at the top represent Dansgaard/Oeschger cycles; YD= Younger Dryas stadial. (After Blunier, 2000, reproduced by permission).

The prerequisites of high hydrostatic pressure (>5 bars) and low bottom-water temperature (< 7(C) for the stability of gas hydrates imply that hydrates occur mostly on the continental slope and rise, below 530 m water depth in the low latitudes, and generally below 250 m depth in the high latitudes. Hydrated sediments may extend from these depths to as much as 1100 m sub-seafloor. Global estimates of methane sequestered in gas-hydrate reservoirs (both in the hydrate-stability zone and as free gas beneath it) vary widely. For example, the Arctic permafrost is estimated to hold anywhere between 7.6 and 1.8 x 104 Gt (Gigatons = 1015 grams) of methane carbon, while marine sediments are extrapolated to hold between 1.7 x 103 and 4.1 x 105 Gt of methane carbon globally (Kvenvolden, 1998). Recently, Buffet et al. (2000) have revised the estimate of total methane hydrate carbon to 1.5 x 104 Gt. They used numerical models to predict the vertical distribution of hydrate as a function of organic matter in the sediments and sedimentation rate.

These models were tested using empirical parameter values from the Blake Ridge. Comparisons between predicted and observed hydrate content, free gas below the stability zone and chlorinity profiles showed reasonable correspondence. Obtaining more precise global estimates of methane trapped in clathrate reservoirs remains one of the top priorities in gas-hydrate research. More meaningful estimates of the amount of gas sequestered in and below the hydrates are essential if we are to resolve the real quantitative impact of the methane emissions from these sources in our climate-change models, as well as in determining the efficacy of hydrates as a future energy resource.

A key unknown, especially for climatic implications, is the nature and fate of methane removal from the dissociated hydrate. How and what quantities of the gas escapes from the hydrate zone and what percentage is dissolved in the water column versus escaping into the atmosphere? Methane flux rates from the seafloor and the hydrated sediments into the water column remain difficult to quantify due to highly variable, transient and diffuse nature of the flux. Oxidation rates of methane in the water column are even more poorly understood.

In a steady state much of the methane diffusing from sediments is believed to be oxidized in the surficial sediment and the water column above (Cranston, 1996). One modeling study (Harvey and Huang, 1995) concluded that in the modeled future global-warming scenarios, the potential of methane from gas-hydrate sources, based on a variety of input assumptions, would seem small compared to the effects of CO_2 emissions, including the envisaged anthropogenic contributions. Thus, a clearer understanding of what happens to the significant quantities of gas that might be catastrophically released from the hydrates becomes imperative. How much of this methane makes it to the atmosphere to force additional greenhouse warming? In the abrupt climate-change scenarios it is assumed that much of this methane is emitted to the atmosphere.

2. Abrupt Climate Change Scenarios

It has been suggested that a sea-level drop of nearly 120 m during the last glacial maximum reduced the hydrostatic pressure sufficiently to raise the lower limit of gas-hydrate stability by about 20 m in the low latitudes (Dillon and Paull, 1983). When a hydrate dissociates, its consistency changes from a solid to a mixture of sediment, water and free gas. Experiments on the mechanical strength behavior of hydrates has shown that the hydrated sediment is markedly stronger than water ice (10 times stronger than ice at 260° K) (Zhang et al., 1999). Thus, a change in the physical properties of the sediment at the base of the hydrate stability field could produce a zone of weakness where sedimentary failure could take place, encouraging low-angle faulting and mass wasting along the continental slope. The common occurrence of Pleistocene slumps on the seafloor have been ascribed to such a catastrophic mechanism and major slumps have been identified in sediments of this age in widely separated margins of the world (Kvenvolden, 1998).

Whether mass wasting at the hydrated-sediment depths is due to hydrate dissociation or other causes, such as, earthquakes or gravity failure, it can potentially release significant quantity of methane trapped below the level of the slump, in addition to the gas emitted from the dissociated hydrate itself. During the glacial periods, these emissions are envisaged to increase together with the incidence of slumping and the easing of hydrostatic pressure on the as the sea level falls. This may eventually trigger a negative response to advancing glaciation and encourage a reversal and termination of the glacial cycle. Thus, there may be a built-in terminator to glaciation, via the hydrate connection (Paull et al., 1991).

Negative response to glaciation in this scenario can initially function efficiently only in the lower latitudes. At higher latitudes glacially-induced freezing could delay the reversal, but once deglaciation begins, even relatively small increases in the mean temperature of the higher latitudes could set off additional emissions of methane from near-surface sources and enhanced greenhouse warming. One model suggests that a small triggering event and liberation of one or more Arctic gas pools could initiate massive release of methane frozen in the permafrost, leading to accelerated warming (Nisbet, 1990). The abrupt termination of the Younger Dryas glaciation (now dated at ca. 11.6 k.y. ago) has been ascribed to such a mechanism (Nisbet, 1990).

To test this idea, Thorpe et al. (1998) modeled the effect of an abrupt release of a "realistic" amount of methane into the atmosphere at the end of a glacial cycle, as constrained by the ice core records. It was concluded that the direct radiative effects of such a pulse of methane emission alone would be too small to account for the reversal of the glacial cycle. However, with certain combinations of methane, carbon dioxide and heat transport inputs, it was possible to simulate changes of the same magnitude as those indicated by empirical data.

3. Climatic Feedback Loop

Paleoclimatic records of the recent geological past gleaned from ice cores from Antarctica and Greenland (see, e.g., Jouzel et al., 1993; Petit et al., 1999) show that at the onset of glacial cycles there is a relatively gradual decrease in atmospheric carbon dioxide and methane (Fig.1). Climatic ameliorations at the end of the glacial cycles, on the contrary, tend to be more abrupt and are accompanied by equally rapid increases in greenhouse gases. While the onset of glaciations is now widely thought to be brought on by Milankovitch forcing (a combination of variations in Earth's orbital eccentricity, obliquity and precession), a mechanism that also can explain the broad variations in glacial cycles, their relatively abrupt terminations remain enigmatic and less well understood. CO_2 degassing from the ocean surface has been suggested as one mechanism, but it alone can not explain the relatively rapid shift from stadials to interstadials (Nisbet, 1990).

The systems response to glacially-induced sea-level fall, felt first in the lower latitudes and then spreading to the higher latitudes can be thought of as a climatic feedback loop that could be an effective mechanism for explaining the rapid warmings at the transitions from the stadial to interstadial cycles in the late Quaternary (also known as the Dansgaard/Oeschger events). These transitions often occur only on decadal to centennial time scales. In this scenario it is envisioned that the lowstand-induced slumping and methane emissions in lower latitudes lead to accelerated greenhouse warming and trigger a negative feedback to glaciation. This also leads to an increase in carbon-dioxide degassing for the ocean. Once the higher latitudes are warmed by these effects, further release of methane from near-surface sources of the permafrost could provide a positive feedback to warming. The former (methane emissions in the low latitudes) would help force a reversal of the glacial cooling, and the latter (additional release of methane from higher latitudes) could reinforce the trend, resulting in apparent rapid warming observed at the end of the glacial cycles (Haq, 1993, 1998).

For the optimal functioning of the climatic feedback model discussed above, methane would have to be constantly replenished from new and larger hydrate pools during the switchover. Although as a greenhouse gas methane is nearly ten times more potent than carbon dioxide, its residence time in the atmosphere is relatively short (on the order of a decade and a half), after which it reacts with the hydroxyl radical and oxidizes to carbon dioxide and water. The atmospheric retention of carbon dioxide is somewhat more complex than methane because it is readily transferred to other reservoirs, such as the oceans and the biota, from which it can re-enter the atmosphere.

Carbon dioxide accounts for up to 80% of the contribution to greenhouse forcing in the atmosphere. Lashof and Ahuja (1990) estimate an effective residence time of about 230 years for carbon dioxide. These retention times are short enough that for cumulative impact of methane and carbon dioxide through the negative-positive feedback loop to be effective methane emission to the atmosphere would have to be continuously sustained from gas-hydrate and permafrost sources. The feedback loop would close when a threshold is reached where sea level is once again high enough that it can stabilize the residual clathrates and encourage the genesis of new ones.

Several unresolved problems remain with the gas-hydrate climate feedback model. The feedback model assumes a certain amount of time lag between events as they shift from lower to higher latitudes, but the duration of the lag remains unresolved, although a short duration (on decadal to centennial time scales) is implied by the ice-core records. Also, it is not clear whether hydrate dissociation leads to initial warming, or warming caused by other factors leads to increased methane emissions from hydrates. Recent data presented by Severinghaus and Brook (1999) implies a time lag of ca. 50 (±10) years between abrupt warming and the peak in methane values at the Bølling transition (around 14.5 k.y.B.P.), although an increase in methane emissions seems to have begun almost simultaneously with the warming trend (± 5 years). High-resolution record of four climatic transitions from stadials to interstadials in the GISP2 ice core from central Greenland shows (Brook et al., 2000) that while the warming events are rapid, methane concentrations increase by 200-300 ppb relatively slowly over a period of 100-300 years.

This, however, does not detract from the notion that there may be a built-in feedback between increased methane emissions from gas-hydrate sources and accelerated warming. If smaller quantities of methane released from hydrated sediments are oxidized in the water column, initial releases of methane from dissociated hydrates may not produce a significant positive shift in the atmospheric content of methane. However, as the frequency of catastrophic releases from this source increases, more methane is expected to make it to the atmosphere. And, although the atmospheric residence time of methane itself is relatively short, when oxidized, it adds to the greenhouse forcing of carbon dioxide. This may explain the more gradual increase in methane, and is not inconsistent with the a very short time lag between the initiation of the warming trend and methane increase, as well as the time lag between the height of warming trend and the peak in methane values. Although there is still no direct evidence to suggest that the main forcing for the initiation of deglaciation is to be found in hydrate dissociation, once begun, a positive feedback of methane emissions from hydrate sources (and its byproduct, carbon dioxide) can only help accelerate the warming trend.

4. Modulation of Gas-Hydrate Stability by Bottom-Water Temperature: The Santa Barbara Basin

Greenland ice-core data has revealed that rapid increase in atmospheric methane coincided with an abrupt warming event at the end of Younger Dryas glacial episode (at ca. 11.6 k.y. B.P.) and occurred over relatively short interval of a few decades (Severinghaus et al., 1998). High-resolution nitrogen- and argon-isotope data from the Bølling transition (from stadial to interstadial) recorded in the Greenland ice also shows that methane concentrations rose over a period of about 50 years, but began increasing 20 to 30 years after the onset of rapid warming (Severinghaus and Brook, 1999). Additional sources of methane have often been attributed to the spread of wetlands during the wet and warm periods (e.g., Blunier et al., 1995).

Kennett et al. (2000) argue that gas-hydrate dissociation is a more likely source of such rapid increases in methane at the onset of interstadials. In a detailed study of the $\delta^{13}C$ record from sediments of the Santa Barbara Basin, off California, these authors, for the first time, implicate temperature increase in the upper intermediate waters (400-1000 m), rather than reduced hydrostatic pressure, as the potential cause for hydrate dissociation. This record displays millennial-scale Dansgaard-Oeschger (D-O) cycles over the past 60 ky that are synchronous with rapid warmings and can be linked to warmings in the ice records from Greenland. These can also be tied to variations in bottom-water ventilation and temperature and changes in benthic foram assemblages. The record also implies that the warming of the surface water and atmosphere lagged behind the warming of intermediate waters.

Kennett et al. (2000) maintain that the energy needed for these rapid warmings could have come from dissociation of methane hydrates. Relatively large excursions of $\delta 1^{3}C$ (up to -5‰) in benthic foraminifera are associated with the D-O events. Though the planktonic-foram $\delta^{13}C$ record remained relative static, during several brief intervals the planktonics also show notable negative shifts in values (up to -3‰), implying that the entire water column may have experienced rapid $\delta^{12}C$ enrichment during these intervals. They conclude that the large negative excursions in benthic $\delta^{13}C$ during interstadials most likely represent injections of biogenic methane in the system from clathrate sources and are consistent with poor basin ventilation, low oxygen levels, low faunal diversity and occurrence of lamination in these intervals.

Figure 2 summarizes the Kennett et al. (2000) model of changes in the methane flux controlled by gas-hydrate instability that is modulated by increase in bottom-water temperature in the Santa Barbara Basin. They argue that temperature increases in the bottom water of 2 - 3.5° C from stadials to interstadials are large enough to cause the dissociation and deroofing of gas hydrates, leading to increased upward flux of methane. Hydrate dissociation probably occurred close to the beginning of interstadials, when bottom water had warmed up but the sea level was still relatively low. Simultaneous large negative shifts of $\delta^{13}C$ in both the benthics and planktonics probably represent brief but massive and localized injections of methane from hydrate sources. Kennett et al. (2000) identify at least four such episodes of massive methane emissions when the entire water column suffered prominent negative $\delta^{13}C$ shifts.

The Kennett et al. (2000) data suggests a time lag between the warming of bottom waters and warming of the sea surface and the atmosphere. Thus, this model would imply that hydrate dissociation, which would be a consequence of bottom-water warming, could lead to catastrophic injections of methane into the atmosphere and accelerated greenhouse warming. Other data (e.g., Severinghaus et al., 1998; Severinghaus and Brook, 1999) suggests that warming occurs a few decades before the peak in atmospheric methane values, implying that warming causes methane increase and may not a consequence of it. Thus, some authors (e.g., Raynaud et al., 1998; Blunier, 2000; Brook et al., 2000) are not entirely convinced of an active role for hydrate methane in stimulating rapid climate change.

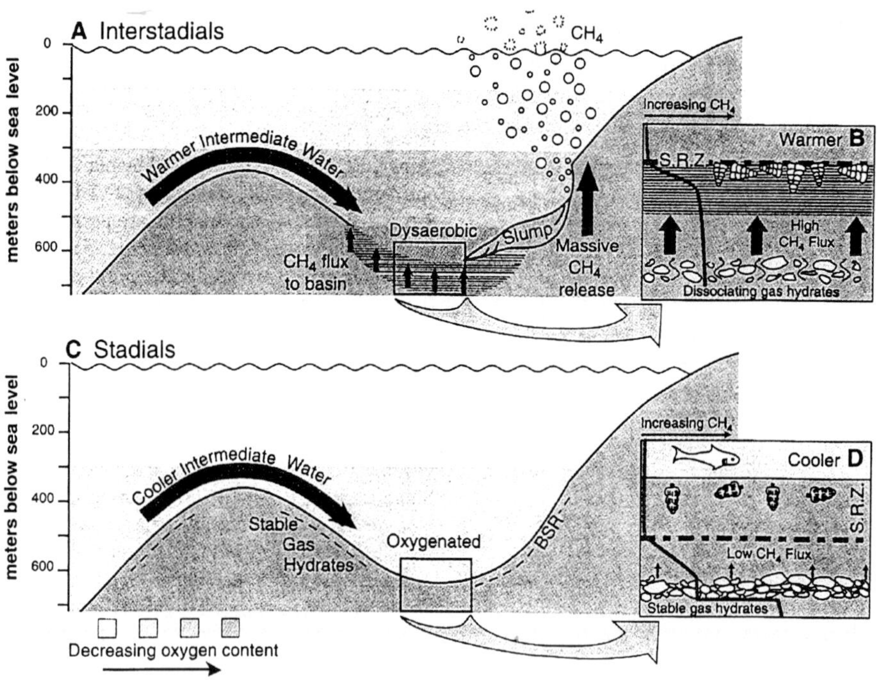

Figure 2. The Kennett et al. model of the influence of changes in bottom-water temperature on methane flux in Santa Barbara Basin. Warmer intermediate waters that can destabilize the hydrate and increase the methane flux to the ocean water and atmosphere (A) characterize the interstadials. During these intervals the sulfate reduction zone (S.R.Z.) is narrow (B). During the stadials (C) the intermediate waters are cooler and the hydrate is stable, with low upward methane flux. The oxygen content of the water increases while S.R.Z. expands (D). (After Kennett et al., 2000, reproduced by permission).

5. Gas Hydrates and the long-term Record of Environmental Change

One potential proxy for the injection of significant volumes of methane into the ocean waters is the changes in carbon-isotopic composition of the carbon reservoir. The $\delta^{13}C$ of biogenic methane in hydrate averages about -60‰ (Kvenvolden, 1993), perhaps the lightest (most enriched in $\delta^{12}C$) carbon anywhere in the Earth system. It has been suggested that massive release of methane from gas-hydrate sources is the most likely mechanism for a pronounced enrichment of $\delta^{12}C$ seen in late Paleocene marine sediments which was also a period of rapid bottom-water warming (Dickens et al., 1995). The dissolution of methane (and its oxidative byproduct, CO_2) in the seawater also led to increased dissolution of carbonate on the seafloor. Thus, major negative shifts in $\delta^{13}C$ that occur together with an increase in benthic temperature (bottom-water warming, see the Santa Barbara example discussed earlier) or a sea-level fall event (reduction of hydrostatic pressure) may provide clues to past behavior of gas hydrates.

6. Massive Methane Injection during the Late Paleocene Thermal Maximum

A prominent negative shift in global carbonate and organic matter (13C during the latest Paleocene warming maximum (around 55.5 m.y. ago) has been recorded from sediments worldwide. The late Paleocene-early Eocene interval was a period of peak warming, and overall the warmest interval in the Cenozoic when latitudinal thermal gradients were also became greatly muted (Haq, 1984). At the close of the Paleocene Epoch ocean bottom waters warmed rapidly by as much as 4° C, with a concurrent rapid shift of -2 to -3‰ in $\delta^{13}C$ values of all carbon reservoirs in the global carbon cycle (Kennett and Stott, 1991, Koch et al., 1992, Bralower et al., 1995). High-resolution data from a sediment core (on Blake Nose in the western North Atlantic) that straddles the Paleocene-Eocene boundary by Norris and Röhl (1999) suggests that much of this isotopic shift occurred within no more than a few k.y. indicating a catastrophic infusion of $\delta^{12}C$-enriched carbon, probably from a methane source. These authors used physical (color reflectance and magnetic susceptibility) chemical (iron cycles) and biotic events to establish the chronology. Their study concluded that the isotopic excursion was synchronous in the oceans and on land.

The late Paleocene thermal maximum was also coincident with a major benthic foraminiferal mass extinction and widespread carbonate dissolution and low oxygen conditions on the seafloor (Thomas, 1998). Earlier Dickens et al. (1995) had suggested a rapid $\delta^{13}C$ excursion of this magnitude could not bc explained by mechanisms that are normally invoked for $\delta^{13}C$ changes (i.e., increased volcanic emissions of carbon dioxide, changes in oceanic circulation and/or terrestrial and marine productivity). A rapid warming of bottom waters in the late Paleocene by 4 °C (from 11 °C to 15 °C) could have abruptly altered the sediment thermal gradients, encouraging catastrophic release of methane from gas hydrates. Increased flux of methane into the ocean and atmosphere and its subsequent oxidation is modeled to be sufficient to explain the -2.5 ‰ excursion in $\delta^{13}C$ in the inorganic carbon reservoir. Dickens et al. hold explosive volcanism and rapid release of carbon dioxide and changes in the sources of bottom water during this time as a plausible triggering mechanism for the peak warming leading to hydrate dissociation. Modeling results of Huber and Sloan (1999) as well as extensive empirical data suggest warm and wet climatic conditions with less vigorous atmospheric circulation during the late Paleocene thermal maximum.

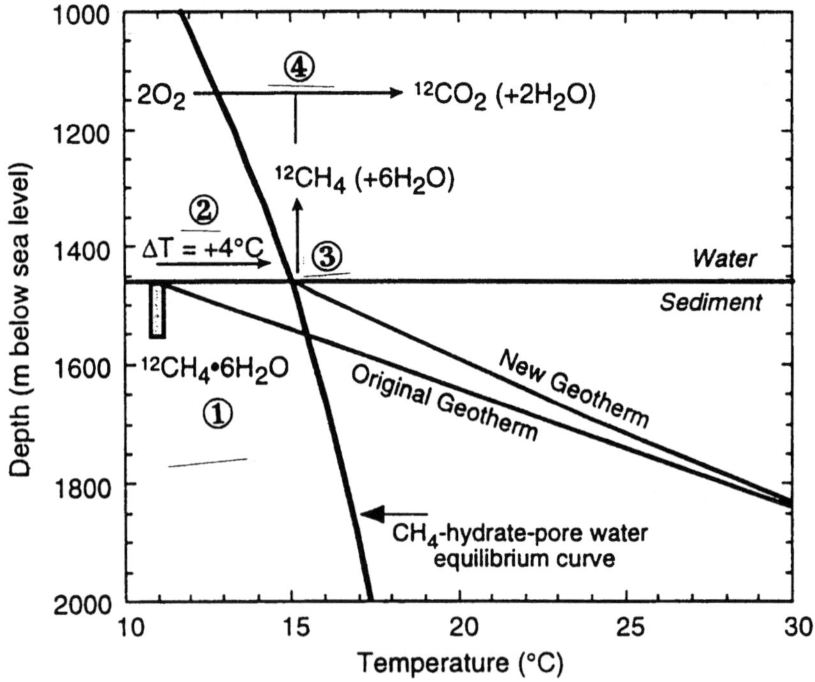

Figure 3. The Dickens et al. model of rapid warming of bottom waters and changes in methane flux during the late Paleocene thermal maximum (LPTM). 1: Before the LPTM solid hydrate ($^{12}CH_4.6H_20$) is restricted to the zone between the sediment-water interface and the intersection of the sediment geotherm (Original Geotherm) with CH_4-hydrate-water equilibrium curve; 2: Rapid warming of bottom water by 4° C during LPTM causes a shift of the geotherm (New Geotherm); 3: Once heat is transferred to the sediment it induces dissociation of hydrate, sediment failure and greatly increased flux of methane (enriched in to the ^{12}C) to the water above; 4: Methane oxidizes in the water column or the atmosphere, producing CO_2 enriched in ^{12}C, and water. (After Dickens, in press, reproduced by permission).

A recent discovery of sedimentary features indicating widespread slumping associated with the sediments of the late Paleocene thermal maximum drilled on the Blake Nose (Katz et al., 1999) seems consistent with the Dickens et al.'s hypothesis. In addition to the sedimentary and isotopic data, Katz et al's faunal data indicates warming of bottom waters, extinction of benthic foram assemblages and widespread carbonate dissolution.

A high-resolution stable-isotopic study of two widely separated Atlantic sites also seems to support the Dickens et al. proposal that biogenic degassing from gas hydrate may have played a key role in altering Earth's climate in the late Paleocene, but with a twist. In the low- and southern high-latitude Atlantic Ocean sediments, Bains et al. (1999) found that the carbon-isotopic shift occurred in a series of injections of methane, separated by periods of stasis, with global consequences during the relatively short interval at the end of Paleocene. Dickens et al. (1995) had proposed that during this time the latitudinal thermal gradients had flattened enough to allow sinking of warm waters to push the hydrates beyond a threshold and to dissociation. Bains et al's (1999) data indicates that there was virtually no change in surface-water temperatures of the low-latitude site for ca. 150 ky before the carbon-isotopic shift, while both the surface and deep waters at the high-latitude site show cooling. Thus, crossing a thermal threshold may not have been the initial cause of hydrate dissociation. Earthquakes, volcanism and slope failure may have been contributory agents, and data also suggests a cooling of higher latitudes (Bains et al. 1999). Could this cooling have been severe enough to cause significant ice accumulation and lowering of sea level?

The late Paleocene-early Eocene eustatic record could offer clues to whether reduced hydrostatic pressure during lowstand may have contributed to gas-hydrate dissociation. The overall trend for this interval shows a rising sea level through the latest Paleocene and early Eocene (Haq et al., 1987), punctuated by several shorter-term sea-level falls throughout and one prominent drop straddling the Paleocene-Eocene boundary. Could the latter have provided the forcing needed for hydrate dissociation at the close of the Paleocene? Early Eocene is also replete with high-frequency sea-level drops of several 10s of meters (Haq et al., 1987). Could these events have maintained the continued warm climates of this interval? These ideas seem testable if detailed chronological and faunal and isotopic data for the interval were available with a resolution that could resolve the leads and lags in the system.

7. First Expansion of Gas Hydrates

One question of historical interest is the timing of the first gas hydrate development in the geological past? The specific low temperature-high pressure constraint for the survival of gas hydrates suggests that they may have existed widely at least since the latest Eocene, the timing of the first development of the oceanic psychrosphere and cold bottom waters (Haq, 1984). Theoretical considerations indicate that hydrates can exist in the continental slope and rise sediments when bottom-water temperatures approach those estimated for late Cretaceous and Paleocene-Eocene (ca. 7-15 °C), though they would occur deeper within the sedimentary column and the stability zone would be relatively thinner. Dickens et al. (1995) estimated a depth of about 900 m below sea level for the top of the hydrate stability zone in the late Paleocene. Only when the bottom waters warmed up to 22 °C would most margins of the world be free of gas hydrate accumulation. The narrower hydrate-stability zones during warm bottom-water periods, however, do not necessarily imply a much reduced methane reservoir, since it also follows that the sub-hydrate free gas zone could be larger, partially making up for the hydrate deficiency.

The Cenozoic eustatic record indicates that prior to late Eocene there is little geological evidence of large polar ice caps, and the mechanism for shorter-term sea-level changes remains uncertain. Nevertheless, the Mesozoic-Paleogene eustatic history shows major sea-level fluctuations of 100 m or more that are comparable in magnitude, if not in frequency, to glacially-induced eustatic changes of the late Neogene (Haq et al., 1987). If gas hydrates existed in the pre-glacial world, major sea-level falls would imply that hydrate dissociation could have contributed significantly to continental margin tectonics and climate change. Massive methane emissions, however, should also be accompanied by prominent (13C excursions, as exemplified by the terminal Paleocene climatic maximum (Kennett and Stott, 1991, Dickens et al., 1995).

8. Concluding Remarks

A significant role for methane hydrates in the longer-term greenhouse forcing and abrupt climate change scenarios and as a major contributor of carbon in global carbon cycle remains contentious. These issues can only be resolved with high-resolution studies of the hydrated sediments, in conjunction with studies of the ice cores, preferably with decadal time resolution. Because of the relatively short residence time of methane in the atmosphere, only decadal resolution (< 50 years) will be capable of resolving the causal relationships and leads and lags in the system. Such insights may well lead to recognition of an important role for methane hydrates in global climate change, and through it, as agents of biotic evolution.

Acknowledgments The views expressed here are those of the author and do not necessarily reflect the views of the National Science Foundation or any other agency of the U.S. Government.

Chapter 12

Potential Role of Gas Hydrate Decomposition in Generating Submarine Slope Failures

Charles K. Paull and William Ussler III
Monterey Bay Aquarium Research Institute
Moss Landing, CA 95039-0628

William P. Dillon
U.S. Geological Survey
Woods Hole, MA 02543

1. INTRODUCTION

Gas hydrate decomposition is hypothesized to be a factor in generating weakness in continental margin sediments that may help explain some of the observed patterns of continental margin sediment instability. The processes associated with formation and decomposition of gas hydrate can cause the strengthening of sediments in which gas hydrate grow and the weakening of sediments in which gas hydrate decomposes. The weakened sediments may form horizons along which the potential for sediment failure is increased. While a causal relationship between slope failures and gas hydrate decomposition has not been proven, a number of empirical observations support their potential connection.

1.1. Causes of Slope Failure

Landslides, whether subaerial or submarine, occur because the stresses exerted on the sediment exceed the sediment's strength. The threshold, when the total stress exceeds the sediment's strength, for any individual slope failure is ultimately the result of a combination of factors.

Some of the more important processes that weaken the sediment result from changes that are external to the sediments that ultimately fail. These factors include over-steepening of slopes by continuing sediment deposition, tectonic changes, or erosive undercutting. Other external events, such as earthquakes and wave pumping, act as transient triggering mechanisms.

M.D. Max (ed.), Natural Gas Hydrate in Oceanic and Permafrost Environments, 149–156.

Internal alterations of the sediments caused by gas hydrate processes or other factors also affect sediment strength. While the majority of the processes associated with early diagenesis act to increase sediment strength, some processes work to decrease it. In particular, increasing the internal pore pressure, increasing the sediment porosity and the formation of gas bubbles will reduce slope stability (Prior and Coleman, 1984).

2. EFFECT OF GAS HYDRATE FORMATION AND DECOMPOSITION ON SEDIMENTS

2.1. Changes in Sediment Strength

The formation and decomposition of gas hydrate is believed to exert a significant influence on the mechanical properties of marine sediments. The formation of methane hydrate results in the extraction of water and methane from the pore space. Thus, liquid water and dissolved (and in some cases gaseous) methane are extracted from the pore space and converted into solid gas hydrate crystals. Replacement of the liquid water by solid gas hydrate will increase the shear strength of the sediments. Moreover, there will also be a decrease in the porosity and permeability of the sediments. The formation of gas hydrate within continental margin sediments may be similar to the effects of water ice formation in permafrost areas.

Conversely, when gas hydrates become unstable, they decompose into water plus gas. The effect of transforming solid materials into a liquid (and possibly a free gas phase) will be to decrease the shear strength of the sediments, making the sediments more prone to failure. If gas saturation is exceeded, gas bubbles will be produced as a consequence of gas hydrate decomposition and their presence is likely to further decrease the strength of the sediment.

The process of gas hydrate decomposition will also affect the pore pressures of the sediments (Kayen and Lee, 1991 and 1993). When methane hydrate decomposes in sediments in which the pore waters are already saturated with methane, a volume of both water and methane will be released into the pore spaces that exceed the volume that was previously occupied by the gas hydrate. The net effect is either a pressure increase (if the sediments are well sealed) or a volume increase (if pressure is allowed to dissipate by fluid flow). Thus the decomposition of gas hydrate can cause pressure greater than hydrostatic pressures (the pressure produced by the weight of a column of water above a location, Kayen and Lee, 1991) especially if low-permeability sedimentary layers are available to seal pressure in. The associated increases in pore pressure, dilation (expansion) of sediment volume, and the development of interstitial gas bubbles all have the potential to weaken the sediment (Prior and Coleman, 1984).

2.2. Sensitivity to Gas Hydrate Related Deformation

The potential for gas hydrate to alter the mechanical properties of sediment is not uniformly distributed with water depth. Obviously, the proper temperature and pressure conditions and the presence of gas are required. Gas hydrates are typically stable in continental margin sediments in a zone that can be viewed as a seaward-thickening prism (Figure 1). This prism has a distinct up-slope limit. The water column temperature structure associated with most mid- and low-latitude regions is such that the upper limit of methane hydrate stability occurs at 500-700 m water depths. Below this level methane hydrates are stable at the seafloor and remain stable throughout a zone that increases in thickness with increasing water depth. However, where there is the presence of even small amounts of other gas hydrate-forming gases that leak up from great depths (e.g. ethane, propane, hydrogen sulfide, carbon dioxide, etc.), significant shallowing of the minimum hydrate forming depth is possible. Such chemically driven shallowing of the gas hydrate occurrence zone is particularly relevant in known petroleum generating areas like Gulf of Mexico.

The sediments that are close to the up-dip limit of the gas hydrate occurrence zone have the greatest potential to be altered by the effects of gas hydrate decomposition. Gas hydrates in this area are sensitive to short term transient thermal perturbations in bottom water temperature, and thus more likely to have experienced repeated episodes of gas hydrate formation and decomposition. However, as the water depths increase, the thickness of the sediments that overly the gas hydrate phases boundary also increases (figure 1). Because temperature changes are primarily transferred through the sediments by thermal conduction, it takes thousands of years for seafloor-warming events to propagate through the sediments and alter the position of the phase boundary. Thus, the effects of transient thermal events become dampened and/or cancelled with increased water depth.

The sediments that are at the up-dip end of the gas hydrate stability field, near where they outcrop at the seafloor, are also subject to the largest potential pore pressure alterations when gas hydrate decompose. This is because the ambient pressure controls the relative volume of gas that will be generated when gas hydrate decomposes (Figure 1). For example, at a water depth of 650-m along the US Atlantic seaboard, which is close to the seafloor intercept of the gas hydrate phase boundary, there is a 2.84-fold net volume increase going from gas hydrate to gas bubbles plus water. In contrast, when gas hydrate decomposes at a total depth of 4-km below sea level, the net increase in volume is only 1.16 fold.

If gas hydrate decomposition is a significant factor in generating slope instabilities, we predict that the frequency of sediment failures should be focused at or just below the up-dip limit of gas stability within seafloor sediments for three reasons: (1). Neither gas hydrate nor gas hydrate-induced deformation will occur at shallower depths. (2) The sediments near and below the seafloor intercept of the gas hydrate phase boundary may have experienced the greatest number of gas hydrate growth and decomposition events because they are most susceptible to seafloor temperature changes. and (3) The potential volume increases associated with gas hydrate decomposition increases toward the shallow water limit of the gas hydrate occurrence zone.

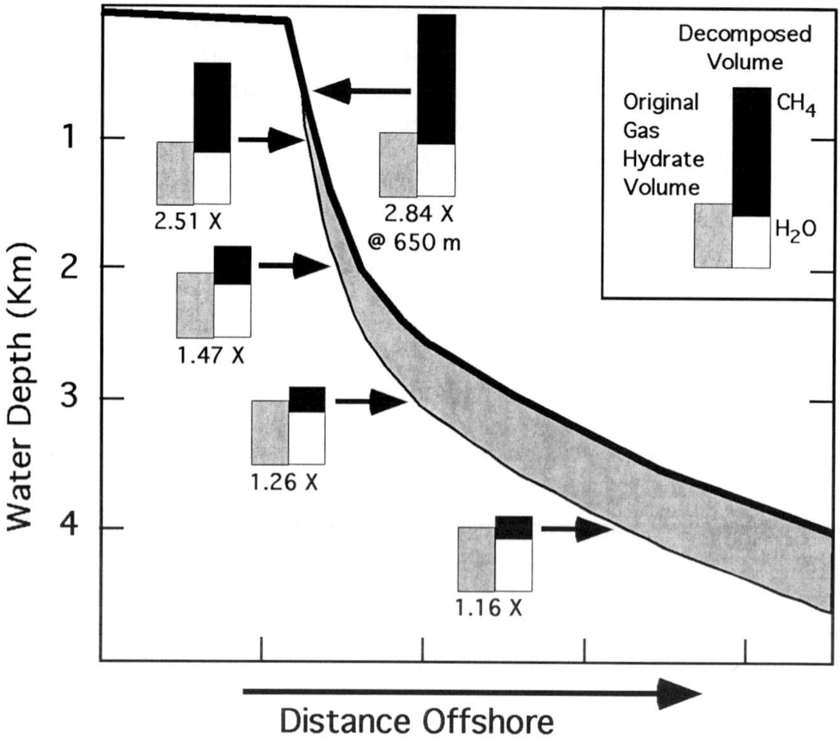

Figure 1. Diagram showing the distribution of gas hydrate stability beneath the ocean floor. The dark line indicates the seafloor. The lighter line is the position of the gas hydrate phase boundary, (which is presumably coincident with the BSR if one is present). The shaded area between the two lines indicates that region where gas hydrate is stable within the sediments. Boxes show the relative volume change associated with a fixed volume of gas hydrate decomposing into water and methane gas(also see Max and Dillon, 1998). Note that the relative volume change decreases with increasing water depth. Diagram is constructed assuming a 30° C/km geothermal gradient and hydrography similar to the U.S. Atlantic margin north of Cape Hatteras.

3. CONNECTION BETWEEN SUBMARINE SLOPE FAILURES AND GAS HYDRATE?

3.1. Submarine Slide Scars

The scars left by submarine slope failures are easily identified in seismic reflection profiles, long-ranged side scan sonar images, and/or multi-beam bathymetry data. Such scars are very common continental margin geomorphic features (Hampton et al., 1996). Many of the mapped submarine slides are significantly larger than the biggest terrestrial landslides (Schwab et al., 1993).

While the existing data document the spatial distribution and scale of the continental margin sediment failures, remote sensing data do not directly reveal the cause of the failure. Moreover, there are a number of challenges that make investigating the causes of submarine slides difficult. Probably most of the known slide scars are not very fresh and important environmental conditions may have changed since the slope failed. The specific conditions responsible for generating the local weaknesses are erased by the failure event. It is also difficult to know whether the surviving sediments on the flanks of the slide are comparable to the section that failed. Marine sediment scars are difficult to investigate in detail because access to the deep-sea is usually limited and always expensive. However, the general patterns of slide scar distribution do provide a geographic context from which we can begin to evaluate the potential role of gas hydrate on slide formation.

3.2. Coincidence of Slide Scars and Known Gas Hydrate Distribution

Many authors have tentatively related major slumps on continental margins to instability associated with the break down of gas hydrate (e.g., Summerhayes et al., 1979; Embley, 1980; Carpenter, 1981; Cashman and Popenoe, 1985; Paull et al., 1991; Rothwell et. al., 1998; Nisbet and Piper, 1998; Cherkis et al., 1999). This inference is usually based on the observation that bottom simulating reflectors (BSR), the only commonly available remote detection indicator for the presence of gas hydrate, occur in the sediments around the slide scar.

Perhaps the best-documented coincidence between slide scars and the inferred distribution of gas hydrate is along the US Atlantic margin. Here, over 200 slump scars have been mapped (Booth et al., 1994). This is the only region where there is a large enough number of mapped slide scars to merit analysis of their spatial distribution. Intensive mapping efforts also show that intermittent BSR's occur along the entire margin, implying that gas hydrates are common within these sediments. The US Atlantic margin data clearly show that the slides are neither randomly distributed nor are they strongly associated with steep slopes (See chapter six, figure 11). Instead, the majority of the slide scars occur at or within the current up-dip limit of gas hydrate stability. In fact the greatest number of scar headwalls occur in the 500-700 meter depth range. Thus, the observed distribution of slide scars is consistent with the distribution that is predicted to occur if gas hydrate decomposition has played a significant role in causing these sediment failures.

3.3. Huge Submarine Slope Failures

The largest of the US Atlantic margin slides is the Cape Fear Slide (Embley, 1980; Carpenter, 1981; Cashman and Popenoe, 1985). Seismic reflection profiler data show that there are numerous normal faults that sole out at or near the level of the BSR near the headwall of the slide and under the sole of the slide (Schmuck and Paull et al., 1989) and that the BSR rises at the edges of the slide scar, but disappears near its center, where gas apparently escaped (Dillon et al, 1993). This suggests that the base of gas hydrate stability (BGHS) is associated with a zone of weakness and failure within the sediments. However, in this case the BGHS does not correspond with the main slide surface.

One of the largest known submarine slides is the Storrega Slide on the Norwegian Continental Margin. The Storrega Slide has a 290-km long headwall scar, extends from over 800 km down slope, and transported over 5,500 km3 of material. The current slide scar is believed to represent at least 3 major sediment failure events (Bugge, et al., 1987 & 1988). The two largest events occurred in the Late Pleistocene. The smallest, but most recent event is known to have been associated with an 11-m high tsunami that washed over the Norwegian coast 7,200 years before present (Bondevik et al., 1997). Like many of the major slump scars on sediment covered continental margins, the Storegga Slide, is flanked by sediments that contain a BSR. In this case the sole of the slide is coincident with the level of the BSR on its sides, suggesting the failure surface was coincident with the base of gas hydrate stability before the slide occurred.

3.4. Connection of Sea Level to Slumping

Hydrostatic pressure changes associated with sea level fluctuations should affect the stability of continental margin gas hydrate. A consequence of lowering sea level is that the decreased seafloor pressure will cause the position of the gas-hydrate phase boundary to move upward in the sediments. This upward shift in the position of the gas hydrate phase boundary within the sediment column will cause gas hydrate decomposition, and should also increase the frequency of continental margin slumping events (Paull et al., 1991).

The general patterns of Pleistocene climate changes are well documented. During the last major ice age, sea level dropped to levels that were about 120 m lower than current sea level at about 18,000 years before present. Thus, if gas hydrate decomposition plays a significant role in generating sediment instabilities, one would predict that there should be an increased incidence of continental margin sediment failures during these sea level low stands.

In principle one could test this prediction by establishing the ages of a number of continental margin sediment failure events to determine whether slumps actually occur more frequently during sea level low stands. Unfortunately, we only know the timing of a few big slides (Embley, 1980; Bugge et al., 1987; Rothwell et al., 1998; Rodriguez and Paull, 2000). While the existing data are consistent with increased failures during sea level low-stands, several of these areas are also known to have higher sedimentation rates during sea level low-stands, which could also cause an increased sediment failure frequency.

Another approach to evaluating this prediction was conducted by examining the frequency of hiatuses in piston cores collected from over the top of the Carolina Rise and Blake Ridge gas hydrate region (Paull et al., 1996). They measured the ^{14}C chronology on 123 samples from within the upper 7 m of 45 piston cores collected throughout the region which were not initially known to be associated with hiatuses. The ^{14}C age distributions from these cores showed that the sedimentation rate over the last 35 thousand years had not changed in this region. However, numerous hiatuses were identified within the period of low sea level between 14 and 25 thousand years ago. This observation is consistent with the prediction that a causative relationship exists between gas hydrate decomposition and slumping frequency.

3.5. Soft-Sediment Deformation in the Geologic Records

We hypothesize that the processes of formation and dissociation of gas hydrate can mobilize sediments by causing radical changes in sediment strength, pore pressure, and volume. So at places where sediments are deposited on slopes and where there is active formation and decomposition of gas hydrate (potentially multiple times), we would expect sediment deformation.

Kennett and Frackler-Adams (2000) and Krause (2000) have recently argued that the widespread soft-sediment deformation effects that are seen in the Neogene Sisquoc Formation in Santa Barbara County, California and at Meiklejohn Peak, Nevada were in part produced by instability produced by gas hydrate decomposition. In California, the inferred bathyal paleoenvironment, the isotopic compositions of the benthic foraminifers, and the associated concretions suggest that gas hydrate once existed in the shallow subsurface of the strata associated with the Sisquoc Formation. Kennett and Frackler-Adams coined the term "clathrites" to describe the type of soft sediment deformation that is seen in these sediments, presumably as a consequence of sediment instabilities associated with gas hydrate decomposition. These clathrates contain intrastratal brecciation, homogenization, diapirism, fluidization pipes, and fluid-escape structures. They also indicate that these lithologies are similar to those that have been found associated with modern gas-hydrate-bearing sediments. In Nevada, classic "frost heave" structures, sheet cracks, and small folds and thrusts are seen in strata from deepwater environments. Because gas hydrate, rather than water ice, is more likely to have formed in this environment gas hydrate formation and decomposition is suggested as the potential mechanism to produce these structures.

4. Conclusions

The mechanical effects of gas hydrate decomposition clearly will reduce the sediment strength and increase the chances of continental slope failure. Although gas hydrate processes within the sediments and associated seafloor slumps have been impossible to directly observe, a great deal of circumstantial evidence strongly supports the concept that gas hydrate breakdown is often instrumental in triggering sediment mass movements on the sea floor.

Editor's note

Following the submission of this contribution, detailed acoustic imagery and bathymetry of the outer continental shelf and slope between 36° N and 38° N (North Carolina-Virginia-Maryland) identified incipient slope failure features (Driscoll et al., 2000). Near the edge of the shelf, these morphological features have the geometry of fault headwalls backing down-to-basin rotational subsidnce. There is little doubt that features such as these are likely related to the initial stages of large-scale slope failure. Marginal shelf collapse has led to the deposition of a number of major sediment slides along the U.S. S.E. continental slope, and these newly recognized incipient collapse scarps may either be failed collapses of a previous episode of shelf instability or more modern.

Although the continental shelf margin, which includes the incipient slope failure features and the uppermost slope, are well above the depth at which methane hydrate is stable (~ 500 m), dissociation of hydrate is regarded as a likely mechanism for slope failure along the continental slope here (Drisoll et al., 2000). It is possible that gas produced by dissociation of hydrate further down the continental slope migrated within the sediments along the slope and was contributary to the mapped incipient slope failure features.

Shelf edge collapse, whether related to hydrate or not, can lead to formation of tsunami that can wash nearby coastlines with little warning. As pointed out by Paull (this Chapter), evidence is preserved in the geological record that hydrate dissociation-related slope failure and large-scale sliding along the Norwegian margin was associated with tsunami on the facing coast. Thus there is also potential for tsunami striking the U.S. SE coast following a similar slope failure (Driscoll et al., 2000). Although slope failure and its relationship to hydrate formation and dissociation is a fundamental geological process that begs study in its own right, there is also an important societal impact that would accompany tsunami striking parts of the U.S. SE coast, which makes research into the gas hydrate system a potentially vital concern to the eastern seaborad of the U.S.

Chapter 13

The U.S. Atlantic Continental Margin; the Best-Known Gas Hydrate Locality

William P. Dillon
U.S. Geological Survey
Woods Hole, MA 02543

Michael D. Max
MDS Research
Suite 302, 1211 Connecticut Ave. NW
Washington DC 20036

1. INTRODUCTION

One of the few attempts to date to map gas hydrate over a large area has been made on the Atlantic continental margin of the United States (Dillon et al., 1993, 1994, 1995). This work has resulted in the production of an extensive data base of seismic reflection lines including both single and multichannel lines, and complete GLORIA sidescan sonar coverage. This work was part of the assessment of the U.S. EEZ and was carried out by the U.S. Geological Survey. Earlier efforts were made by Tucholke et al. (1977) and Shipley, et al. (1979). Research along the U.S. SE continental margin of the U.S. is continuing.

2. DETECTING HYDRATES

How can we locate and map hydrate-bearing sediment in the deep sea? Although hydrates have been recognized in drilled cores, their presence over large areas can be detected much more efficiently by acoustical methods, using seismic reflection profiles. Gas hydrate has a very strong effect on sediment acoustics because it has a very high acoustic velocity (~3.3 km/s — about twice that of seafloor sediments), and thus the presence of gas hydrate in sediments increases their average velocity. Sediments below the gas hydrate stability zone (GHSZ), if water saturated, have lower velocities (water velocity is ~1.5 km/s), and if gas is trapped in the sediments below the hydrate, the velocity is much lower, even with just a few percent of gas. For mapping the hydrate, we used two phenomena observable in seismic profiles that are caused by the velocity effects. The first is the Bottom Simulating Reflection, or BSR, (Chapters 6, 21) and the second is the reduction in reflection amplitudes observed in many places where gas hydrates are known to exist, known as "blanking".

M.D. Max (ed.), Natural Gas Hydrate in Oceanic and Permafrost Environments, 157–170.

2.1. BSR

Reflection strength is proportional to the change of acoustic impedance, which is the product of velocity times density. The base of the hydrate-cemented zone, where it traps free gas bubbles, represents a very large change in velocity, and therefore produces a very strong reflection, the BSR (Fig. 1). This may or may not be a continuous reflection event in seismic profiles because the presence of gas will vary laterally due to variations in trapping configuration, porosity, gas supply, etc. Where it exists, however, the reflection is commonly very sharply defined, because the phase boundary is a distinct, not diffuse, limit to hydrate occurrence. In contrast, the top of gas hydrate within the sediments has no such precisely defined limit created by a phase boundary; actually the top of the gas hydrate stability zone exists above the sea floor in most locations. As a result, even where there is a discrete hydrate accumulation within the sediments, no sharp reflection boundary is observed at its top in seismic profiles because there is no large velocity contrast there. Free gas cannot exist within the GHSZ as long as water exists there because it would spontaneously form gas hydrate.

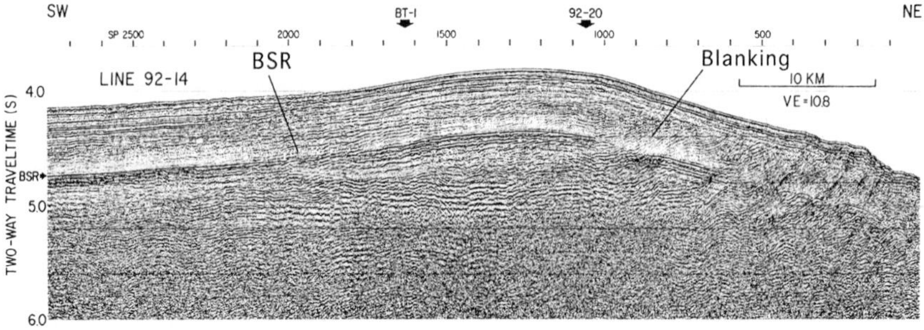

Figure 1. U.S. Geological Survey seismic reflection profile across the Blake Ridge on the continental rise off South Carolina (Fig. 3.), showing clear examples of the type of gas hydrate effects observed on seismic profiles.The base of gas bearing sediments in contact with water- or gas plus water-filled sediments forms a very strong reflecting horizon that parallels the sea floor, known as the "bottom simulating reflection", or BSR. Above the BSR, in the region where gas hydrate is stable, the amplitude of reflections is reduced, a phenomenon known as "blanking" at locations where gas hydrate is concentrated.

The base of hydrate stability, as disclosed by the BSR, occurs at an approximately uniform sub-bottom depth throughout a small area because it is controlled by the temperature; thermal gradients across an area tend to be consistent. In Figure 1 a very strong BSR approximately parallels the sea floor and cuts through reflections from sedimentary strata. The coincidence in depth of the BSR to the theoretical, extrapolated pressure/temperature conditions that define the hydrate phase boundary and the sampling of hydrate above BSRs and gas below give confidence that this seismic indication of the base of hydrates is dependable.

2.2. Blanking

A second significant seismic characteristic of hydrate-cementation, called "blanking" is also displayed in Figure 1; blanking is the reduction of the amplitude (weakening) of seismic reflections probably caused by the presence of hydrate. Many observations of blanking in nature have been associated with gas hydrate accumulations (Lee et al. 1993, 1994, 1996; Lee and Dillon, in press; but the reality of blanking was questioned by Holbrook et al., 1996). Blanking is thought to result from preferential accumulation of gas hydrate in the more porous sedimentary strata, where it would increase the velocity of the more porous, initially lower velocity layers by the introduction of high-velocity, gas-hydrate cement. This would create a more uniform, higher velocity, resulting in reduced acoustic impedance contrasts and thus reduced reflection strengths.

We calculated the amount of hydrate in the most intensively blanked sediments by using known hydrate velocity from published laboratory studies, velocities that we determined for sea floor sediments from multichannel seismic profiles, and porosity in our study area determined in scientific drilling. We assume that such a material is an end member, and, in our computed seismic model, we mathematically "mix" this maximum-concentration, hydrate-bearing material with a similar deposit having no hydrate, in order to model the range of possible blanking effects. The modeled affect on reflection amplitudes of various mixtures of end-member, hydrate-bearing sediment with hydrate-free sediment is shown by computer-generated wiggle traces(Fig. 2).

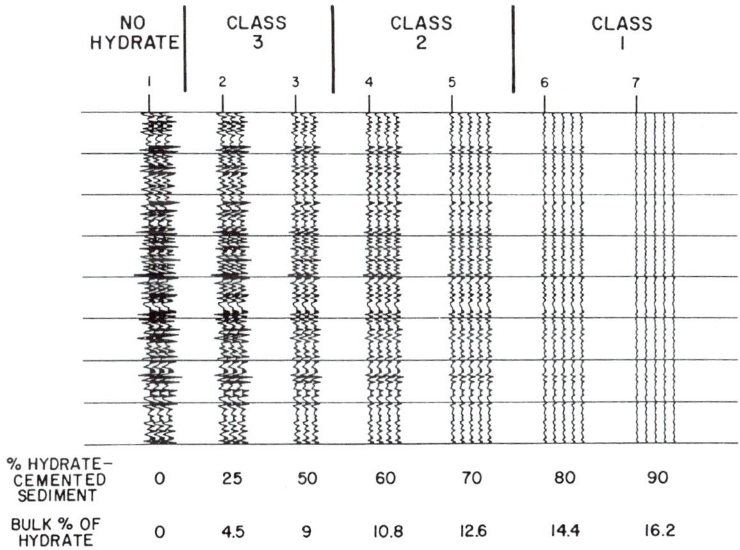

Figure 2. Synthetic seismograms showing the modeled effect of various amounts of gas hydrate in sedimentary strata on the amplitude of reflections from acoustic impedance changes comparable to those of sedimentary strata.

The reduction in reflection amplitude is the parameter of blanking. Computer-modeled increase of blanking (decrease of reflection strength) with increase of hydrate is apparent. Three classes of blanking have been assigned; the boundaries between classes represent a change in the power of reflections by a factor of two. The class boundaries are indicated on Figure 2, and these classes can be related to overall average amounts of hydrate in bulk sediment of approximately 7 percent (class 3), 12 percent (class 2), and 15 percent (class 1) in this example. To recognize classes in a profile, we measure and plot the amount of blanking (the reduction in reflection amplitude) along the profile in a subbottom window within the hydrate zone. These measured values are used as a calibration, and the extent of the three classes is interpreted along the profile. The total thickness for each class is plotted along each profile and the total volume of each class is estimated by interpolating between adjacent profiles in the area being mapped. Once the volumes for each class are estimated, they are multiplied by the appropriate percentages and a volume estimate of hydrate is made.

It must be emphasized that the approach is constructed from a mathematical model that is based on a logical and carefully thought-out concept of how the presence of gas hydrate in a sedimentary structure affects the acoustics of the sedimentary mass. However, at present we have an essentially inadequate understanding of this relationship. Much more work is required in the laboratory to better define this issue, work that is proceeding at present (see Chapters 24 by Winters et al, and 25 by Stern, et al.). Maps that will be presented give an excellent view of the structure of the gas hydrate stability zone and probably a good indication of the relative distribution of gas hydrate in the sediments, but the absolute amounts of gas hydrate indicated in the map probably will need to be adjusted as new modeling capability is developed.

3. DISTRIBUTION OF GAS HYDRATE ON THE U.S. ATLANTIC CONTINENTAL MARGIN

3.1. Volume of Gas Hydrate

Keeping in mind the qualification at the end of the previous section, we will consider a map of the distribution of the relative amount of gas hydrate within the sediments of the U.S. Atlantic continental slope and rise south of Hudson Canyon (Fig. 3). The map indicates the volume of hydrate by isopach contours that show the thickness of hydrate that would appear if all hydrate were extracted from the sediment pores and piled on the sea floor.

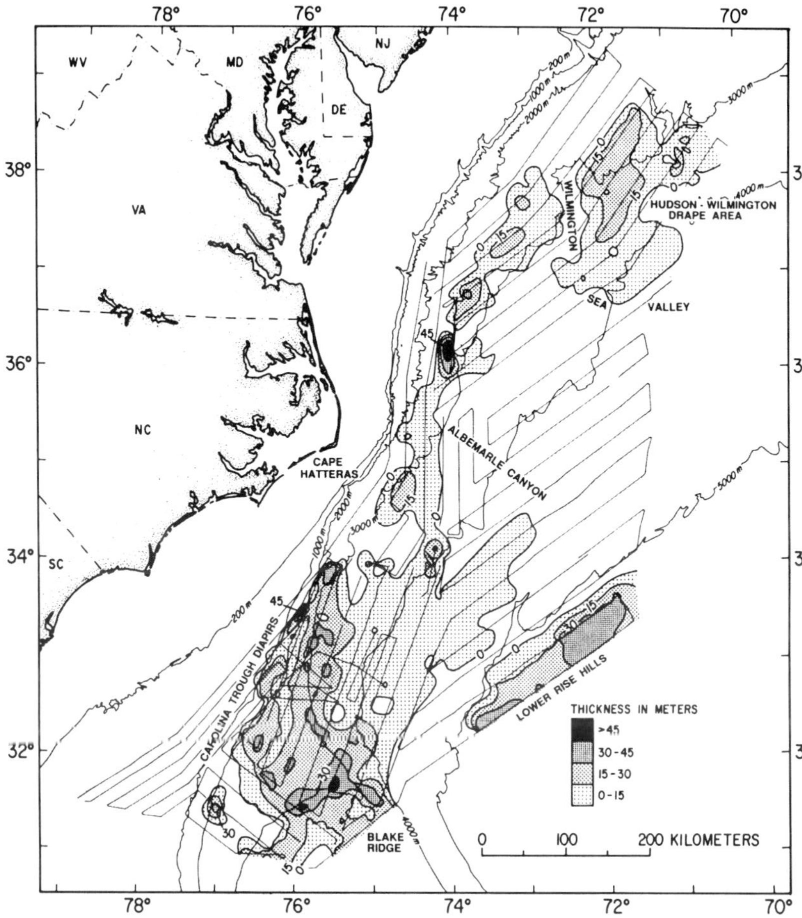

Figure 3. Contours of the volume of gas hydrate existing within the sediments of the U.S. Atlantic continental margin. Volume is indicated as a thickness in meters of the total amount of pure hydrate existing in the pore space. Volumes are estimated by mapping the extent of blanking in seismic profiles in the depth region where hydrate is stable, and using a modeling approach to relate the blanking to proportion of gas hydrate in the sediment.

The map shows that the greatest concentrations of gas hydrate are present south of 34°N, where we identify three major concentrations; a fourth concentration occurs at the northern end of our survey area. One occurs in very deep water, greater than 5000 m, between 71.5° and 74°W on the Lower Rise Hills, which are sediment waves built on the Hatteras Outer Ridge (Mountain and Tucholke, 1985; EEZ Scan 87, 1991). A second concentration occurs on the Blake Ridge, mainly from 75° to 76°W and 31° to 32.5°N (Kvenvolden and Barnard, 1983; Dillon and Popenoe, 1988, Paull et al., 1996). The third southern concentration is contiguous with the Blake Ridge hydrates and extends north-

northeastward from 31.5°N, 77°W to 34° N, 75.5°W; it is labeled 'Carolina Trough Diapirs' (Dillon et al. 1983).

North of 34°N, a series of relatively small, weak concentrations extend along the upper rise to 38°N. At the northeast corner of the map, an extensive, but relatively weak concentration of hydrate, labeled Hudson-Wilmington drape area, covers the region from approximately 37° to 39°N and 71° to 72°W.

The modeling requires considerable development, but in order to provide an example of the possible amount of hydrate gas implied by Figure 3, we calculated the amount of gas hydrate in the Blake Ridge concentration within the small area inside the main 30 m contour on the Blake Ridge (~31.5°-32°N, 75°-76°W) (Dillon et al. 1993). The area within this 30 m contour is approximately 3000 km^2. The calculated volume of hydrate within this area, 1.13×10^{11} m^3, would contain a volume of methane at STP approximately equal to 1.8×10^{13} m^3, or 645 TCF (trillion cubic feet) of methane. For comparison, natural gas consumption of the United States is recently (1997-1999) at a level of about 21-22 TCF/year.

Other estimates of gas volume in the Blake ridge area have been made as indicated in Table 1 (discussed in Collett and Ladd 2000). The methods have varied widely from seismics to sampling to well logging approaches. The assumptions and areas considered vary; Dillon, for example, took a small area where he considered the concentration to be highest. The result is a set of numbers that vary a great deal, but consider the column that indicates volume of gas hydrate per unit area. With our present state of knowledge, the fact that these numbers are all within a factor of 10 is actually rather encouraging.

Total est. gas (Trillion m^3)	Area, (km^2)	Vol./unit area (Billion m^3/km^2)	Reference
18	3,000	6	Dillon et al. 1993
80*	100,000	0.8	Holbrook et al. 1996
70*	26,000	2.7	Dickens et al. 1997
37.7	26,000	1.5	Collett and Ladd 2000
57*	26,000	2.2	Collett and Ladd 2000

* includes gas beneath gas hydrate seal

Table 1. Estimates of gas volume that have been made in the Blake Ridge area.

3.2. Thickness of the Zone of Hydrate Stability A map of the depth of the BSR below the sea floor is shown in Figure 4; this interval - from sea floor to BSR - represents the zone in which gas hydrate is stable. Because the thermal gradient is fairly constant over the continental rise region, and because gas hydrate becomes stable to higher temperatures as pressure increases, one would expect a simple pattern in which the BSR would become uniformly deeper as the water becomes deeper (see Chapter 6, Fig. 3). Clearly the pattern is far more complicated. A general trend of greater BSR depth with deeper water is shown in a plot of data (Fig. 5) from several profiles in the Blake Ridge area (Dillon and Paull, 1983), but the scatter is great. In some places, like the Hudson-Wilmington drape area (Fig. 4), the tendency is reasonably pronounced to proceed from shallow sub-bottom BSR to deep as water depth increases. In mass movement areas, such as the Cape Lookout and Cape Fear slides, a distinct thinning of the hydrate zone is associated with the slide scars. Extreme complexity of the thickness of the hydrate-stable zone occurs in the Carolina Trough diapir area.

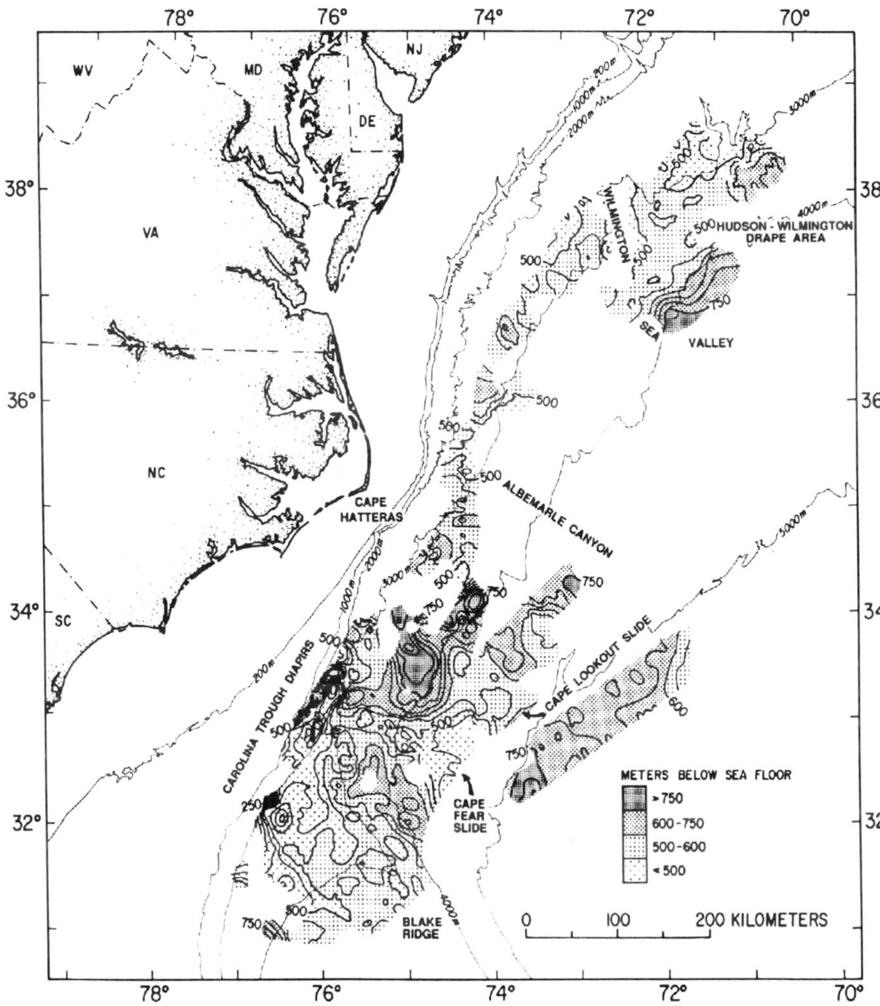

Figure 4. Contours in meters of the thickness of the layer of hydrate stability within the sediments of the U.S. Atlantic continental margin. This is the distance from the sea floor to the BSR. The BSR represents the bottom of the zone of phase stability for gas hydrates.

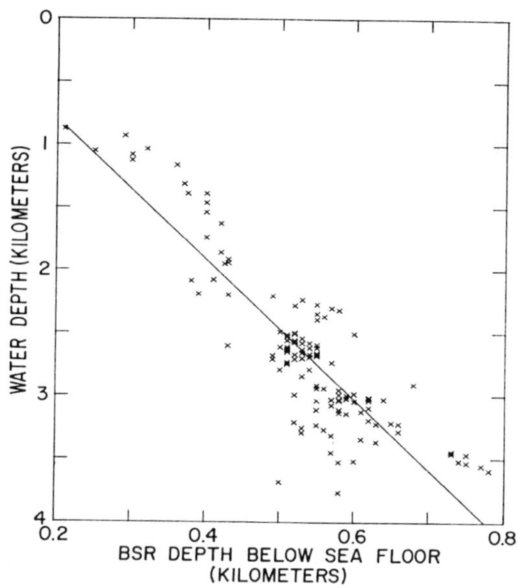

Figure 5. Plot of the depth of the BSR below the sea floor versus water depth for several profiles in the Blake Ridge area (from Dillon and Paull, 1983).

4. GEOLOGICAL CONTROLS ON GAS HYDRATE DISTRIBUTION

4.1. Sediment Deposition

The most significant control on hydrate accumulation appears to be sediment deposition rate. Of the four concentrations of hydrate noted above, three are associated with areas of high deposition. The northernmost (Fig. 3) occurs in a region of undisturbed recent sedimentation (Schlee et al., 1992) between the Hudson and Wilmington sea valleys. The valleys form two major channelways for transportation of sediments across the continental rise. The area between the valleys is a depocenter where more than 300 m of late Pliocene to present sediment accumulated (Mountain and Tucholke, 1985).

The southeastern, deep water hydrate concentration occurs on the Lower Rise Hills, a region of sediment wave deposition that also was active in Late Pliocene to present. Hydrates have not been sampled in either this or the drape area, but their abundance in these areas, as indicated by seismic interpretation, is not surprising. Most marine gas hydrates have been formed from biogenic methane and areas of rapid hemipelagic deposition tend to accumulate considerable amounts of organic detritus and, by rapid burial, preserve it from oxidation at the sea floor, so that it is converted to methane by bacteria within the sediments. Therefore abundance of gas hydrate in areas of rapid deposition would be expected.

The third concentration of gas hydrate in the mapped area is at the Blake

Ridge (Fig. 3), a location that is well-known for high concentrations of gas hydrate, identified both in seismic profiles and scientific drilling (Paull et al. 2000). The source of gas in the Blake Ridge has been bacterial decay of organic material that accumulated in the sediment. A profile across the Blake Ridge (Fig. 1) shows marked blanking of reflections. Blanking is strongest in the lower part of the hydrate zone just above the BSR. The blanking generally seems to be concentrated at the crest of the ridge. The great volume of hydrate within the Blake Ridge sediments, which is the highest concentration mapped, probably results from the shape of the ridge and the capacity of hydrate-cemented sediments to trap gas. The Blake Ridge is a thick (1200 m) sediment drift deposit that accumulated relatively rapidly due to current patterns on the ocean floor; accumulation rates reached 160-190 m/million years (Mountain and Tucholke, 1985). As the sediment accumulated, the ridge built upward and the deeper sediments became warmer; in essence, the surfaces of constant temperature — the isotherms — migrated upward as the sediment surface rose. This warming caused the hydrate in the deeper, older sediments to break down, releasing gas that moved upward through the sediments. The gas accumulated at the relatively impermeable base of hydrate-cemented sediments, which, by its configuration, acted as a trap. The trapped free gas is inferred to have penetrated the GHSZ by diffusion or along pathways provided by small compaction faults and it was immediately converted to hydrate, thus causing a concentration of hydrate at the base of the hydrate-stable zone.

4.2. Diapirism

Off North and South Carolina a linear group of salt diapirs extends along the seaward side of the deep Carolina Trough, one of the four major continental margin basins of the eastern United States (Dillon et al., 1983). Gas hydrates are concentrated around the diapers (Fig. 6). The BSR is observed to rise over each diapir, indicating that the hydrate-stable layer becomes thinner (Paull and Dillon, 1981). The structure of the hydrate-stable layer is very complex in the vicinity of the diapirs (Fig. 7). Actually the correspondence of hydrate-layer structure to diapir location probably is much closer than indicated in Figure 7, as the mapping of diapirs and mapping of hydrate were, of necessity, done on different seismic profiling data sets. The effects on hydrates at salt diapirs is reviewed in detail in Chapter 6 (Dillon and Max). The effects result from the greater thermal conductivity of salt compared to sediment, which produces a warm spot at a diapir, and the presence of salt ions in interstitial fluids, which would act as inhibitors (antifreeze; Taylor et al. 2000). Both effects cause the base of gas hydrate stability to be warped up over the diapir, forming a trap for gas, and the trapped gas tends to nourish the gas hydrate in the vicinity of the diapirs, causing gas hydrate concentrations.

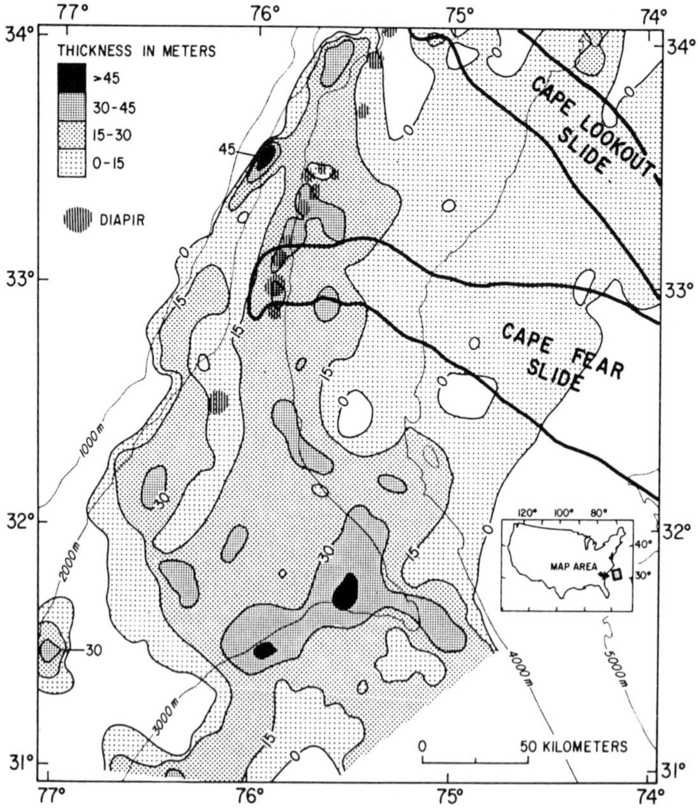

Figure 6. Locations of landslides and diapirs plotted on the map of gas hydrate volume (Fig. 3) for the diapir and Blake Ridge regions. Volume of hydrate is indicated as isopach contours of the amount of hydrate existing in the sedimentary pores.

4.3. Mass Movement

Sea floor mass movements seem to have significant effects on structure of the hydrate layer and on the abundance of hydrate. The broad area where there is a dearth of gas hydrate in the central region of The U.S. Atlantic margin from Wilmington sea valley southward to off Cape Hatters (Fig. 3) is a region marked by overlapping sea floor slide and slump scars. Note the raggedness of the 2000 m contour in this region, which displays the irregularity of the sea floor produced by mass movements of sediment. The continental rise in this region, especially in the depth range of about 3,000 to 4,000+ m, where slides are concentrated, shows almost no indication of gas hydrate in the seismic profiles.

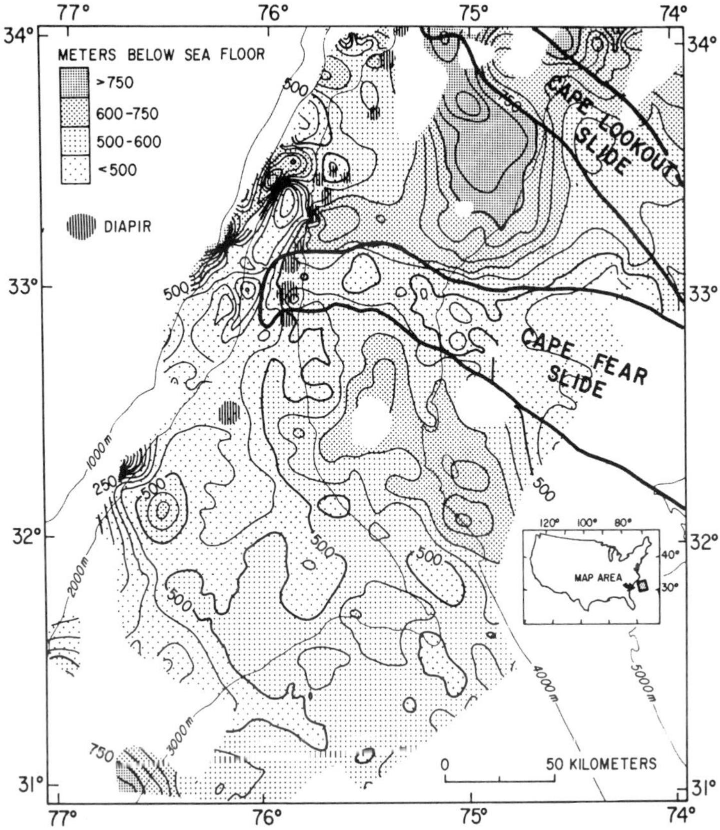

Figure 7. Locations of landslides and diapirs plotted on the map of thickness of the gas hydrate layer (Fig. 4) for the diapir and Blake Ridge regions. Thickness of the hydrate layer is equivalent to the depth of the BSR below the seafloor.

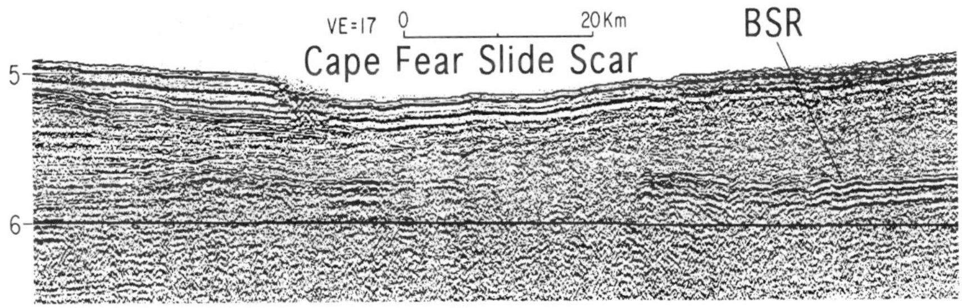

Figure 8. Profile across the Cape Fear slide scar. The BSR becomes shallower and weaker beneath the slide scar.

Fewer slides scars exist south of Cape Hatteras, but those that are present are unusually large (Popenoe and Dillon, 1996). The two major slides are the Cape Fear slide (indicated on Fig. 4) and the Cape Lookout slide. Both slides are mapped in Figure 7 which also shows the details of the structure of the gas hydrate stability zone. Note the marked thinning of the hydrate zone under the slides. As indicated in a seismic profile across the Cape Fear slide (Fig. 8), the BSR becomes indistinct under the center of the slide and the BSR also rises slightly at the sides of the slide scar.

Mapping shows that the BSR-to-seafloor distance commonly decreases by 200 to 300 m from the center of the slide to surrounding, undisturbed areas (Fig. 7), an amount greater than the thickness of slope strata removed by the mass movement. This shallowing of the BSR and the clear association of slide scars to thinning of the gas hydrate-stable zone suggest that displacement of strata by mass movement causes breakdown of hydrate, probably due to pressure reduction resulting from removal of part of the sediment load. The disappearance of the BSR, probably means that both the gas that previously existed there, plus that released when the gas hydrate dissociated, has escaped.

Although mass movement may reduce pressure and cause breakdown of hydrate, the converse is also probably true. That is, if hydrate breaks down for some reason other than mass movement, this can weaken the sediments and foster landslides (Chapter 12, Paull and Dillon). Gas hydrate breakdown (dissociation) will not only remove a cementing medium (that probably is not very important), but also will release both water and gas into sediment pores in volume greater than the volume of gas hydrate that previously existed. If the gas is sealed in, as we would expect, this causes an increase in pressure, known as "overpressure", that reduces the shear strength and essentially buoys up the near-bottom sediments; this can trigger collapses and slides on slopes.

The evidence for such mass movements of sediment tends to be lost when a slide moves off downslope. However, in at least one case on the Blake Ridge (Fig. 4), such a process apparently resulted in a blowout of mobilized sediment and subsequent collapse of the crest of the ridge without causing a slide. The thinning of the GHSZ mapped on the Blake Ridge (Fig. 7) centered at about 31° 55'N, 75° 40'W results from that blowout. A profile across the structure (Fig. 9), a perspective image (Fig. 10), and a series of conceptual diagrams (Fig. 11) indicates how this structure probably formed (Dillon et al., 1998; 2000).

We start with an undisturbed part of the ridge (Fig. 11A), which is comparable to Figure 1. The process passed through episodes of breakdown of gas hydrate, which was perhaps due to the pressure reduction caused by sea level lowering during glacial stages. That dissociation of gas hydrate generated overpressures and caused flow and inflation of the ridge crest (Fig 11B). Eventually rupture occurred and a significant volume (>13 cubic km) of sediment, water and gas escaped (Fig 11C). As this volume escaped, the strata collapsed into the evacuated volume to leave the structure that we see today (Fig. 11D; Fig. 10, profile).

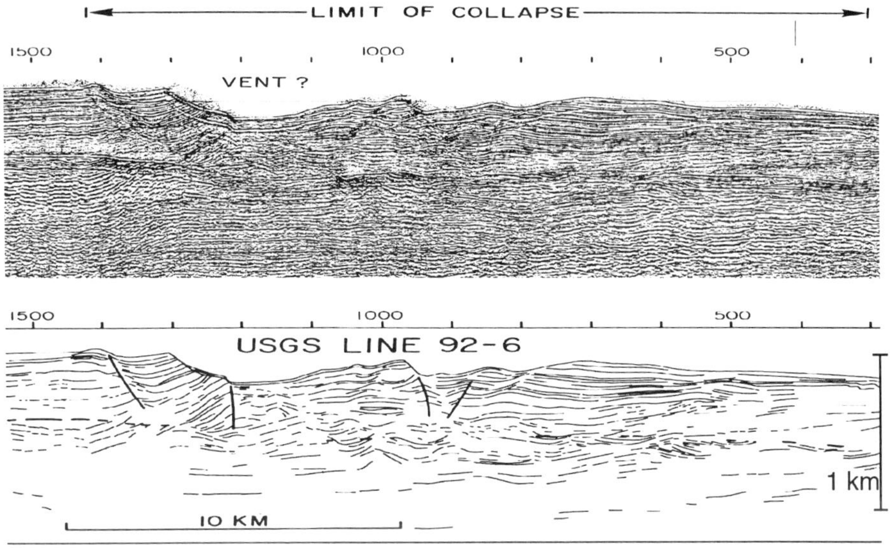

Figure 9. Seismic reflection profiles oriented SW-NE (NE to right) across the collapse structure on the crest of the Blake Ridge.

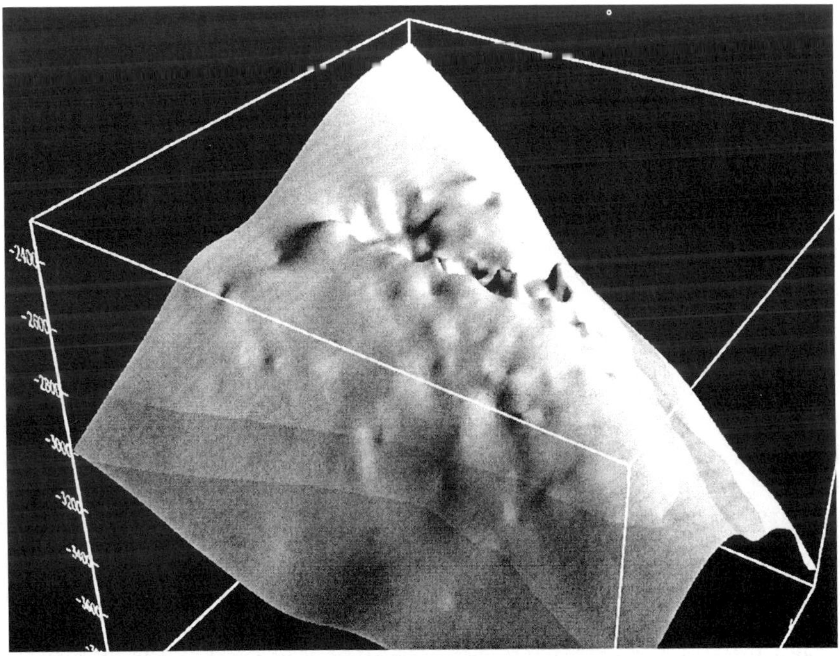

Figure 10. Perspective image of the Blake Ridge, showing the collapse structure at the crest. The area shown is approximately a square, 40 miles on a side and the vertical side of the box extends from 2350 to 3900 m. The profile of Figure 9 crosses the ridge across the center of the collapse.

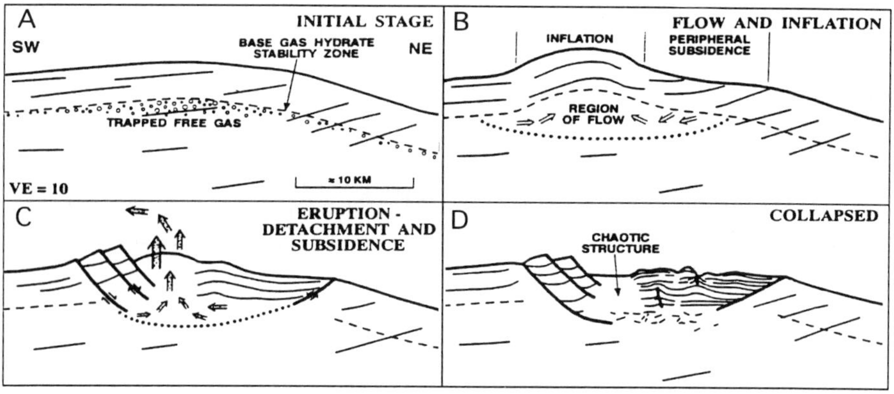

Fig. 11 Conceptual model of the collapse process at the crestal region of the Blake Ridge. The region of flow in the second panel is inferred to be a region in which fluid-rich sediments were mobilized. The zone of trapped free gas indicated in the first panel may have included this entire region of flow.

5. SUMMARY AND CONCLUSIONS

Gas hydrates on the Atlantic continental slope and rise have been sampled in drilling, and they also can be detected in seismic reflection profiles. Characteristics that allow hydrate detection in seismic profiles and mapping of the thickness of the GHSZ and possible volume of gas hydrate are:

1. A bottom simulating reflection, known as a "BSR", which is a reflection that parallels the sea floor and cuts through reflections from strata; the BSR is a reflection from the base of the hydrate-cemented layer where free gas is present beneath it;

2. Reduction in amplitude of reflections within the hydrate-cemented zone, known as "blanking", appears to occur as a function of hydrate development where hydrate displaces water in pore space and causes a diminution of the reflection coefficients of strata.

Four regions of gas hydrate concentration have been mapped on the continental rise in the offshore region between New Jersey and Georgia. Three are correlated with sediment depocenters and one occurs along a concentration of diapirs. Concentration at depocenters probably occurs because most gas in hydrate is biogenic, produced by bacteria that feed on organic material that is preserved from oxidation when it is rapidly buried in fast-depositing marine sediments. A major concentration of hydrate, one not related to a distinct depocenter, is associated with a trend of diapirs that create gas traps resulting in the concentration of gas and gas hydrate. Breakdown of gas hydrate and thinning of the hydrate layer has occurred at sites of mass movement, probably caused by reduction of pressure due to removal of sediment during landsliding. Conversely, breakdown of hydrate may cause mass movements by substituting a weak layer of gassy sediment for a strong layer of hydrate-bearing sediment.

Chapter 14

Gas Hydrate in the Arctic and Northern North Atlantic Oceans

[1]Michael D. Max, [2]Jürgen Mienert, [2]Karin Andreassen, and [2]Christian Berndt

[1]Marine Desalination Systems
Suite 302, 1211 Connecticut Ave NW
Washington, DC
[2]Department of Geology, University of Tromsø
Dramsveien 201, N-9037 Tromsø, Norway

1. INTRODUCTION

The northern North Atlantic and Arctic oceans are morphologically and geologically complex. The constructive axial plate margin of the northern North Atlantic is propagating through Fram Strait, forming a young oceanic crust in the Nansen Basin of the Eurasian end of the deep water Arctic Ocean (Fig. 1). A complex transform along the continental margin of the Laptev Sea is the present termination of this Atlantic-Arctic Ocean spreading center. The North American end of the Arctic Ocean is floored by older oceanic crust carrying a thick sediment prism in the western end of the Canada Basin. The Barents Sea, like the other wide shallow water margins of the Asian Arctic Ocean and narrower continental shelf elsewhere around the Arctic margin, is an epicontinental sea (Eldholm & Talwani, 1977).

The methane generating character of marine sediments is fundamental to the development of hydrate. Any area in which delivery of organic material is high and burial is rapid will lead to formation of biogenic methane. The Arctic and the North Atlantic Oceans are likely to be revealed as major hydrate provinces because the oceanographic conditions (around 0°C) are highly suitable for preservation and burial of organic material (>1.5%). Deeply buried sediments in the Arctic are rich in organic matter. The area comprises one of the largest contiguous sedimentary provinceswith significant amounts of organic carbon (Premuzic, 1980; Romankevich, 1984). The sedimentary framework has been similar in the North American end of the Arctic basin since mid-Mesozoic times and in the Eurasian end since not long after the formation of the plate margin in Magnetic Anomaly 23 times (~ 52 Ma bp) (Vogt, 1986; 1999; Eldholm et al., 1987).

M.D. Max (ed.), Natural Gas Hydrate in Oceanic and Permafrost Environments, 171–182.

Figure 1. Location and geographical names of generalized major basins, ridges and geological features in the Arctic and northern North Atlantic Oceans. From Max and Lowrie (1993). Polar equal area projection. Dot on Lomonosov Ridge is North rotational Pole. AR, Aegir Ridge; BAP, Barents Abyssal Plain; BB, Boreas Basin; BI, Bennet Island; CC, Chukchi Cap; EI, Ellesmere Island; FAP, Fletcher Abyssal Plain; FJI, Franz Joseph Islands; FS, Fram Strait; HM, Håkon Mosby Mud Volcano; GB, Greenland Basin; GFZ, Greenland Fault Zone; GS, Greenland Sea; I, Iceland; JMF, Jan Mayen Fault Zone; KBR, Kolbensey Ridge; KNR, Knipovich Ridge; LB, Lofoten Basin; LS, Labrador Sea; MB, Malene Bukta; NB, Northwind Basin; NOB, Norway Basin; NR, Northwind Ridge; NS, Nares Strait; NOS, Norwegian Sea; NSL, North Slope; PB, Prudhoe Bay; PAP, Pole Abyssal Plain; SAP, Siberian Abyssal Plain; SS, Storegga Slide; SV, Svalbard; V, Vestnesa Ridge; VP, Vøring Plateau; WAP, Wrangel Abyssal Plain; YFZ, Yermak Fault Zone; YP, Yermak plateau.

The Norwegian margin, including the northern North Sea and the Barents Sea (Vorren et al., 1993) along with the North American end of the Arctic Ocean and adjacent land areas (Chapter 5), are proven hydrocarbon provinces. Because the more deeply buried sediments have provided organic material that transformed into gas and oil, sediments in the upper 2 to 3 km of sediment which have not entered the oil window of hydrocarbon maturation (Max and Lowrie, 1993) are likely to provide a rich feedstock for methanogenic bacteria feeding hydrate formation (Chapter 8). Thus, the hydrocarbon-generating potential of marine sediments along this continental margin can be expected to have provided an excellent host for methane generation.

Hydrocarbon exploration in the Arctic has yet to move into many areas where hydrocarbons are likely to be concentrated. However, it is a great challenge because of the economic constraints associated with extreme cold in a remote area, shifting sea ice, environmental concerns, and very difficult logistics. Potentially methane-rich sediments almost certainly underlie large areas. It can be inferred that gas and oil, as well as hydrate deposits, will be more widely found than is currently proven.

Heat flow, which is critical to the thickness of the hydrate stability zone (HSZ) varies considerably in the Arctic and northern North Atlantic Oceans. Sea floor in the age range 100-200 my is 45-55 mW/m^2, younger sea floor has higher heat flows and thinner sediment cover (references to heat flow in Max and Lowrie, 1993). On active ridge sites to 3-4 Ma off-ridge heat flows average about 300 mW/m^2 with some heat flow measurements nearly 400 mW/m^2. The oceanic crust of the northern Greenland-Norwegian Sea has a high heat flow of between 100 and 200 mW/m^2 (Vogt and Sundvor, 1996).

The presence of natural submarine gas hydrates is commonly inferred from seismic reflection data (e.g. Hyndman and Spence, 1992). The base of the stability zone for gas hydrates (HSZ) is geophysically identified by the occurrence of a bottom simulating reflector (BSR) (Stoll et al., 1971). The BSR is a reflection at the boundary between a normal velocity layer or high-velocity gas hydrate cemented sediments and the underlying low-velocity gas-bearing sediments. Whereas compressional velocity values of 1700-2400 m/s are known to be typical for gas-hydrated sediments (Andreassen et al., 1990; Katzman et al., 1994; Lee et al., 1994; Minshull et al., 1994; Andreassen et al., 1995) values below the sound velocity of sea water (SVS) indicate free gas in the pore space. The BSR mimics the shape of the sea floor, often cuts the dominant stratigraphy and is characterized by a high, reversed polarity event (e.g. Lodolo *et al.*, 1993; Katzman *et al.*, 1994; Andreassen *et al.*, 1995).

Oceanic hydrate (Chapter 6) has been geophysically recognized in deep water in a continuous zone along the North Slope of Alaska, in isolated localities in the Barents Sea (Laberg and Andreassen, 1996) and Norwegian continental margins (Mienert et al., 1998, Mienert and Posewang, 1999), and at least one locality in the eastern Labrador Sea (Fig. 1). In addition, hydrate formed independently from the presence of subsea permafrost (Chapter 5) and has been recognized in the Barents Sea (Løvø et al., 1990). Although considerable seismic data exist for the Barents Sea and Norwegian continental margin (Vorren et al., 1993), difficulty in carrying out seismic, bathymetric, and oceanographic surveys have yielded little data for the ice-covered Arctic basin as a whole.

2. Tectonostraticgraphic framework of the Arctic and northern North Atlantic Oceans

2.1. Northern North Atlantic

The Norwegian - Greenland Sea extends from Iceland to the Fram Straits (Fig. 1) and can be divided into a number of tectono-sedimentary provinces controlled by the transform-ridge system. Sediments immediately north of Iceland are up to 1 km thick but thin rapidly to less than 200 m thick in ponded areas along the remainder of the Kolbensey Ridge to the Jan Mayen Fracture Zone. Quaternary sediments on the southeast Greenland margin are up to 2.5 km thick and overlie at least 4 km thick Tertiary and Mesozoic sediments. Sediments on the Jan Mayen Ridge are up to 3 km thick. Sediment thickness variation in the Lofoten and Greenland Basins is characteristic of passive continental margins; thickest sediments along the continental slope and gradually thinning toward the ridge

The Norwegian margin developed during continental rifting between Laurentia and Eurasia, which culminated in Late-Paleocene/Early Eocene break-up (Skogseid et al., in press). Due to the Iceland hotspot mantle temperatures were increased during rifting and break-up, leading to extensive volcanism (Eldholm et al., 1989). The mid-Norwegian margin consists of three rifted margin segments (from south to north): the Møre Margin, the Vøring Margin, and the Lofoten-Vesterålen Margin, and an approximately 200 km long sheared margin segment which constitutes the southern boundary of the Vøring Margin (Figure 1). Extension of the entire area stopped after continental break-up, and apart from minor Tertiary doming the area became tectonically quiet. Doming focussed along the shelf break, and is manifest as the Helland Hansen Arch, and the Naglfar, Vema and Ormen Lange domes. The proposed mechanisms for these domes include ridge-push and differential compaction and asymmetric sedimentation (Doré et al., 1997; Vågnes, 1997). Whereas Eocene to Miocene sedimentation along the margin was moderate, Pliocene and Quaternary glaciations increased sediment input, and led to an up to 2 km thick wedge of clastic sediments east of the dome structures. After the last glacial maximum sedimentation on the mid-Norwegian margin was dominated by submarine mass-wasting (Vorren et al., 1998) potentially triggered by earthquakes and gashydrate dissociation due to climate change (Bugge et al., 1988; Mienert et al., in press).

The Barents Sea is an epicontinental sea extending between Norway and the Svalbard archipelago. It is bounded by Cenozoic passive margins in the north and west. The margin north of Svalbard is a volcanic rifted margin with related igneous intrusions reaching far south into the Barents Sea. The western margin of the Barents Sea is a sheared margin (Eldholm, 1987; Faleide et al., 1991, 1993). Wrench tectonics and an opening component to the predominantly shear setting resulted in complex deformation of the southwestern Barents Sea (Rønnevik & Jacobsen, 1984; Riis et al, 1986, Gabrielsen & Færseth, 1988). The sedimentary strata above the Paleozoic basement comprise an almost complete sequence of sedimentary strata from the Late Paleozoic to the Quaternary (Faleide et al., 1993; Gudlaugsson et al., 1997). However, Neogene uplift and Pliocene-Pleistocene glaciations caused severe erosion of the inner parts of the Barents Sea. Two depocenters developed in the adjacent Norwegian-Greenland: the Bjørnøya Fan south of Bjørnøya and the Storfjorden Fan north of it. More than half of the sediments in these fans was deposited during the last 3 Ma (Eidvin et al., 1993; Sættem et al., 1994; Faleide et al., 1996, Elverhoi et al., 1998).

2.2. Arctic Ocean

The Arctic Ocean floor comprises two distinct major tectonic units. The Amerasian Basin to the Alaskan/Asiatic side of the Lomonosov Ridge (Fig. 1) has a complex internal geological history related to terrane accretion that can be traced into the Pacific Ocean prior to closure of the Bering Sea. The Canada Basin was formed earlier than the Eurasian Basin that lies between the Lomonosov Ridge and the Barents Sea, through which an active constructive plate margin currently passes. The floor of the Arctic Ocean consists of a series of roughly parallel ridges and basins extending from the North American continent to the Eurasian continent. Upper Paleozoic through Cretaceous sediments, including thick shale sequences, have been proven from northern Alaska to northeastern Ellesmere Island, and presumably occur in slope prisms (references in Max and Lowrie, 1993).

The Alaskan/Asiatic end of the Arctic Ocean (Fig. 1) contains the largest single basinal area, the Canada Basin, which is about 1,100 km X 600 km. It is underlain by earliest Jurassic, mid-Jurassic, or mid Cretaceous oceanic crust, and is the oldest part of the Arctic Ocean floor. The abyssal plains, which are underlain by at least 4 to 6 km of sediment, and up to over 10 km of sediment locally, lie at depths of 3 km to 3.9 km. The Makarov Basin lies between the continental fragment of the Lomonosov Ridge and the Alpha-Mendeleyev Ridge. Sediments in the deeper oceanic parts of this basin vary from 1 to 6 km, with greater thicknesses along the Alaska-facing flank of the Alpha-Mendeleyev Ridge.

Up to 4 km of Quaternary sediments occur in the outer sedimentary prism of the Alaskan North Slope overlying up to between 10 km and 12 km of Tertiary through Upper Paleozoic sediments. Smaller sedimentary basins north of Greenland at the juncture with the Alpha Ridge contain sediments between 8 km and 10 km thick. Nearby small sedimentary basins contain sediment between 2 km and 4 km thick. Sediments from 4 km to 8 km occur in the western North Alaska slope, with a major depocenter containing sediment up to 10 km to 12 km thick near the junction with the Northwind Ridge. The continental slope along the East Siberian Sea contains a narrow depocenter from 8 km to 10 km thick that is part of a larger sediment prism extending into the Wrangel Abyssal Plain. The thickest sediments along the junction with the Mendeleyev-Northwind-Chukchi Cap region at the Asian termination of the

Alpha-Mendeleyev Ridge are from 4 km to 6 km thick, but occur mainly within the continental shelf. Sediment thicknesses elsewhere along the Eurasian continental margin appear to be no more than 2 km to 4 km thick.

The Nansen and Amundsen basins are underlain by the youngest oceanic crust zone that represents the propagation of the North Atlantic ridge. These are conjugate basins which are formed from a single axial spreading ridge. Sediments in both basins formed a single prism over extensional crust until establishment of the ridge, at which time the sediments in the Amundsen Basin became more oceanic in character, except at the North American and Eurasian ends where continental sedimentation remains an important influence (Thiede, et al. 1990).

In general terms, the Gulf of Mexico may serve as a well explored analogue for the lesser-known Canada Basin. Both developed at about the same time in the same manner and are characterized by a fixed, major sediment source. Sediment samples from the central Canada Basin have been mainly taken from the shallower water areas of the Alpha-Mendelyev Ridge complex and the Northwind Ridge (Darby et al., 1989) where sedimentary successions in many grabens containing basal Cretaceous beds no more than 1.4 km thick. A few depocenters of sediment thickness > 4 km and >6 km occur in the Chukchi Trough and in the Alpha ridge and along its margins.

3. OCCURRENCE OF BSRs AND HYDRATE LOCALITIES

3.1. Norwegian Margin

Two provinces on the mid-Norwegian Margin show BSR. Whereas the reflectors on the outer Vøring Plateau are related to diagenesis (Skogseid and Eldholm, 1989), the BSR in the vicinity of the Storegga slide are related to gas hydrates (Bugge, 1983; Mienert and Bryn, 1997, Mienert, et al., 1998, Bouriak et al., 2000). These BSRs are located at the seaward termination of a thick wedge of Plio-/Pleistocene sediments, and it is likely that fast sedimentation caused burial of large amounts of organic material. It has been suggested that gas hydrates have destabilized the slope in this area, and that this is one reason for catastrophic slope failure (Bugge et al., 1987, Mienert et al., 1998). Generation of this slide has been related to tsunamis on the facing Norwegian coast (Dawson et al., 1988, Bondevik et al., 1997).

The Storegga Slide is one of the world's largest submarine landslides having moved a total of 5600 km^3 of sediment with an original thickness of the slumped layer of up to 450 m from an area with a size comparable to that of mainland Scotland (Jansen *et al.*, 1987; Bugge *et al.*, 1988, Evans et al., 1996). A coincidence of dissociation of gas hydrates and slope failures exists at the Mid-Norwegian continental margin (Bugge *et al.,* 1987; Jansen *et al.*, 1987; Mienert *et al.*, 1998). High-resolution seismic data allowed to identify two parallel-occuring BSR and associated velocity anomalies in this area (Posewang & Mienert, 1999a) indicating a complex gas and gas hydrate system. Low levels of methane and minor propane from three ODP drill sites on the Vøring Plateau are dominated by biogenic gas (Whiticar and Faber, 1989). This situation suggests that not all hydrocarbons derived from more deeply buried sediments have yet reached the HSZ

North of the northern sidewall of the Storegga Slide, the lower boundary of the HSZ is determined in high-resolution reflection seismic data by a strong BSR which occurs at a depth of approximately 0.35 s two-way travel time (TWT) (Fig. 2). According to the velocity analysis, the corrected depth range of the BSR is 250-285 mbsf (Posewang, 1997, Mienert and Bryn, 1997). The BSR is easily traceable throughout a grid of seismic profiles, indicating the large extent of gas hydrates in this area. The BSR cuts reflections from the dominant strata, mimics the shape of the sea floor and is characterized by high amplitudes. The blanking above the BSR possibly indicates an increase of the hydrate concentration in the sediments. Parts of the BSR along this section are disturbed and exhibit amplitude variations (marked with a,b in Fig. 2). Anomalous high amplitudes of crossed horizons and vertical wipe-out zones are typical indicators for the existence of free gas below the BSR.

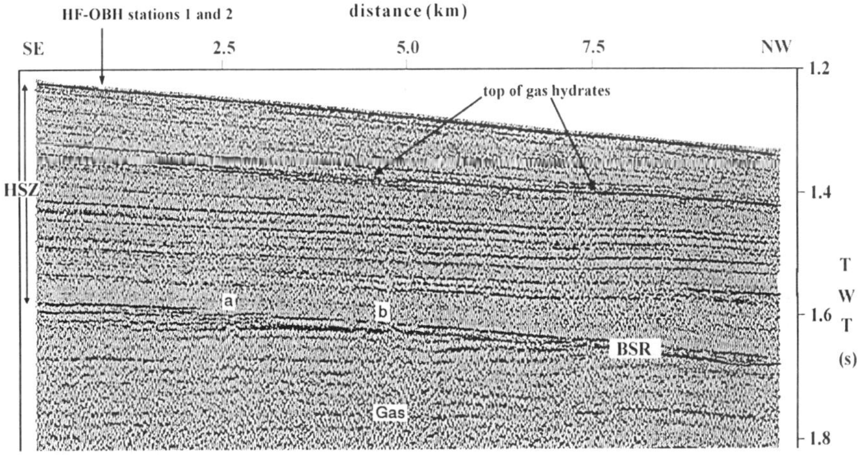

Figure 2. Section of seismic reflection profile acquired with a 2-liter airgun and a 6-channel streamer north of the northern Storegga escarpment. A strong BSR occurs at a depth of 0.35 s TWT bsf. Along this section, parts of the BSR are disturbed and exhibit amplitude variations (marked with a, b). The appearance of a second strong horizon in a depth of 0.125 s TWT is interpreted as the top of gas hydrates (from Posewang and Mienert, 1999a).

The velocity information from the HF-OBH data shows zones of alternating high and low velocity (Posewang and Mienert, 1999a). High velocities with a maximum of 1850 m/s indicate the existence of gas hydrates in a 180 m thick layer. Below these gas hydrate cemented sediments the velocity drops down to a minimum of 1400 m/s, a value lower than the SVS caused by gas-bearing sediments. The thickness of the low-velocity layer is not clearly identified. The transition between gas hydrate cemented and gas-bearing sediments corresponds with the occurrence of the BSR in seismic sections (Posewang and Mienert, 1999a).

A low-velocity zone within the HSZ reflects the complicated hydrate formation mechanism in this area (Posewang and Mienert, 1999a). At a depth of approx. 0.125 s TWT, the velocity drops from 1580 m/s to 1350 m/s, indicating free gas in the sediments. Above the free gas layer, a lithological boundary possibly acts as a seal for rising gas. The significant impedance contrast between the gas-bearing sediments above and gas-hydrated sediments below produces a strong reflection on seismic records in a depth of 0.125 s TWT bsf interpreted as the top of gas hydrates (Fig. 2).

3.2. Barents Sea Margin and Svalbard Margin

The largest contiguous sedimentary prism in the North Atlantic is along the western Barents Sea margin (Myhre and Eldholm, 1988), especially in the Bear Island Fan, where sediments are up to 7 km thick (Vorren et al., 1998; Hjelstuen, et al., 1999). Because rapidly deposited sediments derived primarily from a continental shelf area constitute good source beds, the west Svalbard sediment prism is likely to contain significant amounts of organic material. Moreover, the ocean continent boundary of the western Barents Sea is characterized by high heat flow (Eldholm et al., 1999) facilitating rapid generation of hydrocarbons. Hydrate has been identified in the Håkon Mosby Mud Volcano (HMMV) in the Bear Island Fan (Vogt et al., 1999; Ginsburg et al., 1999), in the Vestnesa Ridge (Vogt et al., 1994) and nearby (Posewang & Mienert, 1999) to the SW of the Yermak Plateau, and by inference within the Malene Bukta embayment (MB) in the Arctic margin immediately to the north of Svalbard (Cherkis et al., 1999) (Figs. 1) Both shallow gas and hydrate have been identified in sediments of the western Barents Sea slope (Eiken and Austegard, 1987; Eiken and Hinz, 1993). Seismic velocity profiles indicate trapped gas below hydrate on the upper continental slope (Austegard, 1982).

The Barents Sea gas hydrate site is at about 350 m water depth. A single shallow seismic reflector of anomalous high amplitudes cuts through dipping layers, interpreted as the base of a gas hydrate cemented layer (Fig. 3) (Andreassen *et al.*, 1990). The BSR that here has the appearance of a "bright spot" occurs in a depth of 0.17s TWT, corresponding to approximately 180 mbsf, one of the shallowest known BSRs observed in north Polar Regions. Velocity analysis from multi-channel seismic data show a high velocity layer above the BSR (>2400 m/s) interpreted as a gas hydrate layer (Andreassen *et al.*, 1990). The strong velocity decrease to values of about 1625 m/s below the BSR, is an indicator of free gas in the sediments.

The Svalbard gas hydrate site ranges in water depth from 860 m to 2350 m. A well-developed BSR exhibits strong amplitude variations, and parts of it show high amplitude reflections below (Fig. 4) (Posewang and Mienert, 1999b). Furthermore, frequency analysis revealed that sedimentary layers below the BSR act as a low-pass filter on seismic signals. Above the BSR, frequencies up to 170 Hz predominate; below the BSR, frequencies of

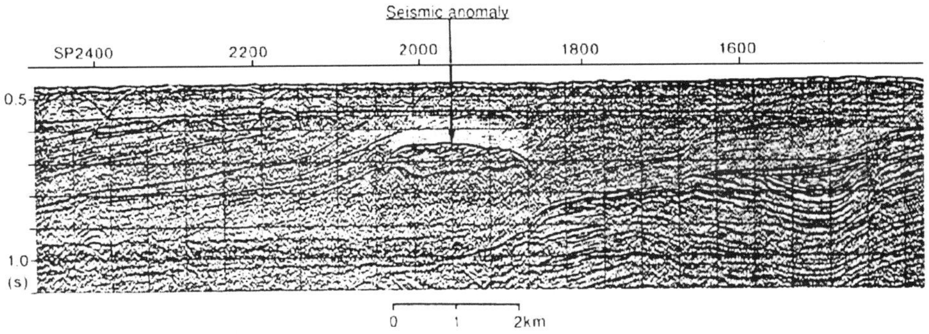

Figure 3. Section of a multi-channel reflection seismic profile located in the Barents Sea. In a depth of 0.17s TWT bsf a seismic anomaly appears interpreted as a BSR. The BSR crosses the dipping sedimentary strata, is characterized by high amplitudes and shows a limited lateral extension above a fault complex (from Andreassen *et al.* 1990).

less than 80 Hz prevail. Free gas, which is trapped below the BSR, might explain the high reflection amplitudes in this subbottom depth and the existence of a low-pass filter on seismic signals. The sealed free gas migrates into the overlaying strata, and therefore, the thickness of the free-gas layer below the BSR varies (Posewang and Mienert, 1999b). Due to the high-resolution character of the data, the so-called 'Base of the Gas Reflection' (BGR) (Camerlenghi and Lodolo, 1994) could be detected and the thickness variation could be calculated (Mienert et al., in press). The reflection amplitude of the BSR also varies according to the thickness of the free gas layer. Due to the short distance to the Vestnesa Ridge, which acts as a heat source (Vogt *et al.*, 1994), the sub sea floor temperature increases downslope. Therefore, the subbottom depth of the BSR decreases with increasing water depth from 0.26 s TWT bsf upslope to 0.22 s TWT bsf downslope (Posewang and Mienert, 1999b). Velocity analysis of multi-channel seismic data shows three distinct layers with different velocities (Andreassen and Hansen, 1995). A nearly 200 m thick layer between the sea floor and the BSR has an average velocity of 1630 m/s followed by a low-velocity layer of 100 m thickness (1450 m/s) and a zone of higher velocities. The high-velocity layer above the BSR is interpreted to contain gas-hydrated sediments, whereas the low-velocity layer indicates free gas beneath the BSR. The interpretation of a gas layer below the BSR was confirmed by velocity-depth models calculated from High-frequency Ocean Bottom Hydrophone (Hf-OBH) data (Posewang and Mienert, 1999b).

3.3. Greenland Margins

The North Greenland continental margin only locally appears to contain enough sediments to either generate sufficient methane or host a HSZ. However, this area is little explored due to difficult environmental conditions and its remoteness. Thick sediments in intracratonic basins on the East Greenland shelf (Jackson et al., 1991) may contain geological petroleum traps similar to those proven by drilling in the Troms-Hammerfest basins of

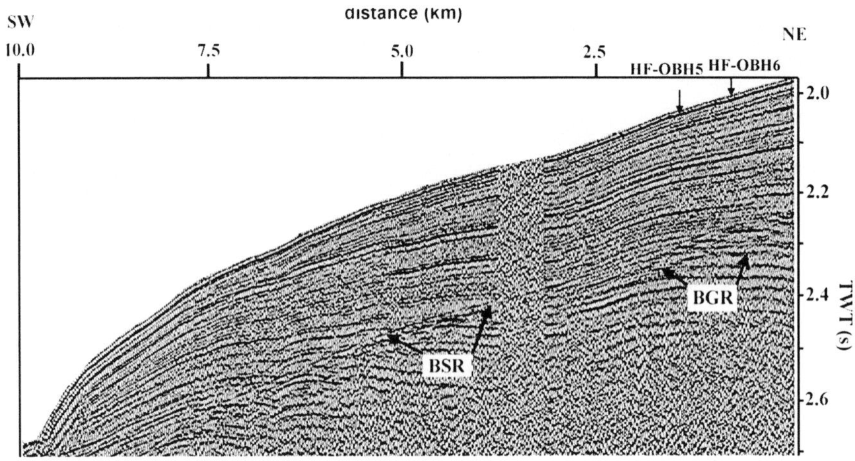

Figure 4. Section of seismic reflection profile based on a 2-liter airgun and a 6-channel streamer located at the Svalbard margin. The arrows mark the locations of the HF-OBH stations 5 and 6 at the sea floor. A strong BSR occurs at a depth of 0.25 s TWT bsf. The BSR parallels the sea floor, crosses the discordant sedimentary strata and is characterized by high amplitudes. Below the BSR the base of gas-bearing sediments is marked by the BGR (from Posewang and Mienert, 1999b).

3.4. Arctic Ocean

Although BSRs have not yet been identified in water depths below 2800 m on the north Alaskan rise (Grantz et al., 1976, 1989; Grantz and May, 1982; Grantz et al., 1989), it is likely that hydrate will be identified in deeper water. Lowrie and Max (1993) discuss a number of indirect identifications of phenomena, such as gas and fluid bursts, which have been used to infer the presence of hydrate in this little known province.

4. AERIAL EXTENT OF HYDRATE FORMATION

Hydrate, often with gas below, has been recognized in continental slopes and rises between about 500 and 3,500 m water depth (Eiken and Hinz, 1989, Mienert et al., in press). All Arctic and northern North Atlantic continental slope areas where sediment thicknesses are greater than about 3 km are potentially areas in which significant gas generation and consequent hydrate formation is possible. Prediction of gas production and the presence of hydrate, however, is uncertain.

A number of criteria are used in determining the area and thickness of potential hydrate development:

1. An average heat flow value of 30°/km for this region.

2. Biogenic gas is produced by the deep biosphere (Chapters 7 & 8).
3. Sedimentary successions throughout the region are regarded as having good gas generation potential.
4. 500 m is a minimum depth for finding hydrate-bearing areas.
5. Bottom water temperature at between the -1.5° C to +1.5° C. Three main areas where hydrate is likely to be found have been identified using these criteria (Fig. 5):

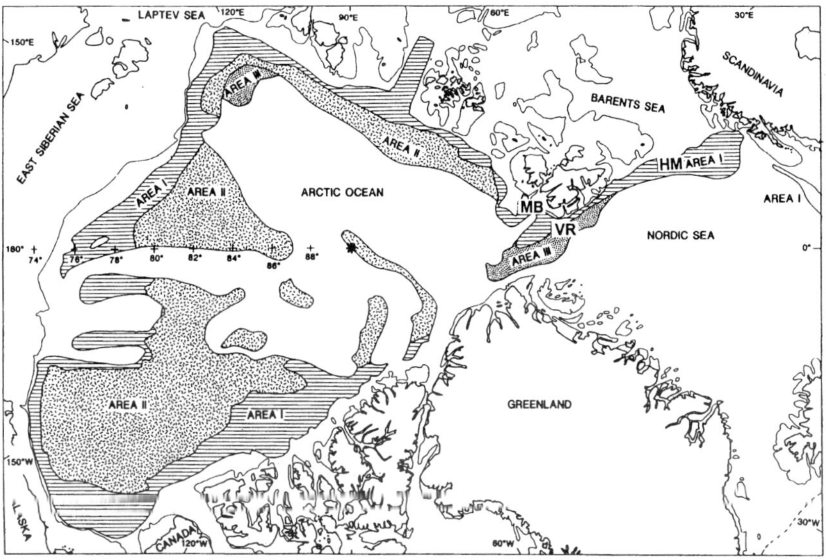

Figure 5. Hydrate likelihood areas, from Max and Lowrie (1993). Slope and abyssal areas separated in Arctic at about the 3,000 m contour. Sediment thickness data not included for areas to the south of dashed line passing across south Iceland. Area I. Continental slope areas between 500 m and 3,000 m follow recognized hydrate development in continental slope north of Alaska (Grantz et al., 1989). Area II. Abyssal areas and sedimentary basins not tied to continental shelf structure where sediment thickness exceeds 3 km. Area III. Areas of abnormally high heat flow associated with plate margin where gas could be sourced from thin sediments or where there is juvenile gas. Area III minimum sediment thickness of between 0.75 km -1 km in ridge vicinities (Knipovich and Laptev).

Area I. This area includes continental slopes between 500 m and 3,000 m water depth where sediment thickness is greater than 3 km. This region the water depth range in which hydrates have been identified along the North Slope of Alaska (Grantz et al., 1989). About 818,000 km^2 of the Arctic Basin and about 154,000 km^2 of the Northern North Atlantic are underlain by areas with these physical characteristics.

Area II. This area includes abyssal regions below 3,000 m (except in the continental slope-ward margin of the Wrangel Abyssal Plain where the 2,500 m contour has been mainly used) where sediment thickness is greater than 3 km. In abyssal areas, no identification of hydrate has yet been made, but if sufficient methane has been generated, there is a thick HSZ in which hydrate could form. About 1,403,000 km^2 of such area occurs in the Arctic.

Area III. Two relatively small areas of thin sediment overlie active transform/ridge systems in which it is known that gas and hot hydrothermal fluids are generated as a by-product of magmatic and volcanic activity. It is possible that hydrothermal fluids would deliver methane to sediments that would otherwise be too thin to generate its own gas. This area consists of about 35,000 km^2 of the Arctic Basin and about 80,000 km^2 of the northern North Atlantic.

The HSZ increases in thickness with increasing depth (Chapter 6) (Miles, 1995). 450 m represents a reasonable average thickness for the HSZ in Area I. For Areas II and III, in deep and very deep water, a thickness of about 700 m is used as a reasonable average thickness.

The thickness of the HSZ may be less important than a mechanism that allows hydrate to concentrate. The HSZ off SE Japan, for instance, is relatively thin (Chapter 18), but the hydrate there is highly concentrated.

Estimations of hydrate and gas were made by Max and Lowrie (1993), but are now regarded as being high (Kvenvolden, 1998).

Conclusions

Gas hydrate has been identified in the North Atlantic and Arctic Oceans. Its extent is not known in detail owing to lack of field data. Geophysical identifications of hydrate along the Norwegian continental margin have been proved by drilling (also see Chapter 23) and seafloor sampling (Vogt et al., 1999).

If the methane generating character of the sediments in the Canada Basin is appreciable, then enormous volumes of hydrate could be present.

Under-ice survey from submarine should resolve the Hydrate Economic Zone (Glossary of Terms) and allow quantification of hydrate.

Acknowledgements MDM thanks P. Vogt, E. Sundvor, O. Eldholm, and Y. Ohta for many discussions. Some of the results have been obtained as part of the SFB 313 of the University of Kiel funded by the Deutsche Forschungsgemeinschaft and the European North Atlantic Margin (ENAM II) project of the MAST III programme of the European Commission (MAS3-CT95-0005).

Chapter 15

Cascadia Margin, Northeast Pacific Ocean: Hydrate Distribution from Geophysical Investigations

G.D. Spence[1], R.D. Hyndman[2,1], N.R. Chapman[1], M. Riedel[1], N. Edwards[3] and J. Yuan[3]

[1]School of Earth and Ocean Sciences
University of Victoria
Victoria, B.C., Canada V6T 1W5

[3]Department of Physics
University of Toronto
Toronto, Ont., Canada

[2]Pacific Geoscience Centre
Geological Survey of Canada
Sidney, B.C., Canada V8L 4B2

1. INTRODUCTION

Natural gas hydrate was first recognized on the Cascadia margin in 1985 through the characteristic bottom-simulating reflector (BSR) on conventional multichannel seismic data (Davis and Hyndman, 1989, Davis et al., 1990). Since then, the Cascadia accretionary margin has received the most intensive studies of any convergent margin for determination of the in-situ properties of marine gas hydrate. Key control for understanding the properties and formation processes of hydrate has been derived from drill holes of the Ocean Drilling Program (ODP) Leg 146, carried out in 1992. Estimates of hydrate concentration were provided through analysis of downhole seismic and resistivity logs and through measurement of chlorinity in pore fluids from recovered sediment core samples.

Most information on the areal and depth distribution of hydrate and on its concentration is obtained using remotely-sensed investigations, primarily geophysical. The region off Vancouver Island has been the target for an exceptionally broad range of such studies, including seismic, heat flow, seafloor electrical sounding, seafloor compliance, and physical property measurements of seafloor piston cores. The primary objective of this chapter is to review briefly the principal geophysical methods used off the northern Cascadia margin, and to provide examples of what we have learned about the broad regional distribution of hydrate and its fine vertical structure.

Although not the focus of this paper, geochemical and biological studies have contributed significantly to understanding the sources of methane in

M.D. Max (ed.), Natural Gas Hydrate in Oceanic and Permafrost Environments, 183–198.

hydrate, as well as hydrate formation and dissociation processes. Examples can be found in the work of Whiticar et al. (1995), who studied the carbon isotope signature of methane dissociated from hydrate and determined that its origin is clearly bacterial and not thermogenic. In a study to determine the nature of the bacterial source of the hydrate, Cragg et al. (1995) found significantly increased bacterial populations and activities in the 10 m interval above the base of hydrate at Sites 889/890.

Seafloor hydrate has been observed at sites off Oregon, particularly in the region known as Hydrate Ridge. In ODP drilling, solid macrocrystalline hydrate was recovered from near the seafloor (2-19 mbsf) at Site 892. It is associated with seafloor venting, or "cold seeps", along with related features such as pockmarks, carbonate pavement and unique biological communities at the vent sites. Over the past 15 years, the region has been the focus of extensive sampling and coring programs, submersible investigations and geochemical analyses. For a review of recent exciting results, the reader is referred to Suess et al. (1999a, 1999b). For information on geophysical investigations on the Cascadia margin off Oregon, see MacKay et al. (1994), Trehu et al. (1995, 1999) and Goldfinger et al. (1999).

2. TECTONIC AND GEOLOGICAL SETTING

The oceanic Juan de Fuca plate has been subducting beneath the North American plate since the Eocene, approximately normal to the Vancouver

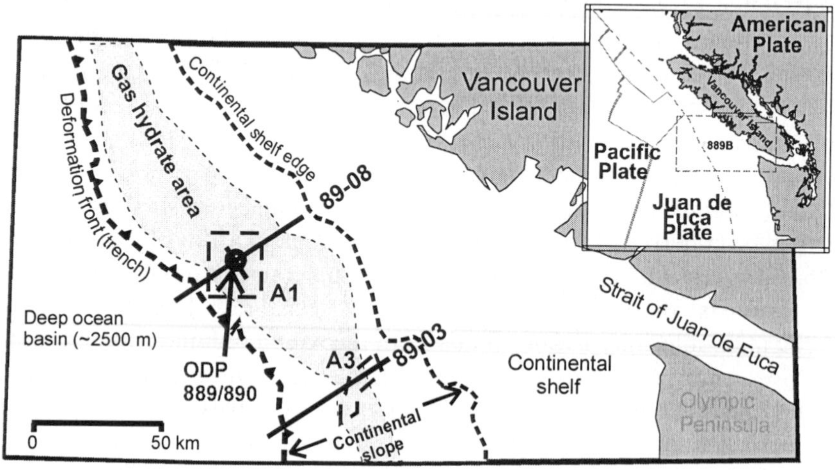

Figure 1. The northern Cascadia margin, showing the region where gas hydrates are found. Detailed grids of lines were collected in regions A1 and A3. Multichannel lines 89-08, 89-03 and 89-10 (perpendicular to 89-08) were collected as part of a site survey for ODP Leg 146.

Islandmargin at a current rate of 45 mm/yr (Fig. 1). This is also the time of emplacement of the Eocene marine volcanic Crescent terrane (and the equivalent Siletz terrane to the south). The Crescent terrane outcrops on southern Vancouver Island and on the Olympic Peninsula. Magnetic, gravity and multichannel seismic data (Hyndman et al., 1990; Dehler and Clowes, 1992) indicate that the Crescent terrane extends northwest of the Olympic Peninsula along the central portion of the continental shelf off Vancouver Island, that it dips landward, and that it acts as the backstop to the accretionary wedge formed by sediments scraped off the underthrust oceanic plate.

2.1 Structure of the Accretionary Prism

Seaward of the deformation front , the Cascadia basin consists of pre-Pleistocene hemipelagic sediments overlain by a rapidly deposited Pleistocene turbidite. Seismic sections image the oceanic crust dipping gently at 3-4° beneath the deformation front (Fig. 2), where the total sediment thickness is 2.5-3 km. The water depths are only about 2500 m due to the young age (5-7 Ma) of the oceanic lithosphere at the deformation front and the thickening sediment section. Initial deformation is accomplished by landward-dipping thrust faults and margin-parallel folds (Hyndman et al., 1994). On MCS Line 85-02 near ODP Site 889 (Fig. 2), a single frontal anticline has grown to a height of over

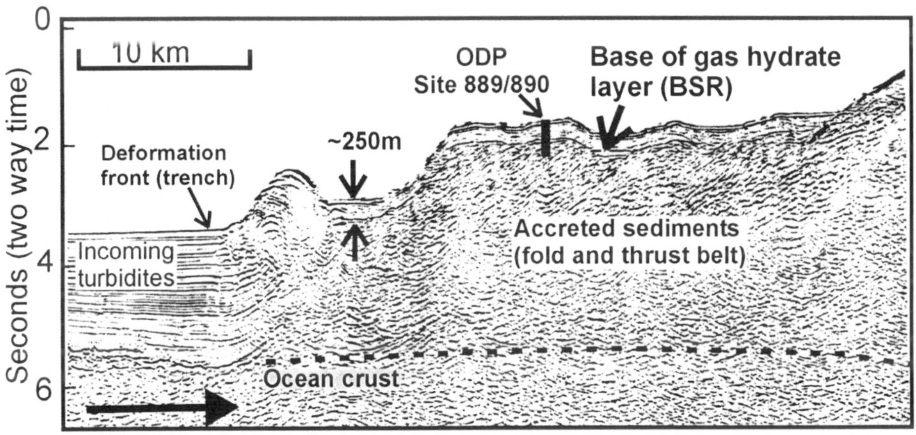

Figure 2. Multichannel seismic line 85-02 which passes near ODP Site 889.

700 m above the basin floor. The anticline developed over a thrust ramp extending down to near the top of the oceanic crust. With similar observations on several seismic profiles, Davis and Hyndman (1989) argue that the present decollement is near the base of the sediment section and most of the sediments are scraped off the incoming oceanic crust. This contrasts with the Oregon margin where the detachment is higher in the section (MacKay et al., 1992).

As the sediment section grows rapidly in thickness farther landward, the accreted sediments undergo severe tectonic compaction and distributed small-scale deformation. From cores recovered at ODP Site 889, sediments recovered in the upper 128 m below the seafloor (mbsf) are silty clays and clayey silts interbedded with fine sand turbidites. These were interpreted as slope basin sediments (Westbrook et al., 1994). Sediments below 128 mbsf are more deformed and compacted, and were interpreted as accreted Cascadia basin sediments. On multichannel seismic sections, slope basin sediments are recognized as well-layered events (e.g., Fig. 3a, CDP 3000-2850), whereas few continuous reflections are observed within the accreted sediments.

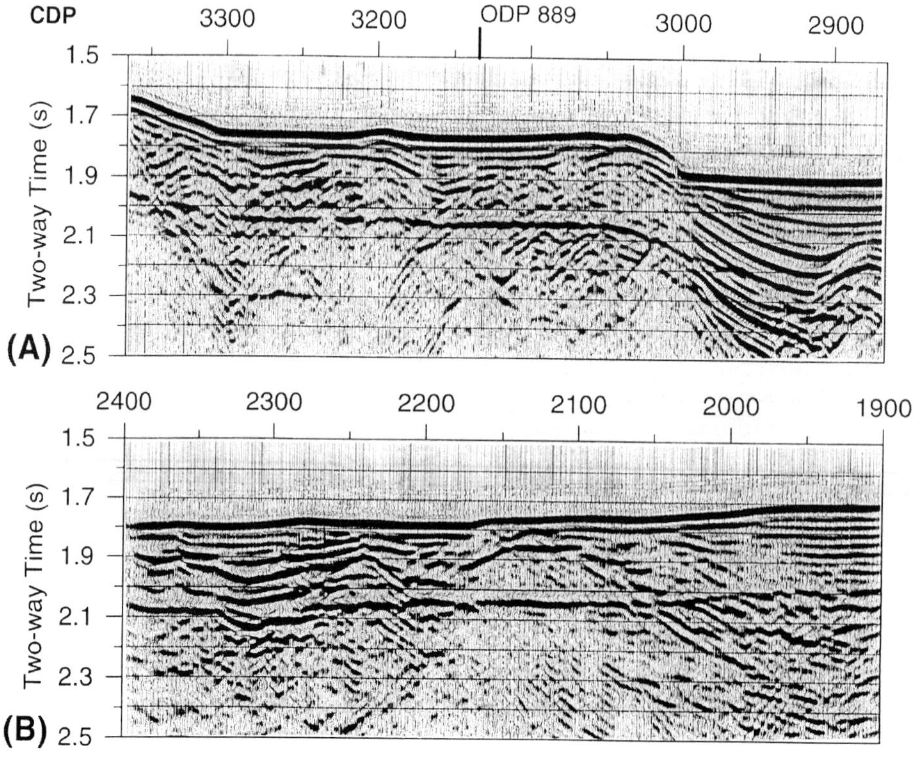

Figure 3. Reflection sections 89-08 (A) and 89-10 (B) near ODP Site 889. The BSR, representing the base of hydrate, is particularly strong in this region.

2.2 Deep-sea Gas Hydrate and Fluid Expulsion

The accretionary wedge sediments thicken landwards by a combination of two processes: vertical thickening and bulk shortening through tectonic compression, and thrust faulting that emplaces initially high-porosity sediments at greater depths. As the equilibrium porosity-depth profile is reestablished, the sediments are consolidated and fluids are expelled (Hyndman and Wang, 1993). Hyndman and Davis (1992) suggested that hydrate layers at and above the BSR are formed when methane is removed from these upward-moving fluids as they pass into the hydrate stability field. This mechanism accounts for the common occurrence of widespread BSRs in sand-dominated accretionary prisms. A similar formation mechanism likely occurs on some passive margins, except that the fluid flow results from rapid deposition of sediments which are initially underconsolidated.

3. MULTICHANNEL SEISMIC LINES AND ODP RESULTS

On regional multichannel seismic (MCS) lines collected in 1985 and 1989, a hydrate BSR was observed in a 20-30 km wide band along much of the 250 km length of the Vancouver Island continental slope (Figs. 1, 2, 3).). A BSR is generally not observed in the well-bedded slope basin sediments (Fig. 3a), but interpretations vary as to whether or not hydrate is present. Hydrate may not be formed because the well-bedded sediments reduce permeability, thus inhibiting vertical fluid and methane flow (Zuehlsdorff et al., 2000); alternatively, hydrate may be present but tectonic subsidence in the basin results in downward movement of the base of the stability field, so that the gas layer is transformed to hydrate and the BSR is much weakened (von Huene and Pecher, 1999).

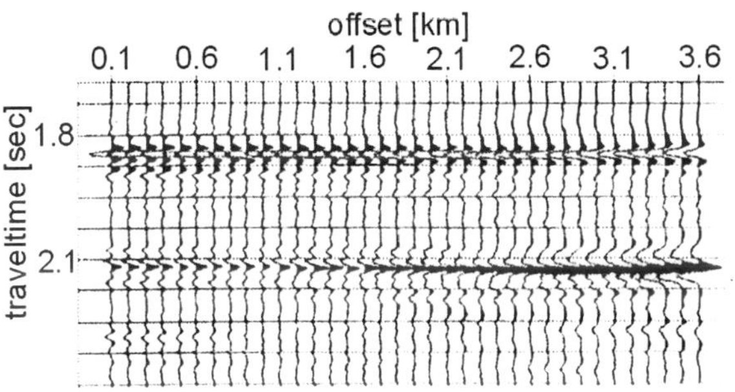

Figure 4. Common-depth-point gather after normal moveout and array directivity corrections from Line 89-08. The gather illustrates the seafloor and BSR amplitude-versus-offset (AVO) behavior.

The BSR signal is generally a single symmetrical wavelet with a reversed polarity relative to the seafloor, indicating a sharp and negative impedance contrast across the BSR (Figs. 3, 4). The seafloor reflection coefficients are typically 0.18-0.24, and the BSR reflection coefficients are about 30% those of the seafloor (Yuan et al., 1996). Careful semblance velocity analyses were carried out on sporadic reflectors down to depths of 2000 mbsf (Fig. 5a). Above the BSR, velocities increase to nearly 1900 m/s, indicating the presence of sediments containing high-velocity hydrate within the pores. However, the critical importance of these measurements is that they provide the only estimate available for velocities below the BSR and gas layer. By extrapolating these deeper velocities upwards, we thus obtain the reference velocity of sediments unaffected by either hydrate or free gas. At the BSR, hydrate produces an increase in velocity of about 250 m/s.

Downhole sonic logs from ODP Site 889 (Westbrook et al., 1994) provide detailed velocity information from about 50 metres below the seafloor (mbsf) to the BSR at 225 mbsf (Fig. 5b). Excellent agreement with the semblance velocity results was obtained. The best constraint for low velocities below the BSR, due to the presence of small quantities of gas, comes from a vertical seismic profile (VSP) at Site 889 (MacKay et al., 1994); unfortunately, these measurements are limited to a depth of only about 30 m below the BSR, and so no conclusions can be made about the thickness of the gas layer.

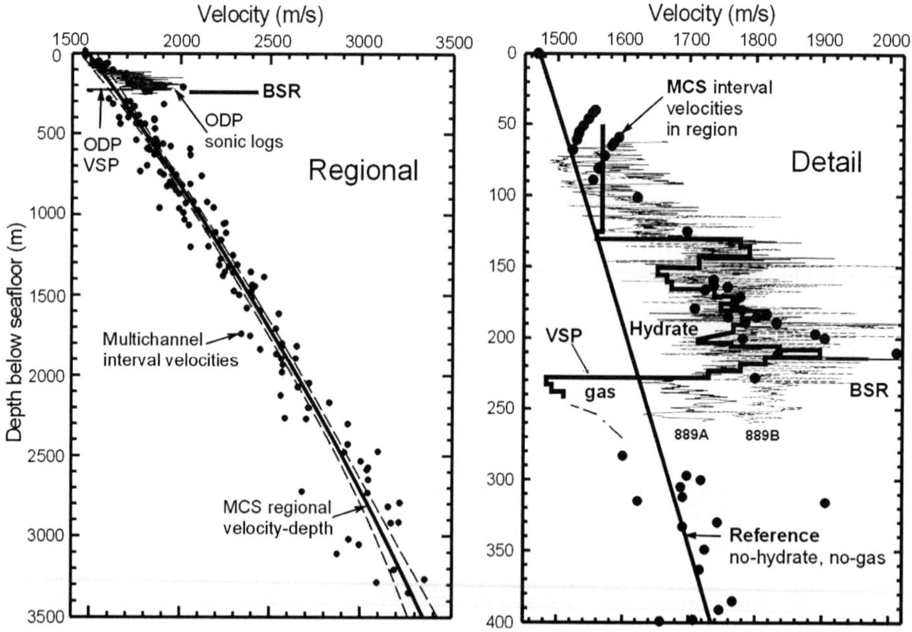

Figure 5. (a) The MCS velocities along L89-10 define the velocity trend down to depths of nearly 3000 mbsf. (b) Sonic logging and vertical seismic profile for Site 889 are shown in comparison with MCS velocities. The solid line fitting the velocities represents the reference velocity profile extrapolated from the deep trend.

Additional information on velocity structure is contained in the phase and amplitude variations with offset for a given common midpoint gather (Fig. 4). The observed amplitude behavior for the BSR, when corrected for the pronounced directivity of the airgun array, shows constant or slightly decreasing amplitudes at near-to-mid offsets and a large amplitude increase at far offsets. Modeling of the amplitude variations indicates that a P wave velocity increase is the main cause of the BSR amplitude increase at large offsets (Yuan et al., 1999). The S wave velocity in hydrate was assumed to increase in proportion to the P wave velocity increase, as may be expected if hydrate cements sediment grains; this is consistent with the increase observed on S wave dipole logs in drilling in the Beaufort Sea (Lee and Collett, 1999) and on the Blake Ridge (Guerin et al., 1999).

Subtle differences in reflection waveforms provide additional constraints on the detailed velocity profile through careful full waveform inversion (Singh and Minshull, 1994; Yuan et al., 1999). The inversion is very dependent on the starting model, which was derived from the sonic, VSP, and semblance velocity profiles. The results indicated that velocities as low as 1500 m/s occur in a 25-50 m thick layer below the BSR (Fig. 5b). Based on the reference velocity estimate at the BSR of 1650 m/s, we concluded that the BSR in this area is consistent with a model described by a velocity increase due to hydrate above the BSR and a velocity decrease due to free gas below the BSR.

4. DETAILED SEISMIC SURVEYS: LATERAL DISTRIBUTION OF HYDRATE AND FLUID FLOW

A drillhole provides only a point measurement of hydrate concentration below the seafloor. Several grids of seismic lines have been collected off Vancouver Island to map in detail the lateral distribution of hydrate or gas away from the drillhole, and to determine if localized concentrations in hydrate or gas can be explained by variation in fluid flow.

4.1 Multifrequency Single Channel Seismic Reflection Data

In 1993, single channel seismic (SCS) data were recorded near ODP Site 889/890 over a tight grid of lines at a nominal spacing of 200 m (Spence et al., 1995; Fink and Spence, 1999). Airgun shots with three distinct dominant frequencies (30, 75, and 150 Hz) were recorded along several coincident lines. Even for the highest-frequency data, the BSR appears as a single reflector. The BSR must thus be due to a single sharp interface, with gradational velocities away from the interface. Mapped over the grid of lines, BSR reflection coefficients reached localized maximum values of 0.15-0.18 beneath topographic highs. This indicates that topography provides a major control on hydrate or gas concentrations, perhaps focussing the flow of methane-bearing fluids from broad areas at depth. The BSR amplitude measurements, when converted to values of velocity above the BSR, were used to estimate hydrate concentrations. Seafloor reflection coefficients were also calculated over the grid. Maximum values of up to 0.5-0.6 were found on the flanks of topographic highs, in marked contrast to typical values of 0.2-0.3 elsewhere in the survey area. Furthermore, the seafloor reflection in these flank regions typically had a much larger amplitude for data recorded with the 75-Hz airgun source, relative to amplitudes from single-fold 30 Hz data (extracted from coincident MCS lines collected in 1989). We interpreted this seafloor amplitude character as evidence for a thin high-velocity layer, with the velocity and density of carbonate; amplitudes for the lower frequency data are reduced because the layer thickness (~2 m) is significantly less than the 30-Hz quarter-wavelength. Off Oregon, Carson et al. (1994) determined reflection coefficients near 0.7 in a region of widespread BSRs near an anticlinal ridge. They observed and sampled carbonate pavement at the seafloor, and attributed its formation to increased fluid flow in the ridge area (Ritger et al., 1987; Kulm and Suess, 1990).

4.2 Heat Flow from Depth Variations in Bottom Simulating Reflector

In 1996, several grids of short-offset multichannel seismic data were acquired in the continental slope region. Piston coring was also carried out at 18 sites, and physical property analyses (including P-wave velocity, density, resistivity and porosity) allowed ground-truth calibration of the seismically-derived seafloor reflection coefficients (Mi, 1998). Over a grid of seismic lines southeast of Site 889 (Area3 in Fig. 1), heat flow was calculated from the depth of the BSR. The regional variation of heat flow, decreasing landwards from about 80 to 65 mW/m^2, was consistent with tectonic thickening of accretionary wedge sediments (Ganguly et al., 2000). Significant local variations in heat flow

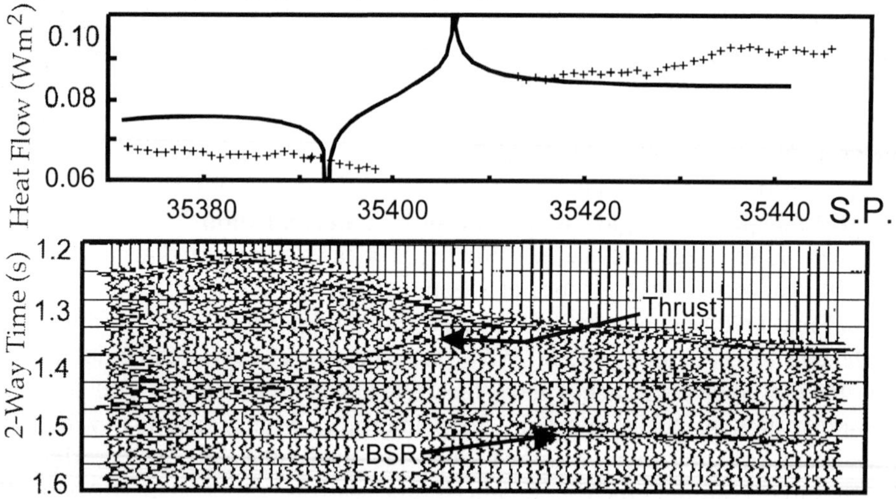

Figure 6. Part of a migrated seismic section from a short-offset multichannel line in area A3 (Fig. 1). Heat flow calculated from the BSR shows a minimum over the topographic high and a maximum on the flanks of the high. A portion of the heat flow behavior is explained by static topographic effects (solid line in *a*). Fluid flow associated with a possible thrust fault may explain the remainder.

were also observed, notably low heat flow values over topographic highs and high heat flow values over the flanks of the highs. As shown in Fig. 6, heat flow at some localities increased by as much as 50% (from 65 to 100 mW/m^2) over a horizontal distance of 1-2 km. Much of this variation may be due to the focussing and defocussing effects of topography alone; applying an analytic solution by Lachenbruch (1968) for the topographic effect, we see that about half of the landward increase in heat flow at SP 35400 (Fig. 6) can be explained by the static effects of topography. The remaining component of this heat flow variation may be produced by dynamic effects. For example, the topographic ridge may be associated with thrust faulting, as identified in Fig. 6. The fault may provide a pathway for the upward migration of warm fluids to produce localized increases in heat flow. Furthermore, regions of reduced heat flow may occur as the thrust fault brings colder near-surface sediments to greater depths.

4.3 Ocean Bottom Seismometer (OBS) Surveys

Several deployments of OBSs have been carried out near ODP Site 889 in 1993, 1997 and 1999. In 1993, wide-angle data were collected during two deployments of OBSs, recording a 120 in^3 airgun with a dominant frequency of 75 Hz (Spence et al., 1995). Simultaneous traveltime inversion of data from four instruments and from coincident normal-incidence lines was carried out to provide a 2D velocity model through the drill site (Hobro et al., 1998). Velocities directly above the BSR ranged from 1.87 to 1.96 km/s. This increase in velocity also appears to be correlated to an increase in BSR reflection strength, which suggests that there may also be a correlation between BSR reflection strength and hydrate concentration. A full 3D velocity inversion from normal-incidence traveltimes and from the full wide-angle OBS dataset has recently been completed (Hobro, 1999).

4.4 Deep Towed Seismic: Detailed Structure at the BSR

High resolution multichannel seismic data were acquired in 1997 using the Deep-Tow Acoustics/Geophysics System (DTAGS) in the vicinity of ODP Site 889/890 (Gettrust et al., 1999). The system was towed at a depth of about 1000 m in water depths averaging 1400 m. Because of its proximity to the seafloor, wide-angle reflection data may be recorded on the DTAGS streamer which has a relatively short length of 622 m. The wide angles facilitate velocity estimation and amplitude-versus-offset analysis. However, to allow velocity analysis and coherent stacking of the data, careful geometry corrections for the varying depth of the streamer were required. Consequently, a method of estimating the depths of the source and hydrophones using sea-surface reflection times was developed (Walia and Hannay, 1999). The frequency band of DTAGS was 250-650 Hz, providing a vertical resolution of ~ 2 m. As well, the near seafloor source and receiver configuration reduces the Fresnel zone size to about 25 m, significantly less than the Fresnel zone width of ~400 m for 30 Hz sea surface data.

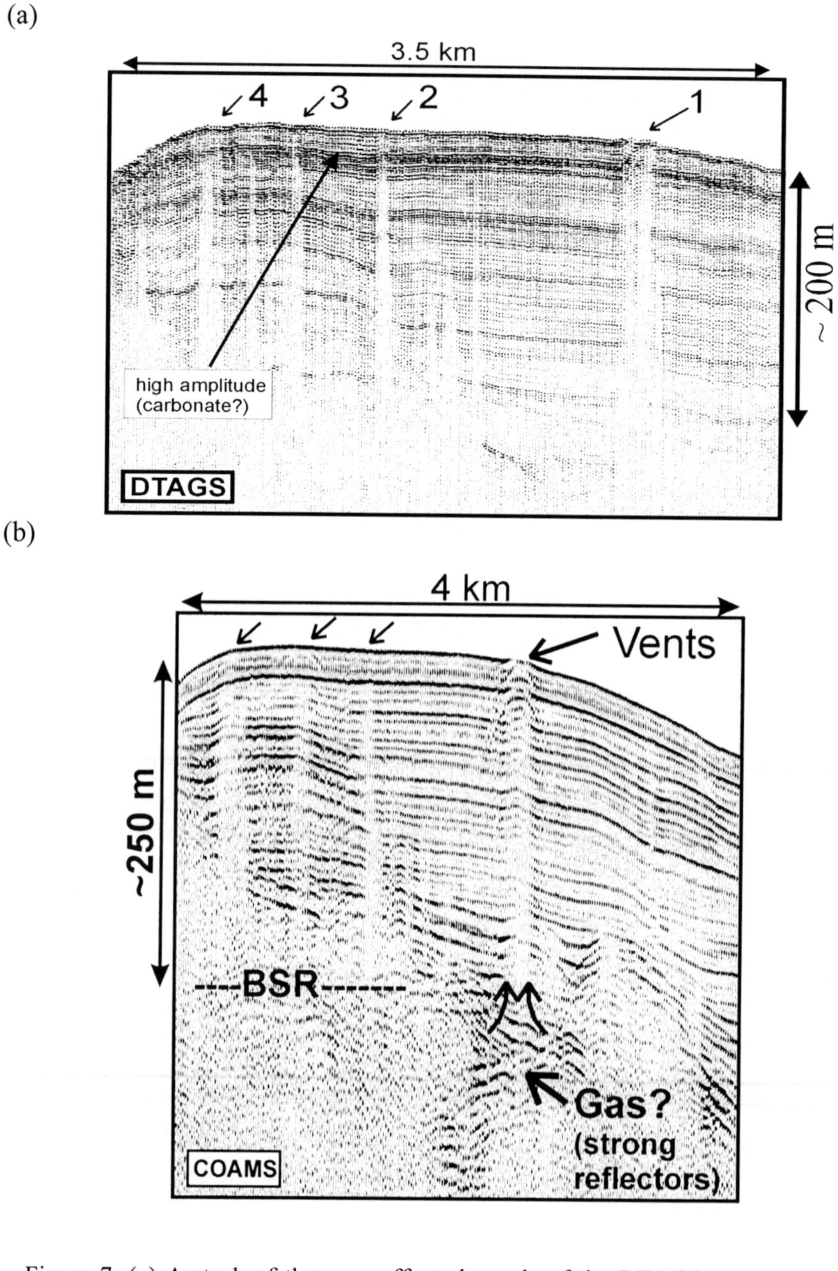

Figure 7. (a) A stack of the near-offset channels of the DTAGS array over same area showing fine structure within the uppermost 200 m of sediments. (b) Portion of an unmigrated multichannel seismic lineover same area as the DTAGS line in (a). Vent 1 is associated with a seafloor pockmark (resulting in a diffraction hyperbola). Note that the sediment horizons change dip at each of the other marked blank zones, indicating the presence of faults.

In a stack of the near-offset DTAGS data, a high amplitude reflector about 10 m below the seafloor can be resolved (Fig. 7a). This near-surface reflector may indicate a carbonate pavement precipitated from upward-advecting fluid (Fink and Spence, 1999).

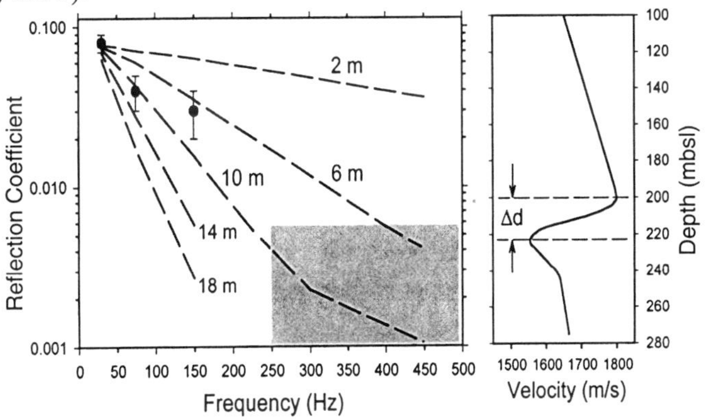

Figure 8. Reflection sections 89-08 (A) and 89-10 (B) near ODP Site 889. The BSR, representing the base of hydrate, is particularly strong in this region.

Combined with 5 previous surveys in the area, the DTAGS data enable us to determine the frequency dependence of hydrate-related reflectors over the broad band from 20-650 Hz (Chapman et al., 2000). The BSR is very strong for low frequency data, but its amplitudes become smaller for the higher frequencies and it is barely discernable at the peak frequency of the DTAGS source (Fig. 8). This behavior suggests that the BSR is produced not by a sharp impedance contrast but rather by a negative velocity gradient. The vertical scale of the gradient is much smaller than the wavelength of the low frequency data, but greater than the wavelength of the DTAGS data. We have modeled this frequency dependence of amplitude using synthetic seismograms, and we infer that the thickness of the velocity gradient layer at the BSR is 6-8 m, with velocity decreasing by 250 m/s (Fig. 8)

4.5 3D Seismic: Seafloor Venting and Possible Pathways for Fluid Flow

A 3D seismic survey was carried out in 1999 using a single 40 cu. in. airgun over a multichannel grid consisting of 40 lines spaced at 100 m plus 8 crosslines. A number of narrow vertical blank zones were observed on many of the seismic sections (Fig. 7b). The region of numerous blank zones imaged at greater detail over a single channel grid with 25 m line spacings. Similar blank zones had been observed on some of the 1997 DTAGS seismic sections, as well as on high frequency data collected by the R.V. Sonne in 1996 (Zuellsdorf et al., 2000. It was suggested that these blank zones represent fluid expulsion channels.

Figure 8. Observed variation of reflection strength with frequency (solid dots). The shaded area gives constraints from DTAGS data. The dashed lines are the modeled variations of reflection coefficient with frequency, for different thicknesses) Δd of the velocity gradient layer which represents the BSR.

The surface expression of Vent 1 (Fig. 7b) was determined in detail by 3.5 kHz sub-bottom profiling, which indicated that the structure was circular with a diameter of about 500 m. Migrated multichannel seismic data clearly show that this blank zone is associated with a seafloor pockmark, and that the polarity of the seafloor reflection changes phase within the pockmark. which indicates the presence of free gas just at or below the seafloor. Vent 1 also bounds a region of anomalously high reflectivity near and below the BSR, which is observed to some extent in Fig. 7b but is particularly prominent in the crossline direction. This high reflectivity can also be explained by the presence of free gas, trapped in sediments beneath the BSR. As observed in Fig. 7b (and in 3D time slices), most of the blank zones can be associated with changes in dip of the sediment horizons. These changes can be traced in 3D across the grid. The blank zones are thus interpreted as faults, which act as conduits for the focussed upward expulsion of fluids and methane gas.

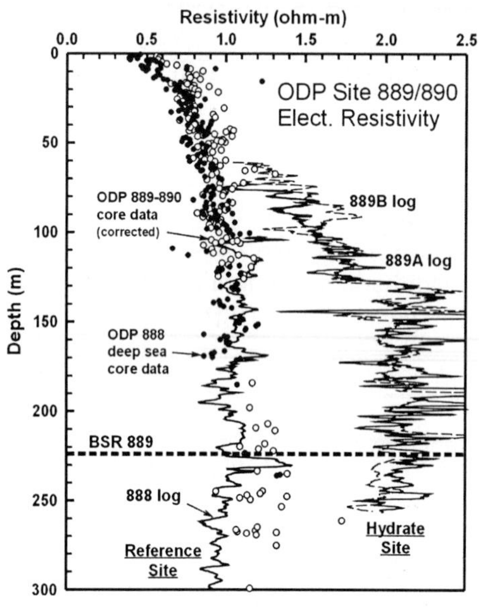

Figure 9. Downhole electrical resistivities from Site 889/890 and the reference site 888. Resistivities of the recovered core samples are also shown.

5. ELECTRICAL RESISTIVITY OF HYDRATES

Electrical resistivity is an alternative parameter that is very sensitive to gas hydrate, because electrically resistive hydrate replaces conductive seawater in the pore spaces. Thus, resistivity measurements have potential as an alternative to seismic analyses for determining hydrate concentrations in deep sea boreholes and for seafloor field mapping of hydrate occurrences.

5.1 Resistivity from ODP Downhole Logs and Core Sample Measurements

At ODP sites off Vancouver Island where a BSR is present, log resistivities have average values of about 2.0 ohm m in the 100 m interval above the BSR (Hyndman et al. 1999; Fig. 9). The abrupt downward increase in log resistivity at 130 mbsf is similar in character to the increase in log velocity (Fig. 5b), and both occur at the base of slope basin sediments overlying accretionary basin sediments. Over the same 100 m interval, the log resistivities average 1.0 ohm m at a nearby deep sea reference site where no gas hydrate is present. As discussed in section 7.2, the difference in resistivity can be used to estimate hydrate concentrations.

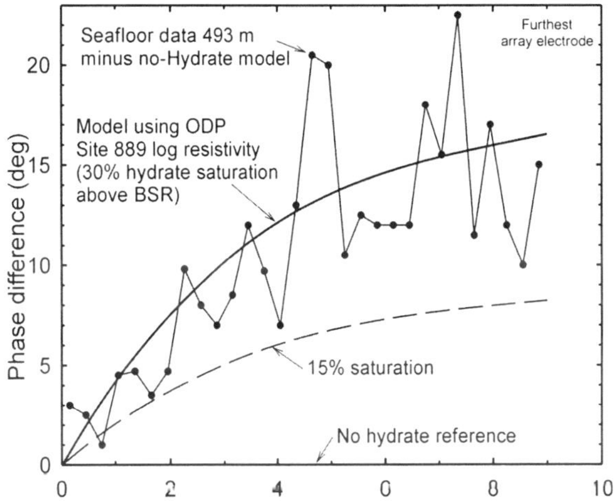

Figure 10. Results of towed seafloor EM system near Site 889. The observed phase differences (dots) best match those for a model in which resistivity above the BSR is 2.0 ohm m, just slightly less than that determined from the Site 889 resistivity logs.

5.2 Seafloor Electrical Resistivity Measurements

A seafloor dipole-dipole electrical system has been developed recently to map the hydrate (Edwards, 1997; Yuan et al., 2000), and several successful surveys have been carried out near Site 889. In this "refraction" electromagnetic method, the traveltime for an electrical signal to diffuse through sediment from a seafloor source to a seafloor receiver is related linearly to the resistivity: the higher the resistivity, the shorter the traveltime. In practice, the phase difference between the transmitted electromagnetic signal and the received signal is viewed as a function of frequency. The measured phase differences are compared to theoretical values calculated for a resistivity-depth model. The general form of the model was based on smoothed resistivity measurements at Site 889 (Fig. 9), in which the 100 m interval above the BSR has constant resistivity while resistivities above this layer decrease uniformly to the seafloor. The seafloor electromagnetic data provide a scaling factor to determine the magnitude of the resistivity relative to a reference. For the best-fit model at a source-receiver separation of 493 m (Fig. 10), resistivities near 1.8 ohm m were obtained for the layer above the BSR., which was within 10% of the log resistivities measured at Site 889.

6. SEAFLOOR COMPLIANCE MEASUREMENTS

Another independent method for estimating hydrate concentration is the measurement of seafloor compliance, since hydrate cementation increases the elastic moduli, especially the rigidity. Thus, this method is particularly sensitive to variations in seismic shear velocity. Compliance is obtained by using a seafloor gravimeter and a precision seafloor pressure gauge to measure the seafloor deformation due to ocean surface gravity waves and longer period internal waves. Successful data have been obtained from eight sites near ODP Site 889 (Willoughby & Edwards, 2000). The measured compliances increased uniformly with frequency, and the basic character of the data could be fit with a model in which seismic velocities increased logarithmically with depth. The addition of a hydrate layer with increased shear and compressional velocity above the BSR improved the fit. However, the absolute resolution of the hydrate layer is limited by the low frequencies of the gravity waves for the water depths found at Site 889. Improved relative resolution may be achieved by comparing compliances near Site 889 with measurements made at locations where no hydrate is expected.

7. ESTIMATES OF HYDRATE AND FREE GAS CONCENTRATION

We conclude this brief summary of geophysical investigations on the northern Cascadia margin by presenting results of various methods to estimate quantitatively the hydrate concentrations within the sediments. Most methods are geophysical, derived from a variety of techniques to determine seismic velocity and electrical resistivity, and these are the only means currently available for estimating concentrations away from the region of the ODP drill holes. However, for completeness, we also present estimates of hydrate concentration determined from geochemical techniques, specifically from measurements of pore fluid chlorinities in recovered core samples. In all estimation methods, we emphasize the importance of determining the reference profile for the parameter being considered, representing its value for sediment for sediment containing no hydrate and no gas.

7.1 Seismic Velocity

We have used two simple model approximations to relate the amount of hydrate concentration in the pore space to sediment velocities. One is to obtain a velocity for the combination of pure hydrate and sediment matrix (i.e., no pore fluid), and then to combine that composite matrix with water-saturated sediment to determine an overall model velocity (Yuan et al. 1996, 1999; Lee et al., 1993). For the observed increase in velocity of about 250 m/s relative to the reference velocity of 1650 m/s, this method suggests that about 10-25% of the sediment pore space is occupied by hydrate.

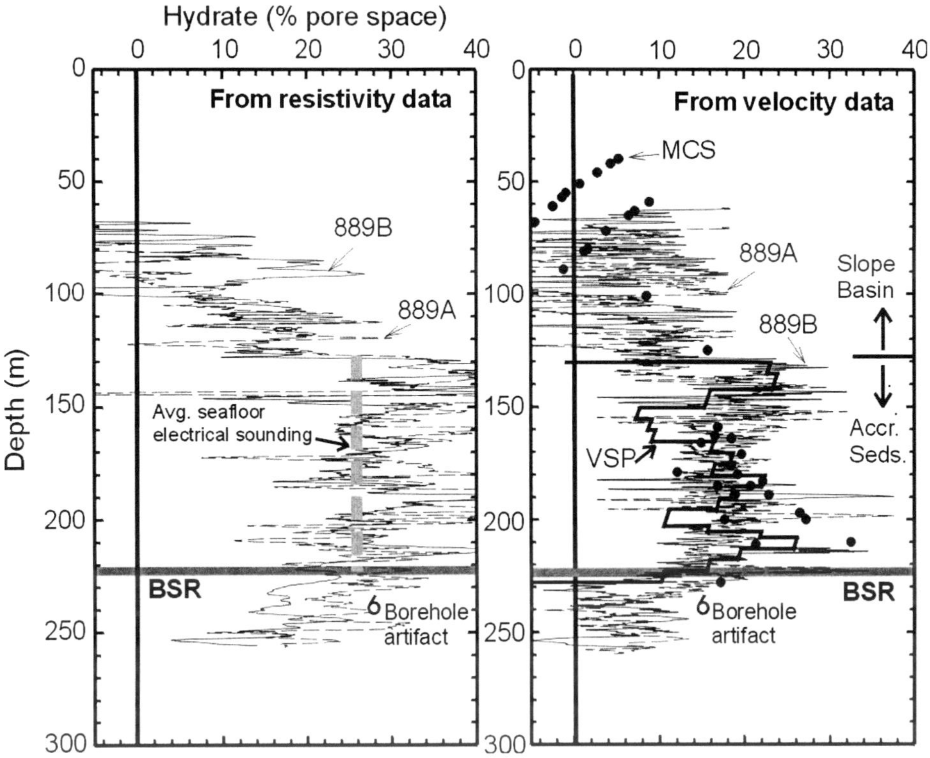

Figure 11. Hydrate concentration near ODP Site 889 estimated from (a) velocity data from sonic logs, VSP and MCS, and (b) log resistivities and core salinities.

The other method assumes that the observed velocity increase relative to the reference (Fig. 5) is due simply to a reduction in porosity as the hydrate fills the pore spaces (Yuan et al., 1996). We have used the velocity-porosity of Hyndman et al. (1993) to calculate inferred porosities and hydrate concentrations for three velocity-depth datasets near Site 889: (1) downhole velocity logs, (2) VSP velocities, and (3) MCS velocities. As shown in Fig. 11a, concentrations reach a maximum of 20-25% of the pore space above the BSR.

To determine the amount of free gas below the BSR, the major constraint is the reduction in P velocity from about1650 m/s to1400-1500 m/s, as determined from VSP data and from full waveform velocity inversions (Yuan et al. 1999). Models and laboratory measurements indicate that a reduction to <1000 m/s is produced with a gas saturation of only 5%, while further increase in saturation makes little change. Thus, a velocity reduction of only ~200 m/s must be produced by a very small quantity of gas, 1% or less of the pore space.

7.2 Electrical Resistivity

An increase in resistivity results from the formation of gas hydrate, which partially fills the pore spaces. It is primarily the saline fluids within the pores that controls the resistivity. Thus, an allowance must be made for the in situ salinity of the pore water. We have shown (Hyndman et al. 1999) that hydrate concentration and in situ pore fluid salinity may be estimated simultaneously from log resistivities and from the pore fluid salinity and porosity measurements made on recovered ODP core samples. The in situ pore fluid resistivities are calculated to be about 80% of seawater; these low salinity fluids are probably produced at greater depths by hydrate dissociation as the base of the stability field has moved upward with time. The hydrate saturation in the 100 m interval above the BSR is calculated to be 25-30% (Fig. 11b). Similar hydrate saturation values of 17-26% were obtained from seafloor electrical sounding experiments (Yuan and Edwards, 2000).

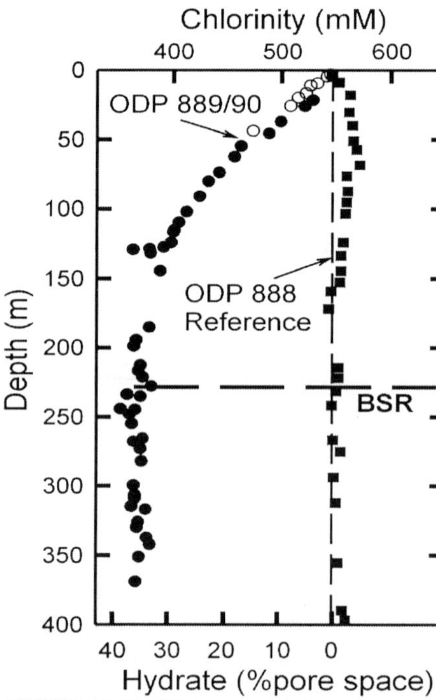

Figure 12. (a) Chloride concentration versus depth at Site 889/890. Average concentration at Site 888 represents the reference chlorinity. (b) Lower horizontal scale gives approximate concentrations of hydrate in pore spaces, estimated from chlorinity dilution.

7.3 Drill Core Chlorinity Dilution

As hydrate is formed, salt is excluded from the crystalline structure As hydrate dissociates upon recovery, the fresh water released produces a reduction in the chlorinity of the pore fluids. The amount of reduction is thus a quantitative estimate of the amount of hydrate present before dissociation. The measured chlorinities in core samples from hydrate Sites 889/890 shows a dramatic decrease with depth, relative to chlorinities at Site 888 where no hydrate was found (Fig. 12). Hydrate concentrations at Sites 889/890 estimated from this chlorinity anomaly (Yuan et al., 1996) rise to about 35% of the pore space at the BSR depth (Fig. 12).

Acknowledgements: This chapter is Geological Survey of Canada Contribution number 2000007.

Chapter 16

The Occurrence of BSRs on the Antarctic Margin

Emanuele Lodolo and Angelo Camerlenghi
Istituto Nazionale di Oceanografia e di Geofisica Sperimentale - OGS
Borgo Grotta Gigante 42/c
I-34010 Sgonico (TS) Italy

1. INTRODUCTION: INVENTORY OF BSRs DISTRIBUTION OFF ANTARCTICA

The Antarctic continent is for the most part surrounded by passive margins, except for a restricted segment along the northern termination of the Antarctic Peninsula. Extensive single-channel and multi-channel seismic reflection surveys carried out in the last two decades, have clarified many aspects of the structure and stratigraphic setting of these margins, and highlighted the importance of the seismic facies analyses for the reconstructions of ice sheet history (i.e., the ANTOSTRAT program, Cooper et al., 1995). Due to the hostile local environment, the knowledge of the Antarctic margins is still uncomplete, allowing for regions adequately covered by seismic surveys to be adjacent to unexplored areas.

The recognition of BSRs on seismic profiles from the Antarctic margin (Fig. 1) has been mostly incidental and only one specific survey has been devoted to the study of BSRs (South Shetland margin, Lodolo et al., 1998). BSRs have been reported in four areas: on the Wilkes Land margin (Kvenvolden et al., 1987; Tanahashi et al., 1994), on the South Shetland margin (Lodolo et al., 1993), on the Pacific Margin of the Antarctic Peninsula (Rebesco et al., 1996), and on the South Orkney continental block (Lonsdale, 1990). Only in two cases, the BSRs represent the base of the HSZ (South Shetland and Wilkes Land margins); the others BSRs appear most likely caused by silica diagenesis or by still unclear compaction effects.

Independently from acoustic evidence of gas hydrates and free gas, abnormally high concentrations of gaseous hydrocarbons have been revealed from core samples recovered by Deep Sea Drilling Project (DSDP) Leg 28 in the Ross Sea (McIver, 1975), the largest circum-Antarctic marginal basin,

M.D. Max (ed.), Natural Gas Hydrate in Oceanic and Permafrost Environments, 199–212.

and this probably has contributed to attract the interest of some commercial oil companies (the French Institute Francaise du Petrole, the Brazilian Petrobras, the Japanese National Oil Corporation, among others) in the extensive exploration of the continental shelves around Antarctica during the early 80's. Nowadays, there is a general consensus among Antarctic researchers that petroleum likely occurs in some of the sedimentary basins both onshore and offshore, but politics has always inhibited active exploration and in particular exploitation, as recently reaffirmed at the Madrid's «Conference for the Preservation of the Antarctic Environment» in 1995.

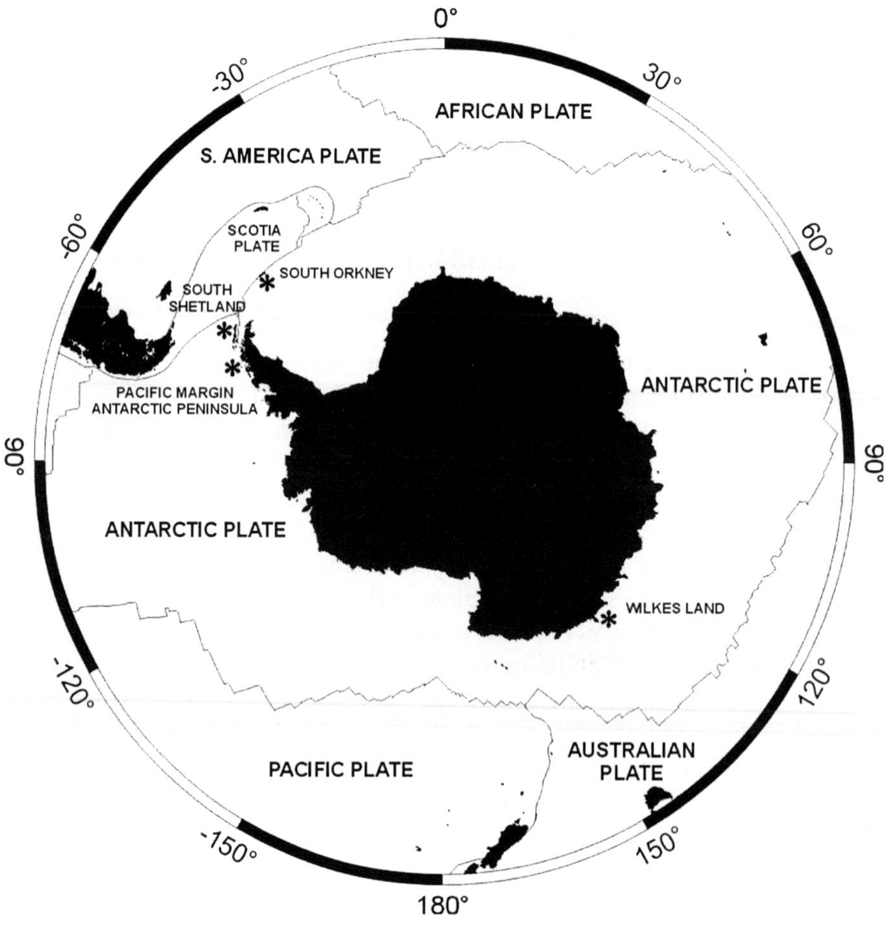

Fig. 1 – Map of the occurrence of BSR's around the Antarctic continent (asterisks indicate the areas where BSR's have been documented from seismic profiles).

2. GENERAL GEOLOGICAL SETTING OF THE AREAS WHERE BSRS HAVE BEEN FOUND

2.1. Wilkes Land margin

The Wilkes Land margin is a passive continental margin (Fig. 2) created by the mid-Cretaceous rift (Eittreim & Smith, 1987; Eittreim, 1994), which definitively separated Australia from Antarctica. Sea-floor magnetic anomalies indicate a sequence of slow, medium, and then fast spreading rates between the two continents (Cande & Mutter, 1982; Veevers, 1987). Traces of rifting events have been preserved in the sedimentary architecture of the margin and in the main unconformities bounding the stratigraphic sequences within the Cenozoic post-rift sequences (Wannesson et al., 1985; Eittreim & Smith, 1987; Tanahashi et al., 1987; Tanahashi et al., 1994).

Cenozoic margin progradation, generating slopes up to 10¡ (Eittreim et al., 1995) is widely observed in the younger sequences overlying aggradational sequences beneath the outer continental shelf, both reflecting advances of the east Antarctic ice sheet to the shelf edge (Cooper et al., 1991). Continental rise deposition is by development of deep-sea fans (trough-mouth fans) and sediment drifts. Finely stratified diatomaceous ooze and mud, reaching about 40 m in thickness on the inner shelf represent Holocene postglacial open marine deposition (Domack et al., 1991) and overlie diamicts of glacial origin. The inner shelf is incised by channels (up to 1000 m relief) carved by the ice (Domack et al., 1989). The outer shelf is characterized by isolated banks (about 200 m below sea level), probably morainal deposits, suggesting former positions of the grounding ice sheet.

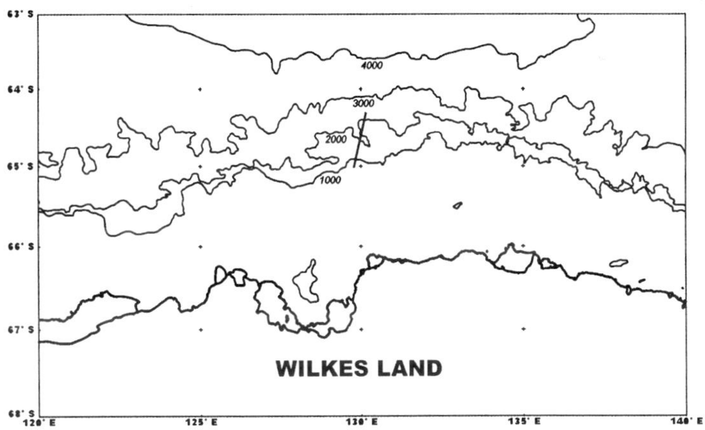

Fig. 2 – Simplified bathymetry (contours every 1000 m) of the Wilkes Land margin. The thick segment indicates the profile presented in Fig. 6.

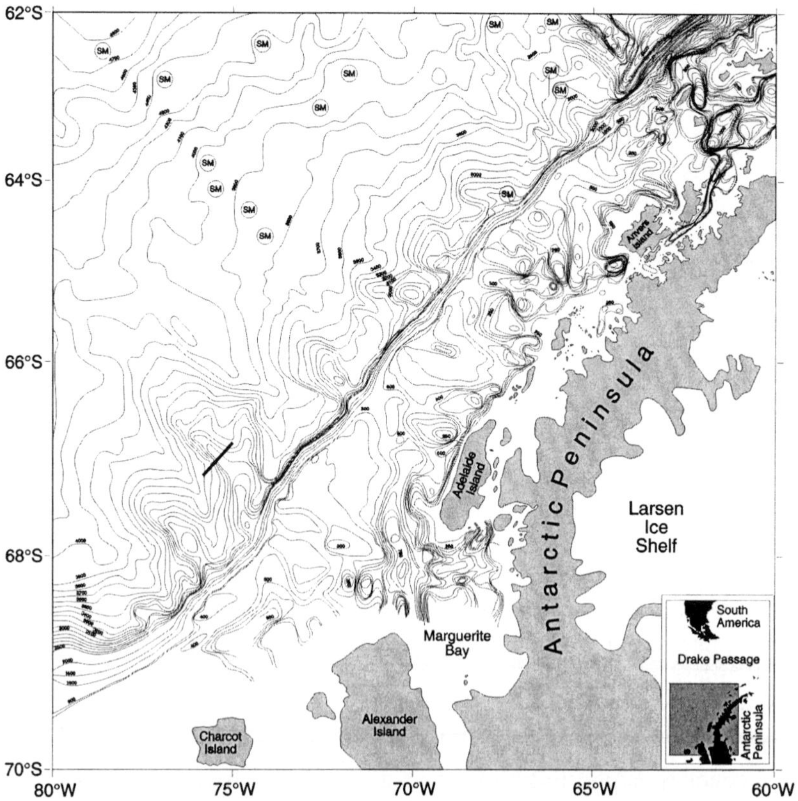

Fig. 3 – Detailed bathymetry of the Pacific margin of the Antarctic Peninsula, obtained combining underway and satellite-derived data (Rebesco et al., 1998). The thick segment indicates the part of profile in Fig. 8.

2.2. Pacific Margin of the Antarctic Peninsula

The Pacific Margin of the Antarctic Peninsula (Fig. 3) was a subducting margin throughout the Mesozoic (Pankhurst, 1990). Subduction stopped during the Cenozoic as a result of north-eastward progression of a ridge crest-trench collisions (Herron & Tucholke, 1976; Barker, 1982; Larter & Barker, 1991a), leaving behind an oceanic basement mid- to late- Cenozoic in age in which the age of magnetic anomalies becomes older moving from the margin toward the open ocean. After ridge crest-trench collision in each segment of the ocean floor delimited by fracture zones, the margin has experienced a local thermal uplift followed by the onset of a subsidence history that can be compared to that of young passive margins (Larter & Barker, 1991b).

The former fore-arc slope (most likely an accretionary wedge) has been buried by thick, mainly prograding sequences, produced by the action of ice sheets grounded to the continental shelf edge at times of glacial maxima (Larter & Barker, 1989, 1991b; Anderson et al., 1990). Margin's progradation is focused into lobate extensions of the shelf edge, separated by broad glacial troughs crossing the shelf (Rebesco et al., 1998; Abreu & Anderson, 1998, and references therein).

Large, up to 5 km wide and 150 m deep, turbidity-channels originate from a dendritic drainage pattern of small feeder channels at the base of a steep continental slope (dip range from 13¡ to 17¡; Tomlinson et al., 1992). Sediment mounds, elongated towards the abyssal plain, were identified on the upper continental rise by Rebesco et al. (1994) and McGinnis & Hayes (1995) and interpreted as asymmetric sediment drifts produced by low-energy bottom currents (Rebesco et al. 1996, 1997). These fine grained hemipelagic sediment drifts, drilled by ODP Leg 178 (Barker,.Camerlenghi, Acton, et al., 1999) are underlain by fine acoustic stratification: planar, parallel or sub-parallel, laterally continuous reflectors conformable to the seafloor on their gentler side; seafloor truncation of reflectors and diffractions associated with irregular seafloor surfaces on their steeper side.

2.3. South Shetland continental margin

The South Shetland continental margin occupies the north-eastern tip of the Pacific margin of the Antarctic Peninsula (Fig. 4), and it represents the only remnant of the former Palaeozoic to Mesozoic Gondwana's subduction margin (de Wit, 1977; Grunow et al., 1992). Because the Antarctic-Phoenix ridge, located seaward of the South Shetlands margin, ceased spreading about 4 Myr ago, present day convergence between the continental domain of the South Shetland platform and the oceanic crust of the Phoenix micro-plate is ensured by a mechanism of roll-back of the hinge of the subducting plate in response to gravitational sinking and opening of the Bransfield strait (Barker & Dalziel, 1983). The margin is delimited by the Hero Fracture Zone to the S-W and by the Shackleton Fracture Zone to the N-E. The Shackleton Fracture Zone, the South Shetland trench, and the South Scotia Ridge, form a triple junction (Klepeis & Lawver, 1996; Aldaya & Maldonado, 1996) in which continental blocks and relatively young oceanic crust become part of a structurally complex geological environment (Gamb™a & Maldonado, 1990; Maldonado et al., 1994; Kim et al., 1995).

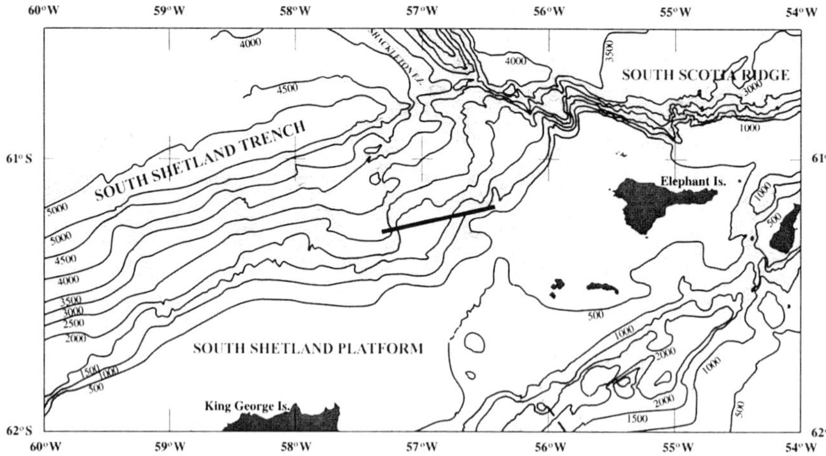

Fig. 4 – Bathymetry of the South Shetland continental margin, with location of the seismic profile presented in Fig. 7 (thick segment). Contours every 250 m. Bathymetric data taken form Klepeis and Lawver (1996).

On a regional scale, the analysis of the seismic profiles has shown the co-existence of two different tectonic styles on the margin, clearly related to distinct deformational regimes. An older regime (active since Late Paleozoic-Early Mesozoic), was responsible of the build-up of the accretionary prism as a consequence subduction of Panthalassa oceanic lithosphere. More recently during the Cenozoic, another tectonic style, superimposed on the older one, affected the margin producing mostly normal faults, which in many cases reach the sea floor. These are particularly well developed in the middle continental slope, where significant extensional basins, with more than 700 m of sediments, are present.

2.4. South Orkney microcontinent

The South Orkney microcontinent (Fig. 5) is the biggest continental element of the South Scotia Ridge (250 x 350 km), an E-W-trending submarine morphological feature along which the present-day Antarctica-Scotia plate boundary is identified. The exposed rocks on the South Orkney Island are mainly metamorphic in origin, closely resembling oucrops of metamorphic complexes of teh paleo-Pacific margin found on the Antarctic Peninsula.

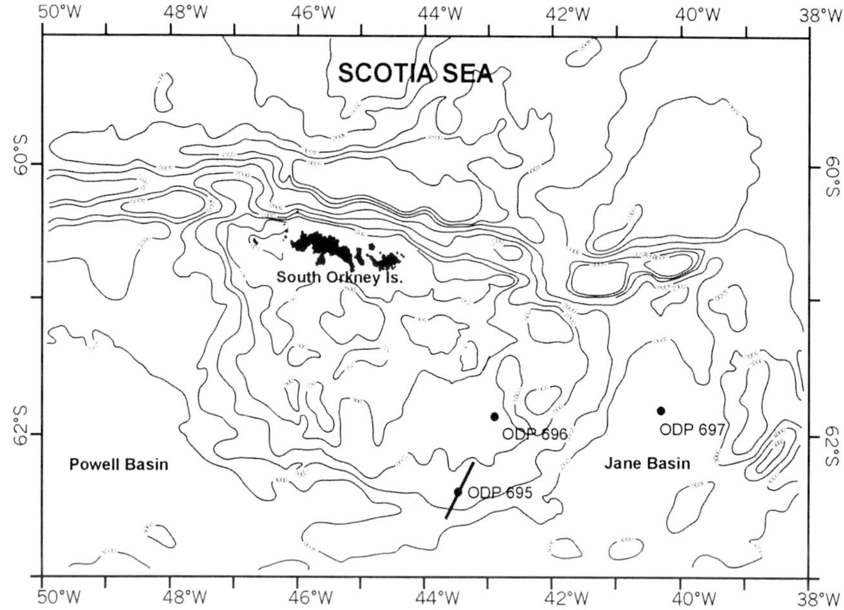

Fig. 5 – General bathymetric map of the South Orkney Islands and surroundings. The thick segment crossing the ODP site represents the part of profile in Fig. 9.

Geodynamic reconstructions have postulated that this microcontinent represents a dispersed fragment of the terrains connecting the southern South America and the northern Antarctic Peninsula, prior to the opening of the Drake Passage during Oligocene time (Barker & Burrell, 1977). It drifted away from the northern tip of the Antarctic Peninsula as a consequence of the development of the Powell Basin (Coren et al., 1997), a restricted oceanic basin presently separating the microcontinent from the northern Antarctic Peninsula. Seismic and gravity data (King & Barker, 1988; Kavoun & Vinnikovskaya, 1994; Maldonado et al., 1998), have documented the continental character of the crust for the South Orkney block, and have reconstructed the deep structure of the fragment, which presents three main N-S-trending epicontinental basins, possibly ancestral structures related to Late Mesozoic rifting.

Two sites drilled during the Ocean Drilling Program (ODP) Leg 113 (Barker, Kennett et al., 1988), on the southeast margin of the South Orkney microcontinent, have allowed the seismostratigraphic correlation with the existing grid of seismic profiles back to the Late Miocene sedimentary sequences, which are of mixed terrigenous, volcanic, and pelagic origin, with the most prominent characteristic the abundance of biosiliceous components.

3. ACOUSTIC CHARACTERS OF THE BSRs

3.1. Wilkes Land margin BSR

An anomalous acoustic reflector (BSR) was firstly observed in the Wilkes Land margin (Kvenvolden et al., 1987), on two multichannel seismic profiles acquired by the U.S. Geological Survey in 1984 (Eittreim & Smith, 1987). They cross-cut the sedimentary configuration that build up a sediment drift (Fig. 6), on water depths ranging from 1600 m to about 3000 m. These reflectors manifest most of the characteristics that suggest they could represent the base of the gas hydrate stability zone, but not the requirement that the sub-bottom depth of the reflectors increase with increasing of water depth. In fact, the depth of the reflector off of Wilkes Land decreases with increasing water depth. Some other areas however, show this similar feature, like in the Nankai Trough (Yamano et al., 1982), where the shallowing seismic reflection with water depth is attributed to oceanward-increase of heat flow. A rough distribution of the BSRs off the western Wilkes Land has been for the first time presented by Ishihara et al. (1996) (see their Fig. 16), and an example of BSR on a seismic profile is shown in their Fig. 17. BSRs in this margin are observed in water depth ranging from 2000 to 3000 m, and occur in 500 to 800 ms two-way traveltime below the seafloor.

Analyses of hydrocarbon gases in near-surface sediments of the continental margin (Kvenvolden et al., 1987), show that methane is the most abundant gas, and likely results from biologic and early diagenetic processes. Heavier hydrocarbons, probably represent terrestrial plant materials that was transported from other continents, or the compounds could be from eroded ancient sediments of Antarctica. Both methane gas and heavy hydrocarbons, however, are in low concentration in the sediments of this margin.

Data collected by the Japan National Oil Corporation in 1982, 1983 and 1990, show the presence of high-amplitude reflectors paralleling the sea-floor on across-slope lines, and one of the seismic profiles has been published (Fig. 2, Tanahashi et al., 1994). Tomographic analyses on seismic data where BSRs have been detected, are planned (Tanahasi, personal communication) with the purpose to univocally attribute the occurrence of BSRs to the base of the gas hydrate stability zone.

3.2. South Shetland margin BSR

The first evidence of BSRs along this margin was documented by Lodolo et al (1993), on multichannel seismic data acquired in 1989. In 1997, a high-resolution multichannel seismic survey was designed to purposely map the BSRs distribution and study its nature (Lodolo et al., 1998; Tinivella et al., 1998). Seismic data analysis has shown that the BSR, which bears the characteristic of reverse polarity with respect to the seafloor reflector, is widespread in NE sector of the South Shetland margin, and is present in water depths ranging from 1000 to 4800 m; sub-bottom two-way travel time of the BSR generally varies from 500 ms at the shallowest water depth surveyed to 900 ms at a depth of 4800 m in the vicinity of the South Shetland trench. The BSR occurs in that part of the margin characterised by a very complex geological setting (see above).

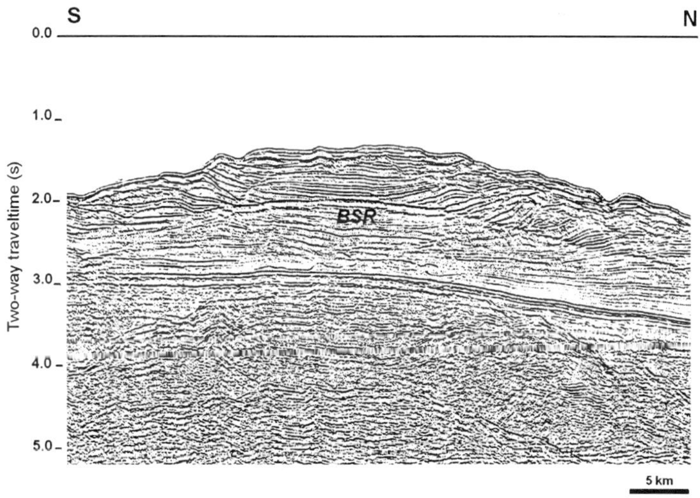

Fig. 6 – Part of multichannel seismic profile (unmigrated) acquired on the Wilkes Land margin showing the presence of a prominent BSR cross-cutting a possible sediment drift (data from the SCAR Antarctic Seismic Data Library System). Location in Fig. 2.

The BSR is superimposed on the sedimentary configuration of the margin (Fig. 7) but it shows a significant difference in strength and lateral continuity in the surveyed area. In areas where the surface geology is dominated by folding, the BSR strength is low, but lateral continuity is preserved. This suggests that the distribution of temperature and pressure has reached an equilibrium to allow for a smoother base of the hydrate stability zone. In areas where faults of the superimposed extensional tectonic regime cut across older compressional structures, the BSR is stronger but less continuous and apparently offset by such faults. A structural discontinuity oriented North to South seems to separate the two zones of BSR with different strength. It is therefore suggested that a direct relationship exists between abundance of gas hydrates and structural complexity, which may in some ways influence the movements of fluids responsible of free gas accumulations and gas hydrate formation.

Reflection tomography techniques applied to some profiles from this region (see Carrion et al., 1993, for a rigorous description of the tomographic method), indicate a velocity trend from the seafloor to the BSR generally consistent with that of normally compacted marine sediments, with an abrupt decrement between the BSR and the underlying normal polarity reflector indicating presence of free gas in the sediment pore spaces (Base of Gas Reflector, BGR, B_hm et al., 1995). The calculated thickness of this gas-bearing layer is approximately 50 m. Conversely, local increments of compressional wave velocity above the BSR can be related either to cementation of normally compacted slope basin sediments by gas hydrate or to overcompaction in accreted sediments (Tinivella et al., 1998).

3.2. Pacific margin of the Antarctic Peninsula BSR

A prominent BSR has been described by Rebesco et al. (1997) on the steeper side of the continental rise sediment drifts, where it is rendered more visible than on the gentler side by the unconformity between seafloor and deeper reflectors (Fig. 8).

The BSR is found consistently at a depth of 600-700 ms twt, approximately coincident with the lower boundary of the 'drift maintenance' depositional stage. The reflector should be regarded as a surface of sharp change of sediment reflectivity, which becomes weaker below. The BSR has not been interpreted as a gas hydrate related reflection by Rebesco et al.(1997) because of the following arguments: 1) The reflector lies too deep (600-800 mbelow seafloor instead of expected 300-500 m) if heat flow is in the normal range for ocean floor of the age that underlies the sediment drifts where it occurs (about 33 Ma). Preliminary observations of the geothermal gradient on this site performed during ODP Leg 178 drilling operations (Barker, Camerlenghi, Acton et al., 1999), confirm this initial statement (Barker, personal communication). 2) The reflector does not show a reverse polarity with respect to the seafloor reflection, 3) The reflector is not associated to a blanking zone above.

During the site survey data analysis prior to drilling ODP Leg 178, a velocity analysis using the same method of acoustic tomography as in Tinivella et al. (1998), revealed that there is no velocity inversion immediately below the BSR, while the velocity profile above the BSR does not show any positive anomaly with respect to the velocity curve of a normally consolidated sediment (Tinivella & Camerlenghi, unpublished

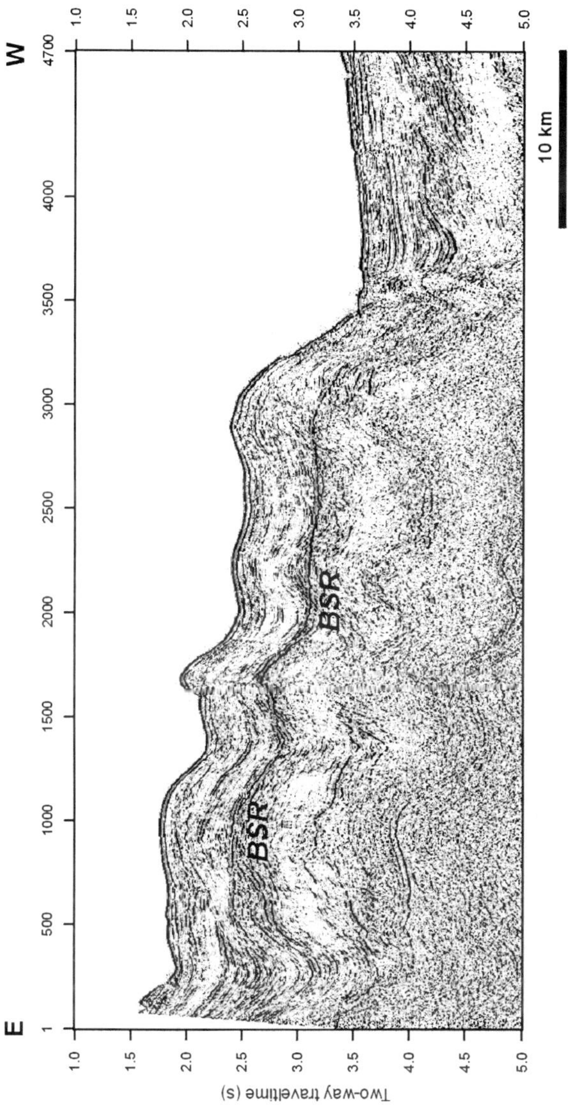

Fig. 7 – Migrated seismic profile acquired along the South Shetland margin, where the BSR strength is particularly pronounced. It crosses a sedimentary anticline without loss in amplitude. Location in Fig. 4.

data). Therefore the seismic data do not suggest any presence of gas in either form above and below the BSR. Additionally, drilling to a few tens of meters above the expected depth of the BSR during ODP Leg 178 did not provide relevant shows of gas, while it confirmed the abundance of siliceous microfossils (diatoms and radiolarians in particular) in the form of opal-A. It is concluded that the BSR in the sediment drifts of the Pacific margin of the Antarctic Peninsula is produced by a diagenetic transition opal-A to -CT.

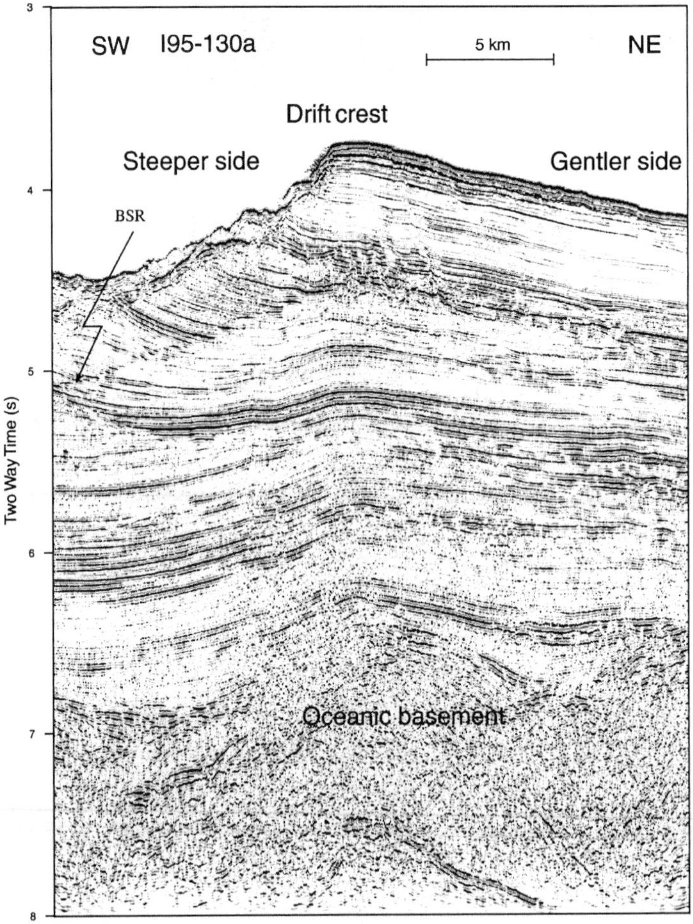

Fig. 8 – Segment of multichannel seismic profile acquired across a prominent sediment drift off the western Antarctic Peninsula margin, where a BSR was identified. Data kindly provided by M. Rebesco, unpublisjed section. Location in Fig. 3.

3.4. South Orkney microcontinent BSR

Seismic data acquired in 1985 over the southeastern continental shelf and margin of the South Orkney microcontinent as a site survey for ODP Leg 113 (Barker et al., 1990), show a high-amplitude reflection lying at sub-bottom two-way traveltime of 500-800 ms (Fig. 9). The widespread cause of the reflection was interpreted as a break-up unconformity associated with the 25-30 Ma opening of the Jane Basin to the east (Barker et al., 1984). In places, the reflector cuts across beddings, and in this case this physical boundary may be either depositional or of secondary origin related to the diagenesis of biogenic silica, possibly combined with a major variation of the detrital input (Lonsdale, 1990). It is in fact well known that the cementation of sediments that occurs with the opal-A to opal-CT transformation, causes an increase in both density and acoustic velocity, producing a downward increase in the acoustic impedance.

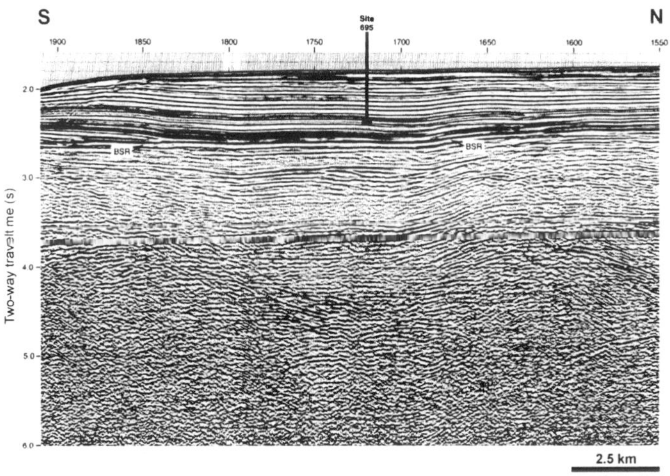

Fig. 9 – Part of the multichannel seismic profile acquired on the South Orkney platform during the ODP site survey (Barker et al., 1990), where a supposed BSR was detected. Location in Fig. 5.

4. POTENTIAL OF THE ANTARCTIC MARGINS FOR HYDRATES FORMATION

A peculiarity of Antarctica relative to other continents is the fact that the source of terrigenous organic carbon (terrestrial vegetation) was eliminated as the climate cooled and ice sheets began to form. However, this cooling was generally accompanied by an increased flux of marine biogenic sediments to the sea-floor. Presently the continental shelves of Antarctica are draped with sediments which have relatively high concentrations of siliceous biogenic material (Anderson et al., 1983), and the content of organic carbon is typically in the range of 1 to 3% (Dunbar et al., 1985). The continental rise sediments, however, contain a much lower amount of organic carbon, as little as 0.1% (Shipboard Scientific party, 1999) owing to a greater dilution of the biogenic component by terrigenous particles.

Extensive scientific seismic exploration surveys have accumulated considerable information to define the geological setting of most of the circum-Antarctic basins (McIver, 1975; Ivanhoe, 1980; Mitchell & Tinker, 1980; Cameron, 1981; Behrendt, 1983b; Aleyeva & Kucheruk, 1985; Davey, 1985; Ivanov, 1985; St. John, 1984; Elliot, 1988; Cook & Davey, 1990; Collen & Barrett, 1990; Anderson et al., 1990, among others).

Scientific drilling has demonstrated that biogenic methane forms in situ in Late Cenozoic deep sea sediments of the Antarctic Offshore, although no data are available on in situ methane saturation with respect to the stability field of methane hydrates. In general, the action of ice sheets on the marine deposition is that of producing a large quantity of unsorted sediments on the continental shelf, in which a very fine grained matrix may become volumetrically important. Downslope re-distribution of sediments tends to sort the material in a proximal mud-dominated slope-rise environment (debris-flows, contourites, and channel-levee deposits), and a coarser grained more distal environment dominated by turbidites.

It could be inferred therefore that hydrocarbon gases formed in situ on the proximal continental margin do not find stratigraphic paths to migrate and concentrate in sufficient proportion to generate a BSR. In addition, the limited tectonic activity of the Antarctic margins, mainly old passive margins, does not create preferential structural paths (along deep faults) to allow deep gas to migrate in solution and concentrate at shallower depth within the hydrate stability zone. In particular, because the seismic appearance of a hydrates (the BSR) depends on the presence of a free gas bearing sediment layer below the hydrate stability zone, a widespread presence of hydrates only on the Antarctic margin would not be in conflict with the few observations of BSRs.

In many other places of the world's margin however, BSRs are found in very fine grained sedimentary successions and with limited or absent deep folding (such as the Blake Ridge as an example). More reasonably, therefore, we are inclined to conclude that the knowledge of the margin is still inadequate in terms of coverage, and adequacy of seismic investigastions to elaborate a firm and definitive statement on the occurrence of gas hydrates along the Antarctic margin.

Chapter 17

Gas Hydrate Potential of the Indian Sector of the NE Arabian Sea and Northern Indian Ocean

Michael D. Max
MDS Research
Suite 302, 1211 Connecticut Ave. NW
Washington DC 20036

1. INTRODUCTION

There is an increasing gap between the demand for natural gas and its availability in India, which is endowed with only limited conventional methane deposits. In 1997 India produced about 70% of its own methane. Today India produces less than 50% of its own methane. In 2005 India it is anticipated that India will produce no more than 36% and by 2010 no more than about 25% of its own methane demands unless new indigenous sources of methane can be identified. Presently the production of gas in India is around 58 million m^3/day, and demand is likely to expand to about 285 million m^3/day. As there is thought not to be a high likelihood finding new conventional methane sources in the foreseeable future, India will have to either considerably scale back plans for industrialization and suppress consumer demand or meet its energy requirements from some other source, such as nuclear energy. India could also import methane or develop indigenous methane from unconventional sources, such as: (i) Coal-bed methane; (ii) Gas hydrate; (iii) In-situ coal gasification.

Exploratory assessment of the likelihood of hydrate in the Indian sea-area by the Oil & Natural Gas Corporation (ONGC) and the National Institute of Oceanography (NIO) established interest in gas hydrate. The topic of Indian hydrate resources was discussed during the Indian Geophysical Union (IGU) conference held in December 1995 at the National Gas Research Institute (NGRI), Hyderabad, which was attended by many senior Indian geoscientists and engineers. An expert hydrate committee was established by the Secretary, Ministry of Petroleum & Natural Gas in 1996 to recommend the steps to be undertaken for the identification and possible development of gas hydrate resource. As recommended by the expert committee, a National Gas Hydrate Program (NGHP) was established. The Gas Authority of India, Ltd. (GAIL) was tasked as the lead agency for coordinating the initial phase of research and

M.D. Max (ed.), Natural Gas Hydrate in Oceanic and Permafrost Environments, 213–224.

efforts of other Indian governmental and research/industrial activities. A wide range of international hydrate experts were consulted.

Seafloor hydrate is currently subject to intensive research and development studies as part of a national program headed by government agencies. India, being in short supply of natural gas and with a growing economy, is making an all-out effort to evaluate the potential of the vast hydrate energy resource. It is hoped that methane from hydrate could help meet the burgeoning demand for safe energy in India.

India has a very large sea area in which hydrate may occur, although much of it is underlain by deep water in excess of 3,000m (Fig. 1). The Indian national sea area is very large as it extends not only offshore India, but across the Bay of Bengal to link with the Island arc in the Indian possessions centered on the Andaman and Nicobar Islands (Fig. 1). In addition, because of the almost unique tectonosedimentary situation of India and the adjacent northern Indian Ocean, the likelihood of methane generation in thick seafloor sediments and the capture of this methane in hydrate is excellent.

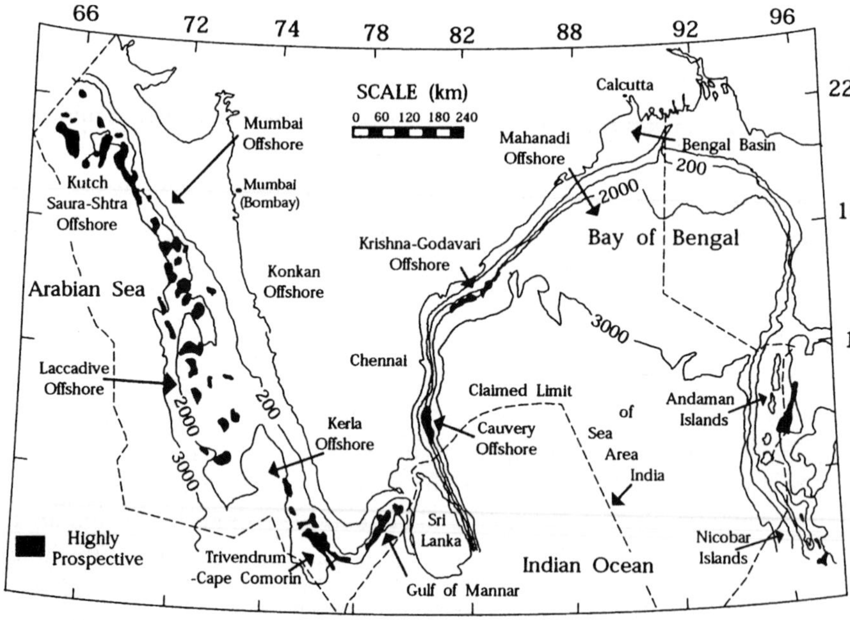

Figure 1. General location map, 200, 2000, and 3000 m bathymetric contours, delineation of the Indian sea area (dashed), and areas of prospective hydrate deposits (black areas) in which hydrate has been identified from seismic interpretation. East longitude, north latitude.

2. GEOLOGICAL FRAMEWORK

India is a cratonic shield area that has drifted northeast from an original position near the SW coast of Africa and Madagascar since the break-up of Pangea at the end of the Paleozoic. The northward overall movement of the oceanic plates in the Indian Ocean have driven the Indian craton into the Asian cratonic landmass and thrust it beneath, tectonically elevating the collision zone and the underplated zone and creating the highest mountains on Earth, the Himalayan mountains, that pass in an arc across the northern margin of India. Erosion of the mountains is extreme and the bulk of the sediments are carried to the south. Most of the major rivers, including the Ganges and Brahmaputra, flow into the Bay of Bengal from the Himalayan Mountains. The Indus River system flows into the Arabian Sea. Both the mouths of these major systems have formed huge deltas and submarine sedimentary fans which are the repository of these Himalayan sediments (Wetzel, 1993).

It is believed that the Indian Plate collided with the Eurasian Plate at about 45 Ma. Major rivers which had started flowing from the Tibetan side toward the south became antecedent rivers flowing across the now lofty Himalayas. Once the collision began and the Himalayas rose, the drainage area of the stream system increased dramatically and the sediments carried by the Brahmaputra system entered the Bay of Bengal. Major clastic wedges began to prograde into the Bengal Basin. Significant volumes of sediments began to bypass the delta and, after merging with the deltas of the Ganges, contributed to the growth of the Bengal deep sea fan. Turbidity currents carried the increased sediment load into deeper waters (Lindsay et al., 1991) as they did in the Indus Fan in the northern Arabian Sea (Kenyon, 1987). The Indus fan appears to contain less sediment volume than the Ganges fan.

The two coasts of India are superficially similar passive margin depositional environments, but are tectonically different in detail. The deep water offshore basins along the east and west coast of India both evolved through a rift and then drift phase in a passive divergent framework (Naini and Talwani, 1982) associated with the northward drift of the Indian ocean plate (Biswas and Singh, 1991). The initial phase of rifting formed a system of NNW-SSE horsts and grabens lying parallel to the western continental margin along a NE-SW trend about parallel to the east coast of India. In the post rift stage, massive carbonates were deposited along the west coast while a series of step faults down throwing toward the basins resulted in growth faults and shale diaprism along the east coast of India. After the mid-Miocene, at which time there was a major tectonic episode and uplift, large scale complexes of prograding clastics began to be deposited along both the east and west Indian paleoshelf margins. There is still active uplift in the Himalayas, which yields heavy sedimentation.

The Andaman-Nicobar region is tectonically very different from the coastal areas of mainland India . This region is associated with an east-dipping subduction margin and island arc system at the junction between the eastern Indian Ocean plate and the complex of plates in the SW pacific. This island arc-plate margin system is the eastward continuation of the Himalayan (Tethyian) confrontation zone to the north of India, which continues from the SW Pacific Ocean north of Australia and along a number of tectonic zones into Europe in the west. Whereas in the passive margin area tectonic folding of the upper several kilometers, and usually more, of the sediment is absent, active tectonism in the Andaman-Nicobar area has formed large folds that provide the possibility of complex hydrate-geological gas traps (Max and Lowrie, 1996; Max and Chandra, 1998).

The continental shelf along the west coast of India is much more broad than the eastern shelf. In addition, the water depth increases sharply at the eastern shelf edge up to water depth of 3,000m along the east coast whereas the western shelf margin is geographically complex owing to presence of islands and bathymetric reentrants. The western margin of the Andaman-Nicobar island arc is sharply defined bathymetrically, but less steep than the eastern, and the northern part of the western Indian continental slope. The Andaman sea itself is a back-arc basin with relatively complex bathymetry reflecting complex internal structure and sedimentation.

3. SEDIMENTATION

In the Bay of Bengal and the Arabian Sea, generation, migration, and accumulation of hydrocarbons commenced with rapid sedimentation related to the orogenic activity marked by formation of the present-day Himalayan Mountains. Thick sediments have been deposited over 3,000 miles from the major river mouths. Organic matter accumulated within the sediments throughout sedimentation, although preservation varies with depth, reflecting tectonic activity, sediments supply, climatic variations, and depositional rates in particular geological strata. A few deep wells have been drilled in relatively shallow water. Sediment thickness data (Fuloria, 1993) shows the thickening from the shield margin into the ocean basins.

3.1. Northern Arabian Sea and Bay of Bengal

The total sediment thickness of the Bay of Bengal and the northern Arabian Sea can conveniently be divided into pre-collision (pre-Eocene) and post-collision (post-Paleocene) sedimentary sections (Curray, 1994; Fuloria, 1993). The younger sedimentary sequence represents the Bengal Fan, which began receiving sediments during the Eocene. The rate of sediment deposition may well have accelerated in the Early Miocene (Curray, 1991; 1994).

The sub-sea Bengal Fan is the largest in the world and has had a high rate of sedimentation. Seismic records (Table 1) in the Bay of Bengal indicate that the maximum thickness of post-collision sediments, which occurs beneath the Bangaladesh shelf, is about 16.5 km. The upper or Neogene section is interpreted as being part of the Bengal Delta, while the lower or Paleogene part is believed to be Bengal Fan that was deposited in deeper water. The 6.1 km of material lying below the Bengal Fan sediments are interpreted as being metamorphosed because ot the pressures and temperatures resulting from the thick overburden (Curray, 1991; 1994). Where this oldest unit is deeper than about 12 km beneath the sea floor, it may have been subjected to low levels of prograde metamorphism that would have 'cooked' and driven off biogenic byproducts such as oil and gas and dramatically dewatered and recrystallized the low-grade metasediment.

The Bengal Fan thins gradually to the south (Rastogi et al., 1999) and it is likely that a conventional proximal to distal pattern in the sediments exists. In the Bay of Bengal, along longitude 85°E, the sediment thickness reduces drastically owing to the presence of a linear basement paleo-high, with about 2 km thick sediments over it for a length of over 600 km, and over 8-km-thick sediments on its west and east (Fig. 1). East of this basement high in the south in the Bay of Bengal, at about latitude 5°N, the thickness of the distal Bengal Fan reduces to between 2 to 4 km.

The geothermal gradients from the deep wells drilled for the exploration of hydrocarbons in the continental shelf areas of India ranges from 2.5 to 4.3 degrees C per 100 m (NGRI, 1996). This is a somewhat higher range of geothermal gradient than is encountered underlain by oceanic crust in the northern Indian Ocean. The sedimentation rate of Pliocene to Recent sediments varies in different basins. It is about 47 cm/1,000yr in the Krishna-Godavari offshore and 13 cm/1,000yr in Kerala. The average geothermal gradient varies between 3-4 degrees C/ 100 m in the Indian offshore, with maximum values being in Kutch area. The geothermal gradient in the Bay of Bengal is on the order of 2 °C per 100 m, which is highly favorable for the formation of a thick hydrate stability zone.

The predicted heat flow values were computed using different procedures based on the genesis of the crustal rocks, such as heat flow versus crustal age relationship for regions of oceanic and hot spot derived crust (Stein and Stein, 1993), the heat flow versus crustal stretching factor relationship (Mckenzie, 1978) for the regions underlain by continental crust.

The organic carbon content of recent sediments varies between 0.14 and 6.18% along the west coast in different water depths (Poropkari et al, 1993). Off the east coast it varies between 0. 5% to 2.1% (Prabhakar and Kumar, 1992). Thus, there is an excellent probability for gas generation formation in sediments below water depths between 1,000-3,000m. The geothermal gradient data (Wright and Louden, 1989; Panda, 1994) have been compiled with geothermal gradient values derived from predictions (Bhattacharya et al., 1997).

3.2. Andaman & Nicobar convergent margin

Andaman-Nicobar offshore, a convergent margin, formed along the eastern subducting margin of the Indian oceanic plate beneath the SE Asian plate. A complex series of Neogene folds zones progressively underthrust from east to west in the N-S Trend. The area around the south Andaman in the fore-arc region was subjected to maximum subsidence in the Neogene. The rate of sedimentation varies between 4-11 cm/1,000yr. The average geothermal gradient is about 1.2-1.8 degrees C/100m, which has been computed using a regression of heat flow and assumed crustal ages for the Andaman Sea region.

The thick Neogene sediments contain a high proportion of unoxidized organic matter but are immature, suggesting that they would be good sources for biogenic methane.

4. HYDRATE FORMATION AND GAS CONCENTRATION

The presence of methane hydrate depends largely upon two key conditions. Firstly, are the pressure-temperature conditions suitable for the formation of hydrate and secondly, can enough methane be produced so as to provide for a strongly positive methane flux that can reach the hydrate stability zone in the upper part of the marine sediments where the methane can be sequestered within hydrate. In addition, inferential evidence is also important. Commercial accumulation of oil and gas in the shelf areas is well established, which can be used to infer formation, migration, and accumulation of hydrocarbons from lateral continuations of the same sediments in the offshore areas.

Virtually all the sediments derived from the Himalayan mountain belt, along with a great amount of biogenic detritus formed from the often lushly vegetative lowlands between the mountains and the marine depositional environment, generally have been deposited together rapidly. These conditions are optimal for the generation of in-situ biogenic methane in the buried sediments. The organic-rich matter within the thick sediments after burial is commonly converted to methane by the action of microorganisms using it as a energy source and generating methane as a by-product.

Gas hydrate is not stable in seafloor sediments of the northern Indian Ocean in water depths less than about 600m water depth. Sediments in both shelves are too shallow for hydrate to form, but both, especially the eastern shelf and upper slope, show good free gas flows and seafloor features such as pockmarks which strongly disturb the seafloor sediments. A high-resolution sparker survey which exhibits widespread acoustic wipe-outs has been carried out in the continental shelves along the western margin of the Bay of Bengal and the Arabian Sea. The wipe-outs represent gassy, chaotic sediments below the top sea-bottom layer (Karisiddaih et al., 1993). Such gas mixed sediments have been observed extensively in the nearshore seabed of the Godavari Basin. Other wipe-outs observed in greater than 500m water depths are mostly located over the faulted continental slope and are probably due to the escape of gas from the subsurface. Thus there is good evidence for strong production of methane from the sediments.

The deep-water areas of India cover the Bay of Bengal and the Arabian Sea from around 600m (6 MPa at about 9°C) to the margin of the jurisdiction of India, to a maximum water depth of approximately 3,500m (35 MPa at about 1°C). This pressure-temperature range is suitable for the formation of methane hydrate over very large areas of sediment that demonstrably is strongly methane productive. India is now undergoing continuation of a post-glacial Holocene transgression which started around 12,500 years ago and tends to stabilize seafloor hydrate. The HSZ lies is between 200-400m thick in western offshore and 400-600m thick in eastern offshore and HSZ varies with heatflow (Rao, 1999).

Where hydrate forms a relatively impermeable layer owing to its replacing sufficient water in porous sediments, a seal can form to the upward movement of methane arriving from depth and can cause gas deposits to form (Dillon et al., 1998; Max and Chandra, 1998). Methane is trapped in the first instance within the hydrate itself, and as subjacent gas where suitable closures occur in which trapped gas can pond (Dillon and Paull, 1983).

Hydrate recycling by the upward migration of the HSZ with continued sedimentation has been advocated as a mechanism for long-term concentration of hydrate (Max and Lowrie, 1996). As sedimentation continues on the sea floor and the underlying sediments subside, the base of the HSZ migrates upward. If hydrate is present, it will dissociate into free This gas can then be reintroduced into the HSZ above by diffusion or migration through fractures. The result of this process is that methane may be recycled between gas below the HSZ and hydrate a number of times.

5. SEISMIC DATA: THE PRIMARY PRELIMINARY METHOD FOR IDENTIFICATION OF HYDRATE SYSTEM DEPOSITS

Over 178,000 line km of seismic data was inspected as part of the preliminary assessment (Table 1).

Institution/ Organization	Data Type	line Km		Total
		Arabian Sea	Bay of Bengal	
ONGC	M	65000	37000	102000
NIO, GOA	M	5000	3000	8000
DGH	M	-	10000	10000
LDEO[1]	S/M	10000	5500	15500
NOAA[2]	S/M	5000	5000	10000
U.S. NAVY	S/M	7500	-	7500
SCRIPPS[3]	S/M	3000	20000	23000
Texas A&M	S/M	2000		2000
Total		97500	80500	178000

Table 1. Table of assessed seismic data. S, single channel seismics, M, Multi channel seismics. 1. Lamont-Doherty Earth Observatory, 2. National Oceanic and Atmospheric Administration, 3. Institute of Oceanography

Starting with the International Indian Expedition in the late 1950s, a large data base of single channel seismic data (SCS) has been acquired in the offshore regions of India (Sehgal et al, 1991). These data sets and some multi-channel seismic data were originally acquired for studies of the structure and tectonics of the continental margins of India and the adjoining deep-sea basins of Arabian sea and the Bay of Bengal. This data, both new marine scientific research (Yuan et al., 1998), seismic data commissioned as part of the gas hydrate assessment, and higher quality commercial data, has been examined for evidence of hydrate and locally reprocessed for interval velocity and precise velocity structure including velocity reversals where gas zones are suspected.

6. AREAS OF HYDRATE LIKELIHOOD

After studying the sediment distribution, bathymetry, hydrothermal data, geothermal data, and other parameters in potential hydrate areas of India, broad estimates of gas hydrate resources have been made with the aim of outlining and grading areas in the Bay of Bengal and the Arabian Sea where gas hydrates may occur. A Geographic Information System (GIS) has been initiated as an archive for hydrate research (Bhattacharya et al., 1997), which can be refined & updated as necessary. Data includes: bathymetry, swath bathymetry (Bhattacharya et. al, 1994), seabed temperature, geothermal gradients, and other geoscientific data such as rate of sedimentation, structural features and sediment distribution, sediment thermal conductivity, TOC, etc. Maps of HSZ thickness have been produced using the analytical approach of Miles (1995) using bathymetry, seabed temperature and geothermal gradient data.

The oceanic areas are around 2.5 million km^2 up to the southern margin of the Indian sea area in the Bay of Bengal and Arabian Sea and have been divided into seven blocks or plays (not shown here) and estimates have been made on the basis of likelihood of prospectively (Table 2). The first order prospects occur between 700 and 3000m water depth, which are drillable, have been selected for further in-house study as the most likely economic targets. In addition, the distance of the various prospects from shore, the width of continental shelves, nearest infrastructure and market to the nearest landing point for the individual prospects, have been taken into consideration at an early stage in an effort to assign overall economic attributes (Fig. 1) .

Seismic attributes of Bottom Simulating Reflector (BSR) such as blanking, velocity inversion and polarity reversal within the estimated hydrate stability zones have been used to infer the presence of hydrate and associated gas deposits. About 80,000 km^2 area out of the entire Indian sea area was found to be favorable for the presence of gas hydrates in water depth ranging from 750m to 3000m. It must be pointed out however, that concentrations of hydrate in shallower water are likely to be more recoverable owing to the greater volume of gas that can be produced, at any given percentage of hydrate in the seafloor sediment, because lower ambient pressures allow the methane produced by dissociation of hydrate to occupy greater volumes of pore space.

Much of the older data is single-channel that was taken for marine research purposes. Most modern seismic data is multi-channel. New multi-channel seismic surveys have recently been completed in the eastern Indian slope area and in the Andaman-Nicobar area. Continued acquisition of seismics is anticipated as well as hydrate-specific processing (Korenaga et al., 1996).

Area Name	Area in Km^2			
	I	II	III	Total
Western Offshore				
Kutch	2,900	4,150	3,250	10,300
DCS	4,105	3,550	600	8,275
Bombay	3,075	800	350	4,225
Konkan	3,800	5,476	-	9,276
Padua-Bank	1,100	3,600	6,700	11,400
Lakshadweep	1,850	3,275	-	5,125
Kerala	275	2,8Q0	1,325	4,400
Cape Comorin	3,500	7,059	-	10,559
	20,625	30,710	1225	36,560
Eastern Offshorc				
Gulf of Mannar	1000	725	-	1,72S
Cauvery	1,900	3,100	-	5,000
Krishna	2,125	750	-	2,875
Godavari	325	150	-	475
	5,350	4,725	-	10,075
Andaman-Nicobar Offshore				
Nicobar	775	475	250	1,500
Andaman	3,125	1,025	92S	5,075
	3,900	1,500	1,175	6,575
Grand Total	29,875	36,935	13,400	80,210

Table 2. Area of gas hydrate prospects (Chandra et al., 1998). Category I - Highly prospective Category II - Moderately prospective, Category III - Low prospective

6.1. Seismic indications of hydrate on the Indian continental slope

BSRs are more prominent in the western offshore region than the eastern offshore area and are distributed discontinuously along the entire length of the western margin, from 8° to 22°N. BSRs and blanking zones originally were identified on nine seismic sections in the western offshore (NIO, 1999) whereas one location in the eastern offshore BSR could be identified in water depths more than 2000 meters. In the western offshore region, the BSR occurrences are confined to a bathymetric window of 2000 to 3000 meters, in most cases along the western flank of the Laccadive Ridge. With one or two exceptions, most of the BSRs are confined to 11-15°N, 71-74°E. In the eastern slope, BSRs are seen at between 1000-3000m water depth. At some localities in the western offshore, for instance offshore Goa, a continuous BSR is 126 km long and is underlain by "wipe out" zone, which is totally devoid of reflections; this and other wipeout zones are most likely highly gas-charged sediments. BSRs commonly vary along the seismic profiles from a minimum of 200 meters to a maximum to over 10 km, between 200 to 900 msec (TWT) below seafloor.

In Kutch slope and in abyssal and slope rise sediments offshore, the strong BSR is, in general, not seen. Velocity inversions and blanking within the predicted HSZ have been used here to demarcate potential areas of hydrate occurrence. Despite the lack of strong BSR, the Kutch offshore area is expected to be favorable for gas hydrate formation.

Good BSR, blanking and velocity inversion are frequently observed at places west of Mumbai (Bombay) and further offshore (NGRI, 1998). The Kerala and Konkan show strong BSR and blanking zone as well as in slope rise sediments immediately to the west. In the Kerala-Konkan area, roughly N-S-trending basement highs and lows appear to provide bathymetric highs that act as hydrate-gas traps (Dillon et al., 1994; 1995; Max and Lowrie, 1996). The slopes of the Laccadive ridge are characterized by discontinuous BSRs.

The same seismic attributes are also seen in Trivandrum-Cape Comorin slope. The lack of prominent BSRs off the west coast may be due to either parallel deposition of younger strata or the absence of free gas beneath the BSR. However, high resolution true amplitude processing, color display of Hilbert amplitude and polarity plots and AVO analysis may help to distinguish the BSR in such areas.

Thick sediments and high rates of sedimentation characterize the eastern margin and the Bengal fan. The offshore basins have potential for oil and conventional natural gas reserves, particularly the Krishna-Godavari basin. This should normally make an ideal situation for the formation of abundant gas hydrates, but clearly defined BSRs are not common. One possible reason for the low degree of BSR occurrence in the eastern offshore could be that the phase equilibrium required for the formation of gas hydrates is disturbed by upwelling fluid movements. Very well defined BSR, clear blanking, polarity reversal are predominately seen in the Cauvery slope and in Krishna-Godavari deep offshore along the east coast. BSR thickness below sea floor should range from 150-350 mbsf in water depths from 750-1900rn. Strong BSRs have been identified in the Krishna-Godavari offshore at about 1300 meters water depth.

The gas hydrate potential of the deep Andaman-Nicobar area appears high owing to the occurrence of well defined BSR, blanking and phase reversal on the rising flank of the fold structures (Chopra, 1985), similar to those seen on the Blake Outer Ridge (Hollister and Ewing, 1972; Dillon et al., 1985; Booth et al., 1996). Only two isolated BSR occurrences have been identified in the northern Andaman sea to date, while the in the southern region BSRs have been identified near the Nicobar Islands (Chopra, 1985) in from 1200 to 2600m water depth. Two prospective areas have been identified in the Andaman area in 850 and 1400m water depths.

In the northern part of the Bay of Bengal, where extremely rapid deposition of organic-rich sediments has taken place during Neogene and Holocene times, large hydrate accumulations are anticipated. No good BSR, blanking, or wipe-out zones, which would provide good evidence for the presence of hydrate and trapped gas, have been identified to date. This is at least in part because the sediment bedding is virtually parallel to the seafloor everywhere in the abyssal fan deposits, and the most obvious feature of BSRs, which is a superposition of the younger BSR across older bedding, cannot develop. Perhaps closer examination or reprocessing of the seismic data may reveal subtle evidence for gas and hydrate similar to those identified from seismic records in the Bering Sea, where large accumulations of thick gas-producing sediments also occur (Scholl and Hart, 1993).

7. ESTIMATE OF METHANE-IN-PLACE IN HYDRATE

For each play area, an estimation of gas hydrate resources has been made (NIO, 1997) by assigning a probability from 5% to 95%. The various attributes have also been graded as a function the probabilities (Collett, 1996). The evaluated gas hydrate resource offshore India is restricted by a severely limited amount of relevant data, particularly in the continental slopes where both commercial data and existing marine geoscience data are minimal (Geoscience, 1998). No evaluation of an Indian hydrate has yet been carried out with the focus of the Japanese hydrate research program, which includes drilling evaluation (JNOC, 1998). Thus, these early estimates are subject to refinement or alteration with the availability of more relevant data in the future.

Estimates range from 95% probability, which is most conservative. These have the highest statistical certainty. The in-place resources of methane in the form of gas hydrates in the Bay of Bengal and the Arabian Sea have been estimated to lie in a range between 1,894 trillion m^3 (66,290 TCF) at 95% probability to around 14,572 trillion m^3 (510,020 TCF) at 5% probability (Table 3). These figures are consistent with an earlier set of estimates for a smaller area carried out by ONGC, which yielded estimates of: 10% probability, 264 TCM; 90% probability, 33 TCM; mean resource, 122 TCM. Although the Andaman Sea area is considered to have considerable gas and hydrate, because it is quite far from the Indian landmass, the problem of transporting gas from this area are considerable; no gas estimates are currently publicly available.

Plays	Probability of equal to or greater than:				
	95%	75%	50%	25%	5%
Bombay Offshore	135	307	454	630	852
Kerala-Konkan Offshore	62	221	1137	1566	2299
N. Arabian Sea	226	440	595	798	1092
S. Arabian Sea	0	0	312	709	1094
Eastern Offshore	1038	1527	2168	3181	4525
N. Bay of Bengal	245	334	468	648	937
S. Bay of Bengal	188	367	1022	2468	3773
Total Reserves (TCM)	1894	3196	6156	10000	14572
Total Reserves (TCF)	66290	111860	215460	350000	510200

Table 3. Estimated gas hydrate resources of India. Equal to or greater than 5% probablility (GAIL, 1996). Editor's note: These figures are speculative and do not represent official government estimates.

What then constitutes the Indian hydrate resource? The best estimates that can be made at this time are very imprecise because of lack of relevant data. Much information must be obtained before a reliable estimate of the "Methane Resource" may be determined. However, in both optimistic and conservative cases, e.g., even at the lowest volumes that can be estimated at present, the absolute size of the potential resource appears to be so large (compared to the 707 billion m^3 of conventional methane available to India presently), that a serious effort is justified to develop means for exploitation of gas hydrates.

8. CONCLUSIONS

1. Regions favorable to the formation of hydrate have been identified using interpretation of existing seismic data in the deep Indian offshore in water depths ranging from 700 to 3,000 m water depth.

2. Abut 80,000 km^2 area is apparently underlain by gas hydrate and subjacent gas deposits.

3. A speculative estimate of about five times the total of conventional Indian gas reserves has been made for methane that may be extracted from hydrate in the Indian offshore. The resource estimates are based on sparse data and are thus only indicative. But the large magnitude of the hydrate resource deserves special attention, and concerted efforts are needed to confirm the figures and draw up a strategy for recoverability, keeping pace with the technological development of deep-water drilling and exploitation techniques.

4. The area of the Bay of Bengal as a potential storehouse of extremely large volumes of hydrates may be unique in consideration of its ideal geological, geochemical, and thermal factors.

5. The Krishna-Godavari offshore is the most favorable area for hydrate in the Indian passive margin in the thick sediments deposited from Miocene to Recent in a growth fault environment. The area lying between 600 and 2,000 m water depth, off the coast of the Mahanadi, Godavari, and Krishna River deltas, is likely to emerge as a huge storehouse of massive hydrate deposits.

6. The Andaman-Nicobar arc basin also displays favorable geological, geochemical, and geophysical attributes favorable for the production of methane and formation of hydrate.

Acknowledgements Thanks to Drs. I.L. Budhiraja and A. Rastogi, and the many Indian geologists who are taking part in the Indian hydrate program. Thanks also to the Indian scientists who originally offered to prepare this contribution but were unable to do so because of administrative difficulties.

Chapter 18

Hydrate as a Future Energy Resource for Japan

Michael D. Max
MDS Research
Suite 302, 1211 Connecticut Ave. NW
Washington DC 20036

1. INTRODUCTION

Japan is heavily industrialized, has a high standard of living, and has a relatively high demand for energy. Yet Japan has never had abundant indigenous energy sources. Because Japan has very limited conventional hydrocarbon resources and potential, the country has traditionally imported virtually all of its energy supplies since the beginning of its industrial revolution in the latter part of the last century. Supplies of energy have always one of the foremost concerns of industry and government at the highest levels.

Japan's present annual consumption of oil is about 2 Million Metric Barrels (MMBBL) while natural gas consumption is 2.3 Trillion Cubic Feet (TCF). Domestic oil production is only 0.25% of annual consumption, and domestic natural gas production is only 3.0% of annual consumption. Japan thus may have the largest energy shortfall of all industrialized nations.

Natural gas consumption in Japan is expected to rise, not only in proportion to oil, but in absolute terms. This is at least in part because of methane's increasing importance as a source of relatively clean exhaust gas as a byproduct of producing the energy (Chapter 28). According to the government forecast of long-term energy supply and demand, natural gas consumption in Japan will reach 2.53 TCF in the year 2,000, 3.10 TCF in the year 2000, 3.53 TCF in the year 2010 and 3.95 TCF in the year 2015, compared with 2.26 TCF in 1997 (Tono, 1998).

Because most of the power produced in Japan uses imported hydrocarbons, and natural gas is widely used in both industry and for private consumption, a secure supply of methane is fundamental to the stability of the economy of Japan. Although methane imported into Japan is widely available currently from a number of producers, principally in the Mid-East, SW Asia, and Russia, any possibility that might exist for the development new types of

M.D. Max (ed.), Natural Gas Hydrate in Oceanic and Permafrost Environments, 225–238.

unconventional hydrocarbon resources in or adjacent to Japan constitutes an attractive proposition. Methane hydrate was identified as one of the most likely of the potential future domestic hydrocarbon resources (Yamazaki, 1997). Considerable expectation has been placed on methane hydrate, a potentially deliverable unconventional natural gas resource because of its promise to contribute to Japan's stable supply of energy in the future.

The head of the energy and natural resources department at the Agency of Industrial Science and Technology (AIST), a unit of the Trade Ministry of Japan, has stated (1995, Reuter, Tokyo), that "Methane hydrates could be a next generation source of energy [for Japan] that can be produced domestically". Japan, at present, is leading the effort to develop commercial technologies to identify hydrate concentrations that are suitable for conversion of hydrate to methane gas and its subsequent extraction.

1.1. Basis and Organization of the Effort to Identify Methane Hydrate Adjacent to Japan

There has been much speculation among scientists and some hydrocarbon industry professionals in Japan about the existence and size of methane hydrate in marine sediments adjacent to Japan. Like speculation elsewhere (Kvenvolden, 1993), one of the biggest stumbling blocks has been the lack of basic knowledge about hydrate. Over the recent past, however, particularly since the mid-1980s, knowledge concerning naturally occurring oceanic hydrate has been greatly improved. In addition, hydrate began to be manufactured in test apparatus for acoustic, thermodynamic, and other material science testing. A considerable body of hydrate data now has documented by many researchers.

Hydrate volumes have been estimated in many places in the world, as the aerial and vertical distribution of methane hydrate is usually estimated primarily on theoretical conditions (Collet, 1998) and by interpretation of seismic studies (Chapter 22). From a resource point of view, previous studies of such hydrates have been mostly restricted to resource assessment (Kvenvolden, 1988; Krason, 1993; Lee et al., 1993). On the other hand, recent studies or discussions are more related to gas recovery (Hovland et al., 1997; Max and Dillon 1998).

Many technical and scientific papers related to methane hydrate, including those dealing with the likelihood of hydrate adjacent to Japan, have now been published. The topic of methane hydrate is apparently attracting more attention because an increased number of papers are being published on hydrate (Preface). In the United States, a program of hydrate research led by the U.S. Department of Energy (Chapter 28) attracted the attention of scientists, especially from the Geological Survey of Japan. The similarity of the tectonostratigraphic position of known hydrate deposits, such as those off the west coast of Central America (Chapter 2), suggested the possibility of significant methane hydrate in the offshore area surrounding Japan.

Preliminary methane hydrate research and development (R/D) studies were begun on a small scale over in the late 1980s by government institutions, such as the Geological Survey of Japan. The objective was to examine and define the likelihood of the occurrence of methane hydrate around Japan. In response to a June, 1994 advisory report of the Petroleum Council of Japan, acting as chief advisor to the Minister of International Trade and Industry, the Ministry of International Trade and Industry of Japan (MITI) initiated a project in 1995 to evaluate methane hydrate around Japan. The report pointed out that methane hydrate, as an unconventional natural gas resource, might be the most important indigenous energy resources remaining to Japan. The advisory report recommended a first five year program of basic survey and research focused on oceanic hydrate around Japan to better understand the resource potential and future commercial possibility. The Japan National Oil Corporation (JNOC) was given the task by MITI of coordinating domestic exploratory work. The study of potential pipeline hazards was carried out by private gas distributors.

MITI-JNOC's exploration surveys on oceanic methane hydrates focused on determining whether such hydrate could be an energy source in the future. MITI and JNOC were given the resources not only to carry out research and survey activities, but also to drill at least one wildcat well on a potential hydrate deposit, not only to evaluate methane hydrate and associated free gas, but also to explore for conventional subjacent hydrocarbons. From a methane hydrate exploration point of view (Tono, 1998), the objectives of drilling must be to clarify the origin of seismic features regarded as revealing the presence of hydrate, to evaluate associated free gas below, and to collect basic data which would help to confirm the possibility of economic concentrations of methane hydrate adjacent to Japan.

To promote this program, in 1995, a Methane Hydrate Development Promotion Committee was established within JNOC. This body consisted of researchers and engineers from universities, national research institutes, oil companies, and gas companies. JNOC's Technical Research Center (TRC) has conducted collaborative studies on methane hydrate with ten private companies in Japan since 1995. In order to carry out initial drilling of at least a stratigraphic exploratory well in 1999 and to examine the possibility of the development of' methane hydrate in the future, a research-industrial consortium was established. The group consists of four Japanese oil exploration companies, two service companies, three gas companies and one electric power development company. The TRC research and development project includes examination of at least six practical fields such as basic properties of methane hydrate, geotechnical modeling and logging technology, seismic survey and processing technology, drilling and coring technology. production modeling and core testing technology.

Four main fields of investigation comprised the basic hydrate research program of Japan. 1. In the first instance, the database for naturally occurring hydrate it very sparse. Data were needed to understand the basic elements of methane hydrate origin, occurrence, and formation, particularly with respect to its effect as a diagenetic material in different types of sediment. 2. Commercialization of the hydrate resource would involve disturbing a natural system that is reactive to changes in temperature and pressure. Therefore, understanding the reaction of the hydrate system as a whole (Max and Chandra, 1998) is important to minimizing hazards and possible effects upon the environment is critical to exploitation of the hydrate resource. 3. Once the methane has been extracted, storage of methane and transport of the methane to the points of use will be on a far greater scale than at present. Because of both the safety and possible impact upon climate of major methane spills, tolerance for mishandling the methane must be minimized. 4. Most important of all, however, to the possible development of oceanic hydrate as a source of methane is the need to develop survey, experimental, and intellectual tools to allow the proper assessment of the three-dimensional disposition of hydrate and determine its volumetric concentration.

At the beginning of 1998, the hydrate consortium also conducted methane hydrate research-well drilling in the Mackenzie Delta in Canada (JAPEX/JNOC/GSC Mallik 2L-38 Research Well) under a collaborative research program with the Geological Survey of Canada (GSC), the U.S. Geological Survey, and universities and research institutions (Chapter 5). Even though the well was on land and in permafrost rather than in an oceanic environment, it was felt to be important that emerging technology and methodology thought to be important to recovering methane from hydrate would be tested in a realistic hydrate environment. In particular, work was carried out to verify new deliverables from the research and development arena, such as the Pressure Temperature Core Sampler (PTCS) and drilling fluid technology, as well as conducting core sampling, wireline logging, VSP and production testing.

The first five year program has been successfully completed in that evidence of substantial hydrate immediately adjacent to Japan has been found and sufficient is now understood about hydrate to continue the hydrate research program. The character of the second five year segment of the program has required altering the structure of the program's organization to focus more on resource development rather than research. It is anticipated that in the near future, during the first decade of 21st century, the term "Recoverable Reserves" may be applied to hydrate deposits adjacent to Japan rather than the initial "Expected Resources" (Tono, 1998).

2. GEOTECTONIC SETTING OF JAPAN'S HYDRATE DEPOSITS

Japan rests on a volcanic island arc of tectonic and magmatic thickening of the crust related to westerly subduction of oceanic crust in the northwest Pacific Ocean. This arc forms the margin of the continent here. However, the tectonic framework is complex and will not be discussed here in detail.

The youngest and most seaward of the accretionary terranes along the N and NW Pacific rim, which is a continuation of the island arc stretching to the

SW from Alaska, is in the process of becoming amalgamated with older subduction complexes that lie along the eastern Asian cratonic cores displayed from China to easternmost Russia. The accretionary prism margin of Japan bifurcates offshore central Honshu, SE Japan, where the younger Izu-Bonin Trench diverges from the northeastern continental margin of Japan to extend along the eastern margin of the Izu-Bonin Ridge, which is itself an accretionary prism and volcanic arc that is emergent from the sea immediately to the south of Japan (Fig. 1).

The Pacific Plate is in collision with both the Eurasian and Philippine Sea Plates. Whereas the Pacific Plate directly abuts Japan in the north, albeit with a possible accretionary relic of the North American Plate or an independent tectonosedimentary terrane similar to those (e.g. Wrangellia-like) accreted to North America along the Canadian and Alaskan Pacific margins, the Pacific Plate does not yet abut Japan directly. The Philippine Sea Plate, or an attenuated relic of it and the sediments riding upon it that have not been consumed in subduction beneath Japan or shortened by tectonic thickening along the eastern margin of the plate that carries the Izu-Bonin Ridge is in collision with Japan along the Nankai Trough (Fig. 1). The subduction of oceanic crust beneath the Izu-Bonin Ridge began in late Middle Miocene to Early Pliocene, and an accretional prism caused by offscarping of incoming sediments on the Philippine Sea Plate has been well developed (e.g., Ashi and Tokuyama, 1997). Subduction along the Nankai Trough confrontation zone is older. Both subduction zones may structurally underlie at least part of the southern part of Japan.

The mechanism of plate collision, subduction, and the formation of accretionary prisms under a compressional stress framework (within which significant lateral and vertical fault movements are commonly part of the tectonic process) is an excellent environment for the migration of methane and associated fluids and other gases, especially from deep sources, and their concentration as hydrate in the upper part of the accretionary prism. The geotectonic situation of Japan is similar to that of the Cascadia Margin (Chapter 15) and the continental shelf margin of the Palmer Peninsula of Antarctica (Chapter 16), where substantial hydrate has also been identified.

The continental margin of Japan, however, is tectonically complex because of the jump of the active north circum-Pacific subduction zone to the Izu-Bonin Trench at the annealing triple point off SE Honshu (Figs. 1, 2). Active subduction in the northernmost part of the Ryukyu Trough, (Nankai Trough) appears to have been retarded by the development of the Izu-Bonin Trench system. The trench in the northern end does not maintain a sharply defined bathymetric depression comparable with the active Japan-Izu-Bonin Trench system, nor is the water depth within the relic trench indicative of strong subduction. The northern part of the Nankai Trough and the Philippine Sea Plate may be subject to tectonic elevation owing to obduction above the west-plunging Japan-Izu-Bonin subduction system.

One of the important features of the eastern continental margin of Japan is that the shallow water shelf is very narrow. In addition, sediment derived from the nearby high relief volcanic mountains of Japan can be swept to the shelf edge and onto the continental slope and remain more immature than sediment winnowed across a wider shelf. Sandy clastic sediment that can be swept rapidly over the shelf edge and deposited on the slope provides for good porosity and permeability sediments in which hydrate may form. Shelf sediment offshore here, for instance, has a well differentiated sand-silty bedding (Tsuji et al., 1998) that can be traced in drill holes and on seismic records from land to the upper slope.

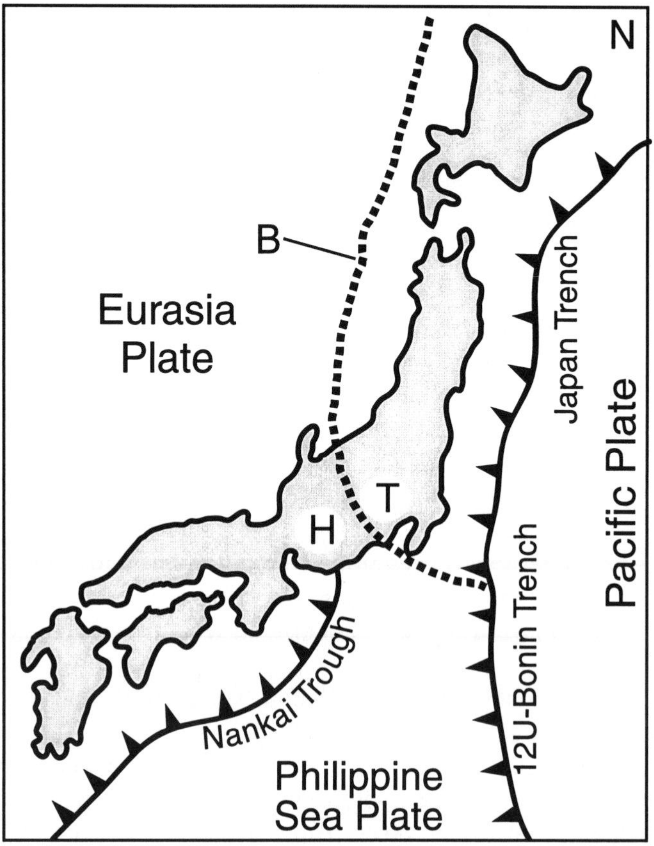

Figure 1. Regional tectonic plate and geomorphic framework in the vicinity of Japan, after Tsuji et al. (1998). B, Possible terrane boundary separating the southern Japan accretionary prism from an amalgamated terrane of North American affinity (N); H, Honshu, the main island of the archipelago of Japan.

3. IDENTIFICATION OF HYDRATE

Identification of areas of hydrate in the vicinity of Japan (Fig. 2) is presently

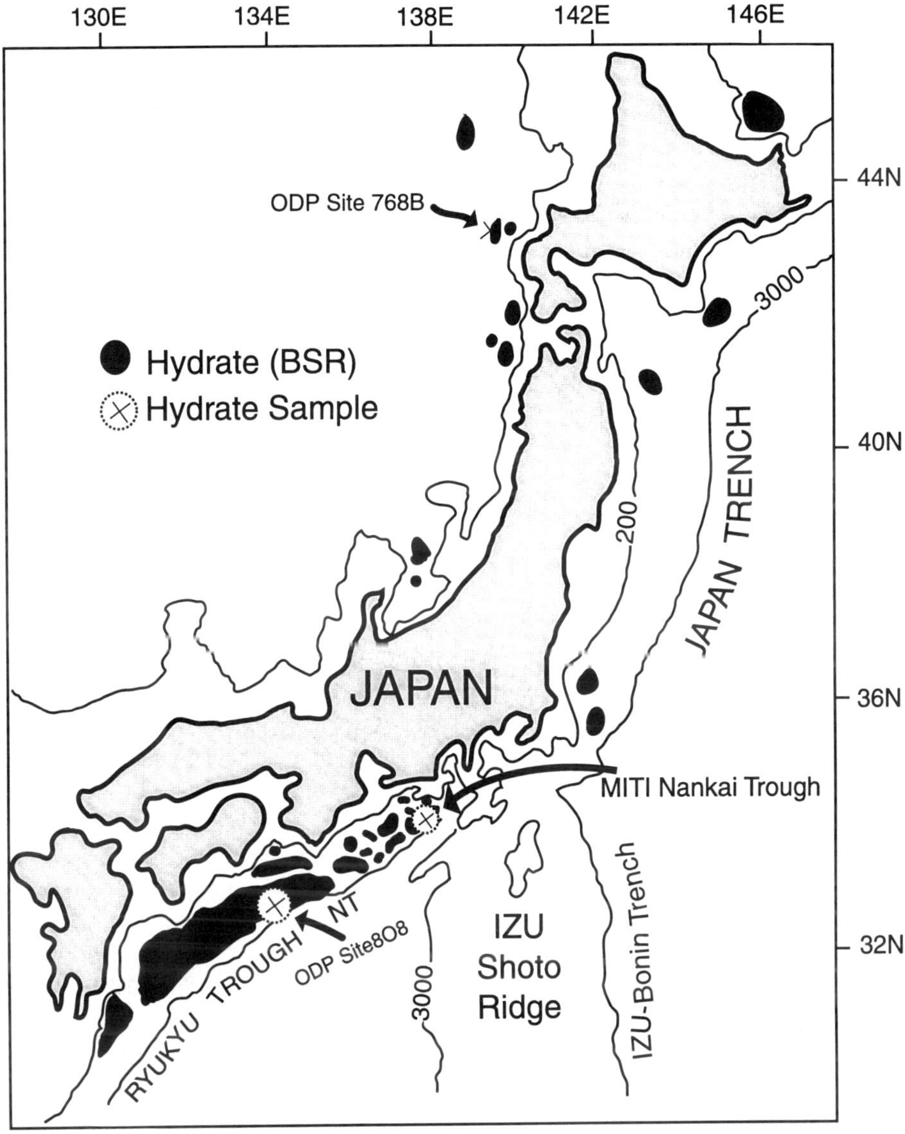

Figure 2. Map showing hydrate deposits around Japan as revealed by presence of hydrate/gas interface BSR. NT, Nankai Trough. Hydrate locations compiled by Mr. Mikio Satoh, Geological Survey of Japan. After: Aoki et al. (1983), Arato et al. (1996), Ashi and Taira (1993), Ashi and Tokuyama (1997), Ashi et al. (1996), Ashi et al. (1999), Baba and Uchida (1998), Honza (1078), Hyndman et al. (1992), Nakamura et al. (1997), Okuda (1996), Sakai (1998), Satoh et al. (1996), Taira et al. (1991), Tamaki et al. (1990), Tsuji et al. (1998), Yamano et al. (1982, 1984, 1992), and unpublished data held by the Geological Survey of Japan.

confined to those areas where well developed Bottom Simulating Reflectors (BSR) have been identified. This interpretation must be carried out with care because apparent BSRs identified on seismics in geological strata in and around northern Japan are known to be due to the presence of Opal A/CT horizons from drilling and stratigraphic analysis.

On high-resolution reflection seismic survey records mapped, at least two of the following gas hydrate BSR characteristics were recorded: reverse polarity (phase reversal) versus sea bottom reflector, relatively high amplitude of the BSR reflection coefficient, cross-cutting of bedding in the sediments, and geoacoustic profile of high velocity hydrate-rich sediment overlying lower velocity gas-rich sediments. BSR regions adjacent to Japan are usually discontinuous over large areas and not clearly defined as a single unbroken seismic feature (detected as "high amplitude zone" not as single continuous reflector), probably because of the bedding causes a strong associated porosity differentiation. Tight sediments do not allow enough gas to pond below hydrate to cause a well defined BSR to form and may not have been porous enough to allow methane to migrate into the bed to form hydrate in the first instance.

Hydrate investigators in the hydrate research program of Japan are aware that using BSR to identify the presence of hydrate yields a minimum distribution indicator. Hydrate is known to extend well beyond BSR (Max and Lowrie, 1996; Max and Dillon, 1998), which marks a hydrate-gas interface. BSR is commonly found below a bathymetric culmination where the base of hydrate forms a closure (= trap) that allows gas that is prevented by hydrate-tightened porosity from migrating upward into the HSZ or in a bedded series where a compound hydrate-stratigraphic trap sequence can form in dipping beds.

The confirmation that very large volumes of methane hydrate occurs in deep marine sediments (Matsumoto et al., 1996) has been one of the major accomplishments of the Ocean Drilling Program (ODP). Hydrate was recovered by deep sea drilling off the SE coast of the U.S. and the western offshore of the U.S. and Central America (Chapter 2). Hydrate has also been sampled in the course of ODP drilling in two localities to the S of Japan along the active front of the Ryukyu accretionary prism and off NW Japan and at one locality to the NW of Japan in the back-arc basin system (Fig. 2) in two ODP wells and one MITI well. Chemical analyses of samples have confirmed that naturally occurring hydrate is composed almost entirely methane hydrate (Chapter 2).

It is yet too soon to say whether the apparently greater concentration of hydrate along the SW margin of the Nankai Trough compared with the identified locations to the NE reflects a longer history of gas migration, and concentration by hydrate formation in seafloor sediments, or whether there was some difference in the timing, rate of elevation, fault history, etc. of the tectonic mechanisms themselves in the two subduction complexes. Perhaps older subduction complexes simply generate more methane later in their history, perhaps the nature of the sediments in the subduction zone or their thermal history was different. On the other hand, identification is at a preliminary stage and it is perhaps be too early to draw unequivocal conclusions. The database is as yet very small as hydrate around Japan remains largely to be quantified; much may occur well away from the present firm identifications (Fig. 2), which are based primarily upon the existence of hydrate/gas BSR. Also, virtually nothing is known about the age of the methane or action of the methane concentration cycle through a process of hydrate formation, dissociation, and formation of subsequent hydrate in any of the deposits adjacent to Japan.

To the NE of the termination of the Nankai Trough margin (Fig. 1), hydrate does not appear to be concentrated near the margin of the accretionary prism but is rather distributed with no apparent pattern that can be directly related to major structures. A number of fairly large hydrate deposits have been identified in the back-arc basins to the NW of Japan as well as between land area of Japan and the continental margin (Fig 2). However, this apparent distribution and relationship may not hold up when more hydrate is identified in the region.

4. SELECTION OF A HYDRATE NATURAL LABORATORY: THE 'NANKAI TROUGH'

Many areas of hydrate have been identified adjacent to the Japanese islands (Fig. 2). One area has been selected as the site at which drilling and testing of sampling and extractive technologies could be carried out. This area was chosen because: 1. It displays well defined BSR; 2. There is a good stratigraphy in dipping clastic sediments in which gas should flow; 3. Conventional hydrocarbons are also anticipated to occur in deeper strata forming potential structural traps; 4. Drilling would be relatively easier in this shallow water site than in deeper, and more remote, sites elsewhere; 5. Operational support is available from ports nearby. The Nankai Trough area is very close (about 200 km) to Tokyo Bay, which is in the center of the Japanese Pacific coast industrial belt; 6. In addition, in the case that the area proves to be a source of commercial methane, it is immediately adjacent to one of the largest hydrocarbon consuming areas in Japan, and the transport distance of the methane from source to market would be minimal. Although this last aspect was not a primary factor in identifying the area in which the first in-situ R&D should be carried out, strongly positive results would allow virtually complete amortization of the R&D by gas production.

The area selected is referred to as the 'Nankai Trough' by Japanese investigators, even though it occurs in relatively shallow water in water depths slightly less than 1 km and not in the nearby geomorphic Nankai Trough, which is a linear bathymetric depression marking the approximate junction between the Japanese island arc and the Pacific oceanic plate (Figs. 1, 2). The term 'Nankai Trough', however, is used here to remain consistent with Japanese hydrate terminology and usage.

The stratigraphy of the nearby land area is characterized by forearc and accretional Cenozoic sediments (Tsuji et al., 1998). Forearc sediments are Mikura, Kurami-Saigo, Sagara, Kakegawa, and Ogasa Groups in ascending order bounded by unconformities, and the sediments consist of clastics which are mostly alternation of sandstone and siltstone with associated conglomerates and tuff. Accretional sediments are Setogawa, Ooigawa, and Takakusayama Groups, made of ocean floor sediments such as basalt, limestone, and chert, and slope sediments including allocthonous limestone blocks. Two MITI wells, "Sagara" and "Omaezaki-oki", were drilled close to the survey area on land immediately to the north. Both reached the Paleogene, the former to the Takakusayama Group and the latter to the Mikura Group.

In the Sagara district on land, the Sagara oil field, which produced small amounts of oil from the alternating sandstones and siltstones of the Sagara Group (late Miocene to Pliocene). The results of two MITI wells suggest that reservoir rocks and matured source rocks can be anticipated nearby offshore. Upper Oligocene strata in the MITI "Omaezaki-oki" (offshore well in about 560 m water depth about 37 km NNE of the test hydrate drill hole) and Upper Oligocene to Lower Miocene in MITI "Sagara" (coastline well about 58 km NE of the test drill hole, Tjsui et al. 1998, Fig. 1) have 0.5 to 1% of TOC (Total Organic Carbon), and have the potential to be hydrocarbon and methane source rocks. Matured source rock (Ro > 0.5) is expected below 3,000 mbsl in MITI "Omaezaki-oki", and most of the Upper Oligocene and Lower Miocene subsided are on the shelf, area are regarded as thermally matured with respect to gas production. Some areas in the 'Nankai Trough' hydrate area contain thick sediments with sufficient organic detritus which would insure significant methane generation and possibly higher gravity hydrocarbons. Sandstones in the Sagara Group have more than 20% of porosity in MITI "Sagara" and demonstrate proven gas production. The Sagara Group and its constituent sandstones is believed to be widely distributed in this region offshore (Tsuji et al., 1998).

Cold fluid seepage containing biogenic hydrocarbons has been sampled by submersible and ROVs. The methane is believed to be produced from dissociation of hydrates (Ashi and Tokuyama, 1997) because of the chlorinity of associated waters and the isotopic character of carbon in gas. These seepages are usually found in the seafloor along faults that lie parallel to the trend of Nankai Trough and are often associated with concentrations of seafloor shellfish and other life. Large-scale pockmark-like depressions are common and thought to be related to movement on faults and associated hydrate dissociation (Kuramoto et al., 1997). Methane-rich gas taken from a JNOC gravity core sample $\delta^{13}C$ values of about 75%, suggests a biogenic origin (Chapter 7).

Although BSR has been recognized in the Nankai Trough research area (Fig. 2) on the basis of observed BSR (Arato et al. 1996; Satoh et al., 1996), hydrate is probably developed more broadly. Estimation of hydrate volumes in the 'Nankai Trough' hydrate study area is on-going (Matsumoto et al. 1996; 1998). Satoh et al. (1996) estimated resources of natural gas hydrates and associated free gas with hydrated layer in the Nankai Trough offshore Shikoku region at from $2.7 \times 10^{12} m^3$ to $1.6 \times 10^{12} m^3$ (Tsuji et al., 1998).

5. DRILLING AND RESULTS

Following the selection of a suitable gas hydrate study or natural laboratory area, and plans to drill and develop the new down-hole and extractive technology required to obtain methane from the hydrate resource, a drilling target was established. The location of the MITI national wildcat well (named MITI "Nankai Trough") is about 50 km southeast of the coast of Japan. Seafloor there exhibits a hummocky morphology and a local relief of about 400-500 meters across the approximately 4,000 km^2 study area. This well was the second drilling experience in a deep water drilling operation around Japan. The purpose of the well was two fold, one for methane hydrate characterization and the other for stratigraphic test and exploration in the deeper part. The well is located in about 950 m water depth, has been selected with reference to the existence of both Miocene prospect and BSR identified by JNOC's seismic surveys. The depth of BSR is estimated to be about 290 m below the sea floor (mbsf). The well depth below sea level is about 3000 meters.

Preliminary to drilling the main hole in 1999, JNOC drilled two holes in 1997 to a depth of about 250 mbsf to collect data for the casing program. Cores were collected intermittently and examined for the existence of methane hydrate. Cores consist of silt-dominated clastic sediments with a small percentage of sand layers. Methane hydrate was not observed in the cores, although chloride concentration of pore water recovered from cores suggests the likelihood of in-situ methane hydrate that had dissociated during its recovery.

Drilling in the 'Nankai Trough' area has now been accomplished and preliminary results of hydrate evaluation are being assessed. JNOC released a summary of the drilling and preliminary hydrate assessment from the drilling on January 20, 2000. The main body of the news release appeared in major Japanese newspapers in the evening editions of January 20, 2000, or in the morning editions of January 21, 2000. The body of the following information is not an official translation but the informal translation of Ko-Ichi Nakamura of the Geological Survey of Japan.

5.1. Operational and Results Summary:

Contract company for operation: JAPEX

Location of the drill: Off Omaezaki Spur (About 50 km off from the mouth of Tenryu River, Shizuoka Prefecture), Depth 945 m

Planned depth of the drill: 2,800 m from the sea surface (1,855 m from the seafloor). (All depths are from the sea surface)

Drilling rig: " M. G. Hulme Jr. ", owned by the American company, Reading & Bates Falcon Drilling.

Total budget (planned): 5,000,000,000 yen

Position: 34-12-56N,137-45-03E (Tokyo datum)
34-13-08N, 137-44-52E (WGS-84)

November 12-14, 1999: The first pilot hole (Water depth: 945 m, Drilled depth: 1,600 m) was drilled to investigate shallow gas. For the purpose of the prediction of possibly sudden expulsion of gas produced by dissociation of hydrate produced from the heat of drilling, the entry point of the drill hole on the seafloor was monitored by an instrumented Remotely Operated Vehicle (ROV). No escaping gas was recognized during the drilling operation.

November 14 to 16, 1999: The second pilot hole was drilled (Water depth: 945 m, Drilled depth: 1,486 m) was also drilled to check the existence of methane hydrate and to forecast the depth of the hydrate based on the data (resistivity, density, etc.) obtained by LWD (Logging While Drilling). High resistivity was recognized at the depth where methane hydrate was anticipated from analysis of the reflection seismic records.

Nov. 16, 1999: The main well was spudded in.

Nov. 19 to Dec. 2, 1999: Conventional coring operations were carried out to obtain cores that might contain methane hydrate. Cores were obtained 5 times between 1,110 m to 1,146 m and 1,151 m to 1,175 m. (Recovery: 35.5 m/60 m, 59 %). Cores using PTCS (Pressure-Temperature Core Sampler, which is designed to maintain the pressure and temperature at the sampled location) were also attempted between 1,254 m to 1,272 m. (Recovery: 5.5 m/18 m, 31 %). The JNOC-developed PTCS and equipment was used for the first time in Japanese waters. However, PTCS cores were not achieved owing to operational difficulties. PTCS coring was attempted 27 times in the interval between 1,175 m to 1,254 m. (Recovery: 29.1 m/79 m, 37 %)

Core samples were obtained from sandy layers distributed in the depth from 1,110 m to 1,272 m from the sea surface by PTCS and other samplers. Based on the analysis of the large amount of gases generated from the samples, anomalous low temperature of the sample and anomalous low chloride ion concentration in the pore water, etc., the existence of the methane-hydrate was proved in three layers (total thickness was about 16 m) between 1,152 m and 1,210 m from the sea surface.

By the time the cores were recovered, no hydrate remained and some of the sediment appeared somewhat distorted, presumably by gas flow and consequent dewatering. Thus, no solid hydrate was recovered. Estimates of gas hydrate volume from all the drilling to date is based on measuring chlorinity and calculating the dilution factor of normal seawater by hydrate dissociation and the volume of hydrate required to provide the dilution effect.

Overall interpretation using core sample analysis data and petrophysical data revealed that gas hydrate occurs in three intervals (net 16 m) at total depths below sea level between 1152 m and 1210 m. It was reported that gas hydrate content within this zone is 20% in bulk rock volume, or about 80% of total pore space, which is almost 10 times more than the samples taken from ODP Leg 164 on the Blake Ridge (Chapters 13, 20, 23). Thus, there is about 525 million m^3 per km^2 methane in this hydrate (at STP), taking a uniform thickness and bulk composition percentage for the purpose of demonstrating a representative hydrate volume only. These drilling results have yielded information that makes the concept of extraction of methane from hydrate promising.

6. CONCLUSIONS

Although the zone in which hydrate is found is thinner than was hoped, the hydrate concentration in the zone was very rich -- about 80% of pore space. These drilling results are regarded as being very promising and substantiate further research and development activities.

To date, no clear and unequivocal identification of recoverable reserves has been achieved. Concentrations of hydrate have been identified, but there has been little quantification as yet.

Dedicated geophysical survey and carefully selected drilling locations that will provide calibration of the survey data are required in the 'Nankai Trough' study area and elsewhere.

No dependable and safe technology for producing methane from seafloor hydrate has so far been proven for use in recovering methane from hydrate deposits around Japan.

The hydrate is present in geological horizons that probably have enough strength and porosity to allow for spontaneous gas flow without collapse of geological strata when hydrate is dissociated, so long as pressures are managed adequately.

Because of the relatively shallow water depth, gas pressures produced during dissociation will be relatively high (Dillon and Max, 1998; 1999). At water depths less than 1.6 km (approximately), the volume of methane concentrated in the hydrate is greater than the compression factor of free gas at the pressure depth of the 'Nankai Trough' hydrate deposit. Dissociation will produce substantial overpressures.

The earliest date that commercial production of methane hydrates is estimated to possibly begin, if sufficient hydrate is identified and extraction methods are perfected, is estimated to be about the year 2010.

Commercial recovery of one-tenth of the methane hydrate is currently regarded as being a fully economic target.

7. FUTURE WORK

Based on the present drilling results, the gas hydrate research program is planning complementary drill holes in the 'Nankai Trough' study area, acquisition of the seismic data (VSP, the translator's guess) and sampling of the cores by improved PTCS to yield samples that can be fully analyzed to answer the many basic questions that should be resolved, such as what is the origin of the methane, what is the history of development of the hydrate, etc. Modeling of the geological and engineering properties of the potential reservoir is necessary to develop extractive methodology. It is anticipated that quantified methodology will be developed that will establish the relationship between the seismic effects of hydrate and associated gas so that wide-area evaluation of the hydrate deposits around Japan can be made with confidence.

Additional surveys (to check the characters of free gas and BSR, to check the continuation of sandy layers, which contain methane hydrate) are to be carried out. The acquisition of the seismic data, probably VSP, and sampling of the cores by improved PTCS (Chapter 22) are planned. To provide ground truth (Chapter 24) for the surveys, additional drill holes and methane extraction experiments are planned.

The northern Canadian hydrate deposits (Chapter 5) will be revisited to carry out tests of new hydrate-specific technologies from drilling to gas processing in a cost-efficient manner. It is critical that developing extractive technologies and learning to deal with the issues surrounding controlled dissociation of hydrate and the safe extraction of the produced methane be achieved as early as possible.

This new technology should be of value at other hydrate localities and as such, may constitute Intellectual Property Rights (Patents) that should have intrinsic value even if the eventual extraction of methane does not involve hydrate deposits around Japan.

It is estimated that about 50 trillion cubic feet (TCF) of oceanic methane hydrates may exist adjacent to Japan in the Nankai Trough(Chapter 10), but it is not clear that all of the identified hydrate (Fig. 2) is included in this figure.

The character of the hydrate research program of Japan is being altered to include considerations necessary for the commercial supply and distribution of abundant indigenous methane supplies.

Acknowledgments Thanks to Dr. Y. Aoki, JAPEX Geoscience Institute and Mr. S. Tono of the Japan National Oil Corporation for discussions.

Chapter 19

A Note on Gas Hydrate in the Northern Sector of the South China Sea

Sheila L. McDonnell & Michael Czarnecki
Naval Research Laboratory
Washington DC 20375-5350

1. INTRODUCTION

In 1998, the Naval Research Laboratory compiled a digital bathymetric database for use in environmental analyses of the seafloor morphology, physiography, and surficial sediments in the northern sector of the South China Sea (McDonnell, et al., 1998). Both detailed bathymetric surveys and random-track bathymetric sounding data were used. Most of this data was obtained from the United States National Geophysical Data Center (NGDC) and the United States National Imaging and Mapping Agency (NIMA), as well as other data not held by these agencies. Compilation created a new set of digital bathymetric contours (Fig. 1).

Existing sediment thickness isopachs (Hayes, 1978, and Lu, et al., 1977) were also digitized. Superposition of the digital bathymetry upon the sediment thickness chart suggested a number of areas that were suitable to host gas hydrate. Publications also were identified that contained figures showing gas hydrate features (Chi et al, 1998; Bochu, 1998; Liu et al., 1993).

Hydrates can be recognized on reflection seismic records because their presence in significant quantities strongly affects the physical properties and thus the seismic response of the sediments in which they occur (Singh et al., 1993) (Chapters 20-22). The Bottom Simulating Reflector (BSR; Stoll et al., 1971) and other gas hydrate indicators, such as blanking (Dillon and Paull, 1983) and wipe-out below the BSR were identified. These indicators of hydrate were found on virtually all the seismic records examined and suggests that hydrate occurs broadly in certain areas of marine sediment to the southwest of Taiwan.

This led to a preliminary evaluation of the likelihood of gas hydrate development in the area using the predictive methodology outlined by Max and Lowrie (1993). Because methane hydrate is a possible future source of methane for fuel (Kvenvolden, 1993; DOE, 1998; Max and Chandra, 1998), identification of potential commercial accumulations of oceanic hydrate adjacent to likely markets is of very specific interest to those markets. The

M.D. Max (ed.), Natural Gas Hydrate in Oceanic and Permafrost Environments, 239–244.

objective of this paper is to note the gas hydrate occurrences in the northern sector of the South China Sea and to show likely areas of hydrate occurrences in relation to the tectonic framework.

2. TECTONOSEDIMENTARY FRAMEWORK AND HOSTING OF GAS HYDRATE

The margins of the northern sector of the South China Sea (SCS) displays a complex, tectonosedimentary framework. The NW margin of the SCS is a passive tectonic margin forming the boundary to the broad Chinese continental shelf. The eastern margin of the SCS is an active collisional margin of a prong of the circum-pacific plate. The Heng-Ch'un Ridge to the south of Taiwan (Fig. 1) is the southern continuation of the Japanese Ryukyu arc which passes through Taiwan.

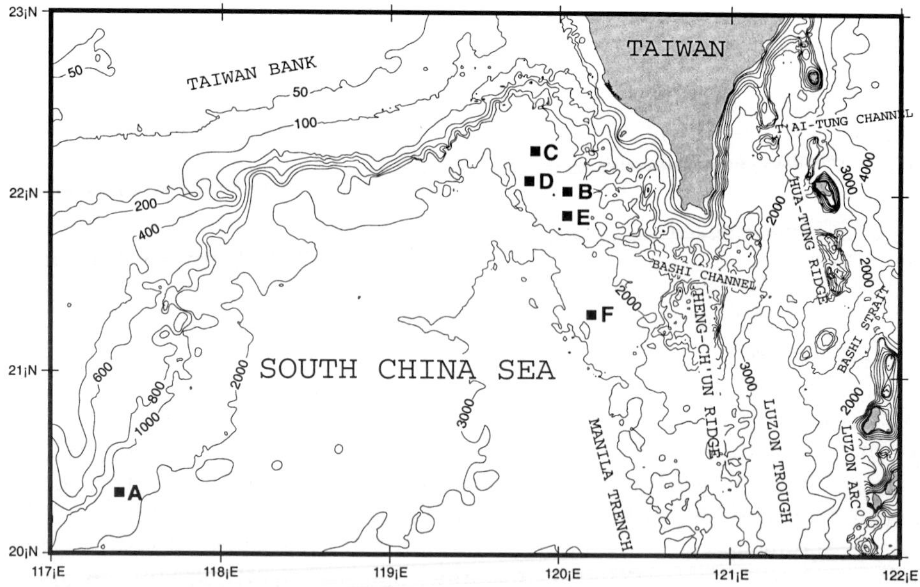

Figure 1. Simplified bathymetric and location map. Contours in m, 50, 100, 200, 400, 600, 800, 1000, 2000, etc. Locations of hydrate from interpretation of seismic reflection lines. A (Bochu, 1998), B (Liu, et al., 1993); C, D, E, F (Chi, et al., 1998).

Sediments along the NW margin of the South China Sea are draped on the continental slope and thickened in complex extensional basins whose linear trends are both parallel and at high angles to the slope trend. Some minor open folding seen mainly at the eastern end of this margin are probably related to compression emanating from the impingement of the collisional arc in Taiwan. Folds appear to be asymmetrically west-vergent, indicating an obducting tectonic framework toward the collisional margin. Indications of hydrate occur in the passive margin sediments of the NW SCS margin (Fig. 1) (Bochu, 1998). Passive margin sediments are likely to be good hosts for hydrate. Similar sediments in the Blake Ridge along the SE U.S. coast retain a porosity on the order of 40-50%, even at more than 500 m depth below the seafloor and below the base of hydrate where hydrate dissociation may have taken place (Dickens et al., 1997).

Sediment thickness contours (Hayes, 1978) show thin sediment in the Heng-Ch'un Ridge. Reflection seismic records (Liu et al., 1993; and Chi, et al., 1998) suggest that the sedimentary succession along the collisional margin is thicker than along the passive margin, which is probably due to tectonic thickening rather than revealing the position of a sediment depocenter.

Regional tectonics along the eastern margin of the South China Sea south of Taiwan are dominated by collisional structures along a structurally young accretionary prism outboard from a volcanic arc. Sediments and the subjacent oceanic crust of the Philippine Sea plate are being actively obducted over the older oceanic crust of the South China Sea basin. Overthrusts and related folding in the accretionary prism are separated from the volcanic arc by a more structurally complex zone containing sediments possibly transported from the deep subduction zone along with possible ophiolites (Huang, 1993) as part of a polycyclic plate history that has not been fully elucidated. Apparent depression of the South China Sea oceanic crust associated owing to impingement of the subduction zone and its consequent loading, is associated with a forearc trench lying upon oceanic crust of the SCS. The forearc trench is filled with relatively thick sediment, on the order of several kilometers thick just west of the deformation front thrust.

Sediments within the accretionary prism of the Heng-Ch'un Ridge appear to be only moderately deformed (Liu et al., 1993), with west verging asymmetrical anticlinal closures between east-dipping imbricate thrusts. Although compaction would be anticipated from the presence of compressional tectonics, the acoustic character of the sediments appears similar where they are folded and unfolded, although this may not be significant with respect to porosity. This similarity could, however, suggest that these dominantly fine-grained sediments contain significant porosity, and probably permeability throughout. These conditions would allow methane to migrate into the hydrate stability zone (HSZ) and be sequestered in gas hydrate.

Gas hydrate has been identified on one multichannel seismic survey line in the area (Liu et al., 1993), but we identify probable hydrate on another four lines of those figured on which hydrate was not identified by the authors. Blanking, transecting of strata by an acoustically blanked zone conforming to the lower HSZ, and discontinuous BSR indicating concentrations of free gas below hydrate, appear to be common. Identification of clear evidence of hydrate on four of the five reflection seismic lines available to us suggests that considerable gas hydrate may occur throughout this area.

3. LOCAL FRAMEWORK FOR METHANE PRODUCTION AND GAS HYDRATE FORMATION

Methane in the oceanic gas hydrate system is confined to the hydrate economic zone (HEZ) which occurs in the upper 1-1.5 km of seafloor sediments where there is methane is concentrated as both hydrate and gas (Max and Chandra, 1998). The character of the sediments, the means of gas generation and migration, and porosity and permeability are all important factors because hydrate forms in the pore space. The ease with which methane and methane-rich fluids can migrate into the HSZ also effects the manner of hydrate accumulation.

Thick sediment successions may develop over considerable periods of time along passive margins where they are unaffected by tectonism and faulting. The character of these sediments is altered principally by compaction which tightens porosity and permeability in a depth-dependent manner. Sediment in active margins, however, are often accumulated as an accretionary prism lying above a subduction zone. The collisional tectonics thicken the sediment pile by thrusting and repetition, as well as focusing sedimentation along a relatively narrow continental slope. Attributes, such as porosity and permeability, can be strongly reduced and highly variable, even in very recently deposited sediments. Fluids carrying thermogenic methane from the subduction zone below the accretionary prism can deliver methane to the HSZ from a source region not available on passive margins.

Where shallow shelves are narrow, organic material in sediment is more likely to arrive in deep water without significant oxidation. The longer sediment is transported before it is buried in water deep enough so that methane generated from its biological detritus can be trapped in a HSZ, the more likely it is that the biological material will be degraded and not be a good source of methane. Sediment can be transported more easily across the very narrow Taiwan continental shelf to the continental slope of the active margin off SW Taiwan than sediment from mainland China can be transported across the broad Chinese continental shelf.

The modern depositional environment has not always existed in this area. During sea level lows, for instance, the shoreline along the Chinese continental shelf was much closer to the slope than it is now. During these times, delivery of terrigenous sediments, likely with significant organic material, whose incorporation in the sediments would be important to the gas-generating character of the passive margin sediments, would be more likely than under present sea level conditions. Thus, sediments along both the active and passive margins probably contain significant organic material that would provide the bacterial feedstock for methane generation.

In the ideal case, good permeability and porosity in the HEZ will yield the greatest concentrations of methane hydrate because greater amounts of methane can migrate and more pore space is available in which the hydrate can form. In low permeability and porosity sediments, methane transport is problematical, except in faults, and the available space in which hydrate can form is much lower (Bolton, 1998).

4. PREDICTIVE AREAS FOR GAS HYDRATE

Three areas of likely methane gas hydrate development have been identified. These areas each have similar interval tectonostratigraphic character, in which sediment thickness and gas-generating capability of sediments are suitably disposed between 1 km and 3.5 km water depth. In addition, seismic reflection data indicate the presence of a BSR in all three areas (Fig. 1).

Area I (Fig. 2) is composed of passive margin sediment built-up to between 1 and 3 km thick. Area II is composed of active margin sediment in a forearc trench on the depressed eastern margin of the South China Sea Plate and in the imbricated sediments of the obducted Heng-Ch'un Ridge. Area III is composed of passive margin sediment that has been encroached upon by the active margin with resultant thick sediment build-up (> 8 km).

Sediments along the Area I and III margins have been accumulating since early Mesozoic (Huang, 1993) during which time considerable hydrocarbon source beds were deposited world-wide (Dow, 1978; Jansa and MacQueen, 1978). Sediments in the Heng-Ch'un Ridge, however, are younger than upper Mesozoic (Huang, 1993). They are more thinly bedded than along the continental margin.

It is in Area III that there is the greatest likelihood of gas generation. The thick sediment build up found here is due to a combination of original sedimentation and tectonic thickening. This zone would have been a depocenter subject to sediment influx from both the NW and the NE.

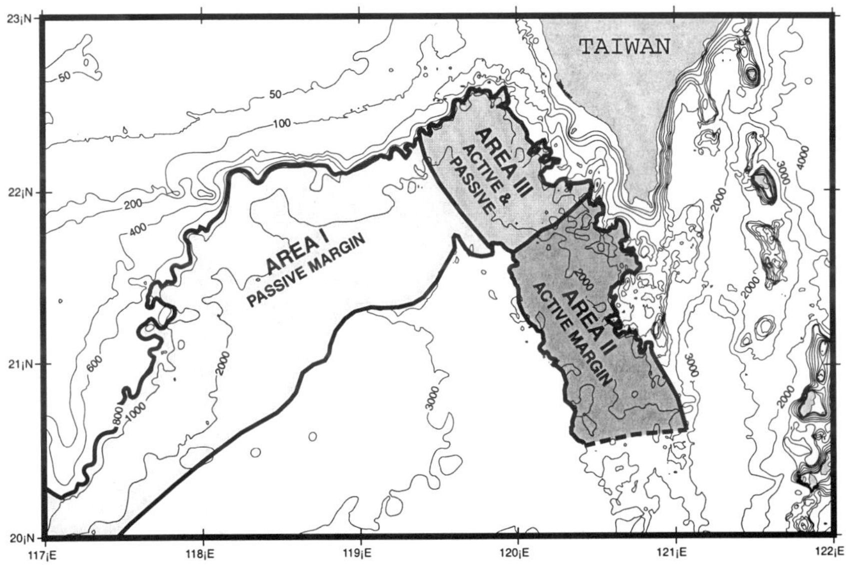

Figure 2. Map of likelihood of hydrate in the sea area immediately south and west of Taiwan. Sediment thickness between 1 km and 3.5 km water depth. Area I, passive margin sediment build-up between 1 and 3 km thick. Area II, active margin sediment and depressed South China Sea Plate trench. Area III, active margin superimposed upon by the active margin.

Uncompacted passive margin sediments may host significant hydrate concentrations. If these sediments are then brought into subduction complexes and accretionary prisms by tectonic encroachment, the tectonism can accentuate hydrate redistribution and methane concentration. The impingement, and apparent SW migration of the western Pacific Plate prong, which passes through Taiwan obliquely into the passive margin sediments should cause concentrations of gas hydrate to form by tectonic accentuation of the cycle of dissociation and reformation of hydrate (Max and Lowrie, 1996).

The hydrate conservation cycle (Max and Lowrie, 1996; 1997) was originally conceived as a response to long-term sedimentation that would result in the base of hydrate rising as a thermogenic response to the elevation of the seafloor. Tectonic activity that raised seafloor, even without the effect of sedimentation and burial, would have the same effect of lowering the seafloor pressure. In a sedimentary prism such as the Heng-Ch'un Ridge, the HSZ of a bathymetrically deep seafloor will thin as a response to lower ambient pressure when the sediment package is tectonically elevated. This causes hydrate stable in the HSZ at depth to dissociate. The concentration of methane in the HSZ by this tectonic process could result in formation of concentrated methane deposits because large amount of methane would be brought into a relatively thin HSZ.

5. CONCLUSIONS

Indications of methane hydrate have been recognized on reflection seismic sections in the northern sector of the South China Sea. A well defined BSR occurs on reflection seismic sections recorded in the sediments of all continental slopes, particularly in active margin sediment prism (Liu et al., 1993; Chi et al., 1998; Bochu, 1998). Because the likelihood of methane generation is locally excellent in the northern sector of the South China Sea on both active and passive margins, gas hydrate is almost certainly more extensively developed than indicated from our limited database.

We have identified three regions in which significant gas hydrate deposits may occur. Hydrate occurs commonly in the accretionary prism sediments to the south of Taiwan. The greatest likelihood of large hydrate deposits occurs in the tectonically reworked sediments of the passive continental margin in the NE corner of the South China Sea, immediately adjacent to Taiwan, across a very narrow continental shelf. This is the ideal case for eventual commercial exploitation of methane from hydrate and possibly trapped gas concentrations.

Acknowledgements Thanks to the Office of Naval Research, PE 0602435N, for funding the bathymetric and geophysical data analysis. The illustrations in this report were created by Irene Jewett of the Marine Physics Branch at the Naval Research Laboratory.

Chapter 20

Introduction to Physical Properties and Elasticity Models

Jack Dvorkin,[1] Michael B. Helgerud,[1] William F. Waite,[1,2] Stephen H. Kirby[2] and Amos Nur[1]
[1]Geophysics Department, Stanford University, Stanford, CA 94305-2215
[2]U.S. Geological Survey, MS/977, Menlo Park, CA 94025

1. INTRODUCTION

Estimating the in situ methane hydrate volume from seismic surveys requires knowledge of the rock physics relations between wave speeds and elastic moduli in hydrate/sediment mixtures. The elastic moduli of hydrate/sediment mixtures depend on the elastic properties of the individual sedimentary particles and the manner in which they are arranged. In this chapter, we present some rock physics data currently available from literature. The unreferenced values in Table I were not measured directly, but were derived from other values in Tables I and II using standard relationships between elastic properties for homogeneous, isotropic material. These derivations allow us to extend the list of physical property estimates, but at the expense of introducing uncertainties due to combining property values measured under different physical conditions. This is most apparent in the case of structure II (sII) hydrate for which very few physical properties have been measured under identical conditions.

Given the inherent variability in field results, the believed relatively small effect of differing guest species will be ignored for most hydrate properties. For this reason, Tables I and II focus primarily on methane and propane hydrate, as examples of structure I (sI) and sII hydrate respectively. Density calculations, however, require specific knowledge of the molecular weight of the guest, and the number of filled cages. Densities are given assuming ideal hydrate cage filling: (guest)•5.75H_20 for sI methane hydrate, and (guest)•17H_20 for sII propane hydrate (only the large cages are occupied). Empty-cage hydrate values, though non-physical, are included for reference. For a more complete listing, see the compilation by Cox (1983).

The elastic property data for ice Ih come primarily from Brillouin spectroscopy data measured by Gammon et al. (1983). From monocrystalline ice Ih data, Gammon et al. (1983) derived wave speed and adiabatic moduli for polycrystalline ice Ih. By combining the adiabatic data with heat capacity and

M.D. Max (ed.), Natural Gas Hydrate in Oceanic and Permafrost Environments, 245–260.

linear expansion data from Table II, isothermal moduli can be calculated using formulations laid out by Anderson (1989).

Waite et al. (in press) used elastic wave pulse transmission to measure the compressional (Vp) and shear (Vs) wave speeds through dense, polycrystalline sI methane hydrate. Waite et al. assumed their material was homogeneous and isotropic and calculated the adiabatic moduli, Vp/Vs and Poisson's Ratio from their simultaneous measurements of Vp and Vs. These values are collected in Table I. Isothermal moduli were derived from these adiabatic values in the same manner as mentioned above for the isothermal ice moduli. These values are also collected in Table I.

Kiefte et al. (1985) and Pandit and King (1982) measured Vp for sII propane hydrate. Kiefte et al. used Brillouin spectroscopy on small volume, relatively pure samples while Pandit and King used elastic-wave pulse transmission on samples that were a mixture of propane hydrate and ice. Pandit and King also measured Vs, and reported values for Vp/Vs and the adiabatic bulk and shear moduli derived from their wave speed measurements. From these reported values, and assuming the material is homogeneous and isotropic, Poisson's ratio and Young's modulus can be derived. The range of tabulated values is given on the low end by Pandit and King's results, and on the high end by combining the Kiefte et al. Vp result with Pandit and King's Vp/Vs ratio. The isothermal moduli are calculated in the same manner as for ice and sI hydrate.

Handa et al. (1984) and Handa (1986) measured the heat capacity at constant pressure for ice Ih and sI and sII hydrate. Their calibration experiments on ice Ih are in agreement with the work by Giauque and Stout (1936). Handa and others publish results as molar heat capacities (J/mol K), but they are reported in Table II in units of J/g K. The nonstoichiometric nature of hydrate makes it uncertain what constitutes a "mole" of hydrate. Handa's results are consistent with a mole of hydrate containing a mole of guest molecules, but even apart from experimental difficulties in establishing the number of guest molecules in a sample, this definition introduces a large apparent difference in the heat capacities of sI and sII hydrate. Because sII with all the large cages filled contains 17 water molecules per guest compared to 5.75 for sI hydrate with all cages filled, the mass of a "mole" of sII hydrate is greater than the mass of a "mole" of sI hydrate as used by Handa et al. Heat capacities are placed on more equal footing if reported per unit mass (J/g K). To convert the Handa et al. molar heat capacity values, the stoichiometry must be known. Handa's assumed stoichiometries, $CH_4{\bullet}6H_20$ and $C_3H_8{\bullet}17H_20$ for methane and propane hydrate, respectively, were used to obtain the values in Table II.

From heat capacity measurements, Handa (1986) deduced the dissociation enthalpies for both hydrate structures. The hydrate enthalpies are given per unit mass for dissociation into ice + gas as well as for liquid + gas. To obtain the liquid + gas enthalpies, Handa added the enthalpy of fusion for water

(Handa et al., 1984) given in Table II to the ice + gas results according to the moles of H_2O present per mole of guest in the hydrate.

Linear thermal expansion coefficients have been measured for ice Ih by LaPlaca and Post (1960), and for sI hydrate by Shpakov et al. (1998) and Tse et al. (in Sloan, 1998). In all cases, the researchers used X-ray diffraction to measure the crystalline unit cell parameter, 'a', for sI and sII hydrate and unit cell parameters 'a' and 'c' for ice Ih. From the variation of the unit cell's volume, V, as a function of temperature, T, the linear thermal expansion coefficient, α, can be calculated according to $\alpha = (1/3)(1/V)(dV/dT)$. It is unclear whether observed differences in α for sI hydrate are due to measurement uncertainties, or real behavioral differences between methane hydrate, measured by Shpakov et al (1998), and ethylene oxide hydrate, measured by Tse et al. (in Sloan, 1998).

A number of researchers have measured the thermal conductivity of gas hydrate and found it to be unexpectedly small compared to ice. The thermal conductivity of hydrate is surprising not only because of its relatively small value, but also because of its anomalous dependence on temperature. Unlike most non-metallic crystalline solids, the thermal conductivity of hydrate rises slightly with temperature (Ross et al., 1981). Tse and White (1981) suggest this deviation from the typical inverse dependence of thermal conductivity on temperature is due to guest molecule vibrations within the hydrate cages. These vibrations can couple with vibrations in the H_20 lattice, reducing the observed thermal conductivity by scattering phonons, the lattice vibration quanta associated with the conduction of heat. In marine and permafrost environments, the hydrate thermal conductivity can be used in estimating the geothermal gradient and heat flow through hydrate bearing sediments (Hyndman et al., 1992; Majorowicz, et al., 1990).

Differences in the dielectric constants between ice and hydrate are due to a lower density of hydrogen bonded water molecules in hydrate relative to ice (Sloan, 1998). In field applications, however, it is the electrical resistivity of hydrate rather than the dielectric constant that is of interest. Though direct measurements of the electrical resistivity of pure hydrate are not currently available, Pearson et al. (1983) and Hyndman et al. (1999) demonstrate a noticeable increase in resistivity for hydrate bearing sediment relative to water saturated sediment. This is due to the replacement of the conductive brine with the resistive hydrate. Through the application of Archie's law, borehole measurements of electrical resistivity have become an important element in estimating the amount of in situ hydrate (Collett, 1998; Hyndman et al., 1999; Helgerud et al., 1999; Guerin et al., 1999). In such calculations, hydrate is assumed to be non-conductive.

Property	Ice (Ih)	sI Hydrate	sII Hydrate
		.79 (empty)	.77 (empty)
Density (g/cc)	.917 [1 f]	.91 (methane) [3 c]	.88 (propane) [3 f]
		1.73 (xenon) [3 c]	.97 (THF) [3 f]
Vp (m/s)	3845 [2 a]	3650 [4 g]	3240, [5 a] 3691 [3 f]
Vs(m/s)	1957 [2 a]	1890 [4 g]	1650, [5 a] 1892
Vp/Vs	1.96	1.93 [4 g]	1.95 [5 e]
Poisson's Ratio	.325 [2 a]	.317 [4 g]	.32
Moduli (GPa):Shear	3.5 [2 a]	3.2 [4 g]	2.4, [5 e] 3.2
Bulk (Adiabatic)	8.9 [2 a]	7.7 [4 g]	5.6, [5 e] 7.8
Bulk (Isotherm.)	8.6	7.2	5.4, 7.5
Young's (Adiabatic)	9.3 [2 a]	8.5 [4 g]	6.3, 8.3
Young's (Isothermal)	9.0	7.9	6.1, 8.1

Table I: Mechanical Properties (Also see Chapter 1)

Property	Ice (Ih)	sI Hydrate	sII Hydrate
Heat Capacity (J/g-K)	2.014 [5 d] 2.097 [5 b]	2.077 [6 d] 2.003 [6 b]	— 2.029 [6 b]
Enthalpy (J/g)	334 [5] —	146 (h→i+g) [6] 437 (h→l+g) [6]	77 (h→i+g) [6] 369 (h→l+g) [6]
Linear Thermal Expansion (10^{-6}1/K)	53 [7 f] 56 [8 f]	87 [9 g] 104 [8 g]	— 64 [8 f]
Thermal Conductivity (W/m-K) [c]	2.23 [8]	.49 [10]	.51 [10]
Dielectric Constant [8]	94	58	58

Table II: Thermal and Electrical Properties (Also see Chapter 1)

a -16°C   d -3°C   g 4°C
b -13°C   e -1.1°C
c -10°C   f 0°C

1. Ginnings et al. (1947); 2. Gammon et al. (1983); 3. Kiefte et al. (1985); 4. Waite et al. (in press); 5. Handa et al. (1984); 6. Handa (1986); 7. LaPlaca and Post (1960); 8. Sloan (1998); 9. Shpakov et al. (1998), 10. Cook and Laubitz (1981).

2. Modeling Wave Speeds in Sediment with Hydrate

2.1. Introduction

Gas hydrates have been found in two general sedimentary environments: (1) clay rich, high porosity ocean and sea bottom sediments; and (2) arctic, onshore sands. In this chapter we present physics based effective medium models for the elastic properties of these two hydrate-sediment systems.

In both environments, we first model the elastic properties of the brine saturated sediment as a function of porosity, mineral and fluid moduli, and effective pressure. Then we modify the model by placing gas hydrate in the pore fluid, in the sediment frame or, for sands, as a grain contact cement. In this chapter, we describe the models in detail and then apply them to well log data.

2.2. Baseline Model for Water Saturated Sediments

The model for water saturated sediments is based on the rock physics model of Dvorkin et al. (1999). This model relates the stiffness of the sediment dry frame to porosity, mineralogy and effective pressure. The effect of water saturation is modeled by Gassmann's (1951) equations. The porosity at which a granular composite ceases to be a suspension and becomes grain supported is called the critical porosity ϕ_c. For a dense random packing of nearly identical spheres, ϕ_c is approximately 0.36-0.40 (Nur et al., 1998). Laboratory experiments have shown that the elastic properties of porous materials are best modeled as mixtures with endmembers of critical porosity and solid material or critical porosity and void space instead of simple mixtures of solid material and void space (Nur et al., 1998). The baseline model for water saturated sediments uses the effective moduli of a dense random packing of identical elastic spheres at critical porosity as its starting point.

The effective bulk K_{HM} and shear G_{HM} moduli of the dry rock frame at ϕ_c are calculated from the Hertz-Mindlin contact theory (Mindlin, 1949):

$$K_{HM} = [\frac{n^2(1-\phi_c)^2 G^2}{18\pi^2(1-\nu)^2} P]^{\frac{1}{3}},$$
$$G_{HM} = \frac{5-4\nu}{5(2-\nu)}[\frac{3n^2(1-\phi_c)^2 G^2}{2\pi^2(1-\nu)^2} P]^{\frac{1}{3}}, \quad (1)$$

where P is the effective pressure; G and ν are the shear modulus and Poisson's ratio of the solid phase, respectively; and n is the average number of contacts per grain in the sphere pack (about 8-9: Dvorkin and Nur, 1996; Murphy, 1985). The effective pressure is calculated as the difference between the lithostatic and hydrostatic pressure:

$$P = (\rho_b - \rho_w) g D \quad (2)$$

where ρ_b is sediment bulk density; ρ_w is water density; g is the acceleration due to gravity; and D is depth below sea floor.

For porosity ϕ less than ϕ_c, the bulk (K_{Dry}) and shear (G_{Dry}) moduli of the dry frame are calculated via the modified lower Hashin-Shtrikman bound (Dvorkin and Nur, 1996):

$$K_{Dry} = [\frac{\phi / \phi_c}{K_{HM} + \frac{4}{3}G_{HM}} + \frac{1 - \phi / \phi_c}{K + \frac{4}{3}G_{HM}}]^{-1} - \frac{4}{3}G_{HM},$$

$$G_{Dry} = [\frac{\phi / \phi_c}{G_{HM} + Z} + \frac{1 - \phi / \phi_c}{G + Z}]^{-1} - Z, \qquad (3)$$

$$Z = \frac{G_{HM}}{6}\left(\frac{9K_{HM} + 8G_{HM}}{K_{HM} + 2G_{HM}}\right),$$

which represents the weakest possible combination of solid and critical porosity material (see Figure 1). K is the bulk modulus of the solid phase. For porosity above critical, K_{Dry} and G_{Dry} are calculated via the modified upper H-S bound:

$$K_{Dry} = [\frac{(1 - \phi) / (1 - \phi_c)}{K_{HM} + \frac{4}{3}G_{HM}} + \frac{(\phi - \phi_c) / (1 - \phi_c)}{\frac{4}{3}G_{HM}}]^{-1} - \frac{4}{3}G_{HM},$$

$$G_{Dry} = [\frac{(1 - \phi) / (1 - \phi_c)}{G_{HM} + Z} + \frac{(\phi - \phi_c) / (1 - \phi_c)}{Z}]^{-1} - Z, \qquad (4)$$

which represents the strongest possible combination of critical porosity material and void space (Figure 1).

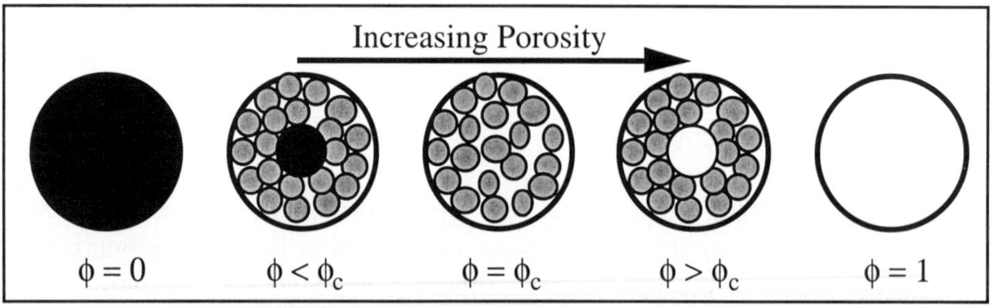

Figure 1. Schematically going from zero to 100% porosity through critical porosity.

If the sediment is saturated with pore fluid of bulk modulus K_f, the shear modulus G_{Sat} and the bulk modulus K_{Sat} are calculated from Gassmann's (1951) equations as

$$G_{Sat} = G_{Dry}, \quad K_{Sat} = K\frac{\phi K_{Dry} - (1 + \phi)K_f K_{Dry} / K + K_f}{(1 - \phi)K_f + \phi K - K_f K_{Dry} / K}. \qquad (5)$$

Once the elastic moduli are known, the elastic wave speeds are calculated from

$$V_p = \sqrt{(K_{Sat} + 4/3\, G_{Sat}) / \rho_B}, \quad V_s = \sqrt{G_{Sat} / \rho_B}. \tag{6}$$

where ρ_B is bulk density.

In the common case of mixed mineralogy, the effective elastic constants of the solid phase can be calculated from those of the individual mineral constituents using Hill's (1952) average formula:

$$K = \frac{1}{2}[\sum_{i=1}^{m} f_i K_i + (\sum_{i=1}^{m} f_i / K_i)^{-1}],$$
$$G = \frac{1}{2}[\sum_{i=1}^{m} f_i G_i + (\sum_{i=1}^{m} f_i / G_i)^{-1}]; \tag{7}$$

where m is the number of mineral constituents; f_i is the volumetric fraction of the i-th constituent of the solid phase; and K_i and G_i are the bulk and shear moduli of the i-th constituent, respectively. The solid phase density is calculated as

$$\rho = \sum_{i=1}^{m} f_i \rho_i, \tag{8}$$

where ρ_i is the density of the i-th constituent.

2.3. Gas Hydrate Part of Pore Fluid Model

If gas hydrate forms in the pore spaces of sediments such that it is suspended in the pore fluid, then its presence affects only the elastic moduli of the fluid and the bulk density of the sample. Therefore, the elastic moduli of the dry sediment are those given by the baseline model. The volumetric concentration of gas hydrate in the pore space is given by $S_h = C_h / \phi$, where C_h is the volumetric concentration of the gas hydrate in the rock. If we assume that gas hydrate and water are homogeneously mixed throughout the pore space, the effective bulk modulus of the water/hydrate pore fluid mixture is the Reuss (1929) isostress average of the water and gas hydrate bulk moduli (K_f and K_h, respectively):

$$\overline{K}_f = \left[\frac{S_h}{K_h} + \frac{1 - S_h}{K_f} \right]^{-1}. \tag{9}$$

The bulk modulus of the saturated sediment is calculated from those of the dry frame using $\overline{K}_f$ as the pore fluid bulk modulus in Gassmann's equations (5).

The shear modulus of the saturated sediment is that of the dry frame. The shear modulus of the pore fluid mixture is zero. This is the limiting factor on this model. It can only be used when the pore fluid does not affect the shear modulus of the overall sediment. This means the model is only appropriate when the gas hydrate does not fill the entire pore space (e.g., $S_h < 0.9$).

2.4. Gas Hydrate Part of the Solid Phase Model

If gas hydrate is instead acting as a component of the load bearing sediment framework, then the original dry sediment calculations must be altered to account for the changes in the effective solid phase (mineral plus hydrate) moduli and sediment porosity. The presence of gas hydrate reduces the porosity of the original sediment ϕ to a new value $\bar{\phi} = \phi - C_h$.

The effective mineral modulus for the gas hydrate/sediment solid phase is calculated from Equation (7) where f_i should be replaced by

$$\bar{f}_i = \frac{f_i(1-\phi)}{1-\phi+C_h}, \tag{10}$$

and gas hydrate should be treated as an additional mineral component with fraction $\bar{f}_h$ given by

$$\bar{f}_h = \frac{C_h}{1-\phi+C_h}. \tag{11}$$

The elastic moduli and velocities of the water saturated gas hydrate bearing sediments are then calculated using Gassmann's equations (5) where porosity and bulk and shear moduli of the solid phase are replaced with the model-appropriate values.

This model is appropriate for any porosity and any fractional filling of the pore space by gas hydrate.

2.5. Gas Hydrate as Contact Cement

The third modeling possibility is the case where gas hydrate forms as an intergranular cement. The basis for this model is the contact cement theory (CCT) of Dvorkin et al. (1994), which calculates the bulk and shear moduli of a dense, random packing of spherical elastic grains with small amounts of elastic cement at the grain contacts. This model is appropriate for granular sediments whose original porosity is approximately that of a sphere pack, namely 36 to 40%. The original CCT theory was only valid for small concentrations of cement or, equivalently, residual porosities greater than approximately 25%. Recently, Dvorkin et al. (1999) extended CCT to high cement concentrations (i.e., residual porosities < 25%) where even the entire pore space may be filled with cement. Together, these theories provide a method for calculating the elastic moduli of dry sediment comprised of sand grains cemented by gas hydrate. The remainder of the pore space is assumed filled with water and Gassmann's equation is used to calculate the bulk modulus of the saturated sediment. The inputs for CCT and its extension to high cement concentration are the porosity of the original (without gas hydrate) sediment; the amount of cement (gas hydrate); the elastic moduli of the original mineral phase; the elastic moduli of the cement (gas hydrate); and the coordination number of the grain pack.

The CCT model is mathematically realized for two specific distributions of gas hydrate (Figure 2): (1) Scheme 1, gas hydrate forms at the grain contacts and (2) Scheme 2, gas hydrate evenly coats the grains.

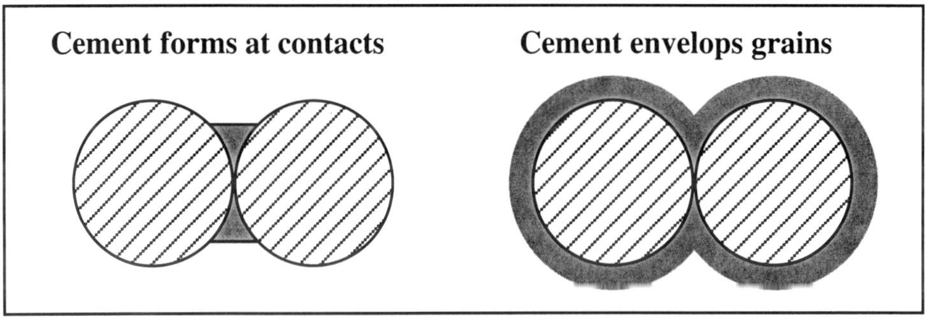

Figure 2. Hydrate as contact cement. Scheme 1 and 2.

The arrangement represented by Scheme 1 is much stiffer than that given by Scheme 2 for the same amount of cement. Also, the residual pore spaces have very different shapes. Scheme 1 produces roughly equidimensional residual porosity while Scheme 2 is star-shaped. The modeling at high cement concentrations is greatly simplified by assuming equidimensional residual porosity (Dvorkin et al., 1999). For this reason, only Scheme 1 is used to model high gas hydrate cement concentration.

To construct the general solution, we begin with the CCT model which describes the elastic moduli of a dense, random pack of identical elastic spheres with a small amount of gas hydrate elastic cement. The effective bulk and shear moduli of the dry cemented sphere pack are predicted to be (Dvorkin et al., 1994; Dvorkin and Nur, 1996):

$$\begin{aligned} K_{CCT} &= \frac{n(1-\phi_c)}{6}(K_c + \frac{4}{3}G_c)S_n, \\ G_{CCT} &= \frac{3}{5}K_{CCT} + \frac{3n(1-\phi_c)}{20}G_c S_\tau, \end{aligned} \tag{12}$$

where K_c and G_c are the bulk and shear moduli of the cement, respectively; ϕ_c is the (critical) porosity of the uncemented grain pack (ϕ_c = 36 to 40%); and n (about 8.5) is the average number of contacts per grain.

Parameters S_n and S_τ are the solutions to integral equations. Dvorkin and Nur (1996) supply the following approximate solution:

$$S_n = A_n(\Lambda_n)\alpha^2 + B_n(\Lambda_n)\alpha + C_n(\Lambda_n),\ A_n(\Lambda_n) = -0.024153 \cdot \Lambda_n^{-1.3646},$$
$$B_n(\Lambda_n) = 0.20405 \cdot \Lambda_n^{-0.89008},\ C_n(\Lambda_n) = 0.00024649 \cdot \Lambda_n^{-1.9864};$$
$$S_\tau = A_\tau(\Lambda_\tau, \nu)\alpha^2 + B_\tau(\Lambda_\tau, \nu)\alpha + C_\tau(\Lambda_\tau, \nu),$$
$$A_\tau(\Lambda_\tau, \nu) = -10^{-2} \cdot (2.26\nu^2 + 2.07\nu + 2.3) \cdot \Lambda_\tau^{0.079\nu^2 + 0.1754\nu - 1.342},$$
$$B_\tau(\Lambda_\tau, \nu) = (0.0573\nu^2 + 0.0937\nu + 0.202) \cdot \Lambda_\tau^{0.0274\nu^2 + 0.0529\nu - 0.8765},$$
$$C_\tau(\Lambda_\tau, \nu) = 10^{-4} \cdot (9.654\nu^2 + 4.945\nu + 3.1) \cdot \Lambda_\tau^{0.01867\nu^2 + 0.4011\nu - 1.8186};$$
$$\Lambda_n = \frac{2G_c}{\pi G}\frac{(1-\nu)(1-\nu_c)}{1-2\nu_c},\ \Lambda_\tau = \frac{G_c}{\pi G};$$

where G and ν are the shear modulus and Poisson's ratio of the grain material, respectively; and G_c and ν_c are the shear modulus and Poisson's ratio of the cement, respectively.

Parameter α depends on the cement distribution. For Scheme 1 (Figure 2) α has the form:

$$\alpha = 2[\frac{\phi_c - \bar{\phi}}{3n(1-\phi_c)}]^{0.25}.$$

For Scheme 2 α has the form:

$$\alpha = [\frac{2(\phi_c - \bar{\phi})}{3(1-\phi_c)}]^{0.5}.$$

The details of calculating the elastic moduli of the pack at low residual porosities are given in Dvorkin et al. (1999).

3. Testing Models with Data

In this section, we apply the models to real world data sets obtained on ODP Leg 164 for gas hydrates in ocean bottom sediments off the southeastern coast of the United States and Northwest Eileen State Well, No. 2 drilled onshore in the Alaskan Arctic. Both sites have a fairly complete set of well logs. The ODP data were acquired in November and December of 1995. The Eileen data are considerably older (1972).

3.2. Northwest Eileen State Well No. 2

We apply the above theoretical models to the well log data collected in Northwest Eileen State Well #2, located onshore North Slope of Alaska. The interval 550-830 m was determined to contain three methane gas hydrate bearing sand intervals (discussed in Mathews, 1986). The gamma ray, resistivity, neutron porosity (with sandstone correction, Schlumberger, 1989), and sonic velocity log curves are plotted versus depth in Figures 3a-d. The correlation of high velocity with high resistivity in the sand intervals, along with gas shows in the mud log (Mathews, 1986), is consistent with the presence of gas hydrate in this well.

In order to provide inputs for the models, we assume, for simplicity, that the mineral phase in the well is pure quartz, and the pore fluid is brine with a salinity of 32,000 ppm and average temperature and pore pressure of 10 °C and 7 MPa, respectively. The density and bulk modulus of the brine are calculated according to Batzle and Wang (1992) and do not vary much in our temperature and pressure regions of interest. The values given in Table III are used for the entire section under investigation. We also assume that the gas hydrate-free porosity of the sand (ϕ) is given by the corrected neutron porosity log. This assumption is based on the fact that the hydrogen density in methane hydrate is essentially the same as that in liquid water (~0.1 mol H/cm^3). Therefore, to first order, the neutron tool should not be able to distinguish between gas hydrate and water in the pore space. The effective pressure, 6 to 9 MPa, is calculated as the difference between the lithostatic and hydrostatic pressures where the average rock bulk density is taken as 2.12 g/cm^3 and the density of brine as 1.024 g/cm^3.

Constituent	Bulk Modulus (GPa)	Shear Modulus (GPa)	Density (g/cm^3)
Quartz	36.6	45.0	2.65
Clay	20.9	6.85	2.58
Calcite	77.8	32	2.71
Methane Hydrate	7.7	3.2	0.90
Brine (Eileen)	2.29	0	1.024
Brine (ODP 995)	2.5	0	1.032

Table III. Input parameters

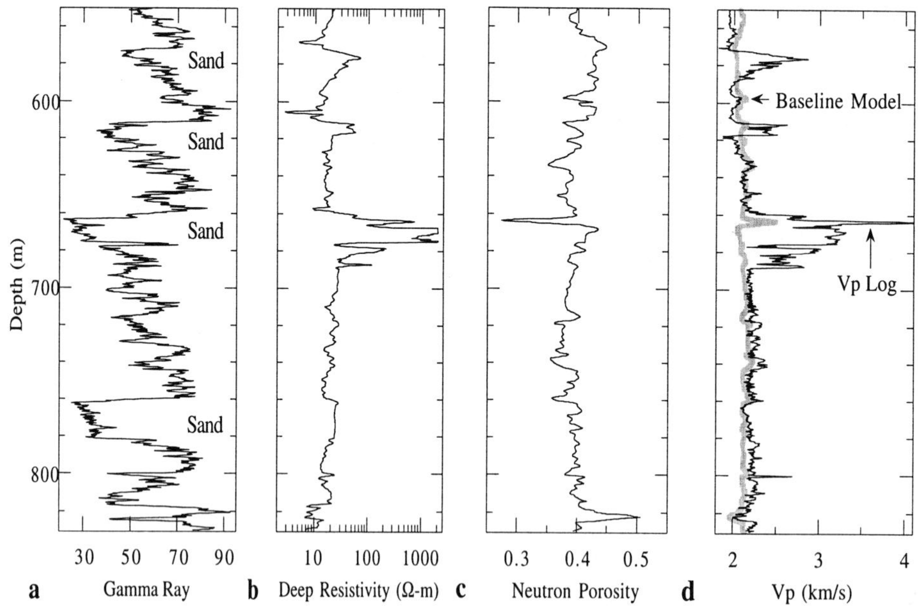

Figure 3. Physical property logs versus depth from Northwest Eileen State Well #2. (a) Gamma ray (b) Deep resistivity. (c) Neutron porosity. (d) Compressional-wave velocity (black) and the baseline model result (gray).

First, we use these inputs to calculate the baseline velocity in the well, assuming that the pore space is fully saturated with brine and does not contain hydrate. The coordination number is fixed at 8.3 corresponding to an average porosity of 0.4 (Murphy, 1982), which we use as the critical porosity. The calculated baseline velocity (Figure 3d) closely matches the measured background velocity. This justifies our choice of background model and input parameters. Next, to model the effect of hydrate on the sediment, we need an estimate of the amount of gas hydrate in the pore space. We calculate the non-water saturation of the pore space from resistivity using the "quick-look" Archie method (e.g., Collett, 1998), which is based on the equation

$$1 - S_w = 1 - \left(\frac{R_0}{R_t} \right)^{1/n}, \tag{13}$$

where S_w is water saturation of the pore space; R_0 is the resistivity of the formation at S_w = 100%; R_t is the formation's true (i.e., measured) resistivity; and n is an empirical constant (about 1.94, Pearson et al., 1986). It is assumed that the R_0 versus depth trend is the same as the background trend of the resistivity log. The choice of background trend is subjective. The trend is supposed to follow the data where gas hydrate is presumably absent and thus highlight the resistivity peaks. In Figure 4a, we offer three linear background fits to the data above 700 m and a single linear fit to the rest of the interval. The corresponding non-water saturation is shown in Figure 4b. We assume that all non-water saturation is methane hydrate. Easily seen in Figure 4b is the high estimated gas hydrate saturation in three of the sandy intervals. Also evident is the lack of precision (±20%) inherent in this technique. However, within the range of saturation values predicted (~20-80%), the models make easily distinguishable velocity predictions.

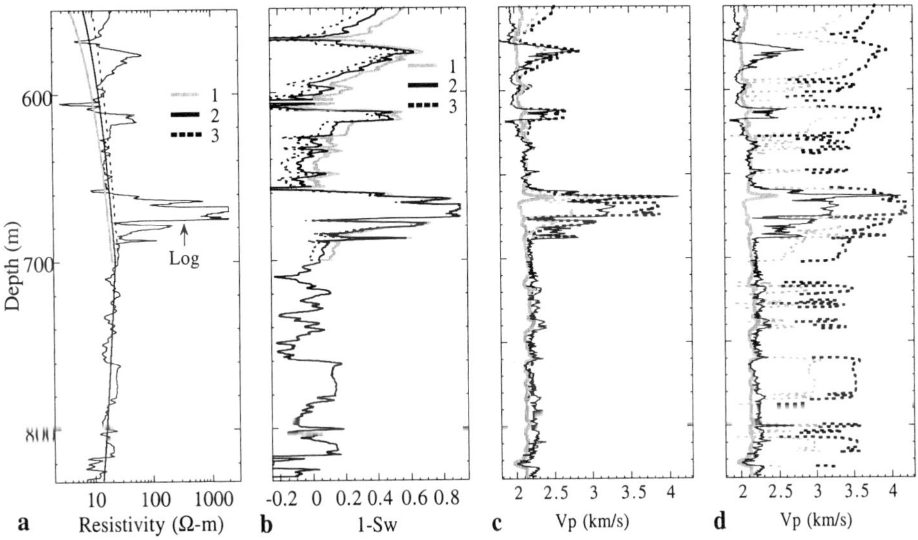

Figure 4. (a) The deep resistivity curve vs. depth and three "quick-look" Archie background resistivity estimates. (b) Non-water saturation from "quick-look" Archie method for the background curves shown in (a). The negative saturation values are artifacts and those intervals are ignored in the elastic moduli calculations. (c) Velocity versus depth using curve 2 in (b). Black: log data; gray: baseline model; gray dashed: gas hydrate in fluid; black dashed: gas hydrate in frame. (d) Contact cement model. Black dashed: Scheme "1"; gray dashed: Scheme "2". The gaps in the velocity curves are where the non-water saturation estimate is negative or where a model is not applicable (see Figure 2 and text).

In our further modeling, we assume the hydrate saturation is given by the middle curve "2" in Figure 4b. The results of applying the "gas hydrate in fluid" and "gas hydrate in solid" models are shown in Figure 4c. Both closely match the velocity data. In contrast, the contact-cement models significantly overestimate the measured velocities (Figure 4d). These modeling results show that in spite of the uncertainty in hydrate saturation estimates, the vastly different predictions of the models allow us to conclude that gas hydrate does not cement the grain contacts at this site.

3.3. Offshore ODP holes 994, 995 and 997

We apply our theory to calculating compressional-wave velocity at ODP Leg 164, Site 995. This site is located off the coast of the southeastern United States on the Blake-Bahama Ridge in a water depth of 2800m. Available data include well-logs, core analyses, and VSP data as well as independent hydrate concentration estimates from chlorinity anomalies, resistivity logs and methane gas volumes evolved from a pressure core sampler.

The interval of interest extends from 190 meters below the sea floor (mbsf) to the bottom of the hole. This interval is lithologically uniform and is predominantly comprised of clay, calcite and quartz (ODP 164, 1996). For the purpose of calculating the effective moduli of the solid component we estimate volume percentages of clay, calcite and quartz as 60, 35 and 5 percent, respectively, consistent with smear slide and XRD data from the site. The elastic properties and densities used in this study for clay, calcite and quartz are shown in Table III.

The bulk modulus and density of sea water in the formation were calculated versus depth as a function of temperature and pressure following Batzle and Wang (1992). The average bulk modulus was 2.5 GPa. Density was practically constant at 1.032 g/cm^3. Effective pressure was calculated as the difference between the overburden and hydrostatic pressure. Porosity was taken from core measurements. Two other input parameters needed in our model, coordination number and critical porosity, were taken at 9 and 0.39, respectively. The modeling results do not depend strongly on these parameters, provided physically reasonable values are used.

In Figure 5a, we plot sonic log data (smoothed with a 61-point median filter) together with the calculated velocity profiles for 0% bulk hydrate (100% water saturation), 2% bulk hydrate, 4% bulk hydrate and 8% bulk hydrate. These calculations are made using the hydrate-part-of-fluid model. Figure 5b shows modeling results using the hydrate-part-of-solid model.

Chlorinity data and pressure core sampler gas volumes suggest that hydrate is located in the interval from 190 to 450 mbsf (Dickens et al., 1997; ODP 164, 1996). The 100% water saturation line in Figures 5a and b intersects the velocity curve at approximately these depths, confirming the validity of our baseline model. The constant gas hydrate saturation lines in Figures 5a and 5b are level curves for estimating bulk hydrate percent from sonic logs. These estimates are similar to independent hydrate estimates from resistivity (Figure 5c), chlorinity and PCS evolved gas data.

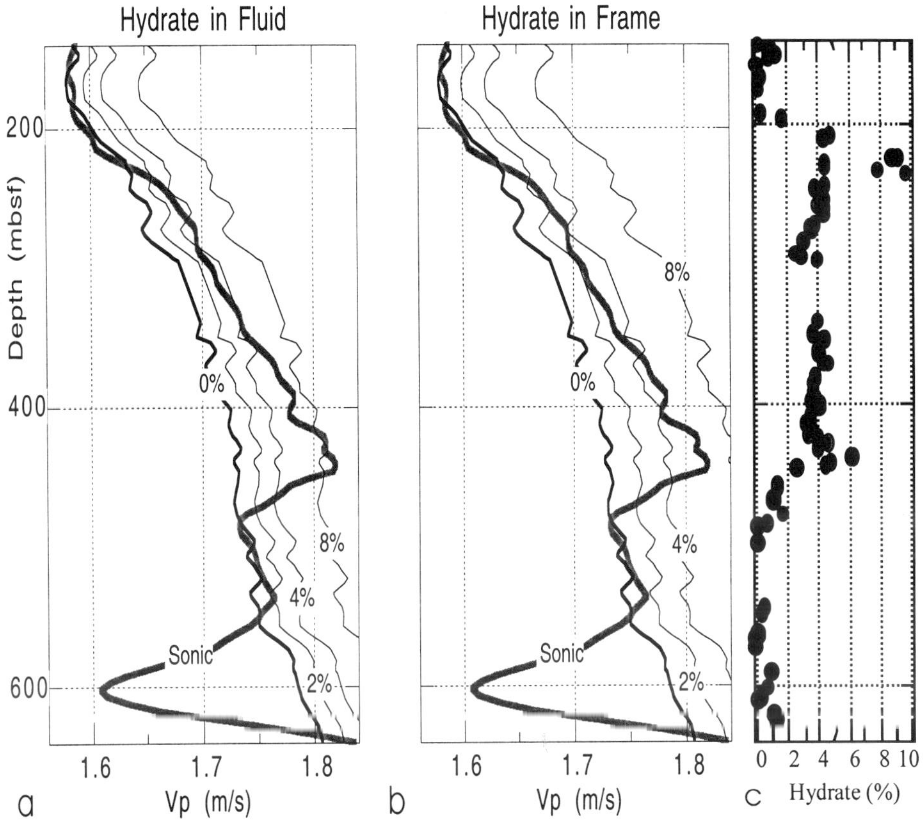

Figure 5. Hydrate concentration from compressional wave velocity log data and resistivity logs. (a) Comparison of well log *Vp* with model results assuming that hydrate is part of the pore fluid. (b) Comparison of well log *Vp* and model results assuming hydrate is part of the sediment frame. For both (a) and (b), model values were calculated at core depths and the results fit with smoothed curves. (c) Hydrate concentration estimate from resistivity log.

This approach to quantifying the effect of gas hydrate on the geoacoustic properties of marine sediment can also be used with seismic data (Ecker et al., 2000). Other similar approaches are presented in Chapters 9 and 25.

4. Conclusion

Rock physics links the elastic properties of the sediment to the amount of gas hydrate present, its location in the pore space, and other parameters (porosity, mineral properties, effective pressure, and pore-fluid compressibility). These elastic properties, in turn, determine the wave speeds and, eventually, seismic signatures of natural gas hydrate reservoirs. As a result, seismic data acquired over prospective gas hydrate deposits can be interpreted in terms of gas hydrate concentration and location in the pore space. The rock physics approach described in this chapter has been used by Ecker et al. (2000) to predict the gas hydrate saturation of the pore space from seismic in the Blake Outer Ridge. Applying quantitative rock physics models to well log and seismic data acquired from sediments containing gas hydrate is the future of gas hydrate reservoir characterization.

Chapter 21

Geophysical Sensing and Hydrate

Peter R. Miles
Southampton Oceanography Centre
Southampton SO14 3ZH, UK.

1. INTRODUCTION

The geographic extent of hydrate deposits and their distribution within the sediment column are still relatively undefined. Both single and multi-channel seismic reflection surveys remain the principle methods of identifying the presence of hydrate but the effects of methane flux through the seabed is best visualized with several other acoustic sensors, namely side-scan. The strong acoustic impedance contrast reflector at the base of the layer - the bottom simulating reflector or BSR - is normally seen where free methane is present beneath the hydrate. Hydrate formation may also cause blanking of the sediment acoustic stratigraphy through cementation of the sediment structure.

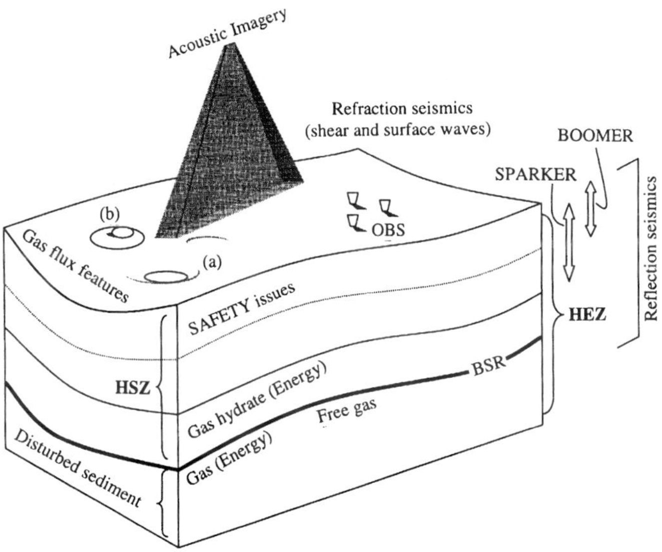

Figure 1. Schematic of geophysical tool targets for characterising hydrate. (a) pockmark; (b) mud volcanoe; HSZ = hydrate stability zone; HEZ = hydrate economic zone.

M.D. Max (ed.), Natural Gas Hydrate in Oceanic and Permafrost Environments, 261–274.

Deep ocean surveys using relatively high resolution side-scan sonar acoustic imagery of the seafloor can reveal pockmarks, mud diapirs and mud volcanoes that are often related to methane flux. These surveys have also mapped large-scale slope disturbances and sediment mass wasting attributed to hydrate instability (Kenyon, 1987;Masson *et al.*, 1997). Larger scale failures appear to correlate with regions of higher hydrate occurrence, at least on the European margins (Miles, 1995). However this relationship can be tenuous in some localities, such as in the Mediterranean Sea, where hydrates are not known to occur widely but are still associated with large scale sediment movement (Rothwell *et al.*, 1998). These factors illustrate the need for quantitative measurement of hydrate distribution in both area and depth. The only tools that can achieve this on the large scale at acceptable cost are geophysical.

To date the acoustic systems and processing were usually designed to reveal hydrocarbon exploration targets or crustal features of scientific interest. However hydrates are confined to the uppermost 0.5 to 2 km of marine sediments and these systems need to be re-tuned when mapping hydrate resource issues at between 200 and 800 m depth in the seafloor. The impact of this hydrate on seafloor stability, safety and the environment is probably confined to the top few hundred meters of sediment (Figure 1).

A commercially successful strategy for methane production from hydrate does not yet exist, though Japan has started to assess their resources as reserves. Therefore we consider effective geophysical techniques for identifying and characterizing the various hydrate issues. Technical specifications are listed in Table 1.

2. GEOPHYSICAL ASPECTS OF THE HYDRATE ZONES

The chemistry of methane hydrates is well described in this volume. Geophysically the impact of hydrate formation is that it strengthens the sediment. If this occurs prior to sedimentary lithification, as it often does, then removal of this rigidity on dissociation can affect sediment stability, particularly in the presence of external disturbance. Therefore the main geophysical effect of hydrate formation derives from the rigidity given to the sediment and this is a consequence of the change in physical properties. Both compressional wave velocity (Vp) and shear wave velocity (Vs) depend on the physical properties of the sediment through which they propagate, therefore if:

K is the sediment bulk modulus - the force required to change volume,
μ is the shear modulus – the force required to change shape, and
ρ is the density, then

$$Vp = \sqrt{((K+4\mu/3)/\rho)} \quad \text{and} \quad Vs = \sqrt{(\mu/\rho)}$$

This shows that Vp > Vs and that the existence of μ not only determines Vp but is necessary for Vs to exist.

Vp will increase from ~1.6 km s^{-1} in normal sediment up to 2.5 km s^{-1} or more in the presence of hydrate. Vs can rise by orders of magnitude and dramatically affects the acoustic properties of the sediment, particularly its ability to transmit shear waves.

The base of the Hydrate Stability Zone (HSZ) shown in figure 2 is bounded by Vp > 1.6 km s^{-1} within the hydrate and Vp < 1.6km s^{-1} below, particularly if there is free methane in the pore spaces. This causes a negative acoustic impedance contrast at the BSR that will reflect seismic energy as an inverse wavelet compared to that of the seafloor and most other reflections. This characteristic and the blanking enables seismic profiling to identify the presence of hydrate and the reflector at the base of a hydrate layer as a BSR.

Below about 450 meters water depth, the temperature - pressure conditions in the sediment are appropriate for hydrate formation, as they are in most ocean basins (Miles, 1995). The geochemical conditions for methane production however are most commonly met along continental margins rather than in the deep oceans and these are well documented in other chapters.

The Hydrate Economic Zone (HEZ, figure 1) is the combined hydrate, methane, and deeper sediment zone from which it is important to characterize methane and the geotechnical properties that bear on the methane recovery (Max and Chandra, 1998). It includes the HSZ and deeper methane and pore fluid zones that are methane rich. In an area where sedimentation has continued over a long period of time, hydrate at the base of the HSZ may become unstable through rising geotherms and dissociate. Methane produced in or below sediment from dissociated hydrate will rise through buoyancy into the HSZ and tend to again form hydrate. The actual process will depend on the local sedimentology and is still a subject of discussion. However this is the zone in which the geotechnical parameters must be extremely well known if safe extraction of methane from methane reservoirs is to be engineered. It is presently estimated that this economic zone is no more than 1.5 to 2 times the thickness of the HSZ. Below this zone, where sediment compaction and

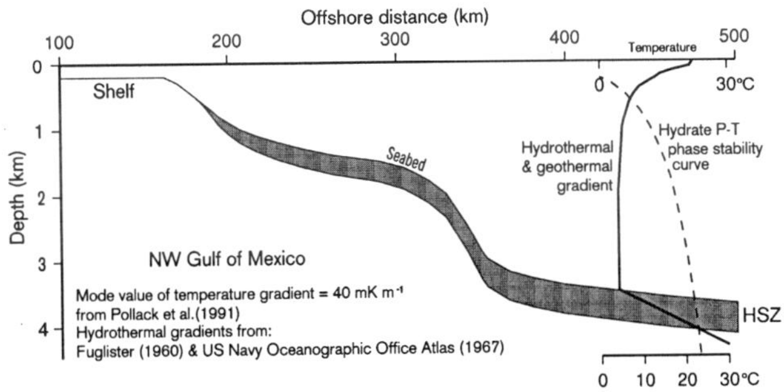

Fig. 2. Hydrate stability P-T phase boundary (dashed) applied to sea temperature and geothermal gradient (solid) offshore Galvaston TX. The calculated thickness of HSZ (Miles, 1995) is the potential maximum depth to that methane hydrate can remain stable, assuming sufficient sediment thickness.

geotechnical properties are more normal, it is likely that no geophysical data need be recovered for economic exploitation of the HEZ.

World-wide the scale and distribution of hydrate deposits appears comparable and controlled by similar circumstances. The sediments in which hydrates form do not appear to exert a significant influence over their nucleation or growth but this has not been quantified. Thus for any geophysical analysis of hydrate the broadest range of sediment types must be anticipated.

3. GEOPHYSICAL SURVEY OBJECTIVES

Concentrated and dispersed hydrate deposits occur in different locations within the seafloor sediments. Hydrate in the lower part of the HSZ could comprise the energy resource potential and create the primary survey interest. This is currently driving research and development interest in both India and Japan. These deposits are near the base of the HSZ where the hydrate will be most concentrated. From a safety aspect, dispersed hydrate at or near the seafloor is seen as a hazard to sediment stability and any engineering operations sited on it. The stability of the seafloor on a geological scale however is important in shaping the continental margin. The redistribution of sediments away from continental margins by sediment mass wasting (Dillon et al., 1995) and the possible formation of Tsunami associated with these events highlight the potential environmental impact. Therefore geophysical surveys will be directed toward the resource, safety or environmental issues, each requiring specific tools determined by the depth of penetration or aerial coverage required.

In order to acquire the basic information required from the HEZ it is necessary to undertake both regional and detailed geophysical surveys that may need to be calibrated by ground-truth sampling. This involves an integrated survey and analysis strategy to bring together the morphology, geology, sedimentology and physical properties of the HEZ.

One of the primary issues for hydrate research is the use and interpretation of acoustic energy to determine the disposition and volume of hydrate and methane in oceanic marine sediments. With a more detailed understanding of the distribution of hydrate and associated methane deposits, a more realistic assessment of the basic process of marine sedimentation and hydrate diagenesis can be made.

4. SEISMIC REFLECTION

This is the most common marine geophysical tool. It involves a discharge seismic source, often an array of airguns and a receiver (Figures 3 & 4). Sediments that contain hydrate will normally show Vp higher than in normal oceanic sediment. Free methane accumulating in the pore spaces beneath the HSZ will cause a significantly lower Vp. Seismic velocity normally increases with depth so this velocity inversion, or negative acoustic impedance contrast, may exhibit the distinctive BSR reflection on seismic records. As the HSZ follows the shape of the seabed, at a near constant depth beneath the seafloor so does the BSR, which marks its base. The BSR is particularly well developed where free methane exists in the pore spacesbeneath the HSZ, the base of which is most likely to contain concentrated hydrate deposits. The BSR will not always be visible and it may vary in amplitude. The Ocean Drilling Program Leg 164 attempted to quantify this variation (Paull et al., 1996). Cementation of the sediment pore spaces also may exhibit seismic blanking on record sections as the inherent acoustic structure within the sediment becomes masked. Velocity analysis from multichannel reflection seismic data, especially where controlled by drilling, is the main interpretive method for estimating methane volume in oceanic hydrate (Holbrook et al, 1996).

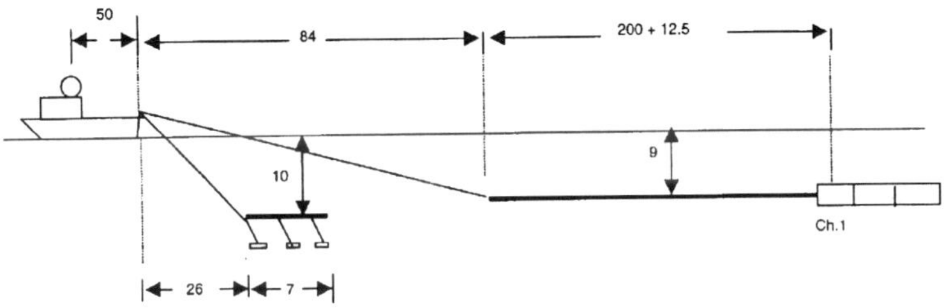

Figure 3 Standard seismic reflection configuration showing relationship between airgun array and streamer. Distances are in metres.

Single channel data may be adequate for the primary task of identifying the presence of hydrate through recognition of hydrate blanking and BSR distribution. In fact single-channel data may produce a more accurate graphic impression of the seismic structure of hydrate and associated methane deposits than processed multi-channel data. The principle advantages of single channel systems are that it is less expensive to operate, can be acquired by none multi-channel specialist vessels and surveyed at higher speeds. It is also noted that many surveys of these data already exist for many continental margin areas. These data can be valuable for regional desk studies. Some data of this type is currently archived by the National Geophysical Data Center in Boulder and the European Commission SEISCAN project (Miles et al., 1997). Figure 4 shows a typical air gun source beam array being deployed. Gun sources can be used singly or as multiple arrays depending on the complexity of the desired source signature (frequency) and energy requirement for depth penetration.

Multi-channel seismic data enables velocity analysis for sequence depth, thickness estimates, and the potential location of hydrate concentrations. However, sedimentary sequence velocity measurements are most accurately measured using expendable sonobuoys – these provide wide angle information to 15 km rather than being restricted to the NMO dictated by the receiver array length. In practice multi-channel velocities are used for stacking and extrapolating sonobuoy results. Normal industry standard data has limited resolution along the shot line and is probably unable to identify significant horizontal velocity changes with a resolution of less than 0.25 - 0.5 km at best. This is because the separation angles between incident and reflected energy are very small in the 1-5 km water depths where hydrate occurs.

More complex acquisition (3D) and processing of multi-channel data, such as amplitude variation with offset (AVO) may also help estimate hydrate density and subjacent methane concentrations at shallow depths below sea level (Andreassen et al, 1997).

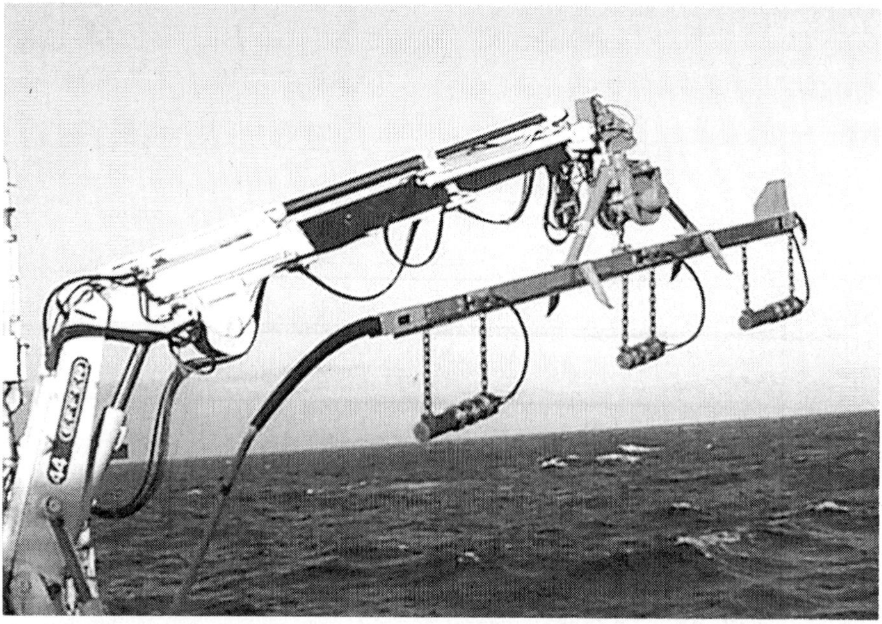

Figure 4. Three-airgun beam array being deployed (D. Booth, SOC).

4.1. High-resolution Boomer and Sparker

These are also impulse discharge sources. Examples are low energy, high-resolution systems (EG&G model 265 UNIBOOM), and medium energy, medium resolution (SPARKER) systems. These can be towed at the same time, with alternate energy sources both received on a single towed array of hydrophones. This technique allows the characteristics of both the uppermost and somewhat deeper strata and structure to be imaged together. Considerable quantities of paper records of single channel data exist, but this is from equipment of an older type that is being replaced by digitally recorded and processed multi-frequency and multi-channel seismic systems.

4.2. Chirp sonar

Chirp systems are sub-bottom profilers so named because of their wideband FM sound sources (e.g., 2-16 kHz over 20 ms). These systems belong to the linear driven source group driving an amplifier with a known voltage waveform. At frequencies at the low end of the linear source range, the acoustic return received at the hydrophone is matched filtered with the outgoing FM pulse to generate a high-resolution image of the stratigraphy. These systems are most successful at higher frequencies (3.5kHz and up), can produce sediment classification based on sediment reflection coefficients, but have restricted depth penetration (about 60m).

Another alternative technology involves the flextensional transducer, a hollow elliptical cylinder (Class IV). This is driven by a stack of ceramic elements mounted internally across the major diameter of the cylinder causing a large volume change. They have been most widely used in the frequency range 1-3 kHz, but designs exist for 300-500 Hz systems (Oswin and Dunn, 1988) although they have a limited operational depth range to about 300 m.

Figure 5. Depth controller 'bird' being affixed to multi-channel streamer on deployment (D. Booth, SOC).

4.3. Special seismic systems

Special seismic systems are those with a source and receiver array geometry different to the conventional position near the sea surface. Several systems have been used to investigate marine sediments containing hydrate. They produce multi-channel seismic datasets that provide very detailed velocity analyses. These analyses can be processed to generate estimates of hydrate and methane occurrence.

A high frequency instrument for seafloor characterization of fine-scale velocity structure in near-seafloor sediments is the Deep-Tow Acoustic

Geophysical System (DTAGS). It consists of a Helmholtz resonator with a broad frequency output between 250 and 650 Hz, and a 24 channel (42 m spacing) towed array, both are towed 500 to 700 m above the seafloor (Rowe and Gettrust, 1993). Penetration depths have proven to be on the order of 700 m. The higher frequency and low attenuation from volume allows 2 to 3 m scale resolution of small geologic features in the HEZ while the larger angles between transmitted and received energy in this deep-tow configuration allows for fine scale 2-D velocity analyses. DTAGS has been used successfully to map hydrate BSRs.

Other specialised systems with either shallow-deep configurations (Pasisar) or narrow beam higher frequency sub bottom profilers (Parasound) can be developed for hydrate applications. They may make a significant contribution to detecting and quantifying the near-surface occurrence of hydrate and be particularly applicable where geotechnical information is required.

4.4. Hydrate specific seismic objectives

Apart from identifying hydrate we need to be able to measure its concentration and physical properties. Collecting seismic data in both single and multi-channel modes permits a wide variety of processing and visualization techniques to be applied. These can reveal the geological and acoustic impedance structures in sufficient detail to identify drilling targets, determine depths, and provide input to economic assessment. Because the economic targets of the oceanic hydrate system occur in the upper 0.25-1.5 km the precise nature of the mechanical strengthening conferred by the formation of hydrate is very important. In addition to the increase in strength, the deformational character of the hydrate is important for determining the level of stress, and the resultant adaptation to strain, that can be tolerated during methane extraction operations (Stern et al., 1996). These can create abnormal thermal and pressure gradients that are important in developing safe extraction strategies.

Although simple pore filling increases the bulk modulus of the sediment, the shear modulus and sediment strength is much less affected than where cementation takes place. It is likely that analyses of the relationship of pressure to Vs and S-wave attenuation in the HEZ may be required to accurately classify the geotechnical properties of hydrate and associated methane deposits. S-wave velocities have been determined from incident P-wave source acoustic experiments (Caiti et al., 1991) and conventional seismic surveys, but more fine-scale characterization is probably necessary. The ideal survey would consist of concurrent P-wave and S-wave measurement using a hydrophone array close to or at the seafloor. This would allow measurement and calibration of incident energy levels at a variety of frequencies and the associated attenuation necessary for estimating methane concentrations and percentages.

4.5. Choosing seismic systems for hydrate assessment

Seismic sources and hydrophone arrays have been developed for the deeper targets in the hydrocarbon industry. The receiver system is generally an array of point receivers up to several km in length controlled at operational depth by 'birds' (Figure 5). This surface-surface system becomes less effective in deep-water because the geometry reduces the CDP advantages, but it is very practicable for a deep-tow system, if slow to operate. However the surface-surface system will remain a primary survey tool because of its availability and practicality.

Deep-tow operations are slow. Therefore in order to better monitor the acoustic energy entering the sediment from surface-surface systems it is desirable to measure the incident wavelet. This can be achieved by locating a hydrophone near the seafloor and will enable attenuation to be used for the analysis of both hydrate and methane charged sediments. This monitoring can also be achieved with a single hydrophone suspended at manageable depths or by using stationary ocean bottom seismometers (OBS), although these are an expensive addition to the operations.

An advantage of OBSs is that additional seismic information such as shear-wave arrivals and surface waves can be observed. Both provide valuable information on Vs at different depths beneath the seabed. For the top 50-100 m Scholte waves, a solid-liquid interface wave similar to a Raleigh wave, shows frequency dispersion from seabed impulse sources. In the presence of a positive S-wave velocity gradient the lower frequency energy travels faster. The seismogram can be inverted to give the velocity structure. Additional S-wave data can be obtained by OBS operated in the conventional refraction mode but owing to the expense of operation should only be considered when specific problems need to be addressed.

Both compressional-wave (Vp) and shear-wave (Vs) velocities should be used to characterize (pore filling) bulk modulus and (cementation) shear modulus properties. Normal seismic reflection profiling (single or multi-channel) operates principally with P-wave arrivals. However S-waves are often generated by the incident wavelet at the seabed or some depth below it. The problem in recording these S-wave paths using a streamer array is that they have to be re-converted back to P-waves prior to re-entering the water layer which generates some unknowns in the interpretation. Hence the desirability of seabed monitoring.

4.6. Seismic processing for hydrate imaging

The information required from the seismic data requires reassessment of the processing that will be most effective for interpretation. The relative importance of seismic data at different levels of resolution has yet to be determined for accurate measurement of the hydrate resource. Most work published on seismic imaging of hydrate has involved data acquired during hydrocarbon exploration. These were designed for deeper targets and may not be most suitable for hydrate confined to the upper 1.5 - 2 km. Whether multi-channel seismic processing is required for information in addition to velocity analyses is uncertain, but single channel data may effectively image the BSR. This will result in records with the least likelihood of processing artifacts that may be miss-interpreted as real impedance horizons related to the hydrate cycle. Thus, acquiring seismic data with the least filtering is desirable. In practical terms concurrent single and multi-channel modes is desirable, if optimistic.

Another aspect concerns the seismic data that already exists. Much of this multi- and single channel data is suitable for reprocessing to re-interpret the relevant targets. This does not involve the expense of data acquisition and is particularly in line with current thinking on environmental information usage. Data obtained for site surveys falls into this category. Regional appraisal can also utilize the vast amount of older data that may exit from surveys beyond those covered by contemporary prospects. Databases for these sections do exist and even those not available in digital form can be transcribed for re-processing, analysis and presentation. These data searches should form part of an initial desk study.

5. ACOUSTIC IMAGERY

5.1. Side-Scan Sonar

Side-scan sonar systems use variations in the intensity and travel time of back-scattered acoustic energy to produce seafloor acoustic imagery. These acoustic intensities are dependent on the scattering strength of the seafloor, distributions of scatterers (e.g., hydrates), degree of bottom penetration, sub-bottom volume scattering, water column characteristics and insonification angle. This is the main tool for imaging seafloor morphology and texture. Acoustic imagery is critical to assessing methane flux from the seafloor. Ocean floor targets for high resolution studies need to be selected from wide-area studies, which can be achieved with very broad swaths and hence, minimum survey time. Higher resolution systems can be used for greater resolution in detailed study areas.

The seafloor imagery contains information about seafloor stability, fluid movement, and structure. Although multibeam bathymetric mapping of hydrate study areas will resolve the bathymetry better than that from the side-scan sonar survey, its high resolution may not be useful for hydrate research characterization without complimentary seafloor imagery. Bathymetric side-scan systems are currently capable of measuring the topography of the seafloor to a precision of approximately 3% of the water depth.

The backscatter effect associated with methane passage through seafloor, or with the presence of dispersed hydrate at or near the seafloor is not known. This type of acoustic analysis is a new research area. Indeed, not enough is yet known about sediment response without the presence of hydrate to predict its application to quantitative dispersed hydrate analysis. It will be necessary to study the in-situ geotechnical target, available equipment optimized for dispersed hydrate research, and the matrix of information required to support the side scan to allow calibration of imagery and backscatter.

5.2. GLORIA

The GLORIA system maps the seabed in swaths up to 60 kms wide, depending on water depth. Operating at 6.5 kHz, the resolution of the system is 50 m. GLORIA is an important regional reconnaissance tool in economic surveys, not only for the quality of the images it produces, but also in its 6-8 knot survey speed. This enables wide area surveys such as an Exclusive Economic Zone (EEZ) or a complete continental margin segment to be assessed prior to site survey selection. The lower operating frequency of the system is known to be capable of penetrating several meters into unconsolidated sediments (Hugget et al, 1992). This enables the system to image structures beneath the masking sediment veneers not seen with higher frequency systems.

Figure 6. Schematic of TOBI deep-tow system.

5.3. TOBI

TOBI provides high quality three-dimensional images of the sea floor. It is a deep-towed highly flexible instrument operational in water depths to 6,000 m at 1.5-3 knots. It incorporates a 30 kHz, 6 km swath, 6 m resolution side-scan sonar with a swath bathymetry. It also has a 3 component fluxgate magnetometer and 7.5 kHz sub-bottom profiler (Figure 6).

The tow cable required to maintain a survey altitude of 300-400 m above the seabed is about 1.5 times the water depth. It has proven useful for both exploration and engineering surveys. The Deep-tow platform which houses TOBI also has a 7.5 kHz sub-bottom profiler which penetrates up to 40 m in soft sediment but much less in sand.

5.4. SEAMARC:

This is a deep seafloor mapping system with a number of varyants. SeaMarc II consists of two sets of acoustic arrays mounted in a five meter long neutrally buoyant vehicle that is shallow-towed below the top of the seasonal thermocline approximately 400 meters behind the ship. The acoustic arrays transmit a signal at 11 & 12 kHz. Bathymetric information is generally reliable out past the first surface-bottom multiple (equivalent to 1.7 times the water depth per side of the tow-fish). Typically a signal pulse produces an average swath width of 10 km (5 km per side) of back-scatter imagery. The system is towed 50-100 m below the sea surface at an average speed of 8 knots. Horizontal resolutions of ca 20-50 m for side-scan and 25-100 m for bathymetry are typical in 500 to 2500 m water depths. Advanced data processing allows for production of high resolution, equal area, essentially geo-referenced datasets.

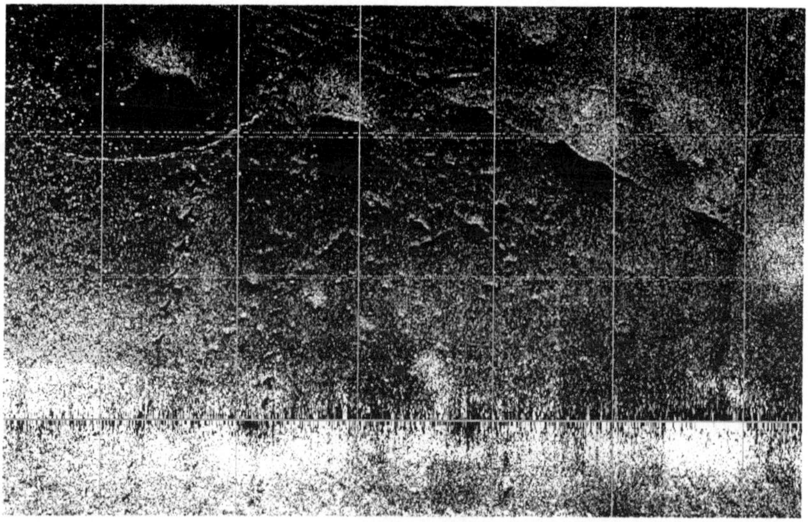

Figure 7. TOBI image of methane pockmarks at 350 metres in the Barents Sea (B. Murton, SOC).

5.5. High resolution side-scan

There are a large number of relatively inexpensive, small and easy to handle side-scans available. Most are usually restricted to shallow water less than 250 m. A frequency of 100 kHz is commonly used, which results in a wavelength of 1.5 cm and a minimum image resolution of about 0.25 m. The frequency can be changed without altering tow configuration or ship speed. Automatic, real-time data processing and equal-area registration of pixels commonly results in a very detailed acoustic image of the seafloor in shallow water.

Synthetic Aperture Sonar (SAS) systems are capable of collecting acoustic backscatter imagery with a high azimuthal resolution. The geometric operational requirements of the system demand a relatively slow tow speed or a reduction in the effective swath width. Not commercially available at present they have the potential to provide very high resolution seafloor images that could avoid optical surveys.

The parametric side-scan sonar is a dual frequency interferometric side-scan sub-bottom swath system. It can penetrate to tens of metres into the seafloor and is ideal for discerning dispersed or localised hydrate in the near seafloor zone.

5.6. Multi-beam bathymetry

Bathymetry is a primary exploration parameter. Low amplitude ridges can trap methane beneath the HSZ as it follows the seafloor. These systems are often hull mounted but smaller systems are designed for temporary installation. They can be operated at 10-12 knots at high resolution. Other wide-beam bathymetry is available from swath mapping systems such as TOBI.

Multibeam bathymetry can be processed as imagery and merged with side-scan and seismic data to provide a fully morphological 3-D representation of the HEZ. This is particularly useful in mapping methane flux features such as pock marks and mud volcanoes. Processing multibeam bathymetric data for its complete time-series signal content also may provide a new method for determining seafloor and immediately sub-seafloor physical properties from which the presence and character of disseminated hydrate might be determined

6. HYDRATE SURVEY DESIGN

Geophysical sensing of hydrate has many similarities to other large- scale issues of marine science. The difference is in the detail – the frequencies used, the configuration of the equipment and in the planning. To date few surveys have been undertaken using equipment optimized for hydrate measurement. A great deal of information relevant to the identification and characterization of oceanic hydrate and associated methane deposits already exists. Much of this can be revisited and reinterpreted, some can be reprocessed or reassessed.

Therefore the first task in assessing hydrate in any area is a desk study. It is imperative to quickly assess all existing information, preferably in a modern GIS-relational database. In those areas where large conventional seismic area surveys of regional reconnaissance character already exist (Booth et al., 1996), these data can be reassessed and reprocessed for detail in the upper 1.5-2 km. The anticipated effects of hydrate in marine sediment should be modeled to identify the best acoustic energy frequency and power to reach selected targets within the HEZ. They can then be studied in detail.

Regional reconnaissance with acoustic imagery will assist in targeting resource potential and minimize cost. Existing geophysical equipment can be re-calibrated, where possible, to produce the best hydrate-sensitive systems for new areas. Detailed deep-tow surveys and analyses are required to follow-up in specific localities.

In-situ ground-truth measurement of seabed physical properties can be made with new instruments such as SAPPA in water depths to 6,000 m. Vp and attenuation can be measured as a function of depth with a vertical resolution of a few cm. . Horizontally and vertically polarised Vs can be measured in surface sediments. Two cone penetrometers can be employed.

Control sites for both concentrated and dispersed hydrate should be calibrated and monitored. Provision should be made to compile all hydrate and methane flux data within a linked map-based, environmental information system using modern relational database and web interface technologies. Much information already exists. The survey strategy will be guided by the information available for a specific locality. This will determine which tools are most appropriate for hydrate sensing and the configuration of their application.

INSTRUMENT	TYPE	SPEED **(kts)**	f **(Hz)**	**S/R**	PENET-RATION
Single channel seismic	impulse	8	30	s-s	>2 km
Multi-channel	impulse	5	30	s-s	>2 km
3-D seismic	impulse	5	30	s-s	>2 km
Boomer	impulse	6	150	s-s	200 m
Sparker	impulse	6	200	s-s	500 m
Sub-bottom profiler	chirp	8	3.5k	s-s	60 m
Flextensional TD	linear	5	500k	ns	200 m
DTAGS	resonator	8	350	dt	700 m
Pasisar	impulse	2	30	s-d	1 km
Parasound	chirp	12	2.5	h	<100 m
GLORIA	side-scan	8	6.5k	ns	~10 m
TOBI TOBI profiler	side-scan	1.5-3	30k 7.5k	dt	- <70 m
SeaMarc	side-scan	8	12k	ns	-
High resolution profiler	side-scan	8	100k	ns	-
Synthetic aperture (SAS)	sonar	1.5		ns	-
Parametric s/s	sonar			ns	~25 m
Multi-beam bathymetry	chirp	8	12k	h	-
SAPPA	direct	-	10k	sb	1 m
OBS (Vp) (Vs) Scholte wave	impulse	-	50 2	sb	>2 km 50 m

Table 1. Instrumentation specification. s-s, surface source and receiver; s-d, surface source deep-tow receiver; ns, near surface; dt, deep-tow; h, hull mounted ; sb, seabed

7. CONCLUSIONS

Many acoustic systems exist that can be used to image the effect of hydrate in marine sediments. It may be necessary to acquire data from more than one system to derive complex datasets that can be used to fully quantify the affects of hydrate and the engineering properties of potential reservoirs.

Chapter 22

Seismic Methods for Detecting and Quantifying Marine Methane Hydrate/Free Gas Reservoirs

Ingo A. Pecher
University of Texas Institute for Geophysics
Austin, TX 78759

W. Steven Holbrook
Department of Geology and Geophysics
University of Wyoming
Laramie, WY 82071

1. INTRODUCTION

Seismic methods are the most widely used approach for indirect detection and quantification of gas hydrate in marine sediments. Historically, the presence of methane hydrate has been inferred on the basis of bottom simulating reflections (BSRs), which mark the phase boundary between hydrate and the underlying free gas zone (e.g. Shipley et al., 1979). In addition to their association with BSRs, hydrates and the underlying free gas affect the elastic properties of the host sediment in ways that are seismically detectable. Partial replacement of pore fluid by rigid gas hydrate causes an increase of both compressional wave velocity (Vp) and shear wave velocity (Vs) (Chapter 20), while the presence of free gas will strongly decrease Vp. Compressional- and shear-wave attenuation (Q_p^{-1}, Q_s^{-1}) may also prove to be hydrate/gas indicators: hydrate may increase both Q_p and Q_s, while gas certainly decreases Q_p (Wood and Ruppel, 2000). Accurate, detailed images of the elastic properties of hydrate deposits therefore hold great promise for remotely quantifying hydrate and gas occurrence and concentrations.

In this chapter, we will address three aspects of the use of seismic experiments for marine gas hydrate quantification, starting with the most indirect approach:

- Detection of conditions that favor gas hydrate occurrence, from regional aspects (stratigraphy, sedimentation rate) to small-scale aspects (migration paths, traps).
- Quantification of gas hydrate regionally, e.g., to enhance global balances, with a focus on large-scale distribution and average saturation.

M.D. Max (ed.), Natural Gas Hydrate in Oceanic and Permafrost Environments, 275–294.

- Quantification of gas hydrate locally, e.g., for safety considerations or perhaps production of methane with a focus on locally high gas hydrate concentrations ("sweet spots").

Although we focus on marine hydrates, many of the concepts we discuss will also apply to permafrost hydrate, with the caveat that BSRs are far less common onshore than offshore. We begin with a brief review of basic seismic principles as they apply to hydrate/gas systems.

2. BASICS

2.1. Wave Types

Both types of body waves, compressional (P-) and shear (S-) waves, are useful in probing hydrate deposits. Shear waves transmit a particle motion that is perpendicular (transverse) to their travel direction. Their speed is controlled by the shear modulus (or "rigidity") and density. Compressional waves have a particle motion parallel to the propagation direction and always travel at a faster speed than S-waves. Their velocity is determined by both compressional and shear modulus, as well as density. Fluids and gases have no "rigidity" (i.e., shear modulus = 0), so they cannot transmit S-waves.

Predicting the P- and S-wave velocities of hydrate- and gas-bearing sediments is not straightforward, since those sediments consist of multi-phase aggregations of minerals (commonly clay, quartz, and calcite), pore fluid (typically saline water), and hydrate or gas. Sediment V_p and V_s are commonly estimated from the properties of the mineral components, pore fill, and the sediment frame (which accounts for the geometry how the grains packed together). Two kinds of approaches have been used to estimate the properties of hydrate- and gas-bearing sediments: empirical approaches, which rely on versions of the Wyllie time-average or Wood equations calibrated to laboratory data [*Lee et al.*, 1996], or physics-based approaches, which use versions of Biot-Gassmann theory [*Helgerud et al.*, 1999]. Uncertainties in both of these approaches arise primarily from the current lack of understanding as to whether hydrates behave as a "cement" between the mineral grains and significantly stiffen the sediment frame, whether they behave as mineral grains as part of the sediment frame, or whether they modify the properties of the pore fill [*Helgerud et al.*, 1999].

With those caveats in mind, several generalizations can be drawn. Because of the low P-wave velocity of gases, sediment Vp is particularly sensitive to even small amounts of gas in the pore space (e.g., (Domenico, 1976)). Sediment Vs, however, is controlled only by grain and frame properties; since the pore fill (gas or fluids) will not transmit shear energy, sediment Vs is relatively unaffected by gas (a small change is predicted due to lower density of gas). Vs is particularly sensitive to changes in the frame "stiffness" due to adding gas hydrates into the frame as additional grains or as cement between the sediment grains. Mathematically, however, the effect of gas hydrates on sediment Vs (and Vp) depends largely on the microscopic details of gas hydrate distribution within the sediments (also see Chapter 20).

2.2. Reflection, Transmission, and P-to-S Conversion

Many of the inferences drawn about the elastic properties of hydrate deposits rely on observations of the traveltimes and amplitudes of reflected and converted waves. In this section we give a brief overview of basic principles of the reflection, transmission, and conversion of seismic waves at a boundary (Sheriff and Geldart, 1995).

2.2.1. Normal incidence

A P-wave that strikes an interface at normal incidence is partitioned into reflected and transmitted P-waves. The precise partitioning of amplitudes depends on the contrast of seismic impedance, Z, which is the product of density and Vp. The reflection coefficient (R) is the fraction of reflected amplitude, the transmission coefficient (T) of transmitted amplitude

$$R = A_r/A_i \,, \quad T = A_t/A_i \qquad (1)$$

where Ai, Ar, and At are the amplitudes of the incident, reflected, and transmitted wave.

$$R = (Z_2 - Z_1)/(Z_2+Z_1) \qquad (2),$$

where 1 and 2 denote the upper and lower layers. This amplitude is lost for the transmitted wave:

$$T = 2\,Z_1/(Z_2+Z_1) \qquad (3)$$

These are the equations for amplitude, not energy; it is possible for the amplitude of a transmitted wave to exceed the amplitude of the incident wave. R can be positive or, if the impedance of the lower layer is less than that of the upper layer, negative. A negative reflection coefficient causes a reversed polarity of the reflected wave from the interface. BSRs, the reflections from the base of the hydrate stability zone, have a reversed polarity, due to the negative impedance contrast from hydrate-bearing sediment overlying gas-charged sediment (Vp is drastically reduced in the presence of gas). Normal-incidence reflection coefficients of BSRs are frequently used as an indicator of the impedance contrast at the hydrate/gas interface (Hyndman and Spence; 1992; Katzman et al., 1994).

2.2.2. Oblique incidence

Obliquely incident waves provide important information on P-wave velocities in hydrate deposits through two observable properties: traveltimes and amplitudes. Wide-angle traveltimes provide the long-wavelength velocity structure of the sediment column through normal-moveout and stacking velocity analysis (e.g.,

(Ecker et al., 2000; Hyndman and Spence, 1992)), forward modeling, or inversion ((Katzman et al., 1994; Tinivella and Accaino, 2000)). The accuracy of velocities determined by traveltime analysis increases with the maximum incidence angle. In deep water, for example, long-offset ocean-bottom seismic data (e.g., (Katzman et al., 1994; Korenaga et al., 1997)) generally provides better velocity determinations than data acquired on a streamer of limited length.

Partitioning of the amplitude of a P-wave that impinges obliquely on a boundary is described by a complex set of equations, the Zoeppritz equations (Zoeppritz, 1919). Assuming an incoming P-wave, two important phenomena occur: (1) The slope of R as function of incidence angle depends on the shear modulus and hence Vs (e.g. Ostrander, 1984). (2) At oblique angles, some amplitude is turned into a S-wave (P-to-S converted wave, PS-wave), due to the horizontal traction imposed by the incident P-wave. Figure 1 shows amplitude partitioning for different contrasts of seismic properties at layer boundaries.

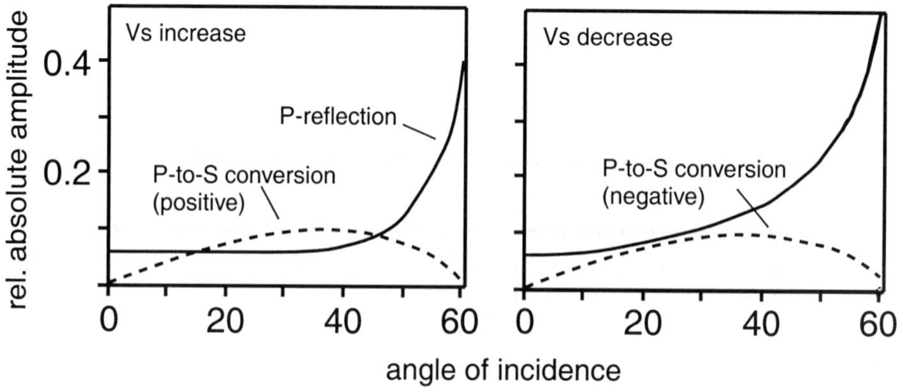

Fig. 1: Amplitude as a function of incidence angle for an incident P-wave, reflected P-wave and reflected P-to-S converted wave. Absolute values normalized to the amplitude of the incoming P-wave. Elastic parameters are realistic for unconsolidated marine sediments close to the seafloor: Upper layer: Vp: 1600 m/s; Vs=200 m/s (left), 400 m/s (right), density=1700 kg/m3. Lower layer: : Vp: 1600 m/s; Vs=400 m/s (left), 200 m/s (right), density=1700 kg/m3. Note the differences in the P-reflection response depending on whether Vs increases or decreases. Reflected P-amplitudes increase until all P-energy is reflected at the critical angle, 62.7° for this velocity contrast. The P-to-S arrival has a negative polarity for a Vs decrease (where the polarity of the arrival for a Vs increase is defined as positive).

Seismic rays that hit a layer boundary at an oblique angle follow Snell's law of optics. The incidence angles of incoming, reflected and transmitted P- and S-waves are inversely related to the layer velocities. Hence

$$(\sin \theta_1)/V_1 = (\sin \theta_2)/V_2 \tag{4}$$

Figure 2 shows schematic raypaths across a layer boundary and the annotation used in this chapter.

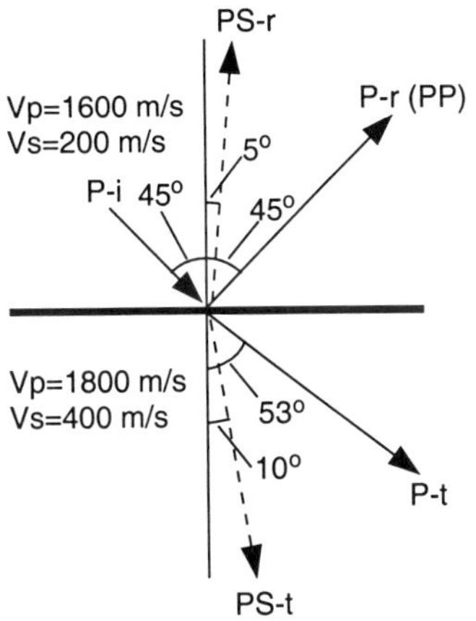

Fig. 2: Ray paths according to Snell's law, for an incoming P-wave (P-i). Elastic model as in Fig. 1, with a Vs increase. P-r: reflected P-wave (usually labeled PP); PS-r: reflected P-to-S converted wave (usually referred to as "PS-wave"); P-t: transmitted P-wave; PS-t: transmitted P-to-S converted wave.

Amplitude analysis provides information that cannot be gleaned from traveltime analysis alone. For example, amplitude-versus-offset modeling is widely used to estimate Vs contrasts from P-wave data. Waveform inversion can provide accurate models of short-wavelength P-velocity variations, as discussed in a later section (Korenaga et al., 1997; Minshull et al., 1994; Pecher et al., 1996a; Singh et al., 1993).

2.2.3. Resolution

The vertical resolution of seismic reflection data depends on the wavelength of the reflected wavelet. It is commonly assumed that without special processing (see e.g. waveform inversion below), 1/4 of a wavelength can be resolved. Assuming a Vp of 1700 m/s (which is typical for sediments in which gas hydrates may occur), and a dominant frequency of 50 Hz (which is at the higher end of conventional seismic surveys), the dominant wavelength is 34 m, implying a vertical resolution of about 8 m. For data with dominant frequency of 400 Hz, in contrast, the dominant wavelength is only 3 m, with resolvable layer thickness of about 1 m (e.g., Fig. 3). The limits of vertical resolution are immediately visible when wiggles from reflections appear to overlap each other. Horizontal resolution is defined by the Fresnel zone, which depends on reflector depth, frequency and velocity (Sheriff and Geldart, 1995). For a velocity of 1700 m/s, a reflector depth of 1000 m, and a frequency of 50 Hz, the Fresnel zone is about 130 m. Thus changes in BSR properties over lateral distances of less than 130 m cannot be resolved. This is often not considered when interpreting seismic data.

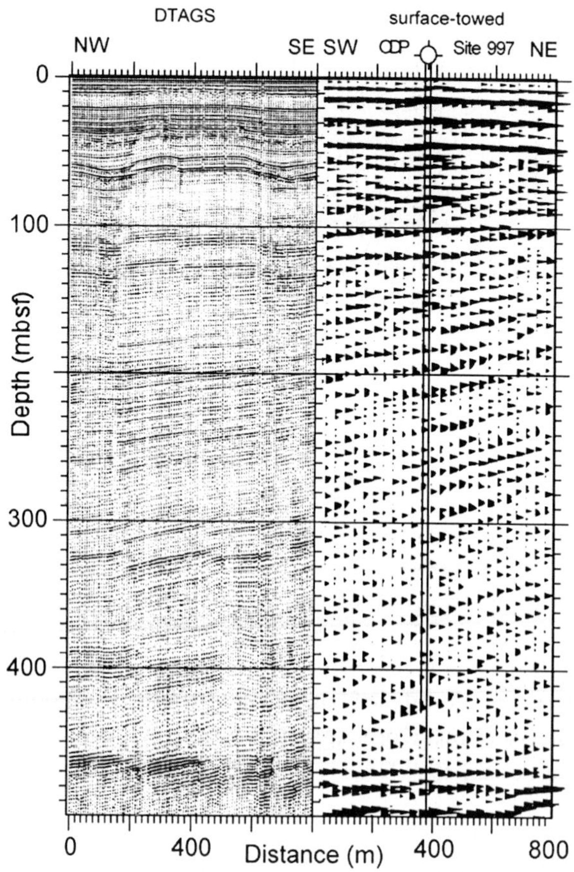

Fig. 3: Conventional seismic reflection profile (right) and DTAGS deep-towed data (left) over ODP Site 997 on the Blake Ridge, after Wood and Ruppel (2000). Data are stretched to depth using velocities measured in the borehole. The BSR is at ~460 meters beneath the seafloor (mbsf). Note that the BSR appears as a continuous reflection in the conventional data, whereas it appears to be "shingled" in the higher-frequency DTAGS data. Also note the much better resolution of near-vertical faults in the DTAGS data.

2.2.4. Attenuation

An obvious approach to improving resolution of seismic data is to use higher frequencies, thus generating waves of shorter wavelength. However, in practice it can be difficult to transmit high frequencies through sediments, since intrinsic attenuation (i.e., attenuation due to intergranular friction) is proportional to the number of wave oscillations and thus increases with frequency. A second form of attenuation is scattering due to small heterogeneities (smaller than a seismic wavelength) or thin layers, which can also be frequency-dependent. Attenuation thus causes an inherent trade-off between resolution and penetration of seismic waves: greater penetration generally requires lower frequencies, which have lower resolution. Seismologists often describe attenuation by referring to its inverse, the quality factor (Q_p or Q_s).

Research into the characteristics of P- and S-wave attenuation in hydrate/gas deposits is in its infancy. Little quantitative information is available about the effects of hydrate on Qp and Qs, though hydrate should increase Qp and Qs (decrease attenuation) if it acts as an intragranular cement. Gas-charged sediments attenuate P-waves, which often precludes imaging beneath gas zones. A recent study of P-wave attenuation in the Blake Ridge hydrate province showed a strong decrease in Qp from the hydrate zone to the gas zone (Wood et al., 2000).

2.2.5. Seismic acquisition

Several types of seismic data are useful in determining the elastic properties and stratigraphic geometries of hydrate/gas deposits.

Seismic reflection data typically have source-receiver offsets that are less than the target depth. At sea, reflection data are typically acquired on a hydrophone streamer towed behind the ship; in single-channel seismic (SCS) data, the streamer has only one channel, while multichannel seismic (MCS) data are acquired on a longer streamer with multiple hydrophones. SCS data can provide excellent images of hydrate/gas deposits (Dillon and Drury, 1996; Katzman, et al., 1994) but provide no information on seismic velocities. MCS data provide both structural images (enhanced by stacking) and velocity information from reflection moveout. Sources used in seismic reflection profiling include airguns and waterguns.

Ocean-bottom seismometers/hydrophones data are acquired by recording surface shots (e.g., airguns) on fixed ocean-bottom seismometers (OBS) or hydrophones (OBH). The longer source-receiver offsets provided by this geometry allow improved resolution of layer velocities, better AVO analysis (Katzman et al., 1994), and more accurate waveform inversions (Korenaga et al., 1997).

- On-bottom refraction data are also recorded on an OBH or OBS, but have the added advantage of moving the seismic source to the seafloor. By placing both source and receiver on the seafloor, the water column is effectively removed from the refraction experiment, allowing refractions from within the hydrate zone to be directly observed. Acquiring such data requires special deep-towed source equipment, such as NOBEL (Christeson et al., 1993), which fires explosive charges on or near the seafloor. NOBEL has not yet been applied to hydrate/gas deposits but holds great promise for improving measurements of velocity in the hydrate zone, especially in deep water (> 2000 m). There are also instruments to generate S-waves on the seafloor, e.g., "shear wave sleds" (Davis et al., 1996), which, despite their limited energy, signal penetration and water depth, may be suitable tools for studying shallow gas hydrates.

Vertical-seismic profile (VSP) data are acquired by placing a seismometer into a borehole, firing a surface seismic source (airgun or watergun), and determining velocity through the hole by measuring the transit time of the seismic wave. The most common form of VSP data is acquired at zero offset, meaning that the surface source is deployed vertically above the receiver, usually from the drilling ship. A second geometry, walkaway (or offset) VSPs, employs shots from a second ship at varying source-receiver distances from the borehole. Zero-offset VSP data provide detailed and accurate seismic velocities and accurate measures of reflector depths, while walkaway VSPs provide improved opportunities for recording converted shear waves. VSP data in hydrate/gas provinces are relatively uncommon (Bangs et al., 1993; Holbrook et al., 1996; MacKay et al., 1994), since they require a drilling ship or specialized re-entry of a pre-existing drill hole.

Near-bottom reflection data are acquired by towing a seismic source and streamer near the seafloor. The Navy's DTAGS (Deep Towed Acoustic Geophysical System) is an example of this kind of system (Rowe and Gettrust, 1993). This system uses a Helmholtz resonator source to achieve a penetration of several hundred meters at high frequencies (250-650 Hz). The resulting vertical resolution and Fresnel zone are considerably improved over surface-towed systems. This system has been successfully used in several hydrate provinces and provides unique images of the hydrate/gas system (Wood and Ruppel, 2000) (Fig. 3).

Ocean-bottom cables (OBCs). Ocean-bottom cables are similar to conventional seismic streamers, but they are deployed on the seafloor and contain 3-componemt geophones at a dense spacing (typically 25-50 m). OBCs are particularly promising for recording converted S-waves (PS-r in Fig. 2). OBC technology alleviates two problems for PS-surveys with OBSs, poor coupling to the seafloor and sparse receiver coverage. The dense receiver spacing in OBCs allows dense two-dimensional coverage of the subsurface with PS-waves. Recent OBCs are also available for deep water (up to 2000 m), which will allow study of Vs in hydrate-bearing sediments. The principal drawback of OBC surveys is their relatively large cost.

3. THE INDIRECT APPROACH: GEOLOGIC FEATURES THAT FAVOR GAS HYDRATE OCCURRENCE

As in oil and gas exploration, a principal role of seismic methods in gas hydrate studies is not necessarily to directly detect the hydrate, but rather to characterize the structure and stratigraphy of the hydrate/gas deposit by obtaining an accurate image of the subsurface structure. Indeed, to date no unambiguous direct hydrate reflections have been observed, so that hydrate occurrence must be inferred indirectly. (Gas, in contrast, has a distinctive seismic signature, and the free gas zone underlying hydrate deposits can often be imaged directly).

Before drilling an oil exploration hole, numerous parameters (e.g., stratigraphic information, porosity, permeability, structure, etc.) are investigated to test the likelihood of hitting an oil reservoir. Research to apply a similar strategy for gas hydrate prediction is underway (*Collett*, 1996). However, our knowledge of gas hydrate formation in nature probably has to improve significantly before we have a reliable set of parameters to investigate the likelihood of gas hydrate occurrences.

Many of these parameters cannot be determined without information from drilling and sampling. Seismic surveys, however, provide detailed 2D and 3D structural images, as well as crucial extensions of stratigraphy away from boreholes. In particular, seismic imaging provides critical maps of:

- layers that are known to contain gas hydrates.
- the "source rock" of the gas, and clues as to whether the gas originates in a deep thermogenic reservoir or in more distributed biogenic source zones.
- conduits that supply fluid and gas into the hydrate stability zone. Layers of high permeability (particularly faults) may allow gas to migrate upward.
- conduits that allow escape of gas through the hydrate stability zone. We still don't know how methane migrates through the GHSZ without getting "trapped" as hydrate. Faults may play a key role in this process.
- possible traps for gas. In particular, it is not clear by what mechanism gas is trapped at the BSR, or whether stratigraphic traps within the hydrate stability zone might focus methane flux, leading to elevated gas hydrate concentration.

4. BOTTOM SIMULATING REFLECTIONS AS REGIONAL GAS HYDRATE PROXIES

4.1 Nature of BSRs

BSRs are the most widely used indicator for gas hydrates in marine sediments. They are located at the base of the hydrate stability zone (Chapter 4), where pressure and temperature conditions are at the phase boundary between hydrates and free gas. Because gas hydrate stability is much more sensitive to temperature than to pressure, BSRs approximately follow isotherms. In undisturbed sediments, isotherms are parallel to the seafloor, which is why BSRs mimic the seafloor (Fig. 4).

The negative polarity of BSRs indicates a Vp decrease across the stability boundary. In principle, this contrast may be caused by elevated Vp in gas-hydrate-bearing sediments above the base of the gas hydrate stability zone (BGHS) (Hyndman and Spence, 1992) and/or by low velocities in gas-charged layers beneath it (Korenaga et al., 1997). Results from various studies indicate that, with some possible exceptions, strong BSRs are principally caused by a drastic decrease of Vp due to free gas, with a relatively small contribution from an overlying "hydrate wedge" (Korenaga et al., 1997). At some locations, however, weaker BSRs may be caused by gas hydrates without underlying free gas (Minshull et al., 1994; Pecher et al., 1996b) (Fig. 4b). Important evidence

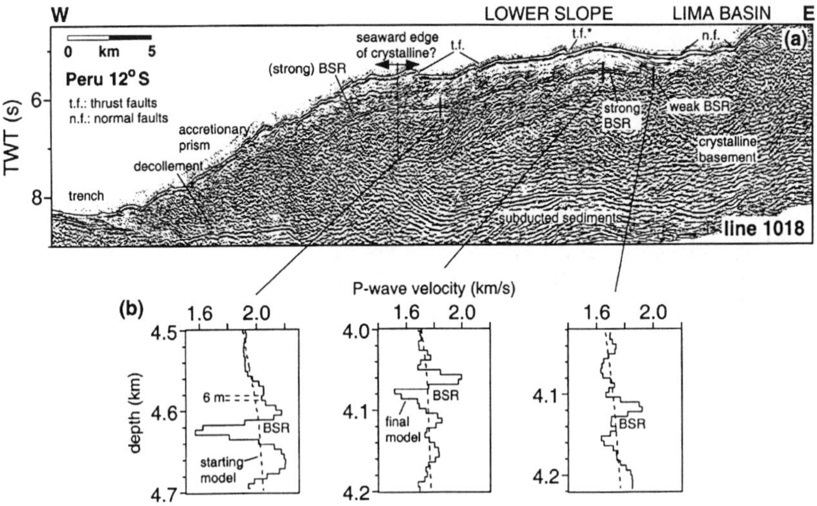

Fig. 4: (a) BSR off Peru. Structural interpretation according to von Huene et al. (1996). TWT: two-way traveltime. The shallowing of the BSR toward the continent is caused primarily by a decrease of pressure and the resulting move of the phase boundary toward lower temperatures, i.e., closer to the seafloor (an additional effect comes from an observed regional increase of heatflow). Note the weakening of the BSR close to its landward termination, which coincides with normal faults. The normal faults may act as conduits that allow gas from the BSR to escape.
(b) Velocity-depth profiles from waveform inversion across the BSR. The right profile at the location where the BSR is weak may indicate that the BSR is mainly caused by elevated Vp above the BGHS, perhaps related to gas hydrates. The other two profiles suggest a thin free gas layer beneath the BSR

for the nature of the BSR comes from waveform inversion, which optimizes a subsurface velocity model by comparing synthetic seismograms to observed data. Waveform inversion is necessary to elucidate the details of the velocity structure around BSRs, because those structures appear to be thinner than a seismic wavelength. The BSR waveform is therefore a complex superposition of reflected signals from the top and base of the layer, which is impossible to model with oversimplified AVO models (e.g., compare the results of Katzman et al. (1994) to those of Korenaga et al. (1997) in the same area).

A major question regarding the operation of the hydrate/gas system is the thickness, origin, and dynamics of the free gas zone. VSPs conducted during Ocean Drilling Program Leg 164 on the Blake Ridge revealed that the hydrate zone there is underlain by a surprisingly thick gas zone (at least 250-m thick) (Holbrook et al., 1996). The free gas zone is characterized by a region of unusually low Vp (Fig. 5) that coincides with a band of high reflectivity that follows stratigraphic layers. This can be explained by slight variations in gas concentrations across layer boundaries, since Vp is very sensitive to gas at low

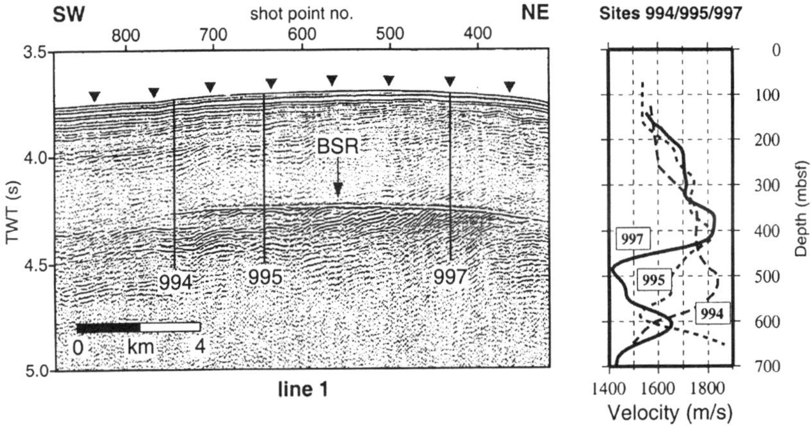

Fig. 5: Reflection seismic section (left) and Vp from VSPs on the Blake Ridge. The BSR at 995 and 997 is at ~440 mbsf. The only reasonable explanation for the low Vp beneath the BSR is a thick free gas zone. Note the zone of high reflectivity beneath the BSR. This is probably caused by slight variations of free gas concentration across layer boundaries. After Holbrook et al. (1996).

concentration (Fig. 6). A similar band of high reflectivity has been reported recently offshore Pakistan (Grevemeyer et al., 2000). Drilling on the Cascadia margin (MacKay et al., 1994) and off Chile (Bangs et al., 1993), on the other hand, showed that the gas zones there are seismically thin (off Cascadia, earlier results from waveform inversion (Singh et al., 1993) were confirmed). The cause of these large differences in the thickness of gas zone are not understood; however, sediment permeability may play a key role.

The character of BSR reflections depends on the frequency content of the data. Recently, BSR reflection strength on the Cascadia margin has been shown to be frequency-dependent, with weaker reflection amplitudes for higher signal frequencies (Yuan et al., 1996). This may be caused by a steep velocity gradient at the BGHS, rather than a first-order step: low-frequency signals "see" a step function, thus generating a strong reflection, while higher-frequency signals see the gradient, leading to weaker reflections. Importantly, DTAGS data, which have unusually high lateral and vertical resolution, show that the BSR on the Blake Ridge is not the continuous reflection observed in conventional surface-towed data, but is rather a set of "shingled" reflections along dipping layers that terminate at the BGHS (Fig. 3). The lower resolution of surface-towed seismic data smears out these shingles, generating a smooth, continuous reflection (Wood and Ruppel, 2000). Thus the very characteristic that distinguishes the BSR as a phase-boundary reflection -- the cross-cutting of strata reflections -- is itself an artifact of the relatively low frequencies of surface seismic data.

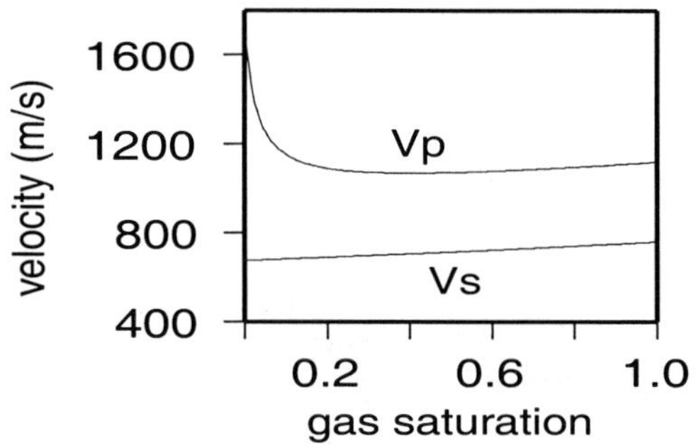

Fig. 6: Vp as a function of free gas saturation, roughly modeled for conditions expected at the BSR on the Blake Ridge. If gas saturation is below 20% of pore space, a slight variation of gas saturation across layer boundaries leads to a strong Vp contrast, and hence, a strong reflection coefficient. These calculations assume uniform distribution of gas in the sediment (i.e., each pore has the same saturation). Uneven distribution of gas moves the minimum in Vp toward higher saturation (Domenico, 1977). Vs is almost unaffected by free gas. A slight increase of Vs is caused by the density decrease associated with gas.

4.2. Possible pitfalls

As the above discussion demonstrates, BSRs are more complex and diverse than previously thought. What matters in the context of gas hydrate quantification, however, is that the presence of BSRs is a reliable indicator that a study area contains gas hydrates. However, there are some pitfalls.

BSRs appear to be obvious features, but they may not be noticed if interpreters don't expect them. In particular, inappropriate seismic processing that focuses on deep sediment sections may destroy BSRs. Also, horizontal layering, by destroying their distinctive cross-strata geometry, may hide BSRs.

In addition, BSRs are absent in many locations where gas hydrates are known to occur. We still do not know what conditions, in addition to the presence of gas hydrates overlying free gas, are necessary to "maintain" BSRs. Two mechanism that may preclude BSR formation are tectonic suppression of BSRs (Fig. 7), proposed for the Lima Basin off Peru, and the escape of gas through faults, which may be responsible for the general lack of BSRs in the hydrate-rich Gulf of Mexico. Moreover, the presence and amplitude of BSRs do not allow any immediate conclusions about whether gas hydrate is present in the shallower sediment section. This is in particular true for shallow-water gas hydrates and gas hydrate outcrops in the Gulf of Mexico.

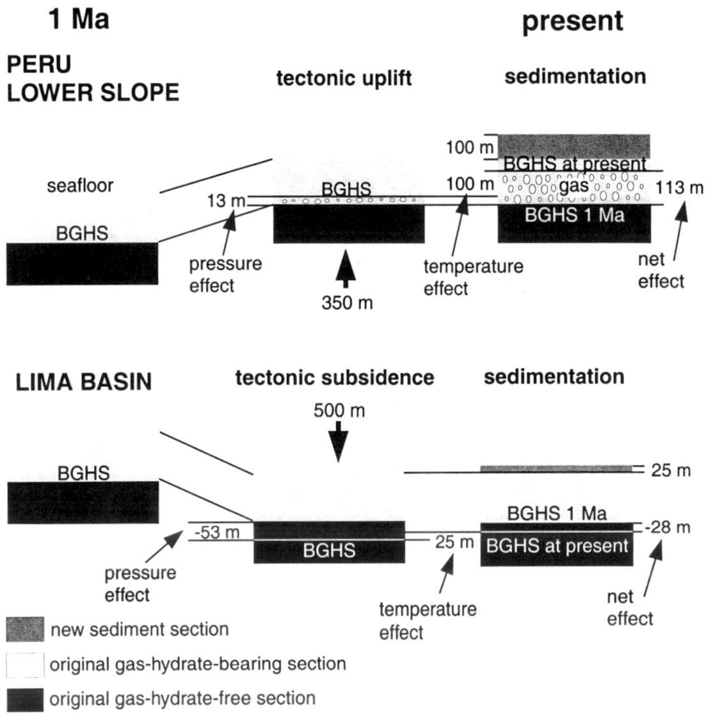

Fig. 7: Tectonic suppression of BSRs as a proposed mechanism to preclude BSR formation. The Lima Basin off Peru (further upslope from the section shown in Fig. 4) is a good candidate for gas hydrate occurrence because of the high organic carbon content in the sediments. However, most of the Lima Basin lacks BSRs. Lima Basin is rapidly subsiding. This leads to an increase of pressure, moving the phase boundary of gas hydrate toward higher temperatures, i.e., downward ("pressure effect"). Sedimentation leads to the opposite effect: Assuming a constant thermal gradient, the BGHS moves upward with respect to the sediment column ("temperature effect"). Both effects have been quantified off Peru (von Huene and Pecher, 1999). On the lower slope (Fig. 4), both uplift and sedimentation lead to an upward movement of the BGHS, causing dissociation of gas hydrates, generation of free gas, and a strong BSR. In Lima Basin, the net effect of subsidence and sedimentation is a downward movement of the BGHS with respect to the sediment section. Eventual gas at the BGHS is predicted to be absorbed to form gas hydrate.

Finally, silica diagenesis from opal-A to opal-CT or opal-CT to quartz also causes BSRs (Hammond and Gaither, 1983). Those phase transitions are associated with a velocity increase and BSRs therefore display a positive polarity. However, apart from that, they look deceptively similar to hydrate BSRs (Fig. 8 – (Scholl and Creager, 1973)).

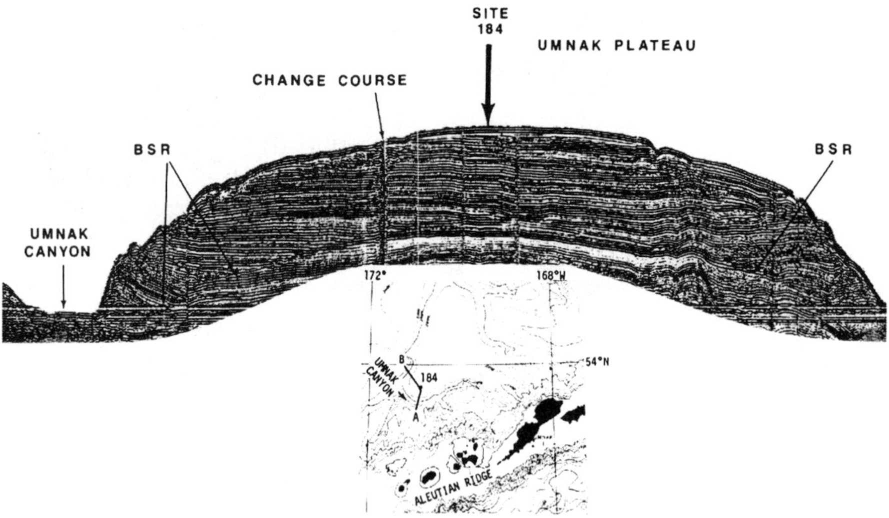

Fig. 8: A "silica BSR" from the Bering Sea. Note the very similar appearance to gas hydrate BSRs (from Scholl and Creager, 1973)

Despite these possible pitfalls, BSRs may be used successfully to constrain regional gas hydrate occurrence, and with proper calibration, concentrations. Calibration from drilling of ODP Leg 164 together with BSR distribution around the Blake Ridge, led to an estimate of gas hydrate amount along the East coast of the U.S (Dickens et al., 1997; Holbrook et al., 1996). Similar studies along the world's margins would be very valuable for global estimates of gas hydrate quantities.

5. THE DIRECT APPROACH: SEISMIC LITHOLOGY

The partial replacement of pore water by solid gas hydrates should affect seismic sediment properties. This expectation fuels research on "seismic lithology" of hydrate-bearing sediments, which attempts to predict the composition of sediments and pore fill (i.e., gas hydrates) from seismic properties.

5.1. P-waves

5.1.1. Amplitude Blanking

Any reader who studies pre-Leg 164 literature of gas hydrates on the Blake Ridge is almost certain to encounter the phenomenon of "amplitude blanking". The blanking concept was developed to explain the striking contrast in reflection amplitude above and beneath the BSR on the Blake Ridge: reflectance above the BSR is generally much lower than beneath it (Fig. 5). According to the blanking hypothesis, this contrast is caused by hydrate-induced suppression of impedance contrasts above the BSR (Lee et al., 1994; Lee et al., 1993), with "normal" reflectance beneath the BSR. However, drilling and VSPs during Leg 164 showed that the situation is exactly reversed: the low reflectance above the

BSR is "normal", due to unusually homogeneous sediments (Paull et al., 1996), while the strong reflections below the BSR are simply bright spots associated with a surprisingly thick free gas zone (Holbrook et al., 1996) (see also chapter 3.1). This observation eliminates the need to invoke blanking to explain reflection amplitudes on the Blake Ridge.

Despite this new understanding of reflectance on the Blake Ridge, it is possible that high concentrations of hydrate may act to suppress reflection coefficients in some areas. The principal idea behind amplitude blanking is that for a constant hydrate concentration, the total amount of gas hydrate is greater in higher-porosity layers. Vp usually decreases with increasing porosity because of the low Vp of porewater compared to that of the frame-building minerals.

If hydrates preferentially form in a high-porosity/low-Vp layer, its Vp will increase more than that of an adjacent low-porosity/high-Vp layer, thus reducing the velocity contrast at the intervening boundary. This is an entirely plausible mechanism for amplitude blanking, although it requires relatively high concentrations of hydrate to effect a significant change in reflection amplitude -- certainly much higher concentrations than are found on the Blake Ridge (Paull and Matsumoto, 2000). Nevertheless, reflection contrasts that unequivocally require blanking have not yet been observed. Thus, although amplitude blanking should not be dismissed entirely, we believe that it should not be considered a reliable gas hydrate indicator until unequivocal evidence for its existence is available.

5.1.2. P-wave velocity

Because gas hydrate has a higher velocity (3270 m/s; (Waite et al., 1998)) than the pore fluid it likely replaces, P-wave velocity should be a useful indicator of hydrate concentration. Indeed, this is the case on the Blake Ridge: the average concentration of gas hydrate increases towards the crest of the ridge, as does average Vp measured in VSP data (Fig. 9). However, the increase is subtle: a lateral increase in velocity of about 50 m/s corresponds to a lateral increase in hydrate saturation from 2% of porosity to 7% of porosity (Holbrook et al., 1996). Such a subtle change was interpretable only because of the uniform sediments of the Blake Ridge, which make it an ideal "natural laboratory" for gas hydrate studies. In a less uniform environment (e.g., an accretionary margin) a change of Vp by about 50 m/s would likely be indistinguishable from lateral sediment variations.

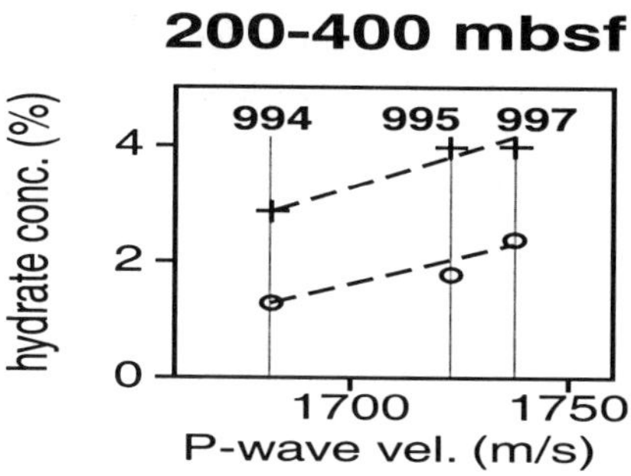

Fig. 9: Average Vp between 200 and 400 mbsf on the Blake Ridge, from VSPs (Fig. 5) vs. average hydrate concentration from chlorinity (circles) and resistivity (crosses). Estimates of hydrate concentration from chlorinity are lower than those from resistivity. However, both methods predict an increase of hydrate concentration toward the crest of the ridge, consonant with the lateral increase in P-wave velocity.

Layers of unusually high hydrate concentration, however, may display a distinct Vp increase. The zone immediately above the BSR may contain relatively large amounts of gas hydrates because of “methane recycling” – gas from dissociated gas hydrates that moves back into the hydrate stability zone to form gas hydrate. Indeed, as mentioned earlier, waveform inversions often show a "wedge" of high velocities associated with BSRs (Fig. 4). Because of the pronounced effect of free gas on Vp, it is in some cases not clear if this "wedge" represents the true Vp structure or if it is an artifact generated by the waveform inversion in order match the high observed reflection coefficients at vertical incidence. In data of sufficiently long offset, however, the wedge seems to be a robust feature (Korenaga et al., 1997).

A general problem for quantifying hydrate concentration is the lack of a hydrate-free "reference” velocity, i.e., the velocity of the host sediment at full water saturation. Vp from below the BSR cannot be used because of the presence of free gas and its pronounced effect on Vp.

5.1.3. Seafloor reflectivity

The reflection coefficient of the seafloor may be a good indicator of gas hydrate outcrops. Seafloor outcrops consist of nearly 100% gas hydrates and are surrounded by sediments that typically have a Vp only slightly above that of water. Hence, the seafloor reflection coefficient should increase. This effect, however, is not as drastic as might be hoped, because pure hydrate has a lower density than sediments, which counteracts the velocity increase. For example, assume a typical gas-hydrate-free soft sediment of 1600 m/s with a density of 1500 kg/m3. Vp in water is about 1500 m/s and density 1000 kg/m3. Hence, the seafloor reflection coefficient is 0.23. The reflection coefficient for gas hydrates with a Vp of about 3270 m/s (Waite et al., 1998) and a density of about 910 kg/m3 (depending on cage occupancy) (Sloan, 1989) would be 0.33. Such an increase should be detectable, but would require diligent data analysis.

Carbonate mats are often observed in the vicinity of gas hydrate outcrops.. The lower end of Vp and density in limestone (as a carbonate) is 3390 m/s and 2000 kg/m3, respectively (Mavko et al., 1998) (these numbers may vary widely). This would lead to a reflection coefficient at the seafloor of 0.64. Carbonates may therefore obscure gas hydrate outcrops. Should carbonate and gas hydrate outcrops be sufficiently separated compared to the resolution of seismic data, however, this difference in reflection coefficients may enable us to distinguish between them.

5.2. S-Waves

5.2.1. Background

The main promise of V_s for hydrate-bearing marine sediments lies in the predicted higher sensitivity of V_s to gas hydrates at lower concentration (Fig. 10). This is mainly due to the low V_s of the hydrate-free unconsolidated marine sediments (typically 100-500 m/s) compared to V_s of about 1750 m/s of gas hydrates [*Waite et al.*, 1998]. Several borehole studies have demonstrated the value of V_s for gas hydrate quantification [*Guerin et al.*, 1999; *Sakai*, 1999].

The main promise of Vs for hydrate-bearing marine sediments lies in the predicted higher sensitivity of Vs to gas hydrates at lower concentration. This is mainly due to the low velocities of the hydrate-free unconsolidated marine sediments (Fig. 10). Several borehole studies have demonstrated the value of Vs for gas hydrate quantification (Guerin et al., 1999; Sakai, 1999)

5.2.2. First field experiments

P-to-S converted waves were first detected in conjunction with gas hydrates during ODP Leg 164 (Pecher et al., 1997a). They were recorded during walkaway-VSPs, for which a three-component geophone was clamped in a borehole. A 13-km long seismic profile was shot over the borehole. At larger offsets, the angles of incidence were large enough to generate P-to-S conversion at a number of horizons. Because of the uniformity of Blake Ridge sediments, P-to-S conversion is likely associated with variations of gas hydrate concentration rather than lithologic boundaries (Pecher et al., 1997b). A second profile was acquired over a three-component OBS at the seafloor. In these data, a clear PS-arrival, apparently from the base of the hydrate stability zone, was recorded (Fig. 11). Gas may cause P-to-S conversion because of the associated density contrast. However, PS-energy would be predicted to be weak, especially given the low gas concentrations encountered on the Blake Ridge. Therefore, it is more likely that PS-conversion is caused by a decrease of sediment Vs at the base of the hydrate-bearing sediments.

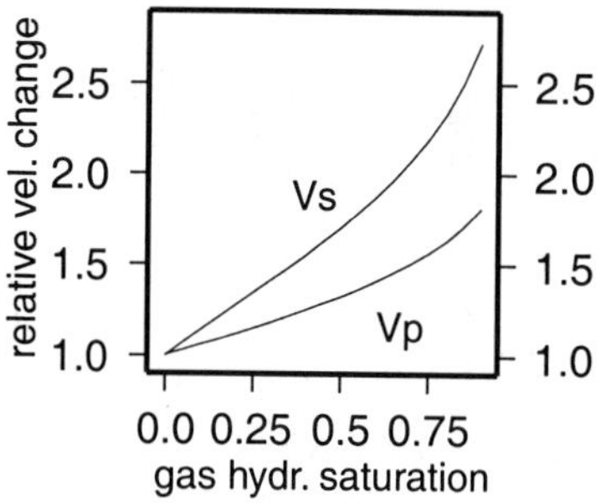

Fig. 10: Predicted relative change of Vs and Vp as a function of gas hydrate saturation of pore space, assuming that gas hydrate replaces part of the sediment frame. Sediment composition and conditions roughly as expected at the BGHS on Hydrate Ridge, offshore Oregon. Vs is more sensitive than Vp mainly because of the low baseline Vs of unconsolidated hydrate-free marine sediments (typically, Vs: 100-500 m/s; Vp: 1500-2000 m/s).

After these encouraging observations, a recent OBC survey off Norway did not show any PS-conversion associated with the BGHS [*Andreassen et al.*, 2000]. It is not yet clear why P-to-S conversion at the BGHS appears to take place on the Blake Ridge but not off Norway. Possible explanations include a possibly low hydrate concentration in the OBC area off Norway, or that strong PS-conversions from densely-spaced ping strata may hide conversion at the base of the hydrate stability zone. This discrepancy is still under investigation, and a new PS-study is planned on the Blake Ridge.

5.2.3. Potential of PS-surveys

The study of shear-wave propagation through gas-hydrate-bearing sediments is still in its infancy, so this topic remains speculative. Much depends on how strongly gas hydrates affect sediment Vs. Assuming that Vs is more sensitive to gas hydrates in the pore space, lateral variations of gas hydrate concentration should have a more pronounced effect on P-S traveltimes than on P-wave traveltimes. This will be tested during a planned PS-survey with OBSs across the Blake Ridge crest, which will allow us to compare PS-traveltimes to P-reflections.

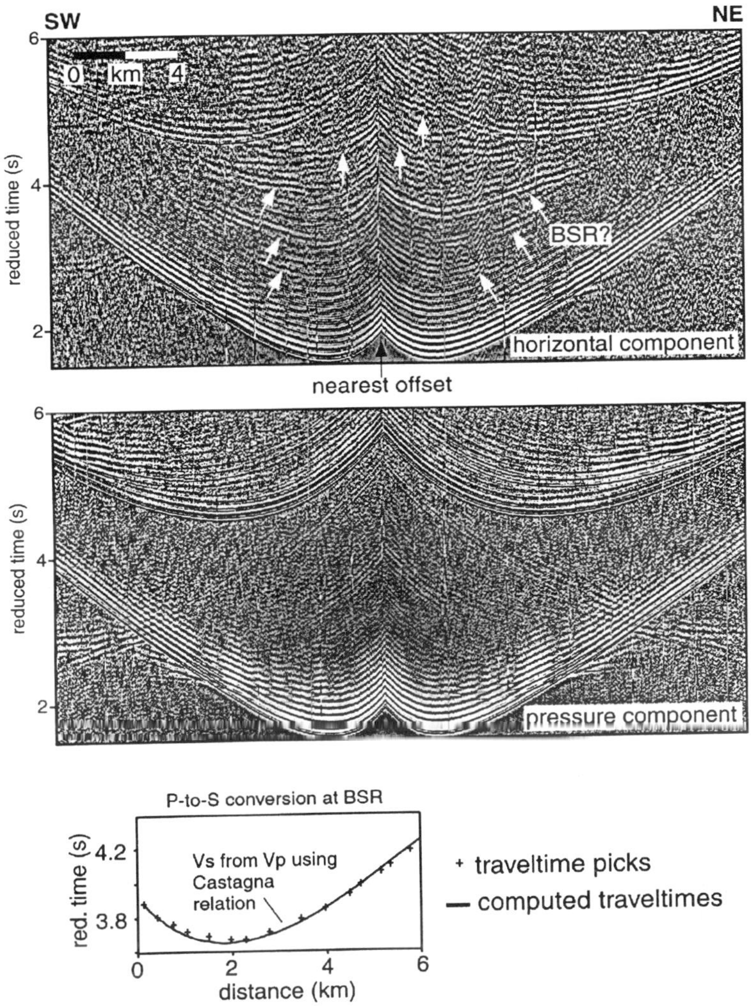

Fig. 11: PS-waves recorded with a three-component OBS on the Blake Ridge (white arrows in horizontal component). Plotted with reduction velocity of 2500 m/s, i.e., source-receiver-offset/(2500 m/s) is subtracted from arrival time. First arrival is the direct wave from the airgun to the OBS. The key in identifying arrivals as PS-waves is that they do not appear in the pressure component recorded by hydrophone a short distance above the OBS, because water does not transmit S-waves. Lower plots: Modeling of traveltimes using a Vp model from the VSP and an empirical relationship between Vp and Vs (Castagna, 1985) to estimate Vs assuming P-to-S conversion at the BSR. Calculated and observed traveltimes match extremely well considering that the Castagna-relation is only empirical. Although this does not prove conversion takes place at the BSR, it demonstrates that this is a realistic suggestion. Considering the uniform sediments beneath most of the Blake Ridge, most likely only parameters that may lead to P-to-S conversion are variations in gas hydrate concentration and free gas occurrence. After Pecher et al. (1997).

The greater potential, however, may be in the detection of layers of high gas hydrate concentration, which should cause P-to-S conversion. Furthermore, because Vs is lower than Vp, shear-waves have shorter wavelengths than P-wave (at a given frequency), so P-to-S surveys often have better resolution than P-wave surveys. PS-arrivals may be particularly well suited to study the BGHS. There, the waveform of the PS-arrival may enable us to determine the fine Vs-structure across the BGHS, just as waveform inversion of P-waves has elucidated the fine Vp-structure (Fig. 4). Unlike Vp, Vs will be almost unaffected by the underlying free gas (Fig. 6). Vs may therefore be directly related to gas hydrate. Vs below the BGHS may thus be useful as a reference value of hydrate-free sediment, since the composition of sediments that host gas hydrates may be identical to that immediately below the BSR. We caution that quantitative PS-analysis has many pitfalls, in particular the high attenuation of S-waves in shallow marine sediments. We nevertheless think that PS-analysis has the potential of leading to a breakthrough in gas hydrate quantification.

6. CONCLUSION

Seismic techniques for detection and quantification are still in the early stages of development. For regional studies, it appears that BSR distribution may be the best proxy for gas hydrate occurrence. However, reconnaissance mapping of BSRs must be calibrated by borehole data and by analysis of factors that may preclude BSR formation even in the presence of gas hydrates.

Detection of layers of highly concentrated gas hydrate, whether for commercial gas production or to assess drilling hazards, requires a more complex and multi-faceted approach. We expect that, as in oil and gas exploration, we will have to develop a set of parameters that define the likelihood of encountering gas hydrates. This will be a multi-disciplinary effort in which seismic surveys will mainly provide accurate images of the subsurface. The other important focus is likely to be on seismic lithology. Extensive research is still needed to understand how gas hydrate affects seismic properties of sediments. We have particularly great hope for PS-wave studies in conjunction with sophisticated rock physics models. This hope is fueled by the breathtaking development of both acquisition hardware and evaluation techniques for marine PS-wave surveys.

Acknowledgments. We thank Warren Wood for providing the DTAGS figure, and David Scholl for providing the silica-BSR figure. This work was partially supported by NSF grant OCE-9504610.

Chapter 23

Ground Truth: In-Situ Properties of Hydrate

David S. Goldberg
Lamont-Doherty Earth Observatory
Borehole Research Group
Palisades, NY 10964

Timothy S. Collett
U.S. Geological Survey
Denver Federal Center
Denver, CO 80225

Roy D. Hyndman
Geological Survey of Canada
Pacific Geoscience Centre
Sidney, BC V8L 4B2

1. INTRODUCTION

1.1. Motivation

The occurrence, distribution, properties, and hydrocarbon reservoir potential of natural gas hydrates in marine sediments continue to be enigmatic questions in marine geoscience. The fact that natural gas hydrate is metastable and affected by changes in pressure and temperature makes its observation and study difficult under laboratory conditions.

As is now well known, gas hydrates are crystalline solids formed of a cage of water molecules surrounding a natural gas molecule under specific conditions of relatively high pressure and low temperature. The supply of gas also must be sufficient to initiate and stabilize the hydrate structure [*Sloan,* 1990; *Kvenvolden,* 1993]. These restrictive thermodynamic conditions are satisfied on many continental slopes and rises around the globe. Most information on the regional distribution and geological environment of hydrate occurrence comes from remote geophysical measurements, especially seismic studies. The most common mapping tool is the characteristic BSR—the large bottom-simulating seismic reflector that is commonly associated with the base of high-velocity hydrate and the top of low-velocity free gas. Some promising initial surveys of sub-seafloor electrical resistivity and seafloor complicance also have been carried out. However, all such remote methods require calibration or "ground truth" with in situ data from drill holes. In addition, since hydrate is unstable

M.D. Max (ed.), Natural Gas Hydrate in Oceanic and Permafrost Environments, 295–310.

under atmospheric conditions it is extremely difficult to recover and retain samples for study in the laboratory. Dissociation of even massive hydrate samples into water and methane gas occurs after only a few minutes of exposure to room temperatures and pressures. Because of this instability, downhole measurements provide the primary ground truth data for the mapping of gas hydrate using surface geophysical surveys. Downhole measurements also provide calibration data for remotely–measured physical properties and the concentrations of hydrate and free gas that variations in these properties represent. Especially critical relations are those between velocity increase and hydrate concentration, velocity decrease and gas concentration, and similar relations for electrical resistivity. Therefore, in situ detection methods that provide ground truth information in drill holes hold an important key to studying gas hydrate occurrence and concentration.

In situ methods have improved our understanding of gas hydrate occurrence and distribution significantly in recent years. Given the continued scientific interest in gas hydrate investigations, the planning and use of new in situ detection methods and technologies will undoubtedly continue to advance in the future. Simply developing a new strategy to measure in situ properties in closely spaced drill holes, for example, could produce a three-dimensional isopach map of the distribution of hydrate below the seafloor. Such maps would provide ground truth information over a significantly wider area than in a single drill hole and more fully constrain the interpretation of a seismic survey over the same area.

1.2. Wireline logging

During the last thirty years, in situ measurements made in drill holes have been increasingly used for scientific applications in marine geology and geophysics, particularly in deep sea applications. Used mostly by the oil industry to map promising formations for exploration and production of hydrocarbons, a variety of instruments have been developed that can be lowered down drill holes to extract information about the subsurface geology. They are essential for measurements like temperature, for example, that must be made in situ. During the previous decade, this class of measurements using sophisticated electronic instruments has become common and is generally referred to as downhole "logging". With this technique, an instrument is lowered down a hole relatively soon after it is drilled and information taken from inside the hole is sent to the surface along a high-speed communications cable or "wireline". A major advantage of logging methods is that they collect data continuously over the depth drilled and with a minimum of disturbance to the natural system. Typically, logs made in a drill hole fall into three general categories: electrical, nuclear, and acoustic. The interested reader is referred to these publications for a more detailed technical discussion of logging instruments (*Doveton* [1986], *Ellis* [1987], and *Goldberg* [1997]).

Complete recovery of continuous log profiles, however, is not always possible with conventional wireline logging technology. The interval immediately below the seabed, and in particular, any sections of the hole that are unstable are not logged because the drill pipe must be lowered 80-100 m below the seafloor to insure hole stability in the softest sediments for logging to begin. This is the typical case in deep sea hydrate environments. To benefit from the use of log data over the entire section, another approach must be taken.

1.3 Logging-while-drilling

Over the last 15-20 years, new technology has been developed to measure in situ properties in oil and gas industry holes that are drilled horizontally where conventional logging with a flexible wireline is not feasible. This innovative technology is called "logging-while-drilling" (LWD) and uses sensors placed just above the drill bit [e.g. *Allen, et al.,* 1989]. The differences between LWD and conventional logging technology, however, must be taken into consideration when using these data in a particular drilling environment [e.g. *Evans,* 1991]. For example, wireline tools record information as they move upward at (approximately) constant speed and sample data at a fixed depth interval; LWD tools record information as they move downward at variable speed (the drilling rate) and sample data at a fixed time interval (Fig. 1). Currently available LWD

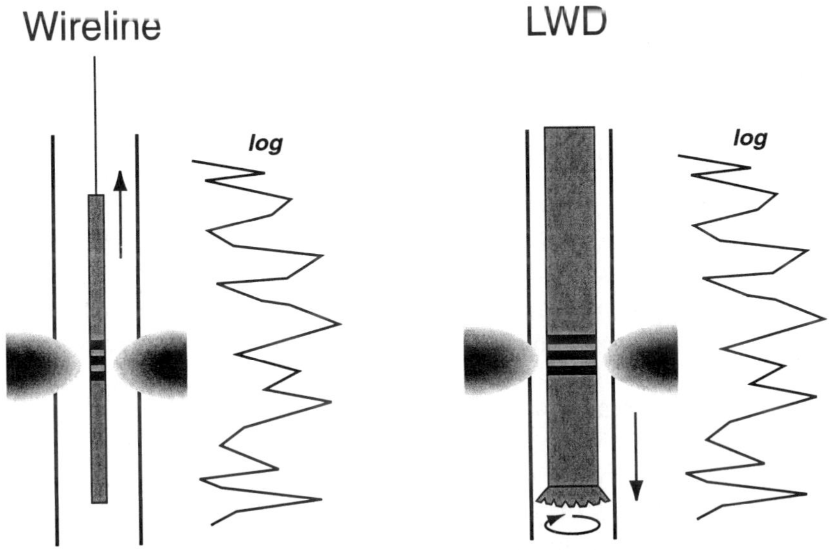

Figure 1. Diagram showing wireline and LWD technologies for in situ logging

devices can make accurate physical measurements including porosity, resistivity, density, and natural gamma radiation, among others, only minutes after the drill bit cuts through a gas- or hydrate-bearing formation. The resolution of these LWD sensors is similar to that of wireline logging tools; neutron porosity measurements have a vertical resolution of about 12 inches, while density and gamma-ray measurements have a vertical resolution of 6 inches, depending in part on maintaining a consistent drilling rate. Unfortunately, the velocity of typical hydrate-bearing formations is less than the minimum value that can be resolved using LWD devices and wireline sonic tools continue to be required for logging in low-velocity sediments [*Goldberg,* 1997].

A primary advantage of LWD over wireline logging in marine environments is that data can be acquired with high resolution without gaps below the seafloor or at the bottom of the drill hole. In gas- and hydrate-bearing sediments, an important additional benefit of LWD is that data is recorded almost immediately after drilling and provides an accurate measurement of in situ properties without allowing time for them to change significantly. The reduced time before the measurement is taken minimizes drilling-induced hydrate dissociation or changes in gas concentration in the vicinity of the drill hole that may seriously affect a wireline log.

2. IN SITU METHODS

2.1. Seismic–downhole data integration

Logs play a crucial role in linking core data with regional geophysical surveys and in providing data where core sections could not be obtained. To study hydrate, researchers typically address the problem with a multidisciplinary strategy, integrating measurements made on core samples with those made using logs and placing both in the context of regional geophysical and seismic studies. The multiple scales of investigation used with seismic, downhole, and core data acquired in the same geological environment complement each other extremely well. Seismic sections are the basis for a regional description and enable a cross-section to be inferred; logs typically have an intermediate resolution of approximately 0.5 m giving continuous information in the region surrounding the borehole; and core samples provide detailed information on physical properties and age. Core, log, and seismic data used jointly also contribute to the confidence in each data set individually. Unlike measurements on core samples, however which are often disturbed during the process of recovery, downhole logs provide a set of continuous information and sample a larger volume of rock than core measurements. In addition, for most hydrate drilling environments, continuous coring does not result in continuous core recovery. Techniques other than piston coring typically results in less than 50% recovery and this proportion is often disturbed by the drilling process. As a result, the true depth of the core becomes ambiguous and the accuracy of measurements made on it may be severely degraded. In situ measurements reduce uncertainty in the sampling depth and eliminate the disturbance of properties that occurs as a sample is taken. Furthermore, because logs have much greater vertical resolution than seismic data, but little lateral resolution, the combination of the two defines subsurface geological structures far better than either data type alone.

The difference in the scale of the physical phenomena affecting each type of measurement may be extreme. The scale ratio from core to log may be greater than 2 x 10^3; the ratio from log to seismic may be 10^6 to 10^7 times larger. In most integrated scientific applications, therefore, downhole logs provide three complementary advantages: 1) data is acquired under in situ conditions, 2) data is acquired in continuous profiles measured throughout the interval with no missing sections, and 3) data is sampled at a larger scale, intermediate between core and seismic measurements. With the expectation that recent advances in technology will continue, research using interdisciplinary strategies that integrate core, log, and seismic data will undoubtedly expand, particularly in studying gas hydrates. Applying complementary techniques to a particular hydrate environment, either in the deep sea or in permafrost areas, is the now the expected scientific approach.

An example of a comparison between surface geophysical data and downhole data is given in Figure 2. Velocity profiles obtained from logs, vertical seismic profiles (VSPs)—another downhole tool measuring the formation velocity—and multichannel seismic experiments are shown as a function of depth in a marine sedimentary environment [data from *Yuan et al.,* 1996]. The increase in velocity above the BSR to about 1800 m/s is attributed to the presence of high-velocity hydrate in all three types of data.

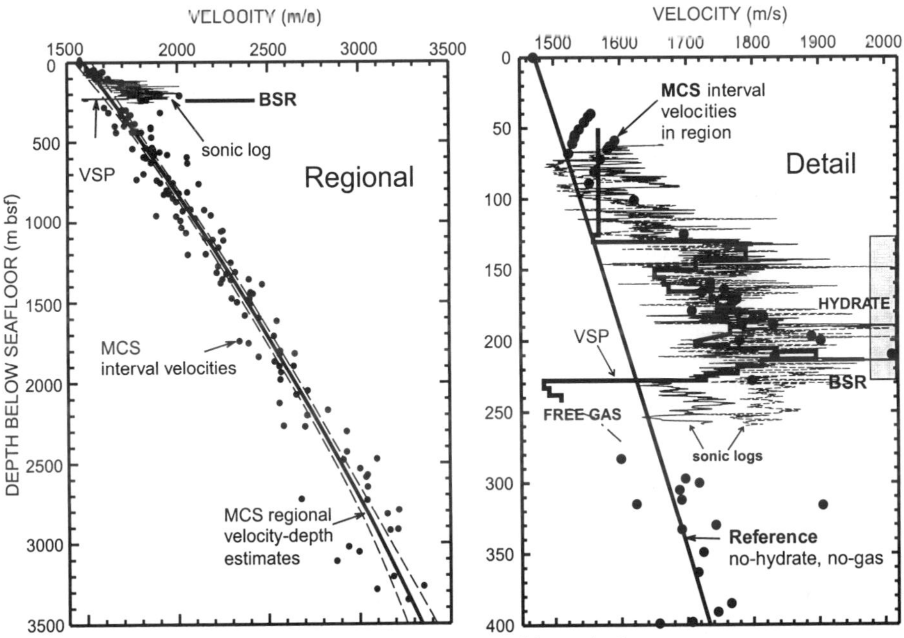

Figure 2. Integration of seismic, VSP, and log velocity measurements.

The detailed diagram on the right highlights the similar response of the three experiments, though each at a different scale, to the presence of hydrate above a reference no-hydrate, no-gas trend; however, significant differences in the resolution of fine-scale variation puts obvious emphasis on the importance of using multiple methods. For example, low velocities below the BSR detected only in the VSP data are interpreted to represent a thin layer of low concentration (<1%) free gas (also see *section 3.3*). This gas layer is not evident in the logs because they have a shallower effective depth of penetration into the formation and do not sense gas that may have been released near the borehole during drilling. The multichannel seismic data also do not have adequate resolution in this area to detect the thin gas layer.

A comparison of downhole resistivity to seafloor electrical sounding surveys in this area similarly agrees over the broad measurement scale of these two experiments. Log-measured resistivity is about 2.1 ohm-m in the hydrate zone, compared to a reference of about 1 ohm-m for no hydrate, and electrical sounding profiles yield similar values of about 1.8-2.0 ohm-m over the same depth interval [*Hyndman et al.*, 1999; *Yuan and Edwards,* 2000].

2.2. Sonic logs

A common approach to identifying hydrate in situ is to directly measure the elastic wave properties of the formation using wireline logs. Distinct changes in the elastic properties are a critical indicator of the presence of hydrate and free gas. Sonic velocity decreases sharply in the presence of free gas, but increases in the presence of hydrate. The shear wave velocity, in particular, allows for the estimation of the shear strength of the formation that increases in the presence of hydrate [*Guerin, et al.,* 1999]. Such observations from sonic logs aid in determining the nature of the BSR, interpreting VSP and seismic data, as well as directly estimating the concentration of methane hydrate and free gas.

Empirical relationships to estimate gas concentrations from *Vp* and *Vs* in a variety of hydrocarbon-bearing formations have been widely used in the oil and gas industry [e.g. *Kuster and Toksoz*, 1974]. If the presence of hydrate and free gas can be accurately identified from VSP and velocity logs, then similar formulations could be used to estimate the volume of methane hydrate from surface seismic data alone. However, such relationships are empirical and a major limitation in using them is our poor understanding of the velocity in massive hydrate and different hydrate-bearing sediments. *Zimmerman and King* [1986], *Lee, et al.* [1996], *Dvorkin and Nur* [1996], and *Guerin, et al.* [1999], among others, all consider theoretical treatments of elastic wave propagation to predict the velocity and pore structure of hydrate-bearing sediments. Most of these models suggest that the microscopic distribution of hydrate infills sediment pores with high-velocity material or acts as an intergranular cement or both, increasing the compressibility and rigidity of the host sediment. Although these models all yield reasonable estimates of the methane saturation and satisfy field observations, they not agree quantitatively. A robust physical model for the velocity of hydrate and hydrate-bearing sediment is therefore still needed.

2.2. Porosity logs

Porosity logs are a fundamental measurement for the in situ detection of gas hydrate. Because the physical properties of the sediment play a critical role in gas migration and the stability of hydrate formation, accurate in situ measurement of porosity is essential. The local distribution of hydrate in the formation may also influence the porosity. Massive, layered, nodular, and disseminated hydrate all have different pore structures and the effects of lithology, such as clay, on the porosity measurement must be considered. A critical factor for quantitative estimate of hydrate and free gas concentrations from logs is therefore the reference properties for no hydrate and no gas. A variety of logs allow this determination. Often, logging methods are combined to differentiate various effects and determine a best estimate of the in situ porosity. The neutron porosity log measures neutron scattering, for example, which is controlled by the total hydrogen content. It provides a direct measure of H^+ but is often adversely affected by the hydrogen bound in clay-rich sediments. The density log measures the electron and bulk density of the formation. Assuming an average grain density of the formation, the bulk density and formation porosity may be computed from the density log though they are highly sensitive to grain density changes. Combining these two measurements and reducing the sensitivity to either bound H^+ or grain density alone, the accuracy of a formation porosity estimate can be significantly improved [e.g. *Schlumberger,* 1989].

Another often reliable estimation of porosity is derived from the resistivity log using Archie's formulation [*Archie*, 1942]. This approach is conventionally used in the oil and gas industry to derive methane gas saturation from the resistivity log. The resistivity of massive hydrate is expected to be about 50 times greater than fresh water, and compared to brine-saturated marine sediment, this is quite high. Disseminated hydrate that replaces more electrically conductive seawater in open sediment pores can thus be detected in situ using logs. *Matthews* [1986], *Collett* [1993], and *Hyndman, et al.* [1999] discuss the approach to estimating hydrate concentration from resistivity-derived porosity in detail. It is important to note, however, that the Archie formation may overestimate hydrate concentration because of an electrical insulation effect—less hydrate than methane gas is necessary to create the same observed resistivity increase—and corrections can be applied to reduce this effect [*Collett, et al.,* 1999]. Moreover, the dissociation of hydrate in the proximity of the borehole due to drilling may reduce the salinity of the pore fluid and consequently increase the measured resistivity [*Prensky,* 1995]. This drilling-induced effect can be minimized by estimating a baseline resistivity for no hydrate and no gas, then using this reference value to calibrate estimates of methane saturation [*Hyndman, et al.,* 1999]. In general, resistivity logs must be carefully corrected in order to estimate methane saturation.

Using LWD, the estimated in situ porosity may be more representative than using laboratory core tests or wireline logs because the measurements are made just above the drill bit with a minimum of drilling disturbance. Similar computations for porosity can be made using neutron porosity and density data from LWD and wireline logs. The standard LWD resistivity tool is an electromagnetic device that responds to the conductivity and the dielectric constant of the formation; it emits a low-frequency electromagnetic wave that can be transformed into shallow- and deeper-penetrating resistivity measurements. The difference in dielectric properties among massive hydrate, pore fluid, and sediment are distinct—gas hydrate behaves like fresh water with respect to the dielectric constant and like sediment with respect to their conductivity. Because of this electromagntic response, however, standard LWD resistivity measurements may be less affected by disseminated hydrate and their signature may not be easy to observe using these devices. Although porosity can be computed using the same Archie formulation from LWD or wireline resistivity logs, examples of LWD-derived porosity in hydrate-bearing sediments have, to date, been limited. Further research into the electromagnetic properties of hydrate-bearing sediments and independent measurement of porosity, pore fluid chemistry, pressurized core samples, and an understanding of the effects of clay are needed to fully evaluate the dielectric properties of hydrate and hydrate-bearing sediment.

3. FIELD EXAMPLES

3.1. The Ocean Drilling Program

The Deep Sea Drilling Project (DSDP) and the Ocean Drilling Program (ODP) have contributed to our basic understanding of marine environments by sampling and experimenting below the seafloor since 1968. Over their history, DSDP and ODP have drilled a wide variety of formations, recovering core samples and measuring in situ logs whenever possible. DSDP and ODP have been the primary source for logging information about the in situ properties of hydrate-bearing deep sea sediments worldwide. During this period, these programs have successfully completed over 190 drilling expeditions, or "legs", around the world's oceans at the time of this writing. More than 1160 drill holes have been drilled and cored, after which many have in turn been logged using downhole instruments. During the DSDP, downhole measurements were conducted in less than 14% of all marine holes drilled, while they have been made in more than 51% of the holes drilled by ODP. This trend has dramatically increased the number of in situ measurements acquired in hydrate environments, yet, to date, DSDP and ODP have cored and logged in fewer than 20 holes in which hydrate has been observed.

Downhole logging at DSDP and ODP sites has provided some of the most representative information known about the properties and distribution of gas hydrate in the subsurface. ODP has recorded logs through hydrate zones in several geographic locations, such as on the Blake Ridge off of the US eastern seaboard [*Paull, et al.,* 1996]. Other hydrate environments near the continental margins of Guatemala, Peru, Cascadia, and Costa Rica have been previously logged [*von Huene, et al.,* 1985; *Suess, et al.,* 1988; *Westbrook,, et al.,* 1994;

Kimura, et al., 1997]. On the Guatemala margin, where the first deep sea log measurements were made, wireline logs were recorded through a 15-m thick massive hydrate layer about 250 m below the sea floor [*Matthews and von Huene,* 1985]. These early data indicated low density, high electrical resistivity, and high compressional velocity in the hydrate layer relative to otherwise uniform clay sediments. Wireline log examples from three other marine locations, the Blake Ridge, Costa Rica, and Cascadia, are discussed in greater detail below. In addition, a considerable amount of knowledge about the in situ properties of hydrate has been gained from permafrost-associated hydrate occurrences on the North Slope of Alaska and from the Mackenzie River Delta of northern Canada. Example logs from these two locations are also shown for comparison. The available logs from these locations include density and neutron porosity, electrical resistivity and sonic velocity logs. Together, these examples document the response of various wireline logging tools to the presence of gas hydrate. Using the above-mentioned methods, it is possible to quantitatively assess the formation porosity and the concentration of methane hydrate under in situ conditions (also see *Collett,* this volume).

3.2 Blake Ridge

The Blake Ridge represents a large area of well-studied gas hydrate on a passive continental margin off the coast of S. Carolina that was drilled during ODP Leg 164 [*Paull, et al.,* 1996]. Most other hydrate areas that have been studied using ODP logs are on accretionary prisms in subduction zone settings. On this passive margin, the hydrate concentrations are relatively low but occur over a considerable depth interval. Sites 994, 995, and 997 comprise a transect of holes that penetrate through a gas hydrate zone within the same stratigraphic interval over a relatively short lateral distance (9.6 km). Prominent BSRs are apparent in seismic profiles across most, but not all, of the length of this transect. The presence of hydrate at Sites 994 and 997 was documented by direct sampling—several pieces of hydrate were recovered from about 260 mbsf at Site 994 and disseminated hydrate was observed at almost the same depth at Site 994. One large, solid piece of hydrate about 15 cm long was recovered from about 331 mbsf at Site 997. No hydrate samples were conclusively identified at Site 995. Instead, based on wireline logging data and geochemical core analyses, disseminated hydrate was inferred to occur from a depth of about 190 to 450 mbsf in all three holes drilled across the Blake Ridge.

Figure 3 shows the wireline logs recorded at Site 995 on the Blake Ridge. Similar to the early observations on the Guatemala margin, disseminated hydrate occurrences on the Blake Ridge are also characterized by increases in log-measured velocity and resistivity. These wireline logs reveal that the hydrate-bearing interval corresponds with an increase in electrical resistivity of 0.1-0.3 ohm-m above background reference levels and with an increase of 0.1-0.2 km/s in compressional velocity. The deep-reading resistivity log reveals several high resistivity intervals of 1.4-1.5 ohm-m near 220 mbsf. Below these anomalies, the resistivity and velocity logs increase with depth to 450 mbsf. Examination of the natural gamma ray and bulk density logs from this hole reveal no apparent lithologic causes for the observed velocity and resistivity increases through the interval. These observations are consistent with the presence of a material having increased resistivity and velocity, but similar density to the pore water

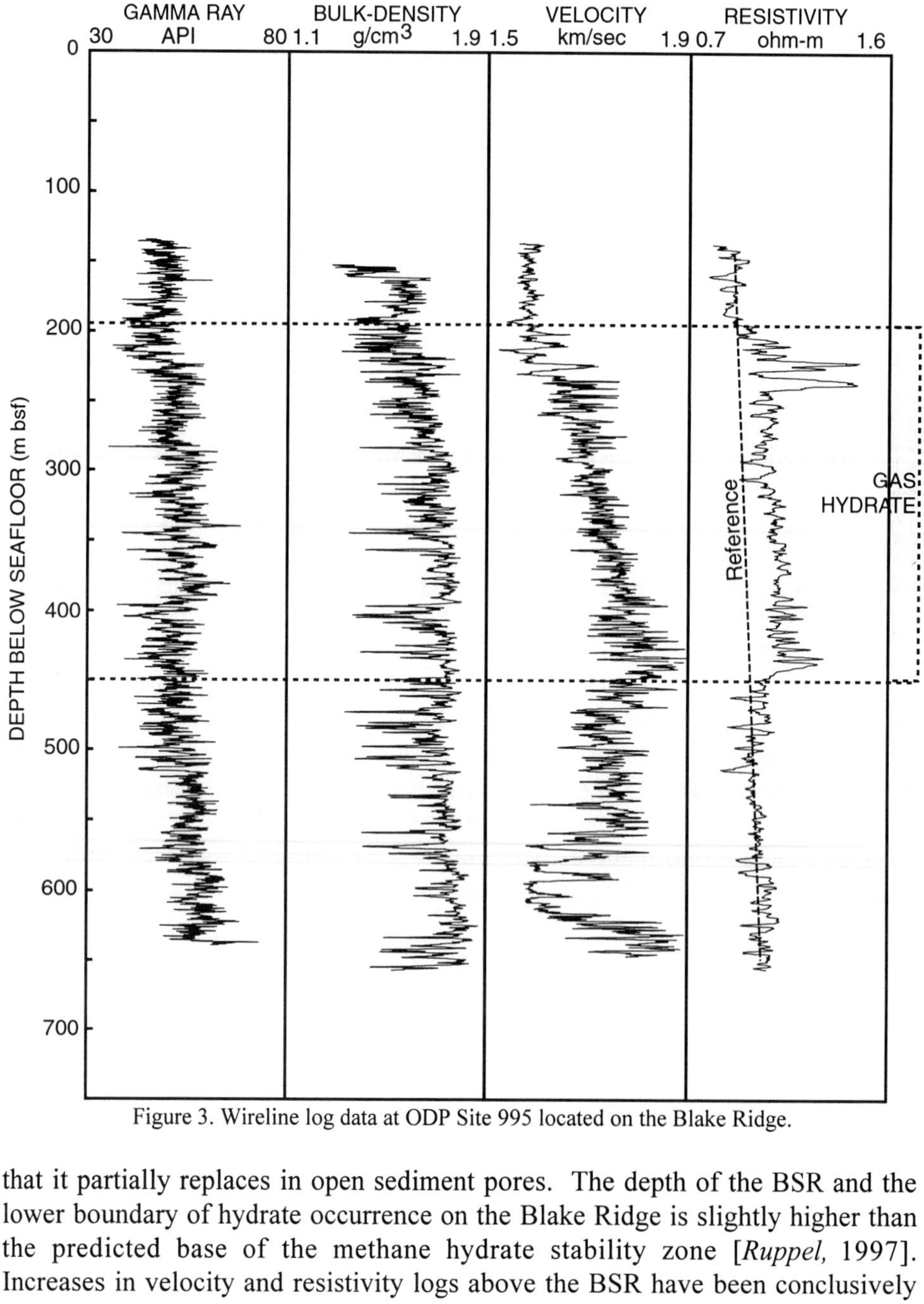

Figure 3. Wireline log data at ODP Site 995 located on the Blake Ridge.

that it partially replaces in open sediment pores. The depth of the BSR and the lower boundary of hydrate occurrence on the Blake Ridge is slightly higher than the predicted base of the methane hydrate stability zone [*Ruppel,* 1997]. Increases in velocity and resistivity logs above the BSR have been conclusively associated with the presence of hydrate, in overall agreement with sampling and other geochemical estimates [*Paull, et al.,* 1996; *Dickens, et al.,* 1997]. The decrease in VSP and sonic log velocity below the BSR is attributed to the accumulation of free gas stratigraphically trapped at this depth [*Holbrook, et al.,* 1996]. Within the interval below the BSR to the depth of the theoretical base of gas hydrate stability, evidence from the sonic logs suggests that hydrate and small amounts of free gas may co-exist at this site [*Guerin et al.,*1999]. The integrated interpretation of in situ logs and seismic data from the Blake Ridge indicates that the velocity contrast across the BSR, though marking a presence of hydrate and free gas over the interval, clearly is not a simple gas–water contact.

3.3. Cascadia Margin

ODP drilling on the Cascadia margin during Leg 146 was designed to study the occurrence of gas hydrate in an accretionary prism associated with an active subduction zone and to examine the role of upward fluid and methane expulsion in hydrate-bearing sediments [*Westbrook et al.,* 1994]. Strong BSRs are widespread over much of the continental slope and provide an opportunity to examine whether they are due to high-velocity hydrate or low-velocity free gas [*Hyndman and Spence,* 1992]. During Leg 146, only minor amounts of hydrate were recovered in the core. At Site 889, drilled off the west coast of Vancouver Island, hydrate is inferred to be present from approximately 127 to 228 mbsf based on geochemical core analysis and wireline logs [*Westbrook, et al.,* 1994]. Only after integration of downhole data and regional seismic velocities was it possible to estimate the concentration of hydrate and free gas above and below the BSR [*Yuan et al.,* 1996]. This velocity contrast was interpreted to be about 2/3 due to high-velocity hydrate and 1/3 due to underlying low-velocity gas.

Figure 4 shows the logs recorded from seafloor to about 250 m bsf at Site 889. From the velocity log, hydrate is interpreted to increase in concentration downward from the seafloor, with an increase at about 130 m associated with the contact between slope basin and underlying accreted sediments. From 130 m to the BSR at about 230 m, the velocity is about 1.8 km/s compared to a reference of about 1.6 km/s. Likewise, the electrical resistivity at Site 889 increases downward from the seafloor with an abrupt increase at about 130 m. From this depth to the BSR, resistivity averages about 2.1 ohm-m compared to the reference of about 1.0 ohm-m [*Hyndman et al.,* 1999]. In addition to the general increasing trend in the logs at Site 889, there are several layers of high velocity and electrical resistivity near 190 and 210 m. There is no corresponding lithologic cause evident in the natural gamma-ray, neutron porosity, and bulk density logs so these thin zones may represent especially high hydrate concentration. The free gas zone underlying the BSR is not expressed in the sonic and resistivity logs, but only as low velocities in the VSP data [*MacKay et al.,* 1994]. *Hyndman, et al.* [1999] interpret the relatively high resistivity log over this interval to indicate that cold drilling fluid has induced hydrate to form from free gas present near the borehole. Alternatively, drilling circulation may have rapidly washed the free gas away from the vicinity of the borehole and beyond the depth of penetration of the logs, but not sufficiently deep to affect the VSP measurements.

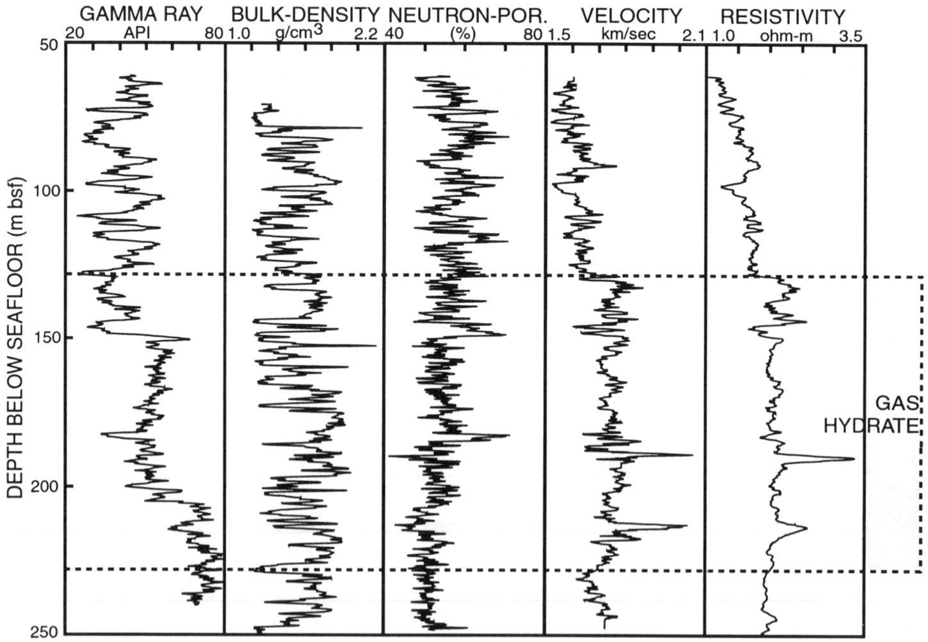

Figure 4. Wireline log data from ODP Site 889 located on the Cascadia Margin.

3.4. Costa Rica Margin

The first LWD data recorded in hydrate-bearing sediments were acquired on the Costa Rica margin during ODP Leg 170. Several holes were drilled into the sedimentary wedge upslope of the Costa Rica trench, penetrating Plio-Pleistocene age sediments that are mostly derived from the Central American peninsula. The wedge was cored, and where significant amounts of material was recovered for sampling, disseminated hydrate was observed in fractures, cemented ashes, and diffuse pore space [*Kimura, et al.,* 1997]. Disseminated hydrate was recovered in these sediment cores, however, their signature was not clearly observed in the LWD resistivity data [*Boissonnas, et al.,* 2000]. Although LWD measurements are made within minutes of drilling and may best represent the undisturbed physical state of the in situ environment, the response of this tool could not distinguish the presence of hydrate in the formations on the Costa Rica margin. A likely explanation is that with only low hydrate concentration, large amounts of seawater and clay in the formation may be too conductive for the hydrate to have a significant effect on the measured resistivity using this electromagnetic device. If hydrate distribution is diffuse it may also rapidly dissociate during drilling and be easily obscured. Additional laboratory experiments and different LWD tools are needed to conclusively identify in situ hydrate occurrences in this environment.

3.5. North Slope of Alaska

The occurrence of gas hydrate on the North Slope of Alaska was confirmed with data from pressurized core samples, wireline logs, and the results of formation production testing in the Northwest Eileen State-2 well located in the northwest part of the Prudhoe Bay oil field. *Collett* [1993] calibrated these log responses to an additional 50 exploratory and production wells in the Alaskan permafrost region and inferred gas hydrate occurrences in six laterally continuous sandstone and conglomerate reservoirs, all of which are geographically restricted to the Kuparuk River-Prudhoe Bay area. The wireline logs from wells in the western part of the area also indicate free gas accumulations trapped below some of the inferred hydrate layers. The Northwest Eileen State-2 well drilled through five such hydrate-bearing layers.

Wireline logs (Fig. 5) from the Northwest Eileen State-2 well. Hydrate occurs in the upper three units and is characterized by high electrical resistivity (ranging from 20 to 70 ohm-m) and high compressional velocity (ranging from 2.0 to 2.8 km/s). *Collett* [1993] explain the increase in electrical resistivity and compressional velocity by the occurrence of gas hydrate.

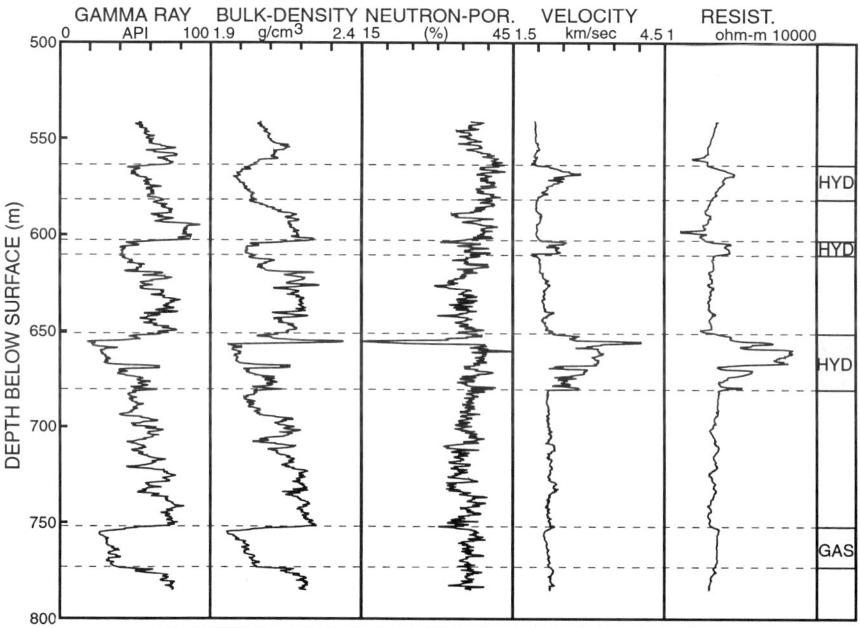

Figure 5. Wireline log data from the Northwest Eileen State-2 well located on the North Slope of Alaska.

The gamma ray, formation density, and neutron porosity logs in these layers are indicative of clean (to slightly shaley) sandstone reservoirs. Of particular note, the low neutron porosity and high density log values from about 654-656 m are likely due to the presence of a "hard", well cemented zone at the top of the sandstone reservoir. The zone characterized by low electrical resistivity, low density, and low velocity from about 751-775 m is interpreted to indicate water and gas-saturated sediment, without the presence of gas hydrate.

3.6. Mackenzie River Delta

The Mallik 2L-38 research hole was drilled to investigate the occurrence of gas hydrate in the permafrost region of the Mackenzie River Delta of Canada [*Dallimore, et al.,* 1999]. A major component of the Mallik drilling program was to apply state-of-the-art wireline logging technology and refine its interpretation in naturally-occurring hydrate formations in a permafrost environment. As the Mallik hole was drilled explicitly for this purpose, the quality of the wireline logs in the hydrate-bearing interval of this formation is excellent. Figure 6 illustrates the logs acquired through a relatively thick hydrate-bearing interval from about 897 to 1,110 m depth. The logs show deep reading resistivity measurements ranging from 10-100 ohm-m and compressional velocities (V_P) ranging from 2.5-3.6 km/s. In addition, a shear-wave velocity log (V_S) was recorded, yielding values from 1.1-2.0 km/s, and V_P/V_S varies between 2.8-2.0 through the hydrate interval. The decrease in velocity at approximately 1120 m depth indicates the presence of a relatively thin free gas layer below the predicted base of the gas hydrate stability zone. High quality VSP data was recorded in the Mallik well and both VSP compressional- and shear-wave velocities are in good general agreement with V_P and V_S from the sonic logs [*Walia, et al.,* 1999]. *Collett, et al.* [1999] estimate porosity from these logs in the range from 20-40%, which suggests that methane hydrate saturation within clean sandstone reservoirs may be especially high in this permafrost environment.

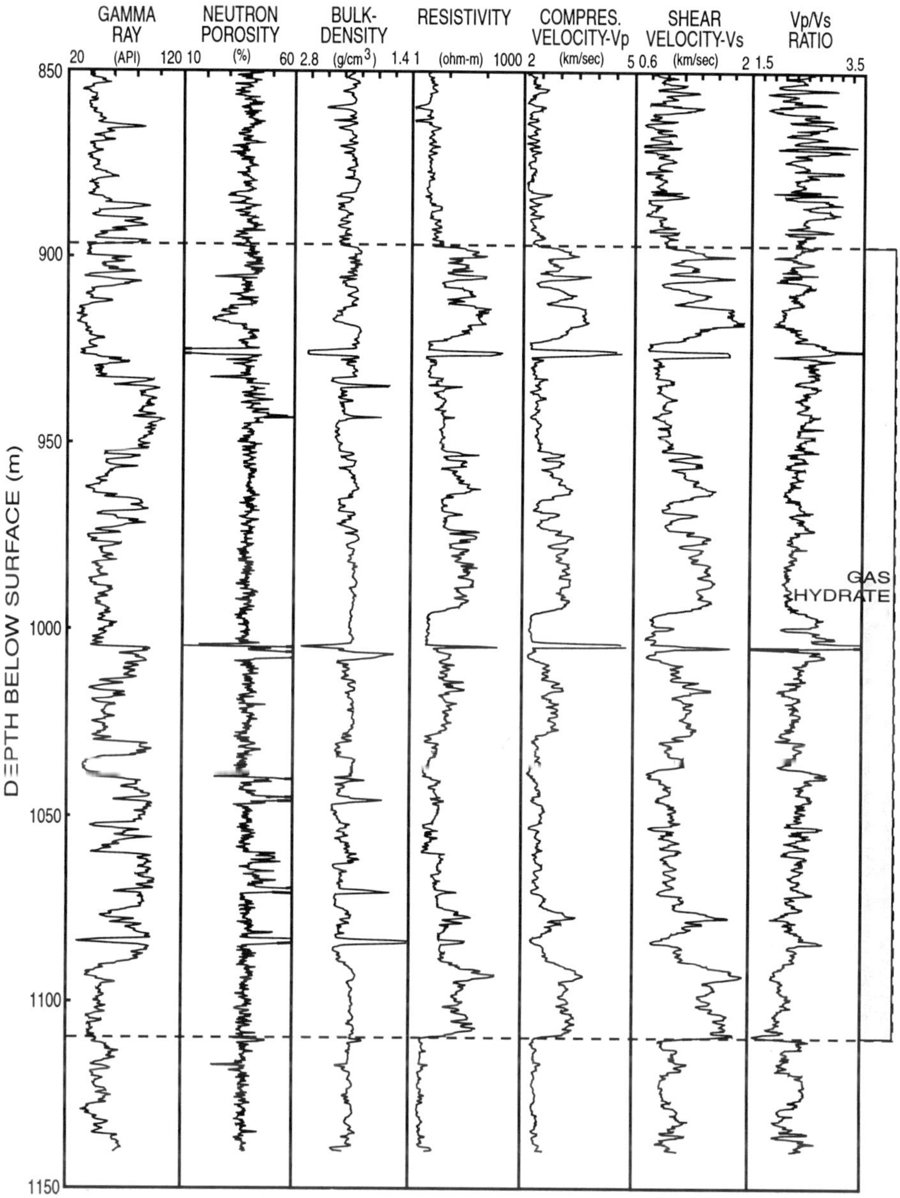

Figure 6. Wireline log data from the Mallik 2L-38 well located on the Mackenzie River Delta of northern Canada.

4. SUMMARY

Because hydrate is difficult to sample and study in the laboratory, in situ detection methods in drill holes are critical in that they provide "ground truth" measurements for constraining surface geophysical data. However, the linkage between downhole and surface information is not necessarily straightforward and their correlation, in turn, must be evaluated in the context of the type and scale of the in situ measurement. Resistivity and velocity logging methods are used most often for this purpose. The dissociation of hydrate or the change in free gas concentration in the vicinity of a borehole due to the drilling process may significantly alter these measurements and the correction or quantitative evaluation of the results is usually required. To minimize such effects, LWD logging methods generally provide higher quality data, but the application of this new technology must be carefully interpreted for the particular host sediment in which gas hydrate may be present.

The measurement of in situ properties using logs is primarily useful for: (1) the identification of hydrate and hydrate-bearing sediment and their distribution with depth, (2) the estimation of porosity and methane saturation, and (3) the calibration on of drill hole information with surface seismic or other remote geophysical data. The table below generalizes the expected response of several common logging measurements in massive hydrate, hydrate-bearing marine sediment, and gas- and water-saturated marine sediments based on the field examples discussed above.

	Massive Hydrate*	Hydrate-bearing Sediment	Water-saturated Sediment	Gas-bearing Sediment
V_p (km/s)	3.2-3.6	1.7-3.5	1.5-2.0	1.4-1.6
V_s (km/s)	1.6-1.7	0.4-1.6	0.75-1.0	0.4-0.7
R (Ω-m)	150-200	1.5-175	1.0-3.0	1.5-3.5
ρ (g/cm^3)	1.04-1.06	1.7-2.0	1.7-2.0	1.1-1.5
Φ (%)	20-50	35-70	35-70	50-90
γ (API)	10-30	30-70	50-80	30-80

Table 1. Common ranges of in situ log properties (*modified from *Matthews,* 1986).

In general, resistivity logs typically have high values relative to seawater-saturated sediment as hydrate forming in open pores replaces the more electrically conductive seawater. Both compressional and shear velocity logs typically increase because hydrate acts as a high-velocity pore filling material and as an intergranular cement, stiffening the bulk compressibility and rigidity of the sediment. Sonic log and VSP velocities tend to decrease in the presence of free gas. Natural gamma ray, neutron porosity, and bulk density logs show little or no apparent decrease in hydrate-bearing sediments. In the future new logging and LWD technologies measuring in situ resistivity, velocity and porosity, all of particular importance for detecting gas hydrate occurrences, will provide greater certainty for many of these key parameters.

Chapter 24

GHASTLI - Determining Physical Properties of Sediment Containing Natural and Laboratory-Formed Gas Hydrate

William J. Winters[1], William P. Dillon[1], Ingo A. Pecher[2] & David H. Mason[1]

[1]U.S. Geological Survey
384 Woods Hole Road
Woods Hole, MA 02543 USA

[2]Institute for Geophysics, Univ. of Texas
4412 Spicewood Springs Road, Bld. 600
Austin, TX 78759 USA

1. INTRODUCTION

Gas-hydrate samples have been recovered at about 16 areas worldwide (Booth et al., 1996). However, gas hydrate is known to occur at about 50 locations on continental margins (Kvenvolden, 1993) and is certainly far more widespread so it may represent a potentially enormous energy resource (Kvenvolden, 1988). But adverse effects related to the presence of hydrate do occur. Gas hydrate appears to have caused slope instabilities along continental margins (Booth et al., 1994; Dillon et al., 1998; Mienert et al., 1998; Paull & Dillon, (Chapter 12; Twichell & Cooper, 2000) and it has also been responsible for drilling accidents (Yakushev and Collett, 1992). Uncontrolled release of methane could affect global climate (Chapter 11), because methane is 15-20 times more effective as a "greenhouse gas" than an equivalent concentration of carbon dioxide. Clearly, a knowledge of gas-hydrate properties is necessary to safely explore the possibility of energy recovery and to understand its past and future impact on the geosphere.

Gas hydrate exists in the natural environment at pressure and temperature conditions that make it difficult to study in situ. For that reason, the U.S. Geological Survey's Woods Hole Field Center developed a laboratory system (GHASTLI - Gas Hydrate And Sediment Test Laboratory Instrument) to simulate natural conditions within the gas-hydrate-stability region. Using this system, gas hydrate can be formed in reconstituted sediment (Winters et al., 2000) and field samples containing gas hydrate can be preserved (Winters et al., 1999) while physical properties are measured.

GHASTLI determines a number of properties of the host sediment prior to gas-hydrate formation, after hydrate has formed, and after dissociation. Currently, four main types of data are measured: (1) acoustic properties (both compression-wave and shear-wave), (2) shear strength and related engineering

M.D. Max (ed.), Natural Gas Hydrate in Oceanic and Permafrost Environments, 311–322.

properties under triaxial test conditions, (3) permeability, and (4) electrical resistivity. These results are used to model and predict the relation between the natural environment, sediment fabric, and the existence of gas hydrate.

Acoustic properties can dramatically change as gas hydrate forms, and this relationship allows remote identification of hydrate. Initial velocities derived from seismic data have been used to quantify the presence of gas hydrate (Lee et al., 1993). This acoustic relationship then has been used as a means of mapping the extent of hydrate and may be a means of estimating resource potential. Also, acoustics have been used to model gas-hydrate interaction with host sediment (Dvorkin and Nur, 1993; Pecher et al., 1999). Currently, a number of theories exist that relate the effect of natural gas-hydrate saturation to acoustic behavior, but testing of these theories has not been possible without a laboratory system capable of simulating in situ conditions. Berge et al. (1999) showed that velocity behavior changes occur when pore-volume saturation exceeded about 35% using sand and R11 (CCl_3F) refrigerant as the hydrate former. It is not yet known if hydrate formed with R11 exhibits behavior similar to natural gas hydrate.

Sediment strength properties can be affected by the presence of gas hydrate, however, measurements of the effects of hydrate on sediment strength are scarce. There is currently concern that offshore drilling in the Ormen Lange gas field in the North Sea could dissociate existing gas hydrate and cause seafloor instabilities (Offshore, 2000). The nearby Storegga slide, which caused tsunamis to impact both Scotland and Norway about 7,000-8,000 ka, is thought to have been influenced by gas-hydrate dissociation. Concerns exist that gas-hydrate dissociation related to mass wasting may release methane, a potent "greenhouse gas", to the atmosphere (Chapter 11). A better understanding of strength properties can be used in stability calculations of: (1) natural slopes (Kayen and Lee, 1991), (2) structures placed on land or the seafloor (pipelines and drilling platforms), and (3) wells drilled to recover natural gas and oil (Briaud and Chaouch, 1997).

Sediment permeability is reduced if significant amounts of gas hydrate occupy intergranular pore space. The changes that are produced are important for slope-stability calculations (Kayen and Lee, 1991) and in determining if gas hydrate can effectively form a seal that traps bubble-phase gas beneath the gas-hydrate-stability zone (GHSZ). The presence of gas beneath the GHSZ is a major contributor to the formation of a bottom-simulating reflector (BSR) present on seismic-profile records.

Because electrical resistivity of gas hydrate is much higher than naturally-existing pore water, measurement of that parameter can be used to indicate when gas hydrate has formed in the laboratory and can be used as a comparison to well-logging measurements. Well logging has provided much information about the presence and properties of gas hydrate in the natural state, for example, Collett et al., 1984, 1999; Collett & Ladd, 2000; and Miyairi et al., 1999. Gas-hydrate saturation has also been calculated from well-log information (Lee and Collett, 1999).

Gas hydrate exists naturally in different materials and under varying environmental conditions. Understanding the factors responsible for hydrate formation, preservation, and concentration may enable prediction of locations where methane from gas hydrate can be economically and safely recovered. GHASTLI provides a means for making measurements related to the above topics under conditions that can be closely controlled. Sediment grain size, specimen bulk density, effective stress, and pore-fluid characteristics can be varied to assess their impact on gas-hydrate formation and behavior.

2. GHASTLI

2.1. Overview

GHASTLI combines a number of separate-pressure-and-temperature-control systems (Fig. 1) to simulate in situ conditions on a 71-mm diameter by 140-mm high right-circular sediment specimen (Fig. 2). The 25 MPa pressure (equivalent to the pressure exerted by a 2,500 m column of water) and temperature (25°C to about -3°C) capabilities of the system provide wide latitude in test configurations and procedures within which gas hydrate can be studied in reconstituted sediment or in field samples containing natural gas hydrate. Instrumentation packages and sensors placed within the different subsystems and in close proximity to the test specimen produce measurements that are logged and displayed by a computer running custom-designed software (Fig. 3). During testing, the sample resides within a silicone-oil filled, main pressure vessel (Figure 1). Separate internal sediment pore pressure and external confining pressure systems are used to adjust isotropic consolidation stress, σ'_c to simulate in situ overburden pressure.

Figure 1: GHASTLI system showing the syringe pumps that control pressure and the main pressure vessel, located just to the right of center in the photo, that contains the sediment test specimen.

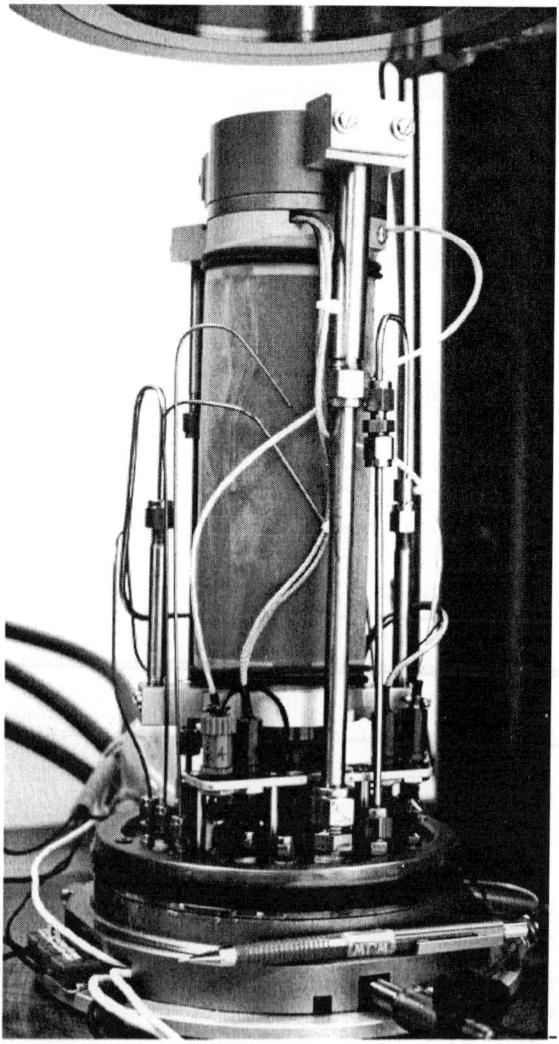

Figure 2: Close-up view of a test specimen about to be raised into the main pressure vessel showing the different pressure and coolant lines. The test specimen (Grey cylinder in the central part of rig) is 14 cm high. The sedimentn sample is vertically raised into the main pressure vessel, located at the very top of the photo, prior to testing.

The test sample is surrounded by flexible membranes. Top and bottom end caps incorporate acoustic transducers and gas or water flow ports. The bottom end cap rests on an interchangeable internal load cell. The top end cap is contacted by a bath-controlled, variable-temperature heat exchanger which imparts a unidirectional cooling front downward through the specimen. To achieve better sample thermal equilibrium the perimeter of the main pressure vessel is also surrounded by cooling coils.

Five 500 ml capacity syringe pumps are used to maintain the confining pressure surrounding the specimen and the internal specimen pressures which include: pore pressure, back pressure, and methane-gas pressure. Flow rates of

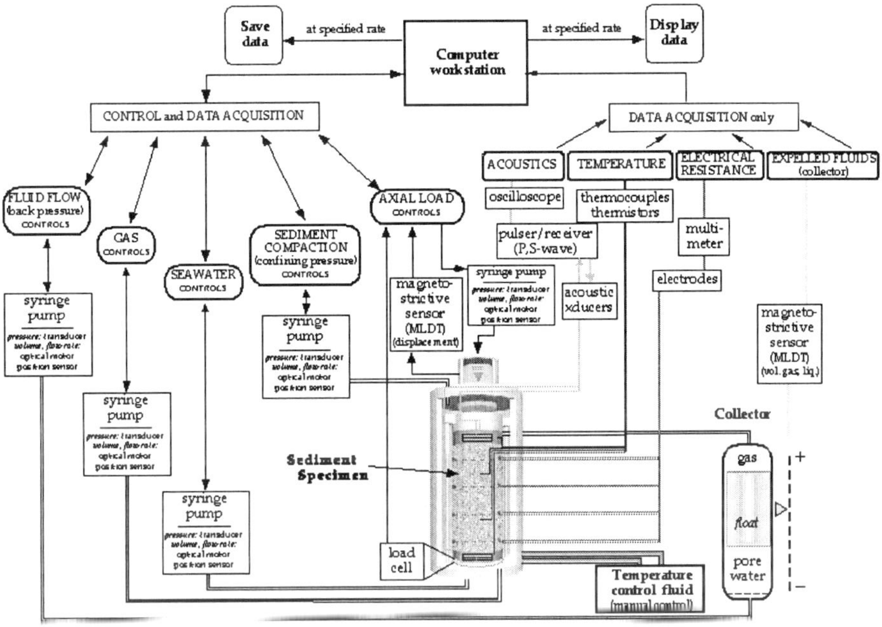

Figure 3: Schematic of main pressure and data acquisition systems (modified from Booth et al., 1999).

0.001 ml/min to over 80 ml/min are controlled by the syringe pumps to a maximum pressure of 25 MPa. The back-pressure system contains a collector that is capable of separating and measuring water and gas volumes that are pushed out of the specimen at test pressures by the dissociation of gas hydrate. A separate syringe pump controls the movement of the load ram during shear tests and is used to determine the height of the specimen (which may change during a test) to increase the accuracy of acoustic-velocity measurements. Four thermocouples and four thermistors are placed against the outside perimeter of the specimen at different heights to measure temperature variations during cooling or warming stages and during gas-hydrate formation and dissociation events.

Because we can reduce specimen temperatures below freezing, ice formation is possible either alone or within the sediment pores. This is important because some physical properties of gas hydrate and ice are similar (Kuustraa et al., 1983; Miyairi et al., 1999) and gas hydrate is present in regions where permafrost is present (Dallimore and Collett, 1995; Collett, 1993).

2.2. Acoustics

P-wave velocity is measured by through transmission using 500 kHz or 1 MHz (natural frequency) wafer-shaped crystals that are located on the back side (away from the specimen) of each end cap. A pulser sends an electrical signal, with a 400 volt pulse height, to the transmitting transducer. The received signal typically has a 200 kHz frequency and is amplified, digitized, displayed on a digital oscilloscope, and recorded by a computer. Acoustic compression-wave and shear-wave velocities, Vp and Vs, are calculated from the specimen length/measured travel time corrected for system delays.

2.3. Strength

Four parameters are measured during triaxial strength tests (Holtz and Kovacs, 1981): load, axial deformation, confining pressure, and pore pressure. Loading is produced by a syringe-pump-controlled ram contacting the heat exchanger which then pushes on the sample at a rate related to the syringe-pump-piston travel. Movement of the ram, which can vary from 0.0001 mm/min to 2 mm/min, is measured using a linear displacement transducer connected to the load ram. Interchangeable load cells can be varied according to anticipated strength of the sediment and gas hydrate.

2.4. Hydraulic conductivity

Hydraulic conductivity measurements are performed (according to procedures in ASTM D 2434-68 (1998), modified for use with syringe pumps) at different phases of testing by flowing water from the bottom end cap up through the specimen and out the top end cap into a back pressure silicone oil/seawater interface chamber. Both constant flow rate and constant head methods are used. Pressures, flow rates, and fluid volumes are recorded from the syringe pumps, and the pressure drop through the specimen is measured with a variable-range differential-pressure transducer or by comparing pressure differences between syringe pumps.

2.5. Electrical resistivity

Four equally-spaced electrical-resistivity probes are placed on opposite sides of the sample. A current is flowed through the outer probes, and the potential is measured between the inner probes at specified times. The direction of current flow is reversed after each reading to reduce the impact of ion concentrations. The electrical-resistivity setup presently being constructed is similar to typical four-point measurements that are performed for geological applications (Chaker, 1996). The measured resistivities are more accurate than those determined with our initial across-the-test-specimen, two-point measurements which were used to map the location of gas hydrate within the test specimen (Booth et al., 1999).

3. PROCEDURES

3.1. Preservation of samples containing natural gas hydrate

We have recovered natural gas hydrate from two field projects: Ocean Drilling Program (ODP) Leg 164 conducted on the Blake Ridge off the coast of South Carolina (Paull et al., 1996) and the Mallik 2L-38 Gas Hydrate Research Well drilled in the Northwest Territories of the Canadian Arctic (Dallimore et al., 1999). Although both projects had pressurized coring systems, gas-hydrate samples preserved for later laboratory testing were recovered using conventional drilling techniques. During the ODP cruise samples containing gas hydrate in finer-grained sediment were brought up through a relatively warm column of water above the thermocline before recovery.

This process contributed to gas hydrate decomposition. By contrast, the coarser-grained samples recovered in the Mackenzie Delta had to be brought up through 640 m of permafrost. The Mallik samples did freeze, but not because of the permafrost. Endothermic cooling produced by gas-hydrate dissociation caused the remaining free water to become ice. Samples from each area were placed into transportation pressure vessels, pressurized with gas, and stored within refrigerators or freezers to maintain the gas hydrate at stable pressure-temperature conditions for the overland trip to Woods Hole, MA.

3.2. Mallik 2L-38 samples

The Mallik 2L-38 test samples were kept frozen prior to placement in GHASTLI. They were removed from the transportation vessels, jacketed, and placed between end caps within a walk-in freezer maintained at about -30°C before being brought into the lab. The presence of ice helped to stabilize the hydrate at atmospheric pressure during the test setup procedures. Although the specimens were still partially frozen at the beginning of testing, they were allowed to warm to a temperature where the ice melted, but the gas hydrate remained solid.

Each GHASTLI laboratory test consists of different steps designed to address different physical-property measurements (Fig. 4). Initial acoustic measurements on gas hydrate and partially-ice-bonded samples (Fig. 5) were followed by converting the ice to water. Temperature was raised to approximately 4°C for at least 24 hours until all ice had melted and acoustic velocity stabilized. After all of the ice (which was not present in situ) melted, temperature was slowly raised in 1 to 2°C increments until dissociation began. The dissociated gas, specimen pore water, and water released from hydrate entered the back pressure collector where water and gas were separated and their volumes measured.

In all cases, the acoustic signal previously established was lost during initial dissociation of gas hydrate in the test samples but was recovered using a "rebuilding" procedure. One strength test was performed on a sample that contained gas hydrate; the rest were sheared after dissociation.

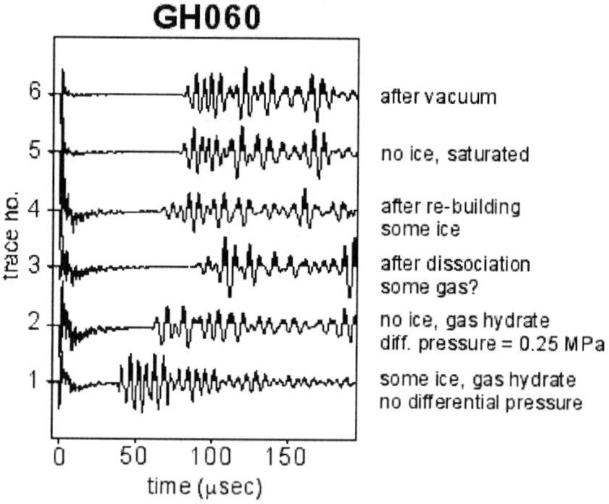

Figure 4: P-wave signals from a Mackenzie Delta sample at different times during the test with sample pores filled with different materials.

Figure 5: Cross section of a black sand sample containing methane gas hydrate formed under simulated in situ conditions in the laboratory. The white gas hydrate occupies almost all sediment pores. The sample diameter is about 70 mm and the original sample length was about 140 mm before sectioning. The wafer at the top of the photo has approximately the same diameter as a test end cap.

3.3. Laboratory-formed gas hydrate

GHASTLI has been used to form gas hydrate within a number of sediment types in the laboratory; for example, Ottawa sand (a standard sediment commonly used for testing), silt, and black sand (Figure 5). Gas hydrate can be formed a number of different ways using GHASTLI. In order to simulate in situ conditions, typically, a water-saturated sediment is initially brought to pressures between 12 and 20 MPa. Confining pressure is increased and consolidation is allowed until excess pore pressures are dissipated. The difference between the internal pore pressure and the confining pressure is related to the simulated overburden stress. Methane is percolated up through the sediment until a pre-determined amount of water, measured by the collector, is pushed out of the sample. Then the temperature of the coolant flowing to the heat exchanger located above the top end cap is lowered until P-T conditions are within the gas-hydrate-stability zone. Physical properties are measured and recorded during these procedures. To dissociate the gas hydrate the procedure is reversed; that is, the temperature is slowly raised until the phase boundary is crossed. Pressure, rather than temperature can also be used as the trigger to form or dissociate gas hydrate. Because the thermal gradient between the top and bottom of the sample can be adjusted from about one degree to many degrees, it is conceivable that a phase boundary can actually be induced at a layer in the sample.

4. RESULTS AND DISCUSSION - SELECTED EXAMPLES

4.1. Index properties and gas hydrate quantity

The index properties (e.g., water content, bulk density, void ratio, porosity) of reconstituted samples typically are calculated from the known mass of solids and water used to form the specimen. These calculations can then be checked with volumetric measurements. The reverse is true for "undisturbed" natural samples containing gas hydrate. Differences between porosity measurements can approximately be accounted for by the expansion of water to ice (about 9%) and water to gas hydrate (about 26%). The amount of methane present, in moles, is determined from the equation-of-state relationships presented by Duan et al. (1992).

4.2. Acoustic properties

P-wave signals are recorded at numerous times during testing (e.g., Fig. 4). P-wave velocities, Vp, are calculated routinely also. Figure 6 shows velocity changes induced in different materials by gas hydrate formation, freezing of the pore water, and by increasing consolidation stress. Knowledge of the material occupying the pore space and the consolidation stress are critical to understanding the implications of the various velocity measurements. Vp values typically increase with consolidation stress because of the greater grain to grain contact stress (primarily) and increased density. This is illustrated by Figure 7 for a Mackenzie Delta sediment, but the rate of velocity change decreases with increasing pressure. Sample shortening during consolidation could also create an apparent velocity increase due to reduced travel distance for the acoustic pulse (if sample height is not accurately measured).

Initial results indicate that samples obtained from the Mackenzie Delta containing natural-gas hydrate acoustically behave as part of the sediment frame or as disseminated throughout the pore fluid, but not as cements between grains.

4.3. Strength properties of samples from Mallik 2L-38 Well

Preserving field samples that contain natural gas hydrate for subsequent laboratory strength testing is inherently difficult. Unless a pressurized coring system that can maintain in situ P-T conditions is used, some gas hydrate may be lost during sample transfer into transportation vessels in the field or later during transfer into a laboratory test system.

If gas hydrate behaves similarly to that of ice in sediment then gas-hydrate-bonded sediment should in most cases be stronger than similar material that does not contain gas hydrate. Andersland and Anderson (1978) demonstrate that the presence of ice can substantially increase the strength of sediment. Frozen sediment exhibits a wide range in strength properties, because strength is influenced by a number of factors: strain rate, temperature, confining pressure, grain size, and density. The strength of sediment containing gas hydrates is probably influenced by these and other factors such as gas hydrate-cage-occupancy.

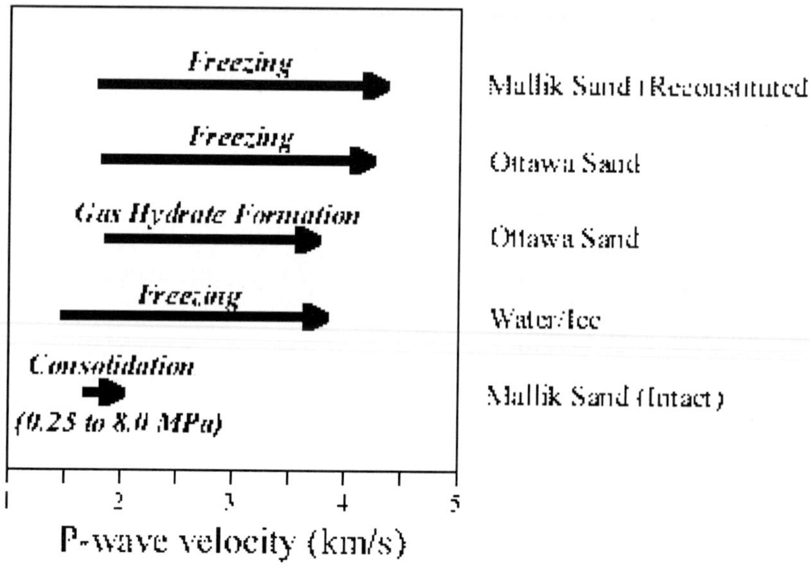

Figure 6: Comparison of changes in P-wave velocity for different materials tested in GHASTLI by forming gas hydrate, freezing pore water, and increasing consolidation stress. The arrows represent an initial material velocity before the indicated process began, followed by a gradual increase in Vp until a final, maximum value is reached.

Undrained triaxial shear tests were performed on samples from the Mallik Well both before and after natural-gas hydrate was dissociated (Figure 8). Plots of the shear stress vs. axial strain for the four test samples show that the strength of the sample containing gas hydrate is indeed much higher. The gas-hydrate-containing sample exhibited higher negative pore pressures during shear. Andersland and Ladanyi (1994) indicate that frozen samples are typically plotted with respect to total stresses because of the difficulty in measuring

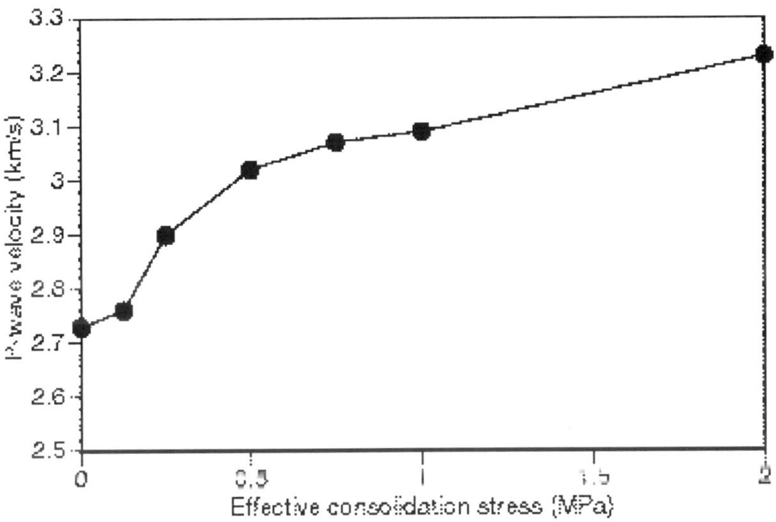

Figure 7: Effective consolidation stress, σ'_c, versus Vp for a sample from the Mallik 2L-38 well.

intergranular stresses. Evidently, in the gas-hydrate-containing Mallik 2L-38 specimen, the pores contained enough free water to transmit pore pressure. The large difference in strength between the specimen containing gas hydrate and the other samples may be related to the fact that many of the pores of the gas-hydrate sample contained a solid that increased the tendency for dilation during shear. Hence the higher negative pore pressure and corresponding strength values.

The maximum effective friction angle (Holtz and Kovacs, 1981) for the hydrate sample was also the highest value of any of the tests (44.4°). The friction angles for the non-hydrate bearing samples (33.8 - 38.6°) are within the typical range of sandy sediment (Hunt, 1984). Depending upon the amount present, gas hydrate has the potential to greatly affect the mechanical properties of the host sediment.

The dissociation of methane hydrate in GHASTLI produces an excess volume of gas which is related to the test pressure and temperature conditions. Typically during a test the internal sample pressure is maintained constant and the additional gas produced by dissociation is measured in the collector.

However, in situ, the production of excess gas may have a destabilizing effect because of the increase in pore pressure accompanied by a subsequent reduction in sediment effective stress. The amount of pressure generated in situ by dissociation depends upon a number of factors, some of which include: P-T conditions, amount and distribution of gas hydrate present, percent of cage occupancy of the hydrate by gas molecules, inherent sediment permeability, rate of dissociation, and faulting present in the region. Measuring sediment and gas hydrate properties is only the first step in predicting in situ behavior.

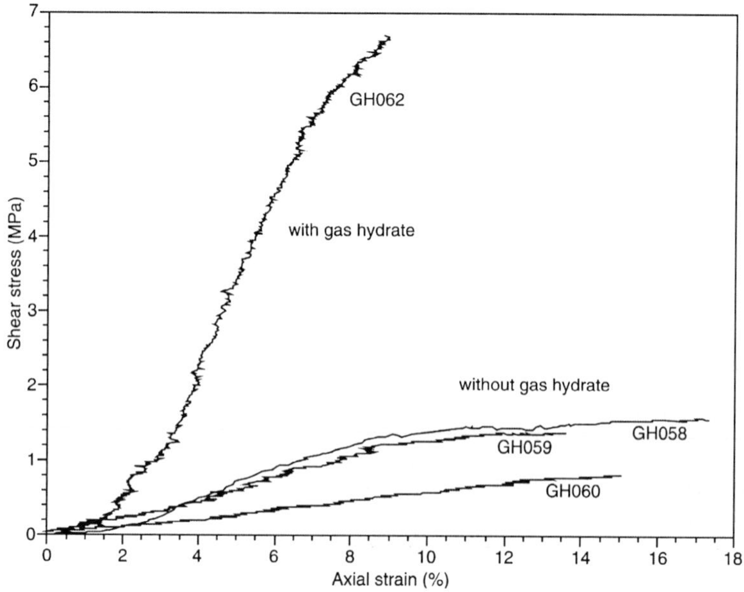

Figure 8: Comparison of triaxial strength plots for a sample containing natural gas hydrate to other samples from the Mallik 2L-38 well after gas hydrate was allowed to dissociate.

5. SUMMARY

GHASTLI is able to produce pressure and temperature conditions that simulate the natural environment. Gas hydrate can either be formed in: (1) undisturbed natural sediment, (2) reconstituted natural sediment, or (3) artificial material; and field samples containing natural-gas hydrate can be preserved. Physical properties, including acoustic velocity, shear strength, permeability, and electrical resistivity are measured at different times during testing to ascertain the effect that gas hydrate has on the bulk physical properties of the host sediment. Those physical properties are used in models that estimate the amount of gas hydrate present in situ, for seafloor and well stability calculations, and for comparison to well logs.

Acknowledgements: The authors thank Dave Twichell and Richie Williams for their helpful reviews.

Chapter 25

Laboratory synthesis of pure methane hydrate suitable for measurement of physical properties and decomposition behavior

Laura A. Stern[1], Stephen H. Kirby[1], William B. Durham[2], Susan Circone[1], and William F. Waite[1]
[1]U..S. Geological Survey, MS/ 977, Menlo Park, CA 94025
[2]U.C. Lawrence Livermore National Laboratory, Livermore, CA 94550

1. INTRODUCTION AND BACKGROUND

1.1. Why study synthetic gas hydrates?

Gas hydrates are an intriguing class of nonstoichiometric compounds that have significant commercial and scientific applications both as an energy resource and as a manufactured material. The last half-century has witnessed a marked escalation in the scope of experimental research on gas hydrates, particularly directed towards the determination of their phase equilibria, formation kinetics, crystallographic and structural properties, transport and thermal properties, effects of inhibitors, and a number of related geochemical topics.

There remains, however, a paucity of reliable experimental measurements of many of the physical, material, thermal, acoustic, and elastic properties of most pure, end-member hydrocarbon hydrates. Instead, either water ice, or hydrates readily formed in the laboratory but rarely occurring in nature (such as ethylene-oxide hydrate and THF hydrate), have commonly been used as analogue material for property measurements. Consequently, there does not exist an accurate and comprehensive database of physical and material properties for end-member gas hydrates, and particularly for those hydrates that are more problematic to form and stabilize in the laboratory. Compounding this problem is the difficulty in retrieving pristine material from natural settings on which to make such measurements, in different laboratories using different methods, and in the general lack of agreement of measurements made on synthetic material.

A wide variety of processes and techniques have been used to synthesize gas hydrates in the laboratory, each yielding a final product that may be highly suitable for some types of experimental testing while clearly unsuitable for others. Here, we focus on laboratory production of pure, polycrystalline methane hydrate and hydrate-sediment aggregates that are suitable for a variety of physical and material properties measurements made on pure, end-member

M.D. Max (ed.), Natural Gas Hydrate in Oceanic and Permafrost Environments, 323–348.

material. The methods described here are based on a materials-science approach that strives to produce final test specimens with highly-reproducible composition, texture, and grain characteristics. Property measurements on such test specimens not only provide end-member material characterization, but also aid in the interpretation of similar measurements made on more complicated hydrate-bearing material retrieved from natural settings.

1.2. Criteria for sample suitability

Historically, bulk quantities of gas hydrates have been produced in many laboratories by a variety of different methods, commonly mimicking either hydrate-forming processes that are thought to occur in nature, or under conditions of pressure and temperature that may occur in nature. Many researchers have grown gas hydrate by bubbling a hydrate-forming gas through seawater- or freshwater-saturated sediment columns, simulating ocean-floor type conditions or processes. Others have produced hydrates from ice + gas mixtures by vigorous agitation, shaking, or rocking procedures that continually renew fresh ice surfaces for hydrate growth. An excellent and comprehensive review of a variety of methods and fabrication apparatus is given by Sloan (1998, chapter 6), to which we refer the reader for further background information.

While the material produced by such dynamic methods may be suitable for some types of measurements, particularly those related to phase equilibria, formation processes, and formation kinetics, there remain some outstanding problems inherent to these procedures which render the resulting material less suitable for those measurements in which precise knowledge and excellent reproducibility of composition and grain characteristics are required. These problems include: (1) The resulting hydrate customarily includes excess or unreacted H_2O, either as liquid water or as ice, which is difficult to separate and remove from the final hydrate product. Not only does excess H_2O contaminate the hydrate and influence all subsequent property measurements, but it also greatly hinders accurate determination of the composition and stoichiometry of the resulting hydrate. (2) Hydrate produced by continuous agitation methods has generally been produced under low to moderate pressures, resulting in low methane content in the resulting hydrate. (3) Synthetic hydrate formed from water or from continually-agitated ice grains does not have well constrained or well controlled grain size, grain texture, or crystallographic orientation, parameters which can strongly affect those properties influenced by grain anisotropy, such as ductile flow behavior or elastic properties.

We present here an alternative method for efficient and routine synthesis of pure methane hydrate in a form that is highly suitable as a standard for properties measurement. We review an *in situ* "seeding" method for hydrate nucleation and growth, a procedure developed from extensive work on other H_2O-based icy compounds in which we demonstrated successful growth of polycrystalline test specimens with controlled and uniform grain size, and with no preferred crystallographic orientation or anisotropy effects (Durham et al.,

1983, 1993). Similarly, gas hydrate samples produced by this method are highly reproducible from sample-to-sample in terms of known purity, composition, stoichiometry, porosity, grain size, and grain orientation. Such sample characteristics are not only necessary for reliable characterization of the intrinsic properties of crystalline solids, but individual parameters can then be changed in a controlled manner for systematic modeling of increasingly complex systems.

This latter point makes the seeding method particularly effective for laboratory modeling of natural systems, as sediment and impurity effects are difficult to isolate in measurements made on natural hydrate-bearing samples due to the inherent complexity of the material and the slow recovery process. Ocean-sediment hydrates recovered from drill core may contain, for example, partially-decomposed hydrate with poorly known stoichiometry, composition, and grain size; excess water or ice produced from partial dissociation; additional pore water; poorly constrained sediment compositions, distributions, grain size and grain orientations; poorly known grain boundary contacts or "cementing" contacts; and possible additional components such as carbonate mineralization or organic detritus. Our seeding procedure allows for easy pre-mixing or layering of a wide variety of particulate matter to the granulated ice prior to converting the ice to gas hydrate, thus enabling fabrication of "tailored" hydrate/sediment aggregates of varying complexity.

1.3. Focus of this chapter: Three basic aspects of gas hydrate synthesis and properties measurements are covered in this chapter. (1) Synthesis apparatus and techniques used for growth of pure, polycrystalline methane hydrate samples, hydrate + sediment ± seawater composites, and hydrates made from liquid hydrocarbons or multi-component systems. Observations of methane hydrate formation from ice grains and a possible model for the reaction are also discussed, as well as key factors for successful synthesis of gas hydrates and for scaling up of sample size. (2) Apparatus and techniques used for measuring stoichiometry, composition, annealing effects, stability, and dissociation rates and kinetics. (3) Applications for property measurements, and apparatus modifications to accommodate compaction and *in situ* measurement of physical, thermal, mechanical, and acoustic properties.

2. SYNTHESIS PROCEDURES AND APPARATUS

2.1. Pure methane hydrate

Samples of polycrystalline methane hydrate can be efficiently synthesized by promoting the general reaction CH_4 (*gas*) + $6H_2O$ (*sol.*→*liq.*) → $CH_4 \bullet 6H_2O$ (*sol.*) (Fig. 1). This product is achieved by the mixing and subsequent slow, regulated heating of sieved, granular, H_2O "seed" ice and cold, pressurized CH_4 gas in an approximately constant-volume reaction vessel (Fig. 2). This seeding method for hydrate nucleation and growth permits successful synthesis of polycrystalline test specimens with controlled and uniform grain size with no detectable preferred crystallographic orientation, which we have verified through sample-replica observation and powder x-ray analysis (Stern et al., 1998a).

Sample fabrication details are as follows: CH_4 gas from a source bottle is initially boosted in pressure (P) by a gas intensifier and routed into sample

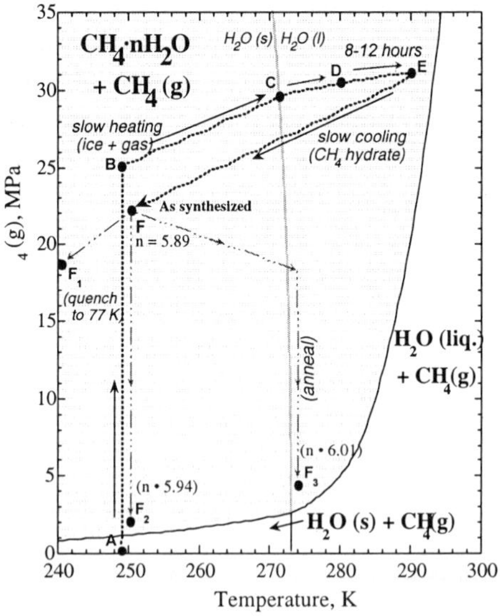

Figure 1: Methane hydrate synthesis and annealing conditions in relation to the CH_4-H_2O phase diagram. Shaded region shows methane hydrate stability field. The metastable extension of the H_2O melting curve is shown by the solid grey curve. Black dotted lines trace the reaction path during methane hydrate synthesis from ice + gas mixtures. Points A-F indicate the P-T path during reaction. For synthesis, H_2O "seed" ice at 250 K (pt. A) is pressurized with CH_4 gas to 25 MPa (B). Heating the mixture through the H_2O melting point (C) and up to 290 K (C-D-E) promotes full and efficient conversion of the ice to gas hydrate. Samples are then cooled to 250 K (F), and can then be quenched (F_1) and removed from the apparatus, or annealed at conditions closer to the equilibrium curve (F_2, F_3). The "n" numbers represent hydrate stoichiometry under various conditions.

molding vessels housed in a conventional deep freezer. The sample assembly (Fig. 2) consists of a manifold of as many as four stainless steel vessels immersed in an ethyl alcohol bath initially held at freezer temperature of ~ 250 K. One vessel serves as a reservoir to store and chill pressurized CH_4 gas, and the others contain the sample molds. Each mold consists of a hollow split-cylinder that encases an indium sleeve filled with a measured mass of H_2O "seed" ice typically packed to 40% porosity. Seed material is made from a gas-free and nearly single-crystal block of ice grown from triply distilled H_2O, crushed, ground, and sieved to 180-250 μm grain size (Durham et al., 1983). The sample vessels with seed ice are initially closed off from the reservoir and evacuated. A disk inserted on top of the packed ice grains prevents displacement of the ice during evacuation. Multiple thermocouples inserted into the base of either sample prior to loading of the ice permit careful monitoring of the sample's thermal history during synthesis and subsequent testing. A Heise pressure gauge and pressure transducers monitor gas pressure on the samples.

The reservoir vessel is first charged with pressurized CH_4 gas to 35 MPa and cooled to 250 K. When fabricating a single sample, the reservoir is opened to the pre-evacuated sample chamber and methane pressure (P_{CH4}) drops to ~ 24 MPa. For fabrication of multiple samples, the reservoir charging, cooling, and opening procedures are repeated to bring the larger volume of the multiple-sample system to about 25 MPa at 250 K. These steps serve to fill the porosity between the ice grains to a molar ratio of CH_4 to H_2O well in excess of that required for complete hydrate reaction. The bath temperature can then be slowly raised by various methods such as by use of a ring immersion heater or a simple hot plate located beneath the alcohol bath. As the samples and gas reservoir warm, they self-pressurize by thermal expansion. Up to 271 K, methane gas pressure (P_{CH4}) increases approximately linearly with increasing temperature, following a slope governed primarily by the equilibrium thermal expansion of free CH_4 in the system. Measurable reaction begins as temperature rises above 271.5 K (the approximate melting point of ice at our synthesis pressure), and consumption of CH_4 gas by hydrate formation slows the rate of P_{CH4} increase (Fig. 3). Progress of the hydrate-forming reaction is monitored by observing the deflection of P from the initially linear P-T curve. Completion of reaction is efficiently achieved by steady heating to ~ 289 ± 1 K over a heating time interval of about 12 to 15 hours after the sample vessel crosses the 271.5 K isotherm (Fig. 3). Data-acquisition software (LabVIEW™, National Instruments) monitors and records the P-T conditions throughout each run. The extent of reaction can be determined mid-run by the measured P_{CH4} offset from the reversible CH_4 expansion curve.

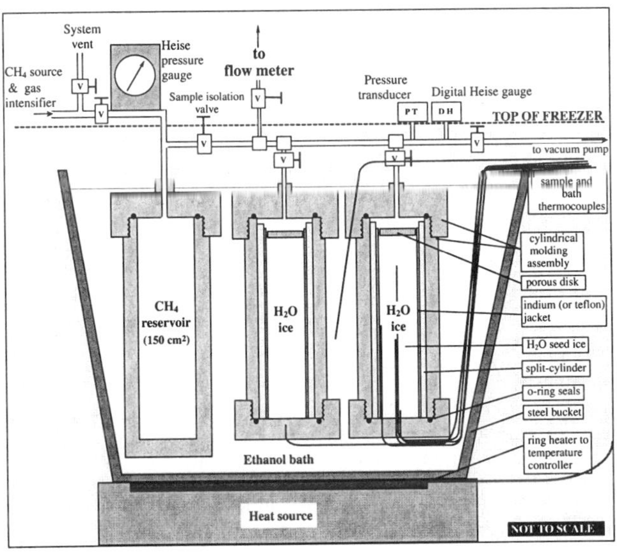

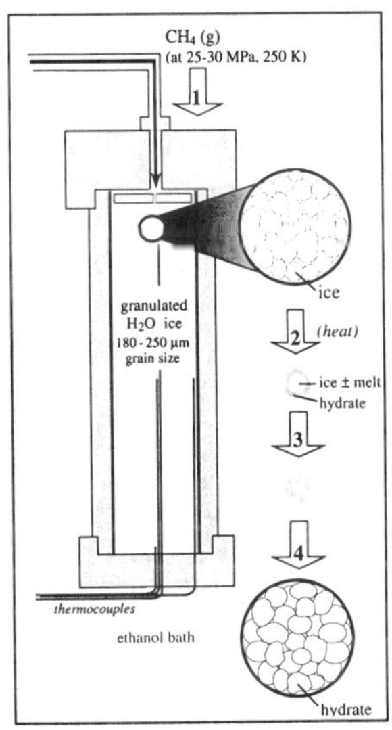

Figure 2: Left: Apparatus for synthesizing test specimens of pure, polycrystalline methane hydrate from CH_4 gas and H_2O ice by methods described in text. **Right**: Postulated synthesis model. Numbered arrows (1-4) are as follows: 1) Cold, pressurized methane gas (35 MPa, 250 K) is admitted to the ice in the sample chamber. The reactants equilibrate to ~25 MPa and 250 K. 2) Warming above the H_2O melting point initiates measurable hydrate formation along the surface of the ice grains, creating composite grains in which a mantle of hydrate envelops an unreacted ice (± melt) core. The reaction rate (diffusion or transport controlled) slows as the hydrate rind grows and thickens, and the inner unreacted core shrinks. 3) Slowly raising the temperature to 290 K promotes further reaction. 4) By the end of the heating cycle, the reaction has reached completion.

Following full reaction, the heat source is turned off, and the system slowly cools through the freezing point of ice and down to freezer temperature

(~250 K). At this stage, the P-T record should be inspected for any indication of a freezing anomaly during the cooling cycle, signifying incomplete reaction. Such an anomaly appears as a discontinuous jump in both the P and T readings at several degrees below 273 K, produced when small amounts of unreacted and supercooled water suddenly freeze to form ice (see Fig. 6 in Stern et al., 1998a, for example). The exothermic process of freezing even very small amounts of water (< 1.5 vol. %), coupled with the associated volume increase, is easily measured by the internal thermocouples and pressure transducers. The presence

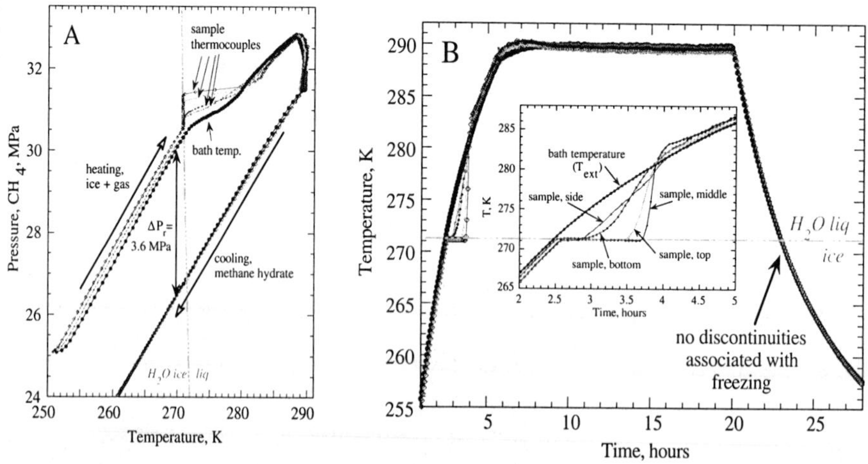

of such discontinuities requires cycling the sample through the melting point of ice a second time to insure completion of reaction.

Figure 3. (**A**) P-T history of a methane hydrate synthesis run (with two samples) during the reaction: CH_4 (g) + 5.9H_2O (ice→melt) → CH_4•5.9H_2O. Warming the ice + gas mixture above the H_2O solidus (dot-dashed line) initiates measurable reaction. Increasing temperature slowly to 290 K, over an 8-15 hour span, promotes full reaction. Complete reaction is marked by a known pressure drop (ΔP_r) from start to finish relative to the extrapolated subsolidus P-T curve. (**B**) Temperature-time profile during hydrate formation. The thermal profile of the sample as it warms through the H_2O melting point is expanded in the inset. Buffering of the sample thermocouples at the H_2O melting point indicates that a measurable fraction of melting of the seed ice occurs over this 1.2 hour stage. Full conversion of the H_2O to hydrate requires 8-12 hours however, and ΔP_r over the early melting stage is not as large as that predicted for complete melting of all residual ice (see Fig. 3A in Stern et al., 1996, for further details).

If no refreezing anomaly appears, samples may then be annealed, tested, or removed from the apparatus. For removal, samples should be vented to low P_{CH4} within the hydrate field (~ 2 MPa at 250 K), then slowly depressurized while cooled with liquid nitrogen. Pressure must be maintained on the sample to keep the gas hydrate within its stability field during the initial phase of the cooling procedure down to 190 K, but the sample should be fully vented before it cools through 109 K, where any residual pressurized methane in the pore space will liquefy. Allowing liquid methane to settle and freeze as the sample is further cooled to liquid nitrogen temperature (77 K) results in a final material that is mildly explosive and very difficult to work with when rewarmed. Following the quenching procedure, the vessels can be disconnected from the apparatus and opened. The inner, hollow split-cylinders containing the jacketed samples are pushed out of the vessels and pried off the samples. Samples are then wrapped tightly in Al foil and stored in or directly above liquid nitrogen.

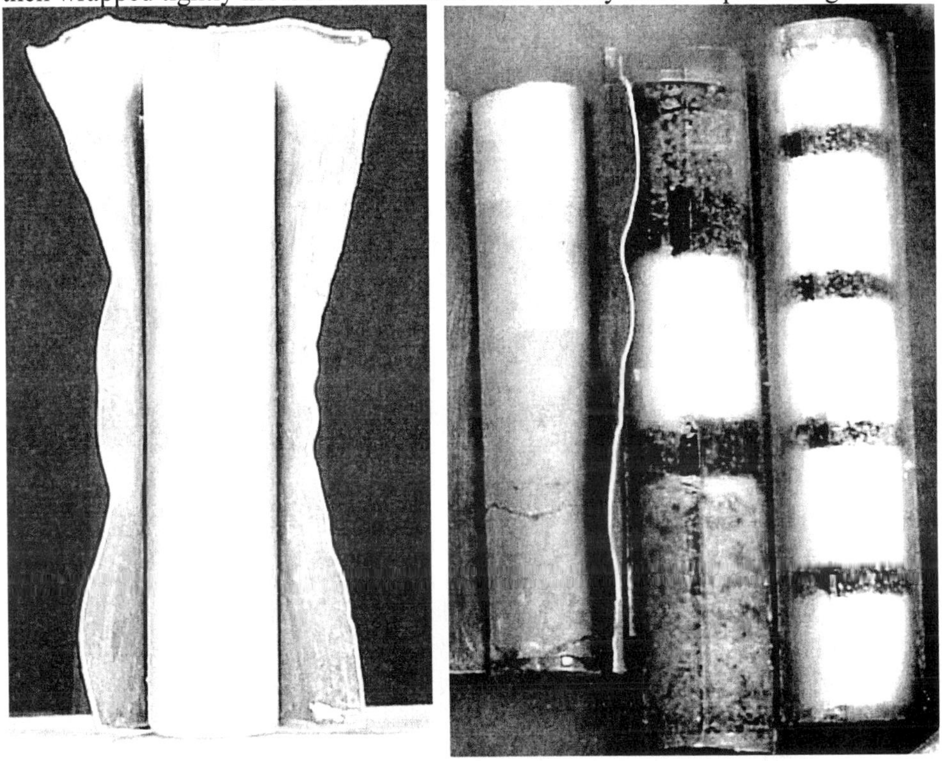

Figure 4. Cylindrical test specimens of polycrystalline methane hydrate and mixed sediment aggregates grown by methods described in the text. **Left**: Sample of pure methane hydrate (CH_4•$5.89H_2O$), 2.54 cm x 11.5 cm, with uniform grain size of ~300 microns and approximately 28% porosity. The encapsulating indium jacket in which the sample was grown has been split and peeled back from the sample. **Right**: Pure methane hydrate + sediment aggregates fabricated with pre-specified characteristics. Samples are 2.54 cm by 10 - 11.5 cm. Sample at left is methane hydrate + pure quartz sand, varying the proportion of sand in each quarter of the sample. Middle and right samples are mixtures of pure methane hydrate and 1.25 ± 0.25 mm black Al_2O_3 particulate. The two samples containing Al_2O_3 were grown in Teflon jackets (split along top surface) rather than in indium jackets.

The resulting material is a cohesive aggregate of uniformly fine, equant, white grains of pure methane hydrate with grain clusters of 250 ± 50 μm grain size (Fig. 4). Samples that were initially packed to 40% porosity will contain 29 ± 1% porosity after full reaction due to the volume increase accompanying conversion of the ice to gas hydrate. Purity of the final material and efficiency of the synthesis are highly dependent on adhering to certain key procedures, discussed below. Careful measurements of gas released from fully reacted samples show that the gas hydrate produced at the prescribed test conditions has a highly-reproducible composition of $CH_4 \bullet nH_2O$, where $n = 5.89 \pm 0.01$.

2.2. Methane hydrate + sediment (± seawater) aggregates

Test specimens of methane hydrate + sediment mixtures are easily produced by pre-mixing or layering sediments with granular ice in each reaction vessel prior to admission of CH_4 gas. With the static growth method, there is no migration of either H_2O or sediment in pre-mixed samples, so samples can be "customized" to any desired mixture or layering sequence. To date, samples have been successfully made with pure methane hydrate inter-layered with discrete horizons of particulate material with grain size as fine as 50 μm (using SiC) and as coarse as 2 mm (using Al_2O_3), with no discernable change in the layering during synthesis (Fig. 4). Alternatively, sediments can be homogeneously mixed with seed ice before synthesis to produce a final product that remains fully mixed.

Saturating the gas hydrate or hydrate/sediment samples with seawater (or fresh water) immediately following synthesis is possible but somewhat problematic, in that it can be difficult to assess the extent and distribution of any secondary gas hydrate growth resulting from the added water. A simple method for saturation involves first depressurizing the sample to P-T conditions that are slightly warmer than the ice point but close to equilibrium conditions (for instance, 275 K and 3 MPa P_{CH4}), then admitting highly pressurized seawater through the upper port of the sample such that the pressurized head space over the water pushes the water into the sample and then gravitationally floods the available pore space. Use of the same hydrate-forming gas to pressurize the water prevents possible contamination of the sample due to entrainment or dissolution of the pressurizing gas within the water. Monitoring the weight changes of both the sample and H_2O reservoir enables determination of the mass of H_2O incorporated into the sample. A more optimal procedure involves a flow-through design to facilitate full saturation of all available pore volume with the added water, easily determined by flowing the water through the base of the sample and observing its emergence from the top end of the sample.

2.3. Methane hydrate formation from melting ice; direct observations, and factors for complete reaction

Hydrate formation and growth processes can be further investigated by optical microscopy, using a simple and versatile optical cell constructed from standard high-pressure valve bodies and SiO_2 capillary tubing (Fig. 1 in Chou et al., 1990). This design was used for observing methane hydrate growth from seed ice, by chilling the cell and loosely filling it with 200 μm ice grains prepared identically to those used in standard samples. The pore volume was then evacuated and flooded with cold CH_4 gas at 23 MPa and 250 K. The cell was then immersed in an insulated dish of cold ethanol, placed under the microscope, and slowly warmed from 250 K to 290 K.

Ice grains subjected to CH_4 hydrate-forming conditions showed visible surface reaction at temperatures well below the H_2O liquidus, almost immediately after exposure to CH_4 gas. This surface appearance did not change appreciably as the grains were subsequently warmed through the H_2O liquidus. Upon further heating, no expulsion of water was observed nor any cracking or collapsing of the hydrate encasement that would be expected to attend the volume contraction associated with bulk melting of the ice interiors. Instead, all grains maintained identifiable shapes and sizes throughout reaction, and changed only by becoming increasingly mottled in appearance as they approached full conversion to hydrate. In separate experiments in which the hydrate was slowly heated and then cycled through the equilibrium curve, however, newly-formed gas hydrate always grew as optically clear material, either coarsely crystalline or as single crystals. Further details and photographs of these experiments are given in Stern et al., (1998b).

At low temperatures in the ice subsolidus field, hydrate formation on ice grains virtually halts (on the laboratory time scale) after initial surface reaction unless the grains are vigorously agitated to renew fresh surfaces for gas hydrate formation. The process of converting the unreacted core to hydrate by raising the temperature, however, is less well understood. We previously hypothesized that the rapid hydrate formation promoted by raising the temperature well above the melting point of ice, in the absence of measurable bulk melting and melt segregation, continues by an essentially solid-state transport or diffusion-controlled reaction (Fig. 2; also Stern et al., 1996, 1998a, 1998b). We speculated that our synthesis conditions promote transport of methane through the outer hydrate rind and inward to the hydrate/ice interface at a rate sufficiently fast such that melt nuclei react to form hydrate faster than they can grow to the critical size necessary for bulk melting and melt segregation. If the rate of bulk melting of the ice can be suppressed by continual removal of melt nuclei, this suggests, by corollary, the possible short-term "superheating" of the ice.

In previous work we noted that while all samples synthesized at 27-32 MPa do in fact show evidence for some melting of the ice cores as the system is initially warmed through the melting point of ice, no evidence was observed to indicate complete melting or melt segregation, even though full conversion of ice to gas hydrate requires about 8 hours by our methods (Stern et al., 1996, 1998a). This early melting stage is now detected with improved accuracy with the multiple-thermocouple design of the reaction vessels that allow significantly improved detection of the thermal lag produced by the latent heat of melting (Fig. 3B). These measurements show that the melt stage is more significant than previously calculated, but still do not indicate either complete or rapid melting of all the unreacted ice in the system immediately upon crossing the ice point.

Further testing is clearly required to better resolve the questions surrounding hydrate formation from slowly-melting ice grains, and whether or not unreacted ice can exist temporarily in a superheated state. Recent NMR imaging reported by Moudrakovski et al., (1999) on hydrate formation from melting ice grains at lower methane pressures (< 12 MPa) showed that in fact particle morphology can be maintained when rinds of hydrate encapsulate an inner core of melt. Similar studies have not yet been conducted at the high P and T conditions of our synthesis procedures, however.

2.3.1. Key factors for successful synthesis of pure methane hydrate The synthesis procedures described here appear to be highly dependent on those aspects that influence the availability, transport, and concentration of gas-hydrate-forming species at the growth front. These factors include elevated P-T conditions, and a high surface-to-volume ratio of the reacting grains to minimize the thickness of the developing hydrate layer through which the reactants must pass.

In terms of the procedures described here, the following three experimental parameters are integral to the full and efficient conversion of ice to gas hydrate without measurable segregation of a bulk melt phase: (1) maintenance of a very high methane overpressure above the methane hydrate equilibrium curve (25-33 MPa), (2) moderate thermal ramping (~ 5-12 K/hr) and subsequent holding of temperature at very warm conditions (288-290 K for 8-12 hours), and (3) using a small initial grain size of seed material (< 300 μm). Synthesis tests conducted with larger grain sizes of seed ice (0.5-2 mm) or lower overpressures of CH_4 (4-11 MPa) resulted in only partial reaction, measurable bulk melting, and significant segregation and pooling of the melt into the lower region of the sample chamber (Stern et al., 1998a, 1998b). Thermal ramping to lower temperatures (< 285 K) or with very slow ramping rates (< 3 K/hr) also reduced the efficiency of the reaction process, and commonly resulted in incomplete reaction and melt segregation. For samples in which melt segregation occurs, it is troublesome and very inefficient to convert the segregated melt phase to gas hydrate under static conditions. Multiple cyclings through the ice point are required, but may still result in a substandard final material. Finally, we note that successful conversion of ice grains to hydrate grains may also be specific to our method for preparing the ice in order to minimize defects, impurities, or grain boundaries that can act as sites for melt nuclei, but this aspect of sample preparation is as yet unverified.

2.3.2. Considerations for scaling up to larger sample sizes

The methods and apparatus discussed in this chapter can be adapted to synthesize material in a variety of sample configurations or scaled up significantly in size. Those factors that affect the rate of reaction and/or the heat flow budget of the system, however, should be given careful consideration to best assess appropriate apparatus modifications to ensure successful fabrication of the desired end-product. Using a gas reservoir chamber that has not been scaled up with an increased sample size, for instance, will affect the relative pressure drop due to the reaction, that in turn can result in partial melting of the sample or affect its thermal state.

Other factors that we have identified in preliminary tests include extremely rapid thermal ramping (>20 K/hr) up to peak temperatures, and synthesizing large samples with low surface-to-volume ratios, both of which can result in heat transfer problems during synthesis due to insufficient dissipation of the exothermic heat of the gas-hydrate-forming reaction. In several cases, the internal sample temperature was observed to significantly overshoot the external bath temperature due to an insufficient rate of heat transfer, resulting from the low thermal conductivity of the newly-forming hydrate and the porous nature of the sample.

Simple methods to circumvent these problems include (1) using a sufficiently large fluid bath surrounding the sample chambers to permit adequate heat transfer and thermal stabilization. (2) Increasing the surface-to-volume ratio of oversized samples to alleviate heat transfer problems. (3) Controlling or moderating the rate of reaction by use of a sufficiently large gas reservoir during synthesis to maintain high gas overpressures during synthesis (the reservoir-to-sample chamber volume ratio should not be less than about 2:1), and by using a moderate thermal ramping rate (5-15 K/hr). (4) Incorporating sediments and/or seawater into large samples to aid in heat transfer and to decrease the total amount of gas hydrate formed in such samples.

Temperature overshoot in the sample interior may not necessarily be undesirable or detrimental, however, as it can be exploited to drive the system to self-buffer along the equilibrium curve at high P-T conditions, and can result in the development of larger and more optically-translucent grains. It is yet to be determined, however, if such material retains a uniform grain size and lack of crystallographic orientation throughout all regions of the sample.

2.4. Synthesis from liquid hydrocarbons or multi-component systems

Synthesis of pure gas hydrates from gas phases that liquefy at our synthesis pressures (CO_2, propane, ethane, etc), or from mixed-gas sources in which components can unmix at synthesis conditions, can be achieved with several modifications to the standard procedures. Certain problems require special consideration, however. These include: (1) the difficulty in achieving high P conditions without unmixing a multi-component source gas, and the inability to access and utilize high T conditions due to the steep equilibrium curves of the phase diagrams for many of these types of gas hydrates. (2) Maintaining a homogeneous distribution of phases in the pore space between the ice grains. This maintenance of phases can be a problem when synthesizing mixed-phase gas hydrates, due to the density differences between the various components and/or phases. Similarly, variations in sample stoichiometry can arise even when working with a single component liquid hydrocarbon, if only the lower portion of the sample remains saturated with the liquid phase while the upper portion is saturated with the gas phase. (3) Quenching the final hydrate product to temperatures below the dissociation temperature of the hydrate, without freezing in hydrocarbon phase in the pore space.

To circumvent these difficulties, it is critical to first establish the P-T conditions at which the source gas liquefies, and at what conditions the gas mixture can be delivered from the source bottle while maintaining its original composition. Many pressurized gas mixtures are guaranteed to be deliverable at room T with known composition, but will partially condense, unmix, and pool in the reservoir at freezer T. In such cases it is necessary to cool and deliver the source gas to the samples by a different means than our standard technique.

One method is to move the gas reservoir down-line from the samples, and to place in its original position a tightly-wound cooling coil that is suspended in the alcohol bath and that connects the source bottle to the samples and reservoir. This configuration allows the source gas to be pre-cooled as it is delivered to the samples, but without letting any liquid settle up-stream of the samples. The samples and reservoir should then all be pressurized in one step.

Alternatively, for single-component systems, a reservoir of pressurized liquid hydrocarbon can be positioned directly above a sample chamber, and the pore space of the ice can then be saturated by using the pressure differential behind the liquid as well as gravitational feed.

Another option for synthesizing mixed-phase gas hydrates that include methane, is to first fully saturate the pore space of the seed ice with the liquid hydrocarbon phase by one of the methods discussed above, and then further pressurizing with methane gas to high pressure (>25 MPa). This method permits accessing higher P-T conditions for synthesis, due to the shift in dissociation conditions to higher P-T conditions (methane-propane hydrate *vs.* pure propane hydrate, for example), and thus enabling more efficient reaction.

Regardless of which method is used, multiple cyclings through the melting point of ice are invariably required to ensure full reaction when using liquid hydrate-forming phases as reactants. Multiple P-T cyclings are particularly necessary for those systems in which the hydrate is stable to only several degrees above the melting point of ice. Our experience has shown that the length of time that such samples spend at warm temperature is less critical than for pure methane hydrate samples, and, instead, maximizing the number of cycles through the ice melting point drives the reaction more efficiently. Internal thermocouples are integral to such tests to establish the point at which melting and/or re-freezing anomalies cease to occur in the synthesis P-T record. Synthesis of methane-propane hydrate, for instance, requires approximately 5-7 cyclings over the duration of several days for complete reaction.

3. MEASUREMENT OF HYDRATE COMPOSITION AND STABILITY

3.1. Gas flow meter and collection apparatus

The volume, and hence mass of gas evolved from a dissociating sample, as well as the rate of gas evolution, can be measured with high precision in the gas flowmeter and collection apparatus shown in Figure 5 that is based on the principles of a Torricelli tube.

The apparatus consists of a closed-ended hollow cylinder made of stainless steel, inverted in a reservoir of distilled water, and suspended from a precision load cell. A column of water is initially drawn up by vacuum into the cylinder, and the weight of the cylinder plus water column is recorded as an initial baseline measurement. Maintenance of a constant drip rate into and out of the reservoir of water in which the cylinder is suspended ensures that the water level in the reservoir remains constant, and hence the buoyancy force on the cylinder remains constant. Methane is bubbled through the water to saturate it prior to dissociation, minimizing further methane solution during testing. Collection of a steady weight baseline is then made on the primed system, after which the hydrate sample can be opened to the flow meter.

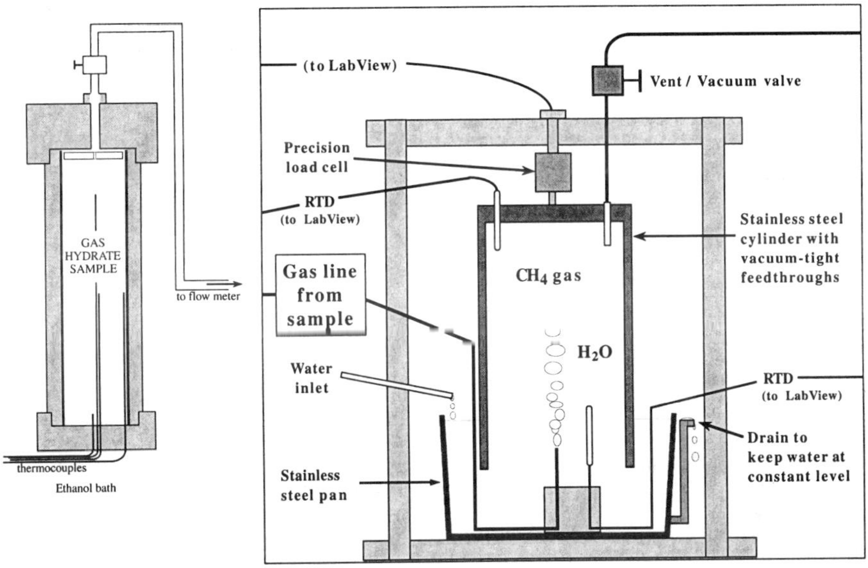

Figure 5. Gas flow meter and collection apparatus for measuring rates and amounts of gas released from dissociating samples. Flow rates may be measured over at least 5-orders of magnitude with excellent precision, while collecting the accumulating gas for mass measurement. Sample size relative to flow meter is not drawn to scale.

As methane gas evolves from the dissociating sample and displaces water in the cylinder, the cylinder weight decreases. The mass of methane collected in the cylinder is then calculated from that weight change by incorporating additional run parameters into the data analysis, including: temperature inside the cylinder, atmospheric pressure, partial pressure of water vapor in methane, and the equation of state of methane gas. Calculation of the number of moles of methane can be simplified by using the ideal gas law, a procedure that accurately predicts the small relative changes of the sample pressure over the range of operation. The capacity of the flow meter shown in Figure 5 is 8 liters, and a typical methane hydrate sample made from 26 g of seed ice will release about 6 liters of gas during dissociation. Samples of the gas can then be collected for subsequent compositional analysis by attaching an evacuated receiving vessel to the vent/vacuum line. The flow meter operates with high accuracy at both high and low rates of gas flow, a result that we have verified over the range 0.01 to 3000 ml/min. Further details are provided by Circone et al., (2000).

3.2. Dissociation procedures

The most accurate measurement of the gas content evolving from dissociating hydrate are those made *in situ*, in which no quenching, depressurization, room-pressure handling, or sample re-installment onto the apparatus is required following synthesis. Either of two general procedures can be followed to take the hydrate samples from post-synthesis conditions of elevated methane pressure down to 0.1 MPa methane pressure prior to dissociation.

The first method, referred to as "temperature-ramping", involves slow cooling of the external bath surrounding the samples while depressurizing the samples, remaining close to but just within the hydrate stability field (Fig. 6A, points 1-2-3). When the hydrate has cooled below 194 K, the remaining gas is vented to 0.1 MPa and the sample can be opened to the gas collection apparatus. Slowly warming the sample (by warming the external bath) above the methane hydrate dissociation temperature (194 K at 0.1 MPa) will then destabilize the hydrate, and the evolved gas is collected in the flow meter (Fig. 5). Warming the sample through the melting point of ice (273.15 K) ensures dissociation of the hydrate as well as melting of the ices. Previously-quenched samples can also be dissociated by this method, by first cooling the external bath (to T < 194 K), and loading the quenched sample into a cold (<194 K) vessel prior to reattachment to the apparatus.

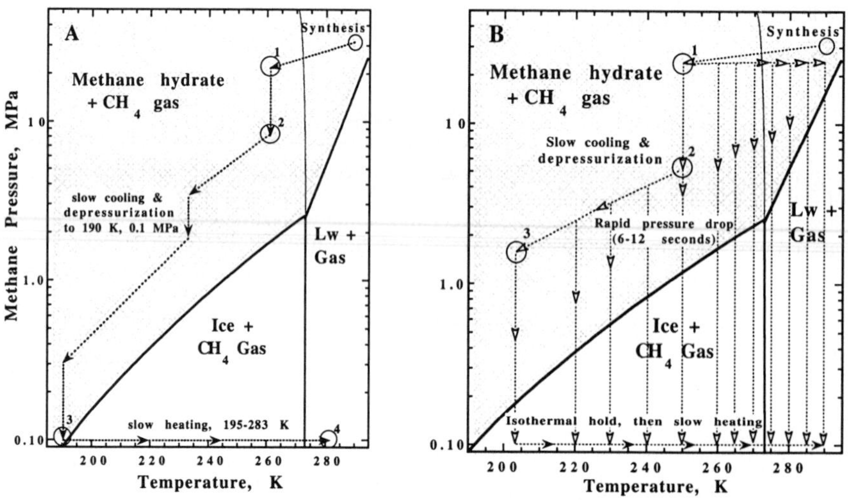

Figure 6. P-T paths for destabilizing methane hydrate for measurement of dissociation rates, stoichiometry, or stability behavior. Samples can be processed in two manners: (**A**) the "temperature ramping" method, or (**B**) the "rapid depressurization" method, each of which are effective for different types of dissociation tests. Procedures are described in the text.

The second dissociation method, termed "rapid depressurization", involves first depressurizing the sample to a smaller overstep of the dissociation curve (Fig. 6B, pts. 1-2 or 3), then rapidly venting the pressure from several MPa above the equilibrium curve down to 0.1 MPa. The vent is then quickly closed while simultaneously opening the valve to the gas collection apparatus. The rapid depressurization is performed over about a 6 to 15 second interval, depending on the magnitude of the initial pressure overstep of the equilibrium curve. This technique is used to measure dissociation rates at a constant external bath temperature, and is particularly effective for isothermal warm-temperature tests to explore the P-T region where hydrate is predicted to dissociate to liquid water + gas. For samples tested at temperatures above 195 K but below 273 K, it is necessary to warm the sample through 273 K after the isothermal portion of dissociation has finished, as in the temperature-ramping tests discussed above.

3.3. Stoichiometry measurements and annealing effects

Stoichiometry measurements are ideally made on samples that are dissociated *in situ* directly after synthesis by the temperature-ramping procedures described above. This method permits stabilization of samples for an extended time at 0.1 MPa and at T < 194 K prior to dissociation, allowing the release of any residual pore gas or adsorbed methane on grain surfaces. Such release is easily detectable by baseline shifts recorded by the flow meter.

Seven samples tested in this manner confirmed the high reproducibility of sample composition produced by the prescribed growth methods; stoichiometry number n of all the test specimens measured at 5.89 ± 0.01 (Fig. 7A). This measured stoichiometry is slightly closer to ideal than that which we reported previously (6.1 ± 0.1, on samples that contained 0 to 3 % unreacted ice, Stern et al., 1996) due to the greatly improved analytical and measurement capabilities provided by the internal thermocouples and the gas collection apparatus, and the current ability to detect very small amounts of unreacted ice. Samples are also now routinely held at the highest P-T conditions during synthesis for several hours longer than previously, to insure reaction of the last several percent of ice to hydrate. (The difference between n of 5.89 vs. 6.1 in methane hydrate samples corresponds to 3.2 vol. % unreacted ice.)

As our synthesis procedures form hydrate at conditions of P and T that are significantly overdriven with respect to the equilibrium curve, it may be desirable to anneal the as-grown hydrate at low P_{CH4} or at equilibrium conditions prior to measurement of certain physical properties. Annealing is easily accomplished through resetting the P-T conditions on the samples to those more relevant to geologic settings, or to conditions close to or directly on the methane hydrate equilibrium curve (Fig. 1). Annealing effectively removes some methane from the hydrate structure (when made by the methods given here), as lower P stabilizes a less ideal stoichiometry. Kinetic factors are also important in reaching an equilibrium composition. On the laboratory time scale, there is a definite trade-off in achieving equilibrium compositions by reducing P at cold temperatures to very low equilibrium pressures, compared to annealing at warmer temperatures to the somewhat higher equilibrium pressures. While this effect has not yet been quantified, preliminary tests show that, as expected, annealing at warmer temperatures promotes a more rapid re-equilibration of n. Allowing a fully-reacted sample (n = 5.89) to anneal at 274 K and 4.5 MPa for 6 days, for instance, results in an increase in the stoichiometry number n to 6.01. Annealing at 250 K for 12 days at 2 MPa, however, only produces an increase in n to 5.94 even though the annealing pressure is lower (see Fig. 1).

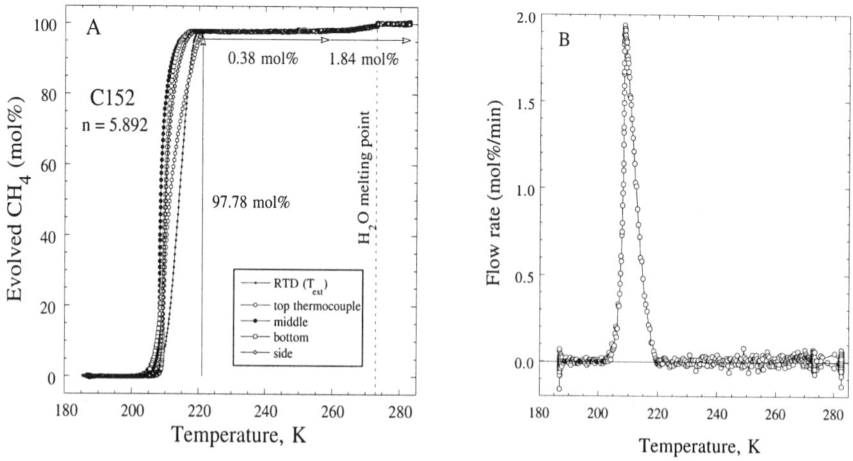

Figure 7. A typical dissociation profile (**A**) and flow-rate profile (**B**) for a sample of methane hydrate grown from 26 g of H_2O ice and dissociated by temperature-ramping procedures (Fig. 6A) by slowly warming the sample above dissociation conditions at a rate of approximately 11 K/hr. This sample yielded 0.244 moles of CH_4 gas, corresponding to a stoichiometry $CH_4 \bullet 5.89H_2O$. Most of the gas evolves over the temperature range 200 to 215 K, and the final 2-5% hydrate is not released until the sample is warmed through the H_2O melting point.

3.4. Dissociation kinetics and phase stability

The methods described above for *in situ* synthesis and stoichiometry measurement are particularly well suited for determination of effects of various pressure-temperature-time (P-T-t) paths on hydrate dissociation kinetics and phase stability, due to the ease of controlling such parameters as pressure, peak temperature, temperature-ramping rate, sample volume, sample composition, and stoichiometry. Below we present some representative results from nearly 100 experiments that demonstrate several applications of the flow meter to the investigation of hydrate dissociation kinetics and phase stability behavior over the temperature range 190 to 290 K at 0.1 MPa, as methane hydrate dissociates to either ice + CH_4 gas (T< 273 K) or liquid water + gas (T > 273 K).

3.4.1. Methane hydrate → ice + gas A suite of methane hydrate samples were removed from initially stable conditions of low T or elevated P by either the temperature-ramping method or the rapid-depressurization method discussed above. Samples decomposed by the temperature-ramping method exhibited highly reproducible decomposition behavior in which approximately 95% of the expected amount of gas evolved over the temperature range 195 K < T < 220 K, as shown in Figure 7. The remaining gas was then only released from the samples (on the laboratory time scale) by warming through 273 K. This resistance to decomposition displayed by the residual 3-5% is likely due to its entrapment along ice grain boundaries or within grain interiors, as it is subsequently released by the melting of the encapsulating ice.

When rapidly depressurized at isothermal conditions ranging from 204 K to 240 K, methane hydrate samples exhibit systematically increasing dissociation rates with increasing temperature, as expected with the increasing thermal overstep of the stability field (Fig. 8A). Samples tested at 204 K dissociated over ~ 3 hours, while those tested at 239 K dissociated within 7 minutes. These times correspond to the duration of the main dissociation event, or about 88-90 % reaction, and the remaining hydrate did not dissociate until the samples were warmed through 273 K and all the ice product had melted.

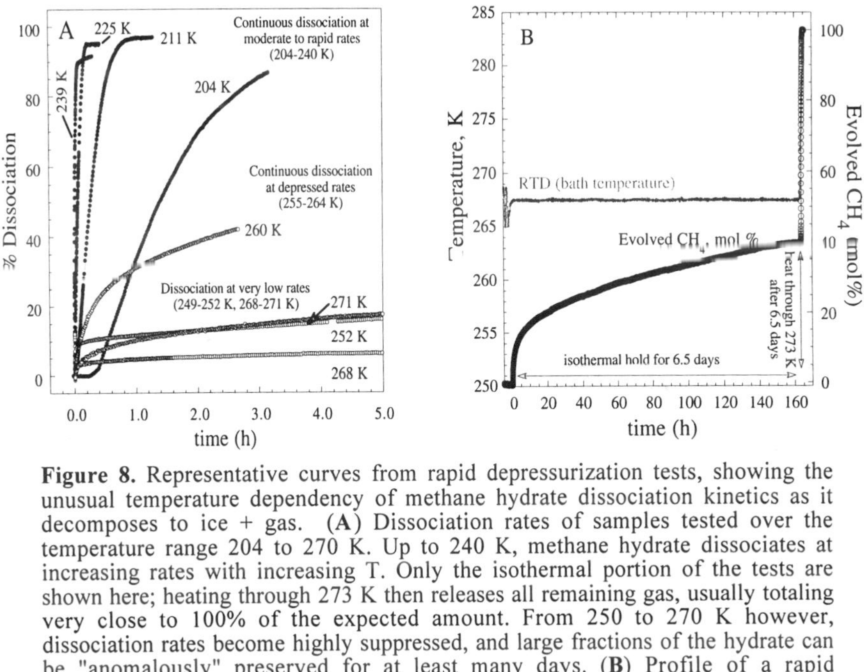

Figure 8. Representative curves from rapid depressurization tests, showing the unusual temperature dependency of methane hydrate dissociation kinetics as it decomposes to ice + gas. (**A**) Dissociation rates of samples tested over the temperature range 204 to 270 K. Up to 240 K, methane hydrate dissociates at increasing rates with increasing T. Only the isothermal portion of the tests are shown here; heating through 273 K then releases all remaining gas, usually totaling very close to 100% of the expected amount. From 250 to 270 K however, dissociation rates become highly suppressed, and large fractions of the hydrate can be "anomalously" preserved for at least many days. (**B**) Profile of a rapid depressurization test at 268 K, showing 60% of the hydrate preserving at 1 atm after 6 days, at temperatures 73 K above the equilibrium dissociation temperature. The expected remainder of gas was then recovered by heating the sample through 273 K.

"Anomalous" preservation" behavior When methane hydrate samples are rapidly depressurized at isothermal conditions between 250 and 270 K, an anomalous preservation effect can be invoked (Fig. 8; also Stern et al., 1998c). This effect is most prominent at temperatures approaching 270 K, where significant amounts of hydrate can remain metastable due to greatly suppressed dissociation rates, even in tests lasting over 6 days (Fig. 8B). Warming of the metastable hydrate through 273 K then promotes full dissociation and rapid release of all remaining gas. Tests in which "preserved" samples were cooled to 190 K and then slowly rewarmed, however, showed that this preservation effect is thermally irreversible, as the samples then dissociated over the interval 198-218 K as observed in the temperature-ramping tests discussed above.

Based on textural observations of quenched samples and the lack of appreciable ice in them, it appears that warm-temperature preservation may be at least partially due to grain surface effects, grain boundary mobility, or structural changes within the hydrate grains. This is in contrast to the residual hydrate in temperature-ramping tests that is released at 273 K, that is expected to be "preserved" by the encapsulating ice. Other tests conducted on unconsolidated methane hydrate and on hydrate + sediment ± seawater mixtures, all samples that were rapidly depressurized at 268 K, demonstrated that while dissociation rates were less suppressed than those from more compacted or pure samples, the rates were still orders of magnitude slower than those predicted by extrapolation of dissociation rates from the lower temperature (204 to 240 K) regime. These results indicate that while the warm-temperature preservation effect is enhanced by grain boundary contacts, it is largely a structural or intrinsic property of methane hydrate.

We still do not fully understand the physical chemistry involved in this preservation effect, but have found it to be highly reproducible. Descriptions of other gas hydrate "self" preservation effects have also been documented by Davidson et al. (1986), Yakushev & Istomin, (1992), Gudmundsson et al. (1994), and Dallimore & Collett (1995).

3.4.2. Methane hydrate → water + gas. Rapid-depressurization tests conducted at temperatures above 273 K show that dissociation proceeds in a systematic manner in which rates increase with increasing external temperatures (Fig. 9). In all tests conducted at T > 273 K, sample interior temperatures plummet and buffer at 272.5 K as dissociation proceeds (Fig. 9; also see Fig. 1 in Circone et al., 2000), an effect similar to that observed on recovered natural hydrate from oceanic drill core material (Kastner et al., 1995). Tests on methane hydrate + sediment ± seawater samples also show that the addition of particulates and seawater both measurably increase the dissociation rate compared to pure hydrate samples. Additional details are provided by Circone et al., (2000). increase systematically (shown by arrow). The two runs conducted at 273.6 K and 273.8 K demonstrate the excellent reproducibility of the test

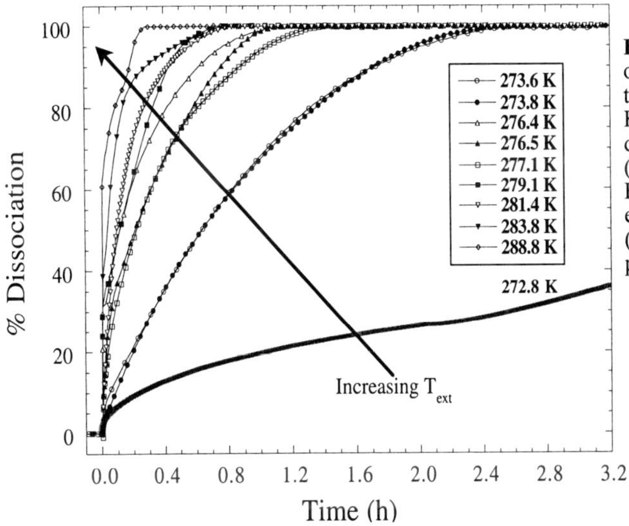

Figure 9. Dissociation rates of rapid depressurization tests conducted at $T > 273$ K, as pure methane hydrate decomposes to water + gas (from Circone et al., 2000, Fig. 2). With increasing external bath temperature (T_{ext}), dissociation rates procedures and results.

4. APPLICATIONS FOR PHYSICAL PROPERTIES MEASUREMENT

The methods described in this chapter for controlled growth of pure methane hydrate are particularly advantageous for further experiments that require either: (1) knowledge of sample purity, good control over P-T conditions and ease of modification of reaction vessel and apparatus specifications (for instance, annealing tests or measurement of gas-solid exchange, thermal conductivity, diffusivity, and acoustic property measurement), or (2) precise knowledge and control of grain characteristics (measurement of fracture strength and ductile flow behavior, for example). We review several modifications to our basic synthesis apparatus (Fig. 2) that we have used to acquire several of these types of measurements. Compaction procedures are also reviewed here, as many property measurements require precise knowledge of the sample porosity and the ability to remove all pore space from the test material.

4.1. Thermal conductivity

The synthesis methods discussed in this chapter are well suited for experiments measuring thermal properties such as conductivity or diffusivity. Techniques developed by von Herzen and Maxwell (1959), for instance, can be adapted to measure thermal conductivity by use of a needle probe design that approximates an infinitely long, continuous line source of heat in an infinite medium. Cylindrical sample geometry can be modeled approximately as an infinite medium, and a needle probe placed along the axis of the sample can be approximated as a continuous line source of heat. Thermal conductivity of a sample can then be calculated directly by measuring the rise of temperature measured by the probe for a given heat input per unit length of wire per unit time, and for a given probe radius.

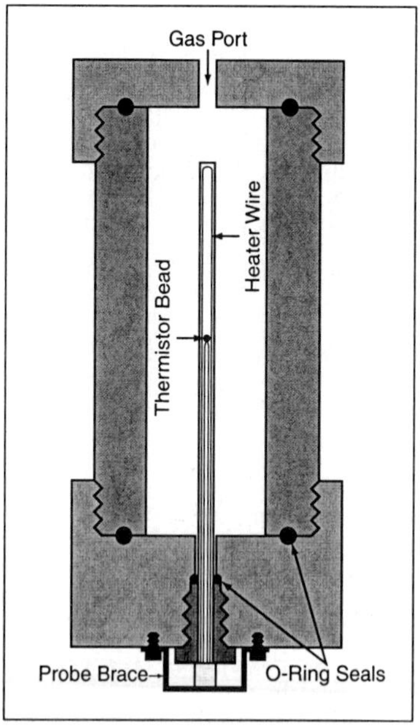

Figure 10. Schematic of sample chamber outfitted for hydrate growth followed by *in situ* thermal conductivity measurement using the needle probe technique. The probe is inserted through the base of the sample chamber prior to packing it with seed ice, and is supported from below by a metal brace to prevent probe expulsion from the chamber at elevated pressure. An O-ring seal around the probe prevents gas leakage through the insertion hole. The needle probe consists of an epoxy-filled hypodermic tube containing a 4 kΩ thermistor and a heater wire running the length of the probe. The probe has essentially the same length as the sample vessel, so departures from ideal cylindrical geometry during the heating phase of the experiment are negligible. Needle probe diameter is not drawn to scale.

The primary adjustment to the standard synthesis apparatus for such measurement involves replacing the standard sample vessel base with an endcap modified to accommodate the thermal probe, as shown in Figure 10. Preliminary success with this design has been reported by deMartin et al., (1999), in measuring thermal conductivity of pure methane hydrate samples and hydrate + quartz sand aggregates. Further discussion of the context of these measurements is provided by Ruppel et al., (this volume). Not only does *in situ* gas hydrate growth with the probe already in place ensure excellent contact between the hydrate and the probe, which is crucial to the success of this technique, but this method also avoids structural and stability problems inherent to drilling holes in quenched material for probe insertion. As this method measures thermal parameters of a porous aggregate, however, it is necessary to either establish the influence of the pore pressure medium on the measurements, or design a hydrostatic compaction capability around the sample to carefully compact and fully densify the material without damaging the probe. Such apparatus and procedural developments are currently in progress.

4.2. Compaction procedures

Compaction of as-molded, porous material can be achieved by either uniaxial or hydrostatic compaction procedures. Our method for uniaxial compaction utilizes an apparatus that permits *in situ* synthesis, compaction, and elastic wave speed measurement (Fig. 11; see also Waite et al., 2000). This apparatus incorporates pistons at each end of the sample chamber, both of which are outfitted with a transducer assembly for wave speed measurement. The moving piston is advanced hydraulically to axially shorten the sample after synthesis, and the length change of the sample is monitored by the linear conductive plastic (LCP) transducer as shown in Figure 11.

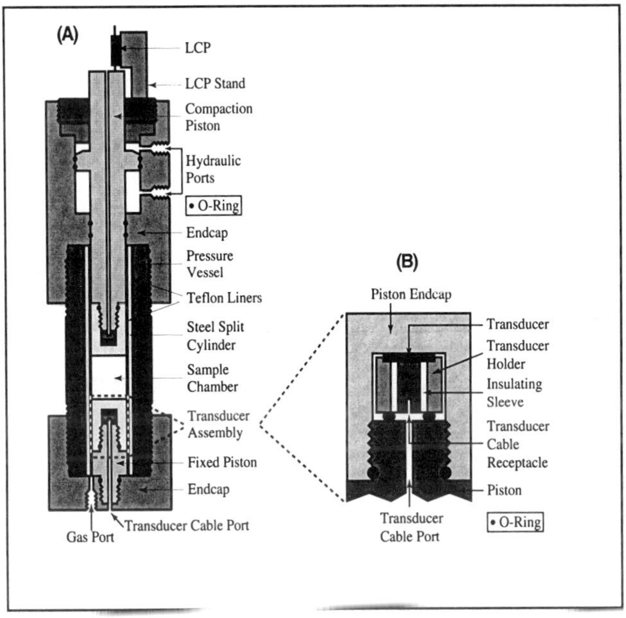

Figure 11. Gas hydrate synthesis apparatus equipped for subsequent *in situ* uniaxial compaction and measurement of acoustic wave speeds (from Waite et al., 2000).
(**A**) Pressure vessel schematic. Polycrystalline gas hydrate is synthesized directly in the sample chamber by the methods described in this chapter, then uniaxially compacted by hydraulically advancing the moving piston.
(**B**) Transducer assembly. Both pressure vessel pistons house a 1 MHz center-frequency piezo-electric transducer (either p- or s- wave), allowing pulse-transmission wave speed measurements to be made throughout the compaction process.

Prior to uniaxial compaction, the sample is vented to a pressure just sufficient to maintain it in its stability field; it is therefore beneficial to begin the compaction sequence at cold temperatures where hydrate is stable at lower pressures. Following this cold compaction, samples are partially repressurized and further compacted while raising the sample's temperature through the ice point. Compaction at high temperature facilitates elimination of the last percent of porosity, although the high strength of hydrate makes it increasingly difficult to achieve full compaction. Monitoring the wave speed profile throughout the compaction process provides independent determination of the presence or lack of any unreacted ice or water in the sample, as a measurable discontinuity in the wave speed can be measured if even trace amounts of H_2O freeze or melt.

For hydrostatic compaction (Fig. 12), a sample of gas hydrate is initially sealed in a soft jacket (such as indium metal) between two end caps or between an end cap and a force gauge, and pressurized from the outside with a confining medium gas. The sealing procedure must be performed at temperatures sufficiently cold to ensure stability of the hydrate. Hydrostatic compaction is best performed in an apparatus in which a piston can be advanced to touch and square the base of the sample, such as with the triaxial deformation apparatus shown in Figure 12. The confining pressure (P_c) is then slowly "stepped" up to 50 or 100 MPa in increments of roughly 20 MPa, taking care not to induce a high-pressure phase transformation in the samples at pressures just above 100

MPa (Chou et al., 2000). Following each compaction step, the piston is advanced to touch and square the bottom of the sample, then advanced just sufficiently to lightly compress the sample in order to compact it with minimal deformation. Optimally, the top end of the sample is attached to a gas line that can be either fully vented or maintained with a low CH_4 back pressure on the sample. Compaction can then be performed at either very cold conditions in the vented configuration, or at warmer temperatures with a regulated pore pressure.

4.3 Acoustic wave speed measurement

Acoustic wave speed, the distance traveled by a wave divided by the wave's travel time, can be measured on hydrate grown directly in the uniaxial compaction apparatus shown in Fig. 11. Ultrasonic waves generated by the transducer in the compaction piston travel through the sample and are received by the transducer in the fixed piston. The travel time through the sample alone is given by the difference in the travel time between a signal passing through the complete system, and a "head-to-head" test signal sent from one transducer to the other when the pistons are in direct contact. The sample length is monitored using the LCP, and checked periodically by directly measuring the piston position relative to the pressure vessel.

In ice or gas hydrate samples, the signal to noise ratio is high enough to observe a distinct and measurable precursor compressional (p) wave generated by the shear (s) transducers. Using a pair of shear wave transducers, the compaction apparatus was successfully used to recover published wave speed values (Vp and Vs) for pure ice Ih, and also provided Vp and the first Vs results for pure methane hydrate (Vp for methane hydrate was measured at 3650 ± 50 m/s, and Vs at 1890 ± 30 m/s; Waite et al., 2000). The simultaneous measurement of Vp and Vs also allows other additional elastic properties to be derived more reliably (see Helgerud et al., this volume).

4.4 Strength measurements

Gas hydrate samples fabricated by the methods given in this chapter are also well suited for deformation testing in the type of apparatus shown in Figure 12. Because the composition and grain characteristics are well known and reproducible from sample to sample, such material is appropriate for rheological measurement and flow law characterization.

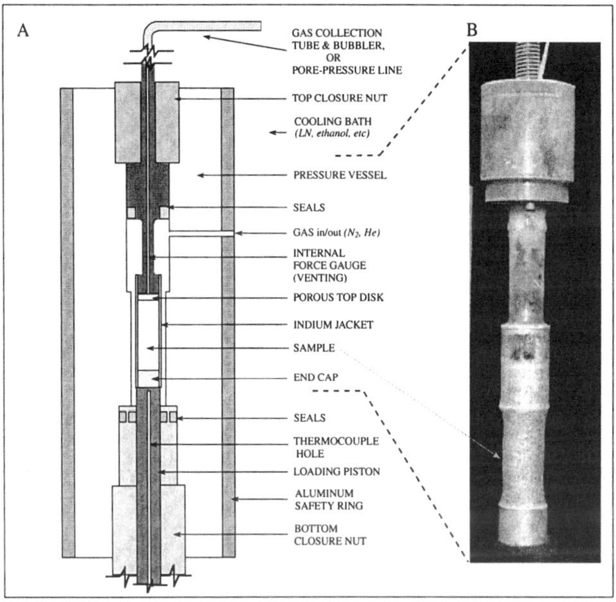

Figure 12. (**A**) Schematic of triaxial gas deformation apparatus used for hydrostatic compaction and subsequent deformation testing of gas hydrate test specimens. A jacketed sample (**B**) can be compacted and then measured for strength in compression tests, over the T range 77 to 300 K, at confining pressures up to 0.6 GPa, and strain rates 10^{-4} to 10^{-8} s^{-1}. The sample is attached to the internal force gauge, and a sliding piston moves through dynamic seals from below to impose axial shortening. A gas collection system or pore pressure line can be attached at the top of the apparatus to monitor possible methane loss during compaction and deformation, or to prevent gas hydrate decomposition.

The strength of methane hydrate samples has been measured in this laboratory in a suite of constant-strain-rates tests in compression, at conditions ranging from T = 140 to 287 K, confining pressure (P_c) of 50 to 100 MPa, and strain rates from 3.5 x 10^{-4} to 10^{-8} s^{-1}. The apparatus shown in Figure 12 is a 0.6-GPa gas deformation apparatus outfitted for cryogenic use (Heard et al., 1990) in which N_2 or He gas provides the P_c medium. The pressurized column within the apparatus consists of an internal force gauge, the jacketed sample, and a moving piston that compresses the sample axially against the internal force gauge at a fixed selected displacement rate. The soft indium jackets in which the samples were grown serve to encapsulate them during compaction and testing to exclude the P_c medium, and provide the additional benefit of superbly replicating the outer surface of the deformed sample, thus enabling subsequent microstructural study at room conditions (Fig. 10 in Stern et al., 1998a, for example). All gas hydrate samples tested in this apparatus were initially subjected to the hydrostatic pressurization and compaction sequence described above.

In tests conducted at T < 200 K, methane hydrate samples displayed strengths that were measurably but not substantially different from steady-state strengths of H_2O ice (Stern et al., 1996, 1998a). Recent tests conducted at warmer test conditions (>250 K), however, show that methane hydrate is enormously strong relative to water ice (Zhang et al., 1999, Durham et al., work in progress), a result which was not previously expected as water ice has commonly been assumed as a general proxy for the mechanical properties of gas hydrates. Figure 13 shows the relative strength differences between methane hydrate and pure water ice deformed in one of our first tests at 260 K, illustrating the dramatic strength differences. Further testing is currently underway to resolve better the strength and flow behavior of methane hydrate, and particularly to determine if possible shear instabilities or other solid-state processes are inherent to the deformation behavior (Stern et al., 1996, Durham et al., work in progress). Such instabilities not only increase the ice content of the sample with increasing strain, but mask the true strength of the pure hydrate due to the increasing ice contamination in the samples.

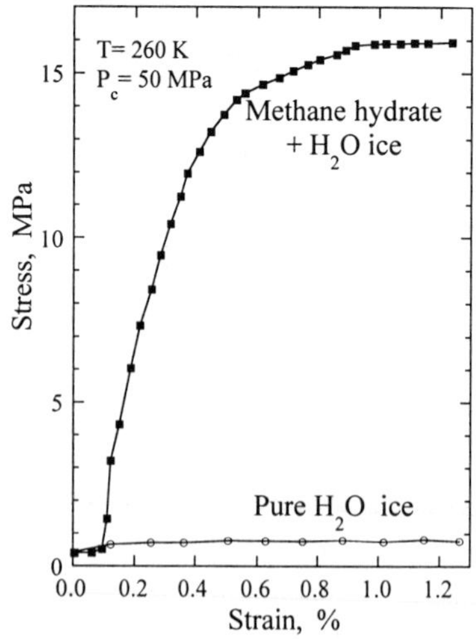

Figure 13. Comparison of the stress-strain histories of a mixed-phase sample of polycrystalline methane hydrate plus H_2O ice (approximately 3:1 volume ratio hydrate to ice), *vs* pure polycrystalline H_2O ice, at comparable test conditions. At lower temperature test conditions (< 200 K), the strength differences between methane hydrate and water ice are not significantly different, due to the high strength of both hydrate and ice at those conditions. At elevated temperatures as shown here, however, the strength contrasts are dramatic. Recent tests show that the strength of pure methane hydrate, even at T > 273, is at least several times stronger than that of the mixed-phase sample shown here (Durham et al., work in progress).

5. Summary and Conclusion

Gas hydrates are challenging materials to investigate for physical properties. Water ice has proven to be a poor analogue material for methane hydrate, indicating that only measurements made directly on end-member gas hydrates will provide reliable results for use in quantitative models that evaluate the amounts of hydrate in sediments, or how gas hydrates may respond to gravitational loading, exchanges of heat, or changes in sea level. Synthesis or procurement of well-characterized and reproducible sample material is therefore crucial to the reliability and accuracy of such measurements.

We describe a simple method for making pure, cohesive, aggregates of methane hydrate that may be easily reproduced in any laboratory. Compositional measurements using a unique precision gas flow meter indicate that the stoichiometry of the resulting material is closely reproducible at $CH_4 \bullet nH_2O$ where $n = 5.89 + 0.01$. Grain size, porosity and aggregate cohesion are also closely reproducible, and the specimen shape is primarily limited by the shape into which granular ice may be packed. Customized aggregates of hydrate-sediment mixtures, or hydrates with mixed hydrocarbon phases, can also be easily manufactured by our *in situ* "seeding" method. Our experience has also shown that by combining methods of synthesis, compaction, and property measurement in a single reaction vessel, it is possible to avoid the structural, compositional, and reproducibility problems that plague experiments that rely on 1-atm sample handing and cold transfer procedures.

Several key findings and results have emerged from this line of research. Direct and precise measurement of the shear and compressional wave speeds have now been made *in situ* on compacted samples on pure, polycrystalline methane hydrate, for example. These end-member values should prove useful in making more quantitative inferences of gas hydrate content in sediment columns as determined from conventional sonic well logs, vertical seismic profiling, or those inferred from arrivals from explosion or other suitably located acoustic sources.

Decomposition rates of pure, porous methane hydrate have also been systematically measured at 1 atm over the temperature range 195 K to 290 K, resulting in the unexpected discovery of a warm-temperature thermal regime (250-270 K) where methane hydrate can remain metastably preserved for extended periods of time. This preservation effect may have important applications to strategies used for retrieval of natural hydrates from drill core, as well as for gas yield rates in hydrate exploitation schemes using pressure release methods to decompose the hydrate.

Another unexpected finding is that pure methane hydrate has a markedly higher plastic flow strength than ice at terrestrially-relevant conditions. The high strength of methane hydrate extends to well above 273 K where water is in liquid form. If gas hydrate serves as a cementing agent in hydrate-bearing sediments, such aggregates should be stronger than permafrost. Hydrate decomposition due to depressurization or heating, as caused by a decrease in sea level or by hot oil extraction from below the gas hydrate interval, could therefore cause large changes in the slope stability of sediments on continental margins, or in the near-surface strength of sediments under drilling platforms.

Lastly, the successful adaptation of the needle-probe method for direct *in situ* measurement of the thermal conductivity of pure, porous methane hydrate confirms that conductivity is extremely low relative to ice and to granular (detrital) sediments. Further measurement on fully dense methane hydrate will provide the additional information on heat transfer properties that is essential for quantitative modeling of the effects of gas hydrate on climate change, and for evaluating the effects of coring and deep oil extraction through gas-hydrate-bearing sediments.

Editor's note:

This Chapter should be read along with Chapter 20, which considers the application of hydrate diagenesis in sediments and the way in which the formation of hydrate alters physical properties of the sediment mass. It is clear from the results of the experiments described here that many of the general assumptions we have been using for hydrate (e.g., water ice is a close analog for gas hydrate) may not be correct in detail. It follows that a number of numerical approaches to quantifying particular effects of hydrate formation and dissociation may also not be yielding correct solutions because the input data is faulty or the assumptions are incorrect.

The Methane Hydrate Research and Development Act of 2000, signed into law by President Clinton on 2 May, 2000, is the spearhead of a broad effort to understand more about gas hydrate and its effect on energy, climate, and other issues affecting the Nation. The preliminary results described in this chapter argue strongly that considerable basic laboratory research remains to be carried out before development of hyrate resources should take place. There is much we need to know.

Chapter 26

Economic Perspective of Methane from Hydrate

Klaas J. Bil
Shell International Exploration and Production B.V.
Rijswijk, The Netherlands

1. INTRODUCTION

Ever since the discovery of natural methane hydrate in the late 1960's (Makogon *et al*, 1972), and the subsequent growing awareness of the enormous quantities of methane locked in them (e.g., Kvenvolden, 1993), research scientists have dreamt about recovering this gas to help satisfy the world's demand for energy. The transition from a petroleum to a methane energy economy would also be environmentally positive because methane burns very cleanly, and produces less CO_2 per unit of energy than any other fossil fuel. In view of the ever increasing concern about the environment, methane will likely become the single most important fuel for many decades to come. Natural methane hydrate potentially holds the promise of (i) energy independence to various countries including the USA, India and Japan, and (ii) enabling extending the use of methane and of the existing gas infrastructure.

Yet methane has never been produced from natural methane hydrate on any significant scale. It has been claimed that part of the gas from the West-Siberian Messoyakha field has been produced from dissociating natural hydrate, but this claim is not regarded by some experts (e.g. Collett and Ginsburg, 1997) as being fully substantiated. Other cases of gas production from natural hydrate have been on a test scale only. Broadly speaking, the problems are in the new technical complexities of developing natural methane hydrate, as compared to conventional gas fields. Such technical complexities include finding and characterizing methane hydrate reservoirs, heat and mass flow in the reservoir, and geomechanical stability issues.

These problem areas are being addressed in a number of comprehensive research programmes around the globe. Large sums of money, in excess of US$ 100 million, have been spent over the last five years and continue to be spent. The lion's share of this work is paid by national governments (Japan, USA, India, Canada and others), as opposed to being commercially driven. Said countries have insufficient indigenous energy production causing concerns about energy independence and stability of supply, and/or their governments have a

M.D. Max (ed.), Natural Gas Hydrate in Oceanic and Permafrost Environments, 349–360.

strong political commitment to environmentally friendly energy. Commercial oil and gas companies, on the other hand, are operating on a worldwide energy market that is mainly driven by price. Indeed, some large commercial energy companies do maintain a research effort in methane hydrate production, but this is on a high risk - high reward basis and the amounts of money spent are nowhere near the US$ 100 million mentioned earlier.

While it would now be technically feasible to produce gas from natural methane hydrate, such an undertaking is indeed not yet economically viable (e.g. BeMent *et al*, 1998). If we want to discuss the future economic perspective of methane from hydrate, we will first have to judge the prospects of its prime competitor: conventional gas, and indeed of the world gas market as a whole.

2. LONG TERM GAS MARKET PROSPECTS

Natural gas has been produced almost since the beginning of commercial oil production, in the early days mostly in the form of associated gas (a by-product). However, for a considerable period gas was seen as a nuisance, having little or no commercial value. Often, it was just flared off. In the 1950's and 1960's, both the volume consumed and the price of gas increased, though the price was still substantially below the average oil price (per unit of energy). The 1973 Arab Oil Embargo changed the picture in a fundamental manner. The first World Energy Crisis resulted in a dramatically increased oil price, and provided the incentive for consumers to look for alternative fuels. As a consequence, gas immediately became a commercially marketable commodity. This resulted in a more than 10-fold increase in gas price within a few years. Since the early 1980's, the average wellhead gas price has hovered around US$2/1000 cu ft (in the US).

There is consensus that the world at large will continue to see a plentiful supply of inexpensive conventional gas until at least 2020-2030. The majority of gas fields could produce at higher rates than they do now with only slightly higher marginal costs. In addition, a significant number of gas fields that have been discovered await development until the current surplus production capacity has been reduced. This results in the anticipation that the price of conventional gas will remain roughly constant (if corrected for monetary inflation) until at least 2020-2030 (Committee J of International Gas Union, 1997; Beck, 1999; Energy Information Administration, 1998; Cochener, 1999; National Petroleum Council, 1999).

The volumes of gas produced (and consumed) will significantly increase over the same period. Most projections agree that by 2030, the world gas consumption will have increased close to three-fold as compared to the volume in 1999. While much of this growth will derive from fuel substitution in the developed world, the main driver will be increased power demand, particularly in developing countries.

Gas supply and demand are to an increasing degree not well-balanced over the globe, and this trend is projected to continue. International gas trade will probably at least quadruple over the next three decades. This expansion in capability will obviously require huge financial investments. It is likely that the most significant exporter will be Eastern Europe/Northern Asia, and the most significant importers will be Western/Central Europe, and Central/Eastern Asia.

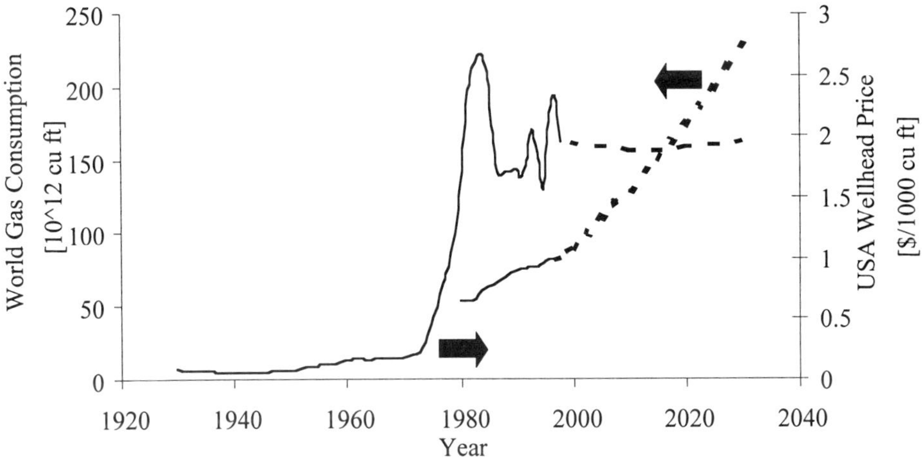

Figure 1. History and average forecast of worldwide annual gas consumption and gas price (USA, at the wellhead).

This assessment of the world gas market in the period 2000-2030 suggests that gas would stay "indefinitely" in demand as long as it is competitively priced. However, it might well take much more than 30 years to develop a mature methane hydrate extraction technology to the degree that hydrate gas production becomes competitive with conventional gas production "at large".

Would the world still need additional unconventional gas resources in, say, 40 years time? How much conventional hydrocarbons would remain and what will be happening to energy demand, alternative fuels / energy sources, fuel cells etc?

There will be special cases and niche markets like Japan or India. These countries deviate from the present global trends of industrial countries in that they have hardly any indigenous sources of hydrocarbons, and pay a premium price for imported gas transported as Liquefied Natural Gas (LNG) or in very long pipelines. Their economic evaluations are obviously entirely different, while also national security related to energy supply becomes an important part of the decision-making process. Hence, these countries have a strong additional incentive to develop methane hydrate off their coasts.

Historically, the trend with most commodities is that continuing technology evolution reduces costs and then makes the resource redundant before it starts to run out. For oil and gas, almost all forecasters have suggested that prices (even in real terms, i.e. corrected for inflation) will rise from the date of the economic forecast. However, when these predictions are considered on a sufficiently long timescale, they have all been incorrect.

Naturally, the further we try and look into the future, the more dominantly are the uncertainties. It is appropriate, therefore, to shift our focus away from the general gas market to the expected timing of methane hydrate production, and issues affecting that timing.

3. THE TIMING OF GAS PRODUCTION FROM HYDRATE

Assuming that gas will continue to be in demand "indefinitely" and to be supplied from conventional sources, at some point in time conventional gas inevitably will begin to become scarce, which will result in increasing prices. At the same time, it can be expected that the production cost of gas from hydrate, if anything, will gradually decrease due to technological development. When the price of the cheapest methane from hydrate matches that of conventional gas, it will then become commercially worthwhile to begin producing methane from methane hydrate (Fig. 2).

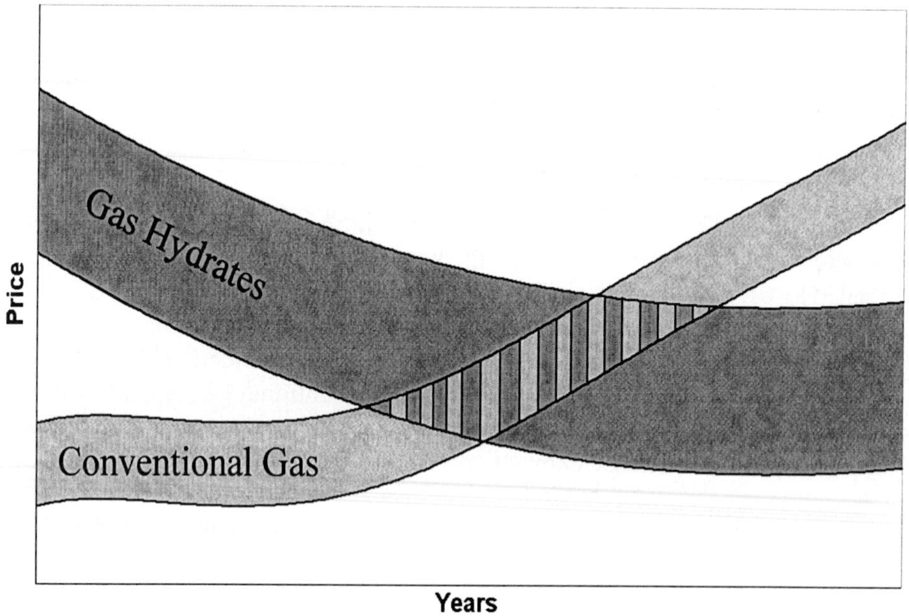

Figure 2. Conceptual cost development of conventional gas, and gas from hydrate. The axes have deliberately been left uncalibrated.

Offshore drilling and production capabilities have been enhanced dramatically over the last decades. Record water depths in offshore drilling have increased from less than 200 m in 1965 to more than 2000 m in the 1990's. Connected to the increasing routine character of offshore operations, drilling and production cost continue to come down significantly. This trend will eventually contribute to offshore hydrate becoming economically viable because much of the technology required can be provided from that developed for conventional deep-water hydrocarbon exploration and production technology.

In addition, horizontal drilling has become a viable technique, and the "reach" of horizontal wells will continue to be extended. In many proposed production schemes for methane hydrate, horizontal wells (e.g. for heat injection) are a key element (e.g., Max and Lowrie, 1997). Without a doubt, the continuing advancements in both offshore operations and horizontal drilling will bring forward the commercial development of methane hydrate.

Recently, I have conducted a simplified "Delphi process" (Turoff and Hiltz, in press) to obtain a judgement from some twelve leading experts working on methane hydrate production research across the globe, both in non-commercial institutes and in commercial enterprises. The question was as to when (if at all) the onset of commercial methane production from natural hydrate will most likely occur. Most experts distinguished between on-shore and offshore hydrate reservoirs, the former being considered as relatively easier targets, because of their generally higher formation permeabilities and hydrate saturations. In addition, the environment on land is less difficult operationally. On the other hand, markets are important and offshore occurrences are likely to be near markets (Japan, India, Canada). All in all, methane production from on-shore hydrate is likely to occur before offshore production.

The experts opinions spanned quite wide ranges, in part because judgements naturally differ, and possibly also because of different views on the implications of the terms "onset" and "commercial". From Fig. 3 it can be seen that the hydrate community on average expects commercial methane production from *on-shore* hydrate occurrences to begin before 2015. (This is consistent with the National Methane Hydrate Multi-Year R&D Program of the US Department of Energy, of which the ultimate goal is commercial methane production from hydrate by 2015.) However, commercial exploitation of *offshore* methane hydrate lies further into the future: the cumulative 50% probability point lies beyond 2060. It must be noted that at least 95% of the methane in hydrate deposits is to be found in continental slope sediments; the offshore represents the fundamental challenge and potentially the greatest reward.

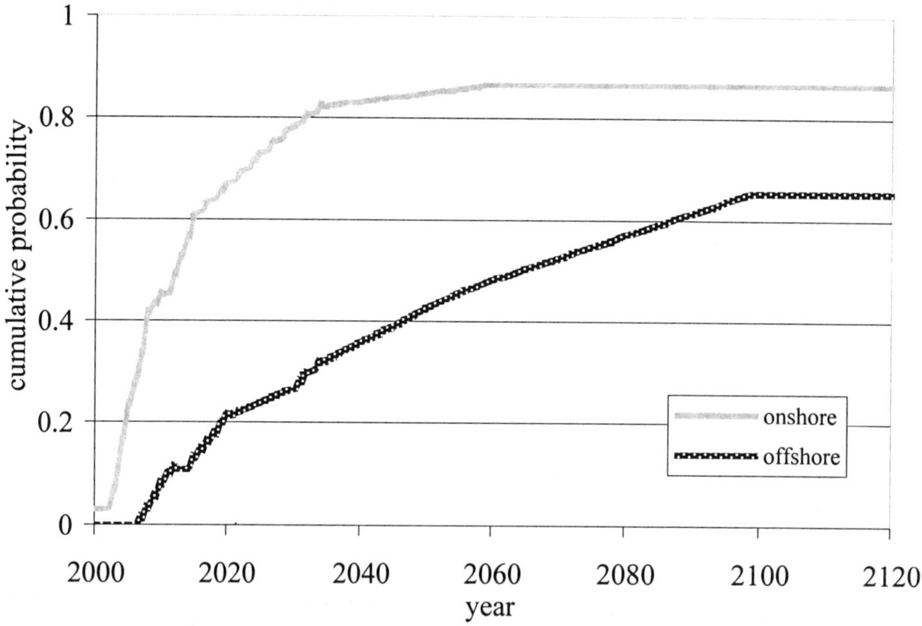

Figure 3. Cumulative compilation of experts' opinions on when the onset of commercial gas production from hydrate will occur.

4. ECONOMIC ANALYSIS

The issue on the economics of methane hydrate can be viewed from several perspectives. One approach pertains to economic calculations in terms of technical costs for exploration, wells, facilities, transport etc, and revenues for produced and marketed gas. This approach will only be dealt with briefly, because to date there are no real-world examples of marketed hydrate gas and hence all calculations must be based on speculative data. The emphasis here is rather on taking a broad-brush, long-term look at the economic justification of spending money on research now, in the hope of realizing future profits.

Several factors currently appear to preclude commercial exploitation of natural methane hydrate. In the first place, it is not known what the best locations are in technical terms like saturation and areal extent. This is partly due to the immaturity of surveying techniques for methane hydrate (including their lack of quantitativeness), and partly to the fact that few dedicated surveys using equipment for hydrate imaging (e.g. Max and Miles, 1999; Miles and Max, 1999) have been carried out to date.

Technically, gas could be produced from "solid" natural methane hydrate occurrences by thermal stimulation (possibly in combination with depressurization), either by using geothermal brine, or by injecting artificially heated hot water or steam. However, with reasonable assumptions on field size and quality, the total gas production costs would range between US$12 and US$30 per barrel of oil equivalent. Hence, this would essentially be a loss-making enterprise. On the other hand, with the current state of technology, the production of free methane from below a hydrate occurrence could probably be done profitably; however, it would not be as profitable as conventional gas production in otherwise similar conditions. This was concluded from a proprietary study by Shell, in which cost engineering and dynamic reservoir models were run on hypothetical, though not unrealistic, natural methane hydrate reservoirs, offshore a large country with little indigenous hydrocarbon resources. More elaborate examples of such calculations can be found in e.g. Kuuskraa (1985) and Godbole and Ehlig-Economides (1985).

In view of the fact that world-wide, large sums of governmental money are spent on research for methane hydrate production, oil and gas companies have addressed the question as to how much budget (if any) they should currently devote to this area. The answer would depend on the view of such a company on uncertainties such as the timing of methane hydrate becoming commercially viable, the size of the potential benefit, etc.

To quantify in economic terms the effects of these uncertainties on the appropriate spending level for the next few (10) years, an analysis such as the following could be useful. This analysis would be most appropriate for a major oil / gas company that normally holds the resources that it produces. It is based on a number of "base case" assumptions that have been listed in Table 1 below. Naturally, these assumptions are very much debatable. The cost and statistically most probable revenues have been calculated (using a discount rate of 10%/year) for three possible spending levels in the period prior to the commercial viability of methane hydrate production, i.e.:

- spending nothing at all (US$ 0/year);
- spending US$ 100,000/year;
- spending US$ 500,000/year.

Probably, by the time that methane hydrate becomes commercially viable, one would want to increase the research effort. Therefore, each of the options includes a "mandatory" step-up effort of US$ 1,000,000 and US$ 2,000,000 in the year before, and the year of commercial viability, respectively.

Number of years (from 2000) before methane hydrate will begin to be commercially viable (years)	10, 20 or 30
Chance of hydrate being commercially feasible in 2010	7.0%
Chance of hydrate being commercially feasible in 2020 (cumulative)	21.0%
Chance of hydrate being commercially feasible in 2030 (cumulative)	27.0%
Chance of the company to be the first if spending US$ 0 pa, including the probability of an adequate step-up timing	1.0%
Chance of the company to be the first if spending US$ 100,000 pa, including the probability of an adequate step-up timing	5.0%
Chance of the company to be the first if spending US$ 500,000 pa, including the probability of an adequate step-up timing	25.0%
Company share of hydrate gas in the first 10 years if they are NOT the first	3.0%
Company share of hydrate gas in the first 10 years if they ARE the first	6.0%
Period that competitive advantage of being the first to be ready fully lasts before falling back to zero (years)	10
Annual increase in market share of hydrate gas in the first 10 years of commercial viability (fraction of total gas production)	0.50%
%age growth of worldwide gas consumption (equivalent to trebling in 30 years)	3.73%
Net revenue for produced gas per barrel of oil equivalent (US$ MOTD)	1.00
Average annual discount rate	10%

Table 1. Base-case assumptions for the economic analysis

In addition to the assumptions in Table 1, for simplicity it has been assumed that:

- Decisions on spending levels pertain to blocks of 10 years.
- Hydrate become commercially viable in a year ending in 0, i.e. 2010, 2020 or 2030. (Obviously, it is possible that hydrate hasn't yet become commercial by 2030, but the analysis has shown that in that case, from a commercial point of view one should certainly not invest money now.) The assumed probabilities in the table have been derived from the offshore curve in Fig. 3.

- The chance of hydrate becoming commercial at any given moment does not depend on the activity level of the company concerned. It will mostly be determined by the market price of gas (driven by supply-demand balance), and by the influence on methane hydrate production cost of technological advances originating from governmentally-paid research programs.
- The probability of success of the company concerned is only determined by what they did in the most recent 10 years, not by what they did before that period. (The probability of success is defined as the probability that hydrate become commercial at a certain time multiplied by the probability that the company can determine this moment long enough ahead to step-up its hydrate research activities timely as described earlier and reap the rewards for its efforts.)

The possible branches from decision points in the decision tree are indicated with uppercase letters, as in Fig. 4. In this figure, the percentages quoted correspond to the base case assumptions as in Table 1. Note that the probabilities of commercial feasibility are stated in a non-cumulative way in this figure, as opposed to the table.

It is assumed that the company will make some money (i.e. 3% market share, see Table 1) from hydrate methane even if they make no research effort now, because at some time in the future there will be an industry-wide capability available to produce methane from hydrate. The additional value for the company is in the possibility of making more money than their competitors by "being the first" (although this strategy also carries greater risk than being the "first to follow"). The additional revenues generated in this way can be accredited to a prior research investment and are hereafter referred to as "revenues".

The potential revenues of each particular path through the decision tree are calculated and then multiplied with its probability of success, and the appropriate costs are subtracted. That way, it is possible to arrive at the Net Present Value for each of the possible decision paths.

From the detailed calculations for above base case, which are not presented here, it is very clear that from a commercial point of view, the best option at this moment is to invest no money at all in hydrate production research (decision path CFI in Fig. 4). This is mainly caused by the relatively low probability of success estimates.

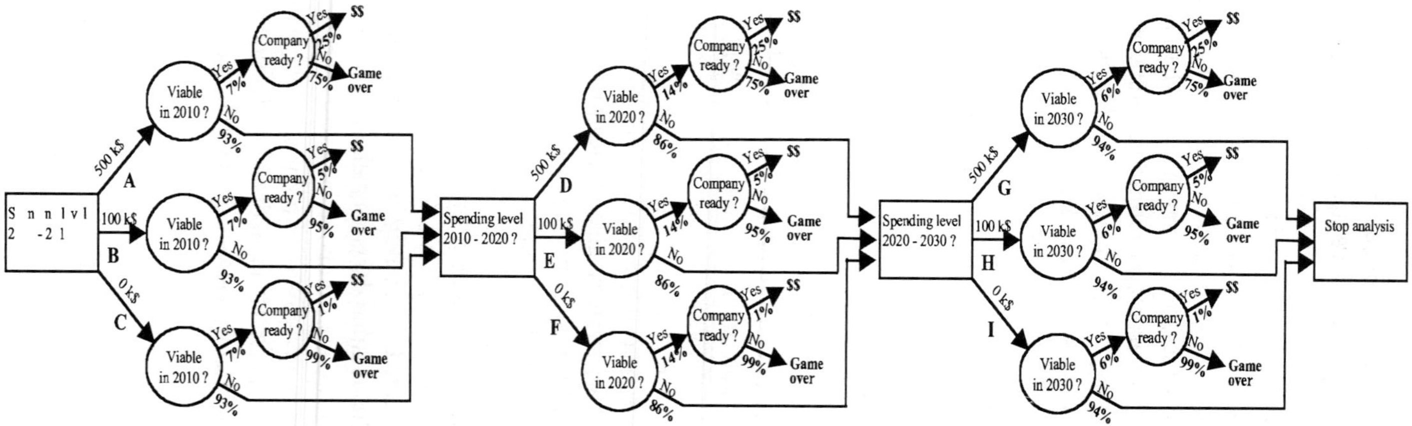

Figure 4. Decision tree showing the possible branches for which cost-revenue calculations were made.

Remarkably, all Net Present Values appear to be negative; i.e. the company would lose money even if the spending level were US$0 per annum. This is, in a way, an artifact of the algorithm that requires a step-up in spending level towards the end of the considered term, as indicated earlier. If the "mandatory" step up in spending level would be omitted, cost and revenues for the path CFI would both be zero, as would therefore the Net Present Value. If we would add more 10-year blocks to the calculation (which was not done to prevent the calculation from becoming overly complicated), the preference for zero-spending mode would probably extend more into the future. So while there is indeed some artificiality in the way the calculations have been set up, the conclusion is robust.

The calculations were carried out in a way that is convenient for perturbation analysis. It appeared possible to tune the input assumptions such that spending US$ 100,000 pa in the period 2000-2010 comes out as the best option within the decision tree. This could be achieved by various sets of assumptions; e.g. when making all of the following changes:

- Chance of hydrate being commercially feasible in 2010, 2020 and 2030 would be 10%, 20% and 30%, respectively;
- Chance of the company to be the first if spending US$ 100,000 and US$ 500,000 would be 20% and 35%, respectively;
- Period that competitive advantage lasts would be 15 years;
- Net revenue for produced gas would be US$ 2 per barrel of oil equivalent.

This set of assumptions can still be viewed as realistic, and leads to a (positive) Net Present Value of US$ 538,000 for option B (higher than any other). Hence, one might be tempted to conclude that this option, i.e. spending US$ 100,000 pa is reasonably justifiable. It must be noted, though, that simply "making money" is not enough. A company would have to weigh this profit-generating option against other investment opportunities, some of which may be in non-energy related areas, that may yield a better return per dollar invested.

It appeared to be very difficult to make investing US$ 500,000 pa from 2000 the best option by tuning the base case assumptions. The only way to achieve this by tuning a *single* assumption is to change the "net revenue for produced gas per barrel of oil equivalent" to more than US$69. Obviously, this is utterly unrealistic because oil at that price would have a serious impact upon the world economy and change the setting of the decision process. Also, simultaneous tuning of several assumptions did not lead to a reasonably realistic set of assumptions that would justify spending US$ 500,000 pa. From this it can also be concluded that spending *more* than US$ 500,000 pa from 2000 would be even less feasible.

All of the foregoing analysis would be most appropriate for relatively large oil or gas companies. A smaller, maybe more entrepreneurial or risk-taking type of company would have a different perspective:

- They probably could move quicker, possibly by taking up a consultant or advisory role.
- They would typically have no ownership of resources, and would run fewer risks because their costs would be recovered as a "fee for service".
- In some countries (including the US) tax laws would allow to fiscally write off research and development costs.
- In some countries tax laws allow distributions to directors of some companies to be free of tax if the research and development was carried out in that country (e.g. Ireland).
- Government provided funding for methane hydrate research in the US and elsewhere may also flow to private companies.

It can be concluded that excellent commercial opportunities may exist for a company that can develop technology that fills a niche or is otherwise a (potential) key component of hydrate recovery schemes. Even though the large oil and gas companies of this world may be reluctant to invest now, they definitely have an interest in the fuel of the future.

Acknowledgements

I would like to thank Owen BeMent and Wim Swinkels for stimulating discussions and for critically reading the manuscript. I also thank Shell International Exploration and Production B.V. for permission to publish this work.

Chapter 27

Hydrate Resource, Methane Fuel, and a Gas-Based Economy?

Michael D. Max
MDS Research
Suite 302, 1211 Connecticut Ave. NW
Washington DC 20036

1. INTRODUCTION

The possibility that hydrate may supply extremely large volumes of methane over a long period of time will be the main motivation for carrying out research into the naturally occurring hydrate system. Despite the possible significance of hydrate to ocean/atmosphere chemistry and its impact upon climate and seafloor stability, the amount of money available for research into that facet of hydrates which may generate economic return is potentially large enough to support very substantial research efforts. Thus, examining economic potential of hydrate is important.

Will purely economic judgements be made with respect to the development of the methane resource of hydrate and can methane be recovered commercially? Economics are different in different countries where displacement of imported fuels by locally derived fuels may be tolerable at costs that are substantially above those of other countries. Economics are also different in countries without large stocks of petrocurrency and in countries where the existing tax regime allows governments to cut the tax on fuels to subsidize local fuel production and yet not raise prices to its consumers. In short, hydrate may be developed first in those countries where it makes economic sense to do so.

There is not a free market in world fuels. For instance, the first 'petroleum shock' and the rise of politically motivated and economically sophisticated petroleum producing countries in the early 1970's, has made obvious the fact that even though very large reserves of petroleum occur around the world, many of these restricted sources are subject to at least partial withholding from the markets of major industrial countries such as the United States for reasons that have nothing to do with the simple economics of supply and demand. In short, the industrial West does not control the majority of the world petroleum supply. With the passing of the Cold War political and economic framework, a new geopolitical situation has arisen, especially with

M.D. Max (ed.), Natural Gas Hydrate in Oceanic and Permafrost Environments, 361.

regard to the world energy market. Producers and consumers are no longer closely tied within well-defined political-economic associations. In addition to the weakening of political restraints that once emanated from competing superpowers, the hydrocarbon producer countries have shown that they can discipline themselves, and via controlling the price of petroleum, establish themselves as an important power block in the emerging new world economy.

The Organization of the Petroleum Exporting Countries (OPEC) is an intergovernmental organization whose stated goal is the stability and prosperity of the petroleum market. Eleven countries take part as full members. The OPEC member countries currently supply more than 40 per cent of the world's oil and they possess about 78 per cent of the world's total proven crude oil reserves. It is in the interest of these OPEC countries to attempt to coordinate their oil production policies in order to achieve a stable market and a higher price for their product. In addition, the OPEC countries established the OPEC Fund for International Development in 1976. The OPEC Fund provides development loans and grants to non-OPEC developing countries. Its loan commitments to date exceed $3.4 billion. Distribution of these funds undoubtedly affects the political and economic decisions of countries which benefit from this new source of financial aid.

Is there a shortage of hydrocarbon fuels? The simple answer, at least in the short term, is no. "There are today more than a trillion barrels of proven reserves [of oil] (recoverable at current prices under current conditions)." (Jaffe and Manning, 2000). There is presently a production capacity in excess of demand and new discoveries, such as those along the eastern South American and west African continental shelves and slopes are adding to the world petroleum reserve. If the total fuel reserves of the world are set against demand, there are many years of conventional hydrocarbons available. The essential issue is not the existence of petroleum, but of production and distribution. If the available production capacity was utilized to capacity and a completely free market set the prices, effective oversupply would drive the price low and hold it there for many years. On the other hand, low prices and rapid depletion of producer's finite natural resources may not be in the best economic interests of producer countries who naturally wish to keep the price of petroleum high.

Although the views of individual consumers who ultimately pay for petroleum products in the industrial world have tended to be given the greatest publicity, there are other considerations and potential consequences. A real superabundance of petroleum may lead to "danger... [that] long-term trends point to a prolonged oil surplus and low oil prices over the next two decades." (Jaffe and Manning, 2000). A truly globalist economic framework, where national boundaries do not constitute economic barriers, raises the possibility of serious political problems arising in producer country societies where income from exporting remains too low to allow significant investment. Instability in the oil-producing countries because of an inevitable decline of oil prices has the potential to introduce societal instability. And it is to be feared that this instability could spill over into First World countries of the industrial world, at least in part in the form of large scale migratory pressures.

Thus, a basic conflict appears to be developing and becoming formalized between producers and consumers in the energy market. This competition between producers and consumers to control, or at least strongly influence, the price of hydrocarbon fuel is almost certainly to be a feature of the world's geopolitical and economic framework for the indefinite future. It will be necessary to fashion an economic framework for world energy that will take into account the objectives of producer countries while insuring the availability of energy supplies to consumer countries.

2. ENERGY SECURITY

What constitutes energy security? Every industrial society is fundamentally based on the uninterrupted supply of sufficient fuel at acceptable prices. The maintenance of dependable fuel supplies therefore underpins economic and political stability of industrial countries and constitutes a basic security issue. The issue of energy security is often expressed as a matter of price, but there is a deeper concern. Industrial countries such as the U.S., which was once energy independent, has now become dependant upon imported energy. However, energy security to the U.S., which prints the world's only petrocurrency, consists of supporting the present economic framework and maintaining the ability to purchase petroleum, albeit at a price that may be somewhat higher than could be obtained in a completely free market. Other countries such as those represented by highly industrialized Japan, which never had significant indigenous energy sources but has sufficient currency reserves to also purchase as necessary, and India, which represents a large number of countries which are not petroleum exporting countries but wish to raise their standard of living and industrialize, view energy security in a different manner.

Energy security can be guaranteed only where a consumer country is in direct control of its energy sources. Only a few industrial, or First World countries, such as Norway and Brazil may be in this enviable situation, although in the past the U.S. was once largely energy independent during its era of greatest industrialization. Most countries are either producers or consumers, yet only a few producers have invested their money in developing infrastructure that will outlast their ability to produce energy. Saudi Arabia, for instance, may have squandered up to 13 trillion dollars income from exporting petroleum without developing a modern commercial economy. What then are the options of countries that wish to insure the security of their energy supplies; that is adequate supplies at adequate costs?

The obvious answer to a potential energy shortage that may be created anywhere in the chain of supply, no matter how artificial or periodic, is to develop indigenous replacement energy sources that would be under the direct control of the consumer country. This may not be an impossible goal for many countries, especially those that border the sea and can exploit their deeper water continental margins for hydrocarbon deposits is that can now be reached using rapidly evolving technology. Israel, for instance, succeeded in developing a Gross National Product that is now larger than Saudi Arabia with no easy source of investment money as would be associated with being a hydrocarbon exporter and with virtually no indigenous energy supplies. But the success of Israel in building its commercial society with no energy resources has not stopped Israel from exploration for its own energy resources. And Israel's luck has recently been good. A new offshore gas field in Israel's territorial sea area could supply Israel's natural gas needs for 15 years without any need for imports and geologists say it is likely that other large sources of natural gas lie close to the current sites off Ashkelon (Ha'aretz, 2000). Brazil, Nigeria, Angola, Australia, etc. are all currently engaged in hydrocarbon exploration, much of which is taking place in their immediately adjacent sea areas.

Where countries are not able to find and develop new conventional hydrocarbon sources, they will have to either considerably scale back their plans for industrialization and suppress consumer demand or meet their energy requirements from some other source, such as nuclear energy. Countries could also import hydrocarbons in volumes beyond those they originally contemplated or develop indigenous gas (methane) or petroleum from unconventional sources, such as: (i) coal-bed methane; (ii) in-situ coal gasification; and (iii) gas hydrate. Of these possible sources of new hydrocarbon energy, identification of gas hydrate and extraction of methane from it has recently, and rapidly, become an attractive exploration topic, especially for countries such as Japan, which has a long-term focus on energy supply and security.

3. COMMERCIALITY OF HYDRATE

Will hydrate be economic?, and, When will hydrate be economic?, with respect to the extraction of methane. Although these are valid questions, there is neither a simple nor a single answer. Although there are world oil prices for different types of natural petroleum, the cost of methane varies from place to place. There are two reasons for this. Firstly, in countries that do not have indigenous gas supplies, the price of gas is the delivered or landed price, which is usually in the form of Liquid Natural Gas (LNG) if no pipeline transport is possible. Secondly, some countries, such as the United States, have low taxes on energy while other countries, for instance in Japan and the European Union, have high taxes. Thus, not only does the floor price of gas vary from country to country, but the cost to consumers vary considerably.

For methane from hydrate to compete on a price basis, it must meet different cost tests in different countries. In North America there are considerable natural gas supplies and the base price of gas is relatively inexpensive (Table 1). In Japan, however, virtually all natural gas is imported (Chapter 18) and the base price, which is the landed cost of LNG, is higher. The retail price of gas in Japan is extremely high (Table 2), largely to reduce the demand for gas.

Cost Factors	Methane from Hydrate		Conventional gas
	Thermal injection	Depressurization	
Investment (M$)	5,084	3,320	3,150
Annual cost (M$)	3,200	2,510	2,000
Total production			
(MMcf/year)**	900	1,100	1,100
Production cost			
($/Mcf)	3.60	2.28	1.82
Break-even wellhead			
price ($/Mcf)	4.50	2.85	2.25

Table 1. Economic study of gas hydrate production after MacDonald (1990) National Petroleum Council (1992), and Collett (1998) for the North Slope of Alaska. * Assumed reservoir properties: h=25 ft, (=40%, k=600 md; ** Assumed process: injection of 30,000 b/d of water at 300 F. Thermal injection involves melting hydrate with hot fluid; depressurization involves lowering gas pressure to induce hydrate dissociation (Chapter 10).

Cost Item	U.S. $ per Mcf
U.S. Futures Price*	1.90-3.00
**U.S. 2Q 2000	4.00+
Japan, LNG imported	3.50-4.50
Japan, Industrial	15.00
Japan, Residential	35.00

Table 2. Comparison costs of methane gas. LNG, Liquid Natural Gas; Mcf, thousand cubic feet. (Collett, 2000, pers. comm.). *1998-1999. ** 2nd Q 2000.

Hydrate production cost projections for the North Slope of Alaska, and by inference hydrate in the Mackenzie Delta area of Canada (Table 1), appear to lie in the commercial range now, although this is a recent development caused by the rising price of gas. The price of gas is liable to increase because indigenous conventional gas from the U.S., Mexico, and Canada (Alberta) will not be able to supply U.S. gas needs by 2005 to 2010 (T. Collett, pers. comm.). Profit margin issues, however, may dictate that conventional gas will be extracted before the higher cost production from hydrate is obtained on a large scale, which would have the effect of delaying methane extraction from hydrate. However, development of permafrost hydrate resources could be profitable soon, but at a lower profit margin than conventional gas.

The hydrate deposits of Japan occur offshore (Chapters 6, 18) and no production cost estimates are publicly available. Nonetheless, the concentrated hydrate deposits offshore Japan occur near shore in relatively shallow water under conditions that are favorable for methane recovery (Max and Dillon, 1998). It is likely that production costs of these Japanese resources can be kept relatively low. Methane from these Japanese hydrate deposits could be commercial there before it is economically feasible in North America because the costs and pricing will allow the maintenance of relatively high profit margins and tax returns. There is also a lack of indigenous gas, and a political imperative to become energy independent if possible.

Japan will almost certainly continue to develop its hydrate resources and produce gas as soon as it can. Some displacement of imported gas supplies could take place within a 4 to 5 year time frame, with significant displacement of imported supplies in a 5 to 10 year time frame. The aim of energy independence is a future goal. India and other countries may also chose to develop their hydrate resources for their own monetary, economic, political, or environmental reasons because any methane they recover will not involve large-scale outflow of petrocurrency. In North America and elsewhere where gas can be transported by pipeline from conventional gas fields, commercial development of hydrate is probably further in the future (Chapters 10, 26), on the order of 25 to 50 years. Energy security or development of hydrate to displace other fuels such as coal and oil on an environmental basis, however, may prove to be more influential - perhaps because of government regulation - than a simple cost comparison. This could cause a telescoping of the world-wide development of hydrate resources into a much nearer term regime.

4. EMERGENCE OF GAS HYDRATE AS A POTENTIAL METHANE RESOURCE

Until recently the very existence of natural hydrate was one of nature's closely held secrets. And it has only been in the last several years, culminating with the Offshore Drilling Program (ODP) leg 164 (referred to in papers in this volume, e.g., 7, 11, 22 & 24), that a quantitative data set has been acquired from direct sampling and measurement. The magnitude of the potential methane hydrate resource is apparently huge.

In terms of the likelihood of commercialization, it is often instructive to examine the overall size of a potential resource or economic target and the percentage of the total which would constitute an economic target. Estimates of the Indian hydrate system resource, for instance (Chapter 17), are made on the basis of volume of hydrate per % likelihood of recovery from the total estimate. As an example, if the statistically derived figure for the volume of methane in gas hydrate within the U.S. EEZ of 200,000 Trillion Cubic Feet (TCF) proves to be near reality (and it is the best technical estimate at present), then an economic target of 1% of the resource would be 2,000 TCF.

This figure is about equivalent to 80 years supply at an average consumption of 25 TCF (current U.S. usage is about 22 TCF [U.S. Department of Energy]). An economic target of only 5 % of the estimated U.S. EEZ resource, which is the greatest statistical probability used in evaluation of the Indian resource (Chapter 10), would result in a U.S. supply about 400 years of natural gas. At an economic target of 10%, gas could displace most other energy sources for a considerable period of time. Because the initial resource appears to be so large, achieving recovery of only a small part of the methane from hydrate would appear to be relatively rewarding, not only on a year-by-year commercial basis, but for the energy security of the U.S. and indeed any country with potential hydrate resources.

A commercially successful strategy for methane production from hydrate, particularly oceanic hydrate, does not yet exist. Indeed, both positive and negative oceanic hydrate resource evaluations have been published. Yet gas hydrate (presumably mainly methane) volume estimates of from 5% to over 80% of pore fill of large volumes of marine sediments have been recognized, and this is encouraging for the eventual development of a hydrate methane resource.

There remain many basic research issues to be resolved prior to the potential exploitation of methane gas from hydrate. We know very little about the naturally occurring material itself (Chapter 25) and how it forms in sediments (Chapter 20). The nature of fluid flow in deep water marine sediments (Chapter 9) is very imperfectly known and vital to reservoir characterization. Current production knowledge based upon conventional gas reservoir performance almost certainly does not describe the conditions of oceanic hydrate deposits.

4.1. Oceanic Hydrate System Economic Targets: The Big Prize

Gas hydrate in permafrost areas thus far have been identified in association with conventional hydrocarbon deposits. They are associated with geological traps that concentrate the subjacent hydrocarbon deposits, and are very different in apparent genesis from oceanic hydrate deposits, which are dominantly formed from biogenic gas and are not associated with geological structure in trap rocks or with petroleum or condensate of any sort. Permafrost hydrate (Chapter 5) is readily accessible from the land, except where deposits may exist in relic permafrost in continental shelves, and as such, constitutes a ready economic target. Permafrost hydrate, however, is estimated to be no more than about 5% of total hydrate, most of which exist in oceanic hydrate deposits (Kvenvolden, 1993; Max and Chandra, 1998).

Oceanic methane hydrate deposits (Chapter 13) exist because the flux of methane into the hydrate economic zone (HEZ) (Max and Chandra, 1998) is greater than the flux out of it (Chapter 8). The methane is part of the biosphere that is in constant flux, even on the scale of local observation and time scales on the order of minutes. Escape of methane into the ocean and atmosphere, biodegradation, and a variety of chemical reactions that consume methane are more than balanced by methane production and concentration of methane in the oceanic hydrate system. Methane is concentrated in both solid hydrate and as gas accumulations, both of which present clear economic targets. Both hydrate and gas occupy primary and secondary porosity in the marine sediments, and both significantly alter the physical properties of the sediment in which they occur. The presence of hydrate may increase the strength of the sediment both by reducing porosity and replacing fluid (and gas) with solid hydrate, while cementing the sediment grains to some extent. Gas weakens the sediment and expels interstitial fluid.

Although large accumulations of hydrate have been recognized in many continental margins, those localities known about today may not turn out to be the primary economic targets for methane recovery. Our knowledge about hydrate occurrence is largely coincidental to the reasons for which most of the surveys on which hydrate has now been identified were carried out. Much more quantification of the resource needs to be carried out. 'Sweet-spots' where hydrate is concentrated must be identified. To date, only the Japanese (Chapter 18) have identified a deposit that may be suitable for development.

Solid hydrate will have to be converted to gas in-situ prior to recovery. Gas deposits associated with hydrate may be the initial economic targets. Methane derived from contiguous hydrate through dissociation, as a response to depressurization of the gas reservoir, will provide a serendipitous partial recharge of the methane gas reservoir.

4.2. Natural Gas as a Fuel

Once the methane potential of the hydrate deposits in the U.S. Exclusive Economic Zone (EEZ) is better described and methodologies for methane recovery developed, it will be reasonable to consider transition from a petroleum-based to a gas-based economy. Methane as a fuel offers clear advantages over oil or coal: immense resource potential, ease of transport via in-place distribution infrastructure, less carbon dioxide release per unit volume burned, minimum release of exhaust pollutants from clean-burning methane, a fuel, which is chemically inert, non-polluting, and biomedically tolerable to humans.

Natural gas, which is primarily methane, is an excellent fuel for combustion, either in open-flame burning to produce heat or in controlled circumstances within fuel cells, for a number of reasons. In comparison with other natural gases, methane contains the highest H:C ratio. That is, there is more hydrogen with respect to carbon in methane than in all the other hydrocarbon gases. When methane is burned it produces less CO_2 per mole than any other fossil fuel. Methane also produces less CO_2 per mole than alcohols, where OH substitutes for one molecule of H, and much less than in liquid petroleum gasoline and oil based fuels.

The recent attention given to global warming, as enshrined in the internationally recognized Kyoto accords, have focused considerable attention upon the volume of CO_2 emissions and the need to curtail or reduce them. An aim of the U.S. government is to achieve a 15% energy efficiency savings goal within the next five years (by 2005) and a reduction of energy use in federal buildings to 30% below 1990 levels by 2010. Switching to natural gas from oil fuels, as is currently underway in California where older oil-fired power stations are being refitted with combined cycle gas turbine generating capability, will have an immediate effect on lowering the CO_2 emissions, at least from that station, with no reduction in power output, and is in accordance with the stated goals of the government for energy use reduction and reduction of CO_2 emissions.

Solid fuels and some oils produces substantial pollutants and CO_2 upon combustion. Natural gas, on the other hand, contains very little sulfur, phosphorus, or nitrogen, and other pollutants that can become aerosols when fuels rich (e.g., coal) in these pollutants are burned. Natural gas (methane) produces almost pure CO_2 and water when burned because of its initial purity. Being a gas, further purification is possible using relatively inexpensive continuous industrial processes. CO_2 and the exhaust gas produced from methane combustion do not produce smog because the pollutants that are cooked and mixed into a photochemical smog by sunlight radiation are not present in the exhaust to begin with, as is the situation with both liquid hydrocarbon and solid fuels without emission control purification.

5. METHANE GAS FROM OCEANIC HYDRATE: THE RACE IS ON

Four developments in the last 10 years have both broadened and deepened interest in developing hydrates as a source of methane gas.

1. Much more is known about hydrate and their disposition because of the sudden rush of marine geological and geophysical research on continental slopes, where most of the oceanic hydrate appears to exist in significant concentrations. The greatest progress in gas-hydrate knowledge in the past five years has probably come from the results of the Ocean Drilling Project (ODP); especially from Leg 164, which took place in November/December, 1995, and is the only ODP major ocean drilling program (other than the drilling being carried out by the Japanese in their waters) devoted solely to the study and open dissemination of information relating to gas hydrate. The region of the Blake Ridge, where four holes were drilled, is considered to have very high concentrations of gas hydrate and to be a potential source of methane, especially where the hydrate concentrations occur in relatively shallow seafloor depths (Max and Dillon, 1998; 1999).

2. Even the most conservative estimates of oceanic hydrate volumes now acknowledge that the potential resource is very large, on the scale of double the volume of the energy content of all known hydrocarbons on Earth combined (Kvenvolden, 1993; this volume).

3. Government agencies (India, Japan, and the U.S. foremost) have developed hydrate research programs to recover methane from oceanic hydrates (Preface).

4. Much of the deep water drilling and gas and oil-handling technology that is required for hydrate recovery has been developed in industry's pursuit of conventional deep water liquid hydrocarbons during the past ten years. Industry is now drilling safely and recovering conventional hydrocarbons in water depths in which gas hydrate concentrations are known to exist (Max, 1999; Max and

Dillon, 2000). The technological base required to explore and develop hydrate has largely been paid for.

6. THE GAS-BASED ECONOMY

Use of methane or a liquid derivative of methane, such as methyl alcohol, as fuels in the place of oil-based fuels would provide for a dramatic reduction in unwelcome byproducts of combustion. Use of methane or methanol as one of a mix of fuels would undoubtedly be environmentally positive (Saricks, 1989). Although the primary focus of economists dealing with fuels has tended to focus on the bottom-line cost of production, transport and end-user cost of fuels, the awareness that environmental concerns and external costs may have to be factored into the fuel cost equation, may cause revision of the present conclusion that oil-based fuels are preferable to methane or one of its derivatives. A strong case can be made that a large number of hidden subsidies support the current low price of oil-based fuels (Hubbard, 1991). For instance, is the deployment of U.S. forces in the Persian Gulf area accountable as a petroleum subsidy?

A number of states in the United States, principally California, New York, and Wisconsin, have already enacted legislation requiring some external costs to be taken into account when proposing new energy generating activities. Fuller external accounting that include costs to society, economies, transportation infrastructure, environment, health, etc., rather than the bottom-line oil company accounting, all tend to favor gas over oil-based fuels. Bringing market prices in line with energy's hidden costs will be one of the great challenges of the coming decades.

Even with no disruptions of petroleum supply to the World economy, the economic energy base will shift from oil to gas in the course of the next quarter century or so. This shift will be dictated by nothing more than the limited amounts of petroleum liquids, tar sands, oil shales, etc. that are available. However, geopolitical considerations, such as significant perturbation of the oil flow from major producing regions like the Mid-East, and environmental considerations that concern emissions from combustion, may force an earlier transition to the gas economy than could be predicted from an analysis of energy cost economics alone. Thus, the shift from oil to gas as the primary non-nuclear source of power is a matter of concern today, and not an obscure energy concern, the exploitation of which will be indefinitely pushed into the future.

What will energy prices be in the near future? We do not know. If there are no disruptions to society or energy supply, it is likely that an inexpensive energy supply will continue to be the industrial base world-wide. No one really knows how competitive any particular type of hydrocarbon is going to be over the coming 10-20 years on a purely commercial basis. How will development of hydrate affect the energy industry? Developing hydrate as a source of methane will undoubtedly stabilize the availability of low-cost energy. It is likely that permafrost hydrate is probably close to being a commercial methane source now, even within North America.

Glossary of hydrate terms

Hydrate science is still in its early stages. Different terminology is often found in the literature for the same gas hydrate related objects and mechanisms. Soon, common use and concensus will determine the accepted nomenclature. Hydrate descriptive terms are strongly science based at present and the use of potential economic hydrate terms in an economic sense (Max and Chandra, 1998) may differ somewhat from scientific usage. 'Hydrate' vs 'hydrates', for instance often appear to be used indescriminately, as does 'hyrate' or 'gas hydrate'.

BGR: Base of Gas Reflector. This is sometimes a strong positive impedance contrast at the base of a gas-rich zone below hydrate, where gas has largely separated from pore fluids, or at the base of a gas zone within the HSZ.

Blanking: Blanking refers to a weaker pattern of reflection traces in reflection seismic records that is regarded by some as being caused by the presence of hydrate. The seismic model of the effect is: hydrate fills the pore spaces and reducing the impedance contrasts of the original bedded sequence. Some geophysicists are not convinced that apparent blanking is evidence of the presence of hydrate.

BOH Base of hydrate. Also: **BGHS**, Base of Gas Hydrate Stability. This refers to the actual base of naturally occurring hydrate, which may be either higher or lower than present calculations may indicate. The gas hydrate physical system is imperfectly understood in nature. The base may be difficult to identify when there is no BSR, especially where the bedding surfaces of marine sediments are parallel bedded with the seafloor.

BSR: Bottom-Simulating Reflector. Seismic term. This is a strong negative impedance contrast at the contact between higher Vp and density hydrate-rich zone and sediments having lower Vp and density below. BSR is usually recognized as a reversed impedance contrast gas zone. The BSR is commonly taken as is a surface marking the phase boundary of and the actual base of the hydrate stability zone.

Dissociation. The 'melting' of gas hydrate owing to rising temperature or diminishing pressure. Pioneered as a methane recovery method from hydrate in Russia (see chapter 6).

Dissociation Feedback: Gas hydrate contains within it a self-moderating or feedback system which act as a rate-controlling mechanism during dissociation. Whatever the cause of dissociation, heat is absorbed. This cools the system and may cause dissociation to slow or cease. Run-away dissociation feedback may take place when hydrate is brought to near-atmospheric pressures which usually will initiate a complete dissociation; the system may absorb enough heat under this condition to freeze the local water.

Depressurization (recovery): Term for artificial dissociation process where pressure in a gas section in contact with hydrate is pumped down and the pressure lowered. Hydrate in contact becomes unstable and will dissociate, absorbing heat.

Inhibitor (recovery): Term for artificial dissociation process where fluids having the effect of an antifreeze with respect to gas hydrate is introduced. The presence of the fluid (e.g. brines, methanol, etc.) changes the position of the phase boundary in pressure-temperature space and causes hydrate lying between the original and the inhibitor-determined position of the phase boundary to dissociate. Water-ice analog is seen where icy streets are melted by the application of salt.

Thermal (recovery): Term for artificial dissociation process where warm fluids are introduced in a hydrate section and the rise in temperature and the availability of heat allows the hydrate to 'melt'.

GAIL. Gas Authority of India, Ltd. Original coordinators of the Indian Gas Hydrate Research Program.

HEZ: Hydrate economic zone. This includes all of the HSZ and sediments below solid hydrate that include both gas-charged and sediment which does not display a significant level of compaction for its depth. This is the zone in which the geotechnical parameters must be extremely well known to allow for safe extraction from gas reservoirs and production of gas from the hydrate. It is presently estimated that this economic zone may be up to 1.5 to 2 times the thickness of the HSZ.

HGZ: Hydrate Gas Zone. Zones of free gas that ocur mainly trapped below the HSZ where the formation of hydrate has lowered permeability enough to trap gas. Gas zones may also occur within the hydrate stability zone.

HSZ: Hydrate Stability Zone. This is the zone in which hydrates are expected to be stable based on calculations of pressure and temperature with depth in the seafloor. Also known as Gas Hydrate Stability Zone (**GHSZ**). It extends from the seafloor downward in the marine sediments to some depth determined by rising temperature. Hydrate in the HSZ is most stable near the surface and progressively less stable downward.

Hydrate, gas hydrate, methane hydrate: Oceanic gas hydrate, is dominantly methane plus water. Hydrate is a diagenetic crystalline solid that forms as a diagenetic material in sediment pore spaces within the local HSZ. Hydrate (singular) is commonly used where one gas species is dominant such as in methane hydrate. Where more than one gas is present in the hydrate lattice and where more than one hydrate structure is present, hydrates (plural) is correct usage.

HZ: Hydrate Zone. This is the zone within the HSZ in which hydrate is actually stable. The upper surface of the HZ may extend to the surface or it may be depressed below a sulfate reduction, or other chemically-active zone. The base of the HZ may occur either higher or lower, or be imbricated and complex, depending on local chemical, structural, and thermal deviations from a simple theoretical model.

JNOC. Japanese National Oil Company. Coordinated the first 5-year program of basic research and development of the Japanese gas hydrate research program.

OBC: Ocean Bottom Cable. Also Seafloor Array or OBA (Ocean Bottom Array). Linear array of geophones resting on the seafloor used primarily for sensing shear and interface waves.

OBH: Ocean Bottom Hydrophone. Used for sensing acoustic energy in the sea, usually near the seafloor in association with OBS, OBH or OBC(A), to allow separation of pressure wave from shear wave energy.

OBS: Ocean Bottom Seismometer. This device records acoustic energy propagating through the bottom; usually on 3-axes so that a full 3-dimensional solution can be found.

Oceanic hydrate system. Economic term referring to all natural gas, which is concentrated as both solid, crystalline hydrate and associated free gas. Useful for describing entire potential economic methane section in which hydrateis or gas is concentrated. Where either hydrate or free gas is the economic target, use of the term 'hydrate' alone is unintentionally misleading.

Ryokyu Trough: Also spelled Ryokyo. Subduction zone collisional margin of the eastern Eurasian plate that can be traced from north of Taiwan to offshore SE Japan. Synonoumos in its northern end with the Nankai Trough.

TGR: Top of Gas Reflector. This is a strong negative impedance contrast that marks the top of a gas-rich zone. Where this occurs immediately below the HSZ, it may be coincident with the BSR. May occur within an HSZ where gas has expelled pore fluids and has precluded further formation of hydrate. Both the top and the base of gas reflectors may not necessarily mark a thermodynamic passage (phase boundary) similar to that defined by the BSR.

Vertical Array: A vertical array of hydrophones used in conjuction with stepped-out sources to acquire an acoustic dataset used primarily for determining physical properties of the upper seafloor for acoustic propagation modeling.

VSP: Vertical Seismic Profile. This is the standard method for carrying out reflection seismic proviles from ships at sea. A horizontal array consisting of a number of hydrophones are towed behind the ship. A source of energy, such as an airgun, watergun, transducer (boomer) or a sparker, is used to privide for regular, short bursts of acoustic energy, which is recorded by the hydrophones after being reflected from the seabottom and from reflectors within the bottom.

Other useful terms

Exothermic: Gas hydrate formation is exothermic, that is, it gives off heat when hydrate crystallizes.

Endothermic: Gas hydrate formation is endothermic, that is, it absorbs heat when it dissociates.

Heat of fusion: Heat given off or absorbed during hydrate formation and dissociation, taken as equal in amount but different in sign for each transformation. This has been measured at about 54 kJ/mol (hydrate) and but has been calculated to be about 57 kJ/mol (Sloan, 1997). This is much higher than the heat of fusion for water-ice, which is about 6 kJ/mol.

Saturation: The state of equilibrium in which no further inert gas is taken up in solution

Common gas laws:

Boyle's Law: Boyle's Law defines the relationship between pressure and volume. It states that at a constant temperature, the volume of a given mass of gas will vary inversely with the absolute pressure, or $P_1V_1 = P_2V_2$.

Charles' Law: Charles' Law concerns the relationship between temperature, volume and pressure. It states that 'if the pressure remains constant, the volume of a given amount of gas is directly proportional to the absolute temperature'.

Dalton's Law: Dalton's Law concerns the composition of air at various pressures. It states that 'the total pressure exerted by a mixture of gases is the sum of the pressures that would be exerted by each of the gases if it alone occupied the total volume'. The total pressure is the sum of the partial pressures of the gases present and as the overall pressure increases, so the partial pressure of the constituent gases increases.

Henry's Law: Henry's Law relates to gas absorption in fluids. It states that 'the amount of gas that will dissolve in a liquid at a given temperature is directly proportional to the partial pressure of that gas over the liquid'. At increased pressures, increased volumes of gas can dissolve in liquid.

Selected References

References for each chapter have been amalgamated here. Because most of the contributors are actively publishing scientists, there are a number of publications referenced that are in press at the time that the book was transferred to the publisher. Many papers are referenced in more than one chapter. In a number of cases the year of publication has a suffic 'a' or 'b' as cited in at least one of the referencing papers. Other papers may reference the same citation without the suffix. The editor has not been able to verify every reference and minor errors in and incomplete citations may be present.

Abegg, F. & Anderson, A. 1997. The acoustic turbidity layer in muddy sediments of Eckernförde Bay, western Baltic: methane concentration, saturation and bubble characteristics. Marine Geology 137, 137-148.

Abreu, V.S. & Anderson, J.B. 1998. Glacial eustasy during the Cenozoic: Sequence stratigraphic implications: American Association Petroleum Geology Bulletin 82, 1385-1400.

Acheson, D. 1997. From Calculus to Chaos. Oxford UP, 269pp.

Aldaya, F. & Maldonado, A. 1996. Tectonics of the triple junction at the southern end of the Shackleton Fracture Zone (Antarctic Peninsula): Geo-Marine Letters 16, 279-286.

Aleyeva, Y.R. & Kucheruk, Y.V. 1985. Geodynamic reconstructions and the prediction of oil and gas potential of poorly known regions: the case of Antarctica: Intetnational Geological Reviews 27, 1383-1395.

Allen, D., Bergt, D., Best, D., Clark, B., Falconer, I., Hache, J.-M., Kienitz, C., Lesage, M., Rasmus, J., Roulet, C. & Wraight, P. 1989. Logging While Drilling, Oilfield Review 1, 4-17.

Andersen, K., Rawlings, C.G., Lunne, T.A. & By, T.H. 1994. Estimation of hydraulic fracture pressure in clay. Canadian Geotechnical Journal 31, 817-828.

Anderson, D.L. 1989. Theory of the Earth, Blackwell Scientific Publications, Brookline Village, MA.

Anderson, J.B., Brake, C., Domack, E.W., Myers, N.C. & Wright, R. 1983. Sedimentary dynamics of the Antarctic continental shelf: In: Oliver, R.L., James, P.R. & Jago, J.B. (eds). Antarctic Earth Sciences, Canberra, Australia, 387-389.

Anderson, J.B., Pope, P.G. & Thomas, M.A. 1990. Evolution and hydrocarbon potential of the northern Antarctic Peninsula continental shelf: In St.John, B. (ed). Antarctica as an Exploration Frontier - Hydrocarbon Potential, Geology and Hazards. American Association Petroleum Geologists Studies in Geology, 31, 1-12

Anderson, R.T., Chapelle, F H. & Lovley, D.R. 1998. Evidence against hydrogen-based microbial ecosystems in basalt aquifers. Science 281, 976-977.

Andersland, O.B. & Anderson, D.M. 1978. Geotechnical Engineering for Cold Regions; McGraw-Hill Book Company, New York, 566pp.

Andersland, O.B. & Ladanyi, B. 1994. An Introduction to Frozen Ground Engineering; Chapman & Hall, New York, 352pp.

Andreassen, K., Bertwussen, K.A., Mienert, J., Sognnes, Hl, Henneberg, K. & Langhammer, J. 2000. Multi-component seismic data in gas hydrate investigations, American Association Petroleum Geologists Annual Meeting, 340.

Andreassen, K., Hart, P.E. & Grantz, A. 1995. Seismic studies of a bottom simulating reflection related to gas hydrate beneath the continental margin of the Beaufort Sea, Journal of Geophysical Research 100 (B7), 12,659-12,673.

Andreassen, K., Hart, P.E. & MacKay, M. 1997. Amplitude versus offset modeling of the bottom simulating reflection associated with submarine gas hydrates. Marine Geology 137, 25-40.

Andreassen, K., Hogstad, K. & Berteussen, K.A. 1990. Gas hydrate in the southern Barents Sea, indicated by a shallow seismic anomaly. First Break 8, 235-245.

Andrews, J.E., Brimblecombe, P., Jickells, T.D. & Liss, P.S. 1996. An introduction to environmental chemistry. Blackwell, Oxford.

Aoki, Y., Tamano, T. & Kato, S. 1983. Detailed structure of the Nankai Trough from migrated seismic sections. In: Watkins, J.S. & Drake, C.L. (eds). Studies in Continental Margin Geology. American Association Petroleum Geology Memoir 34, 309-322.

Arato, H., Akai, H., Uchiyama, S., Kudo, T. & Sekiguchi, K. 1996. Origin and significance of a bottom simulating reflector (BSR) in the Choshi Spur Depression of the Offshore Chiba Sedimentary Basin, central Japan. Journal Geological Society Japan 102, 972-982.

Archie, G.E. 1942. The electrical resistivity log as an aid in determining some reservoir characteristics, Journal Petroleum Technology 5, 1-8.

Ashi, J., Segawa, J., Le Pichon, X., Lallemant, S., Kobayashi, K., Hattori, M., Mazzotti, S. & Aoike, K. 1996. Distribution of cold seepage at the Ryuyo Canyon off Tokai: the 1995 KAIKO-Tokai "Shinkai 2000" dives. JAMSTEC Journal Deep Sea Research 12, 159-173.

Ashi, J. & Taira, A. 1993 Thermal structure of the Nankai accretionary prism as inferred from the distribution of gas hydrate BSRs. In: Underwood, M.B. (ed). Thermal Evolution of the Tertiary Shimanto Belt, Southwest Japan: An Example of Ridge-Trench Interaction., Geological Society Ameica Special Paper 273, 137-149.

Ashi, J. & Tokuyama, H. 1997. Cold seepage and gas hydrate BSR in the Nankai Trough. Proceedings of the International Workshop on Gas Hydrate Studies, the second joint Japan-Canada workshop, Science Technology Agency (JISTEC). Geological Survey Japan, AIST, MITI, 256-273.

Ashi, J., Tokuyama, H., Ujiie, Y. & Taira, A. 1999. Heat flow estimation from gas hydrate BSRs in the Nankai Trough: Implications for thermal structures of the Shikoku Basin. AGU 1999 fall meeting, EOS supplement, F929-930.

ASTM. 1998. Standard test method for permeability of granular soils (constant head) D 2434-68. In: American Society for Testing and Materials, *Annual Book of ASTM Standards*, West Conshohocken, PA, 202-206.

Auzend, J.-M., Dickens, J.R., Van de Beuque, S., Exon, N.F., François, C. & Lafoy, Y. 2000. Thinned crust in southwest Pacific may harbor gas hydrate. EOS 81/17, 182, 185.

Baba, K. & Uchida, T. 1998. Newly discovered BSRs in marginal seas of eastern Japan (in Japanese). Abstracts of the 1999 Technical Meeting of the Japanese Association for Petroleum Technology, 81.

Bains, S., Corfield, R.M. & Norris, R.D. 1999. Mechanisms of climate warming at the end of the Paleocene: Science 285, 724-727.

Balkwill, H.R. 1978. Evolution of Sverdrup basin, Arctic Canada: American Association of Petroleum Geologist Bulletin, 62/6, 1004-1028.

Bangs, N.L., Sawyer, D.S. & Golovchenko, X. 1993. Free gas at the base of the gas hydrate zone in the vicinity of the Chile triple junction. Geology 21, 905-908.

Bard, H.B. & Fairbanks, R.G. 1990. U-Th ages obtained by mass spectrometry in corals from Barbados, sea level during the past 130,000 years: Nature 346, 456-458.

Barker, P.F. & Burrell, J. 1977. The opening of Drake Passage: Marine Geology 25, 15-34.

Barker, P.F. & Dalziel, I.W.D. 1983. Progress in geodynamics of the Scotia Arc region: In: Cabre, R. (ed). Geodynamics of the Eastern Pacific region, Caribbean and Scotia arcs. Geodynamics Series 9 AGU Washington D.C., 137-170.

Barker, P.F. 1982. The Cenozoic subduction history of the Pacific margin of the Antarctic Peninsula: ridge crest-trench interaction: Journal Geological Society London 139, 787-801.

Barker, P.F., Camerlenghi, A., Acton, G.D., et al. 1999. Proceedings ODP Initerum Reports 178 [CD-ROM]. Ocean Drilling Program, Texas A&M University, College Station, TX 77845-9547, U.S.A.

Barker, P.F., Kennett, J.P., et al. 1988. Proceedings ODP Scientific Results 113. College Station, TX (Ocean Drilling Program), 942pp.

Baross, J.A., Lilley, M.D. & Gordon, L.I. 1982. Is the CH_4, H_2 and CO venting from submarine systems produced by thermophilic bacteria? Nature 298, 366-368.

Barrer, R.M. & Stuart, W.I. 1957. Non-stoichiometric clathrate compounds of water. Proceedings of the Royal Society London, Series A243, 172-189.

Batzle, M. & Wang, Z. 1992. Seismic properties of pore fluids. Geophysics 57, 1396-1408.

Bauer, J.E., Druffel, E. R. M., Williams, P.M., Wolgast, D. & Griffin., S. 1998. Time dependent variation in the natural radiocarbon content of dissolved organic matter in the eastern north Pacific Ocean. Journal Geophysical Research 103, 2867-2882.

Bauer, J.E., Reimers, C.E., Druffel, R.M. & Williams, P.M. 1995. Isotopic constraints on carbon exchange between deep ocean sediments and seawater. Nature 373, 686-689.

Bauer, J.E., Spies, R.B., Vogel, J.S., Nelson, D.E. & Southon, J.R. 1990. Radiocarbon evidence of fossil-carbon cycling in sediments of a nearshore hydrocarbon seep. Nature 348, 230-232.

Bauer, J.E., Williams, P.M. & Druffel, E.R.M. 1992. Recovery of submilligram quantities of carbon dioxide from gas streams by molecular sieve for subsequent determination of isotopic (^{13}C and ^{14}C) natural abundances. Analytical Chemistry 64, 824-827.

Beck, R.J. 1999. World Petroleum Industry Outlook: PennWell.

Beeder, J., Torsvik, T. & Lien, T.L. 1995. Thermodesulforhabdus norvegicus gen nov, sp nov, a novel thermophilic sulfate-reducing bacterium from oil field water. Archives of Microbiology 164, 331-336.

Behrendt, J.C. (ed). 1983. Petroleum and mineral resources of Antarctica. U.S. Geological Survey Circular 909, 75pp.

BeMent, W.O., Bil, K.J., Drenth, A.J.J., Klomp, U.C., Roodhart, L.P., Sambell, R.H., Swinkels, W.J.A.M., White, D. & Zoons, C.W. 1998. Are There "Show Stoppers" to Commercial Gas Hydrate Production: International Gas Hydrate Symposium, Chiba, Japan, 197-204.

Berge, L.I., Jacobsen, K.A. & Solstad, A. 1999. Measured acoustic wave velocities of R11 (CCl3F) hydrate samples with and without sand as a function of of hydrate concentration, Journal Geophysical Research 104, 15415-15424.

Bethke, C.M., Reed, J.D. & Oltz, D.F. 1991. Long-range petroluem migration in the Illinois Basin. American Association Petroleum Geologists Bulletin 75, 925-945.

Bhattacharya, G.C., Murty, G.P.S., Srinivas, K., Chaubey, A.K., Sudhakar, T. & Nair, R.R. 1994. Swath Bathymetric Investigation of the Seamounts located in the Laxmi Basin, Eastern Arabian Sea, Marine Geodesy 17, 169-182.

Bhattacharya, G.C., Ramprasad, T., Kamesh Raju, K.A., Srinivas, K., Murty, G.P.S., Chaubey, A.K., Ramana, M.V., Subrahmanyam, V., Sarma, K.V.L.N.S., Desa, M., Paropkari, A.L., Menezes, A.A.A., Murthy, V.S.N., Anthony, M.K., Subba Raju, L.V., Desa, E., Veerayya, M., Rastogi, A. & Deka, B. 1997. Gas Hydrate Stability Zone Thickness Map of Indian Deep Offshore Areas - A GIS Based Approach. Report to National Institute of Oceanography and GAIL, 5pp & figs.

Bidle, K. A., Kastner, M. & Bartlett, D. H. 1999. A phylogenetic analysis of microbial communities associated with methane hydrate containing marine fluids and sediments in the Cascadia margin (ODP site 892B). FEMS Microbiology Letters 177, 101-108.

Bily, C. & Dick, J.W.L. 1974. Naturally occurring gas hydrates in the Mackenzie Delta, N.W.T.: Bulletin of Canadian Petroleum Geology 22, 320-352.

Bird, K.J. & Magoon, L.B. 1987. Petroleum geology of the northern part of the Arctic National Wildlife Refuge, Northeastern Alaska: U.S. Geological Survey Bulletin 1778, 324pp.

Biswas, S.K. & Singh, N.K. 1991. Westem India deep Sea Basin, Exploration thrust area, Bulletin ONGC 25, No. I.

Bjorkum, P.A., Walderhaug, O. & Nadeau, P.H. 1998. Physical constraints on hydrocarbon leakage and trapping revisited. Petroleum Geoscience 4, 237-239. Discussion, replies by authors: Clayton, C.J. ibid. 5, 99-101; Rodgers, S. ibid 5, 421-423.

Bjorlykke, K. & Hoeg, K. 1997. Effects of burial diagenesis on stresses, compaction and fluid flow in sedimentary basins. Marine & Petroleum Geology 14, 267-276.

Blotevogel, K.H. & Fischer, U. 1985. Isolation and characterization of a new thermophilic and autotrophic methane producing bacterium - methanobacterium thermoaggregans spec- nov. Archives of Microbiology 142, 218-222.

Blunier, T. 2000. Frozen methane escapes from the sea floor. Science 288, 68-69.

Bochu, Y. 1998. Preliminary exploration of gas hydrate in the northern margin of the South China Sea. Marine Geology & Quaternary Geology 18, 11-18.

Böhm, G., Camerlenghi, A., Lodolo, E. & Vesnaver, A. 1995. Tomographic analysis and geological contex of a bottom simulating reflector on the South Shetland Margin (Antarctic Peninsula): Boll. Geof. Teor. Appl. 37, 3-23.

Boissonnas, R., Goldberg, D. & Saito, S. 2000. Electromagnetic modeling and in situ measurement of gas hydrates in natural marine environments, In: Gas Hydrates and challenges for the future, G. Holder (ed). New York Academy of Science, New York, 912. paper 14/1180, (in press).

Bolton, A. & Maltman, A. 1998. Fluid-flow pathways in actively deforming sediments: the role of pore fluid pressures and volume change. Marine and Petroleum Geology 13, 281-297.

Bondevik, S., Svendsen, J.I., Johsen, G., Mangerud, J. & Kaland, P.E. 1997. The Storegga tsunami along the Norwegian coast, its age and run-up, Boreas 26, 29-53.

Bonham, L.C. 1980. Migration of hydrocarbons in compacting basins. American Association Petroleum Geologists Bulletin 64, 549-567.

Booth, J.S., O'Leary, D.W., Popenoe, P. & Danforth, W.W. 1993, In: Schwab, W.C., Lee, H.L. & Twichell, D.C. Submarine Landslides: Selective Studies in the U.S. Exclusive Zone, U.S. Geological Survey Bulletin 2002, 1-13.

Booth, J.S., Rowe, M.M. & Fischer, K.M. 1996. Offshore gas hydrate sample database with an overview and preliminary analysis. U.S. Geological Survey Open-File Report 96-272, 23pp.

Booth, J.S., Winters, W.J. & Dillon, W.P. 1994. Circumstantial Evidence of Gas Hydrate and Slope Failure Association on the United States Atlantic Continental Margin, In: Natural Gas Hydrates, Sloan, E.D., Happel, J. & Hnatow, M.A., (eds). Annals of the New York Academy of Sciences 715, 487-489.

Booth, J.S., Winters, W.J. & Dillon, W.P. 1999. Apparatus investigates geological aspects of gas hydrates. Oil & Gas Journal 97, 63-70.

Borowski, W.S., Paull, C.K & Ussler III, W. 1996. Marine pore-water sulfate profiles indicate in situ methane flux from underlying gas hydrate. Geology 24, 655-658.

Borowski, W.S., Paull, C.K, Ussler III, W. 1997. Carbon cycling within the upper methanogenic zone of continental rise sediments: An example from the methane-rich sediments overlying the Blake Ridge gas hydrate deposits. Marine Chemistry 57, 299-311.

Borowski, W.S., Paull, C.K. & Ussler, III, W. 1999. Global and local variations of interstitial sulfater gradients in deep-water, continental margin sediments: Sensitivity to underlying methane and gas hydrates. Marine Geology 159, 131-154.

Boudreau, B.P. 1997. Diagenetic models and their implementation. Springer-Verlag, Berlin, 414 pp.

Bouriak, S., Vanneste, M. & Saoutkine, A. 2000. Inferred gas hydrates and clay diapirs near the Storegga Slide on the southern edge of the Vøring Plateau, off Norway, Marine Geology 163, 125-148.

Bralower, T.J., Zachos, J.C., Thomas, E., Parrow, M., Paull, C.K., Kelly, D.C., Premoli-Silva, I., Sliter, W.V. & Lohmann, K.C. 1995. Late Paleocene to Eocene paleoceanography of the equatorial Pacific Ocean. ODP Site 865, Allison Guyot: Paleoceanography 10, 841-865.

Brereton, G.J., Crilly, R.J. & Spears, J.R. 1998. Nucleation in small capillary tubes. Chemical Physics 230, 253-265.

Brewer, P.G., Orr, F.M., Friederich, G., Kvenvolden, K.A., & Orange, D.A. 1998. Gas hydrate formation in the deep sea: in situ experiments with controlled release of methane, natural gas, and carbon dioxide. Energy & Fuels 12, 183-188.

Briaud, J.-L. & Chaouch, A. 1997. Hydrate melting in soil around hot conductor. Journal of Geotechnical and Geoenvironmental Engineering 123, 645-653.

Brooks, J.M., Kennicutt, M.C., Fisher, C.R., Macko, S.A., Cole, K., Childress, J.J, Bidigare, R.R. & Vetter, R.D. 1987. Deep-Sea Hydrocarbon seep communities: evidence for energy and nutritional carbon sources. Science 238, 1138-1141.

Brooks, J.M., Cox, B.H., Bryant, W.R., Kennicutt, M.C., Mann, R.G., McDonald, T.J. 1986. Association of gas hydrate and oil seepage in the Gulf of Mexico: Organic Geochemistry 10, 221-234.

Brooks, J.M., Field, M.E. & Kennicutt, M.C. 1991. Observations of gas hydrate in marine sediments, offshore northern California: Marine Geology 96, 103-109.

Brooks, J.M., Kennicutt II, M.C., Fay, R.R. & McDonald, T.J. 1984. Thermogenic gas hydrates in the Gulf of Mexico. Science 225, 409-411.

Brooks, J.M., Kennicutt, M.C., III, Fay, R.R., McDonald, T.J. & Sassen, R. 1984. Thermogenic gas hydrates in the Gulf of Mexico: Science 225, 409-411.

Brown, J F. 1962 Inclusion compounds. Scientific American 207, 82-92.

Brown, K.M. 1994. Fluids in Deforming sediments; Chapter 7. In: Maltman, A.J. (ed). Geological Deformation of Sediments. Chapman and Hall, London, 362pp.

Brown, K.M., Bangs, N.L., Froelich P.N. & Kvenvolden, K.A. 1996. The nature, distribution, and origin of gas hydrate in the Chile Triple junction region. Earth & Planetary Science Letters 139, 471-483.

Bryan, K.M. & Markl, R.G. 1966. Microtopography of the Blake-Bahama Region. Columbia University: Lamont Geological Observatory Technical Report 8 (CU-8-66 NOpBSR), 26pp.

Buffet, B.A. 2000. Clathrate hydrates. Annual Reviews of Geophysics (in press).

Buffett, B.A. & Zatsepina, O.Ye. 1999. Metastability of gas hydrate. Geophysical Research Letters 26, 2981-2984.

Bugge, T. 1983. Submarine slides on the Norwegian continental margin with special emphasis on the Storegga area, Continental Shelf Norway. IKU, Trondheim, Norway Publication 10, Section 5.7.3.

Bugge, T., Befring, S., Belderson, R.H., Eidvin, T., Jansen, E., Kenyon, N.H., Holtedahl, H. & Sejrup, H.P. 1987. A Giant three-stage submarine slide off Norway. Geo-Marine-Letters 7, 191-198.

Bugge, T., Belderson, R.H. & Kenyon, N.H. 1988. The Storegga slide, Phil. Transactions of the Royal Society of London 325, 357-388.

Caiti, A., Akal, T. & Stoll, R.D. 1991. Determination of shear-velocity profiles by inversion of interface wave data, in Hovem, J.M., Richardson, M.D. & Stoll, R.D. (eds). Proceedings of the conference on shear waves in marine sediments, La Spezia, Italy 15-19 October 1990: Dordrecht, Kluwer, 557-566.

Camerlenghi, A.& Lodolo, E. 1994. Bottom simulating reflector on the south Shetland margin (Antarctic Peninsula) and implications for the presence of gas hydrates. Terra Antarctica 1, 154-157.

Cameron, P.J. 1981. The petroleum potential of Antarctica and its continental margins: APEA Journal 21, 99-111.

Cande, S.G. & Mutter, J.C. 1982. A revised identification of the oldest seafloor spreading anomalies between Australia and Antarctica: Earth and Planetary Science Letters 58, 151-160.

Capone, D.G. & Kiene, R.P. 1988. Comparison of microbial dynamics in marine and freshwater sediments: Contrasts in anaerobic carbon catabolism. Limnology and Oceanography 33, 725-49.

Cappenberg, T. E. and Prins, R. A. 1974. Interrelationships between sulfate-reducing and methane-producing bacteria in bottom deposits of a freshwater lake. III. Experiments with 14C-labelled substrates. Antonie Leeuwenhoek Journal Microbiology Serology 40, 457-469.

Carpenter, G. 1981. Coincident sediment slump/clathrate complexes on the U.S. Atlantic continental slope. Geo-Marine Letters 1, 29-32.

Carrion, P., Boehm, G., Marchetti, A., Pettenati, F. & Vesnaver, A. 1993. Reconstruction of lateral gradients from reflection tomography: J. Seismic Exploration, 2, 55-67.

Carson, B., . Holmes, M.K., Umstattd, K., Strasser J.C. & Johnson H.P. 1991. Fluid expulsion from the Cascadia accretionary prism; evidence form porosity distribution, direct measurements, and GLORIA imagery. Philosophical Transactions Royal Society London A335, 331-340.

Cashman, K.V. & Popenoe, P. 1985. Slumping and shallow faulting related to the presence of salt on the Continental Slope and Rise off North Carolina. Marine and Petroleum Geology 2, 260-272.

Castagna, J.P., Batzle, M.L. & Eastwood, R.L. 1985. Relationships between compressional-wave and shear-wave velocities in clastic silicate rocks. Geophysics 50, 571-581.

Cha, S.B., Ouar, H., Wildeman, T.R. & Sloan, E.D. 1988. A third-surface effect on hydrate formation. Journal of Physical Chemistry, 6492-6494.

Chaker, V. 1996. Measuring resistivity choosing the right method: ASTM Standardization News, April, 30-33.

Chandra, K., Singh, R.P. & Julka, A.C. 1998. Gas hydrate potential of Indian Offshore Area. In: 2nd conference and exposition on petroleum geophsics SPG-98, Chennai (India), 357-368.

Chanton, J.P., Martens, C.S., Paull, C.K., Coston, J.A. 1993. Sulfur isotope and porewater geochemistry of Florida escarpment seep sediments. Geochim. Cosmochim. Acta. 57, 1253-1266.

Chapelle, F.H. & Bradley, P.M. 1996. Microbial acetogenesis as a source of organic acids in ancient Atlantic coastal plain sediments. Geology 24, 925-928.

Chapman, N.R., Walia R., Gettrust J., Wood W., Hannay D., Spence G.D., Hyndman R.D., MacDonald R. & Rosenberger A. 2000. High resolution deep-towed multichannel seismic survey of deep sea gas hydrates off western Canada: Geophysics (submitted).

Chapman, R.E. 1983. Petroleum Geology. 2nd Edition. Elsevier. 413 pp.

Chasar, L.S., Chanton, J.P., Glaser, P.H., Siegel, D.I. & Rivers, J.S. Radiocarbon and stable carbon isotopic evidence for transport and transformation of DOC, DIC and CH_4 in a northern Minnesota peatland. Global Biogeochemical Cycles (in press).

Cherkis, N.Z., Max, M.D., Vogt, P.R., Crane, K., Midthassel, A. & Sundvor, E. 1999. Large-scale mass wasting on the north Spitsbergen continental margin, Arctic Ocean in: Gardiner, J., Vogt, P. & Crane, K. (eds). Mass wasting in the Arctic. Geomarine Letters Special Issue 19, 131-142.

Cherrier, J., Bauer, J.E., Druffel, E.R.M., Coffin, R.B. & Chanton, J.P. 1999. Radiocarbon in Marine Bacteria: Evidence for the Age of Assimilated Organic Matter. Limnology and Oceanography 44/3, 730-736.

Cherskiy, N. & Makogon, Yu. F. 1970. Solid gas-World reserves are enormous: Oil and Gas International 10/8, 82-84.

Cherskiy, N.V., Tsarev, V.P. & Nikitin, S.P. 1985. Investigation and prediction of conditions of accumulation of gas resources in gas-hydrate pools: Petroleum Geology 21, 65-89.

Chi, W.C., Reed, D.L., Liu, C.-S. & Lundberg, N. 1998. Distribution of the bottom simulating reflector in the offshore Taiwan collision area. Terrestrial, Atmospheric and Ocean Sciences 9, 779-794.

Childress, J.J., Fisher, C.R., Brooks, J.M., Kennicutt II, M.C,. Bidigare & Anderson, A. 1986. A methanotrophic marine mulluscan (Bivalvia, Mytilidae) symbiosis: mussels fueled by gas, Science 233, 1306-1308.

Chopra, N.N. 1985. Gas hydrate—an unconventional trap in forearc regions of Andaman offshore. ONGC Bulletin 22 /1, 41-54.

Chou, I. M., Burress, R., Goncharov, A., Sharma, A., Hemley, R., Stern, L. & Kirby, S. 2000. In situ observations of a new high-pressure methane hydrate phase. EOS, (in press).

Chou, I. M., Pasteris, J. & Seitz, J. 1990. High density volatiles in the system C-O-H-N for the calibration of a laser Raman microprobe. Geochimica Cosmochimica Acta 54, 535-543.

Chou, I.M., Pasteris, J. & Seitz, J. 1990. High density volatiles in the system C-O-H-N for the calibration of a laser Raman microprobe. Geochimica Cosmochimica Acta 54, 535-543.

Christeson, G.L., Purdy, G.M. & Fryer, G.J. 1993. Seismic constraints on shallow crustal emplacement processes at the fast-spreading East Pacific Rise. Journal Geophysical Research.

Circone, S., Stern, L.A., Kirby, S.H., Pinkston, J.C. & Durham, W.B. 2000. Pure methane hydrate dissociation rates at 0.1 MPa and temperatures above 272 K, Annals of the New York Academy of Sciences, 3rd International Conference on Gas Hydrates (in press), 20pp.

Claypool, G.E. & Matava, T.R. 1999. Limits on gas saturation beneath methane gas hdyrates. EOS 80/46 supplement, F503.

Claypool, G.E. 1996. Influence of water solubility, phase equilibria and capillary pressure on methane occurrence in sediments (abstract), American Association Petroleum Geologists Annual Convention, San Diego, Program Abstracts, A27, Tulsa Oklahoma.

Claypool, G.E., Presley, B.J. & Kaplan, I.R. 1973. Gas analysis in sediment samples from Legs 10, 11, 13, 14, 15, 18, and 19, In: Creager, J.S., Scholl, D.W., et al., Initial Reports of the Deep Sea Drilling Project 19. U.S. Government Printing Office, 879-884.

Claypool, G.W. & Kaplan, I.R. 1974. Methane in Marine Sediments. In: Kaplan, I.R., (ed). Gas in Marine Sediments. Plenum, 73-99.

Claypool, G.W. & Kvenvolden, K.A. 1983. Methane and other hydrocarbon gases in marine sediments Ann. Rev. Earth Planetary Science 11, 299-327.

Clayton, C.J. & Hay, S.J. 1994. Gas migration mechanisms from accumulation to surface. Bulletin of the Geological Society of Denmark 41, 12-23.

Clayton, C.J. 1999. Discussion on Bjorkum, P.A. 1998 "Physical constraints on hydrocarbon leakage and trapping" Petroleum Geoscience.

Clennell, M.B., Hovland, M., Booth, J.S., Henry, P. & Winters, W.J. 1999. Formation of natural gas hydrates in marine sediments 1. Conceptual model of gas hydrate growth conditioned by host sediment properties. Journal of Geophysical Research 104, 22,985-23,004.

Cochener, J. (Gas Research Institute) 1999. Drilling into the 21st Century: International Association of Drilling Contractors Annual Meeting.

Coffin, R. Fry, B.B., Peterson, B.J. & Wright, R.T. 1989. Carbon isotopic compositions of estuarine bacteria. Limnology Oceanography 34, 1305-1310.

Coffin, R.B. & Cifuentes, L.A. 1999. Analysis of microbial carbon cycling with stable isotope analysis in the Perdido Estuary, FL. Estuaries 22, 997-1006.

Coffin, R.B. & Cifuentes, L.A. 1993. Approaches for measuring stable carbon and nitrogen isotopes in bacteria,. In: Kemp, P.F, Sherr, B.F., Sherr, E.B. & Cole, J.J. (eds). Current Methods in Aquatic Microbial Ecology. Lewis Publishers, Boca Raton, FL, 663-676.

Coffin, R.B., Cifuentes, L.A. & Pritchard, P.H. 1997. Effect of remedial nitrogen applications on algae and heterotrophic organisms on oil contaminated beaches in Prince William Sound, AK. Marine Environmental Research 1, 27-39.

Coffin, R.B., Velinsky, D., Devereux, R., Price, W.A. & Cifuentes, L.A. 1990. Stable carbon isotope analysis of nucleic acids to trace sources of dissolved substrates used by estuarine bacteria. Applied Environmental Microbiology 56, 2012-2020.

Collen, J.D. & Barrett, P.J. 1990. Petroleum geology from the CIROS-1 Drill Hole, McMurdo Sound: implications for the potential of the Victoria Land, Antarctica: In St.John, B. (ed). Antarctica as an exploration frontier. American Association Petroleum Geologists Studies in Geology, Tulsa, Oklahoma 31, 143-151.

Collett, T.S. 1993. Natural gas hydrates of the Prudhoe Bay and Kuparuk River area, North Slope, Alaska: American Association of Petroleum Geologists Bulletin 77/5, 793-812.

Collett, T.S. 1993. Natural gas production from Arctic gas hydrates, *in* Howell, D.G., ed, The Future of Energy Gases: U.S. Geological Survey Professional Paper 1570, 299-311.

Collett, T.S. 1995. Gas hydrate resources of the United States, in Gautier, D.L., Dolton, G.L., Takahashi, K.I., & Varnes, K.L. (eds). 1995 National assessment of United States oil and gas resources on CD-ROM: U.S. Geological Survey Digital Data Series 30.

Collett, T.S. 1996. Geologic assessment of the natural gas hydrate resources in the onshore and offshore regions of the United States. Proceedings 2nd International Conference on Natural Gas Hydrates, 499-506.

Collett, T.S. 1998. Well log evaluation of gas hydrate saturations: Transactions of the 39th Annual Symposium of the Society of Professional Well Log Analysts, May 26-29, Keystone Resort, Colorado, Paper MM.

Collett, T.S. 1998. Methane hydrate: An unlimited energy resource? In: JNOC-TRC, proceedings of the International Symposium on Methane Hydrates, resources in the near Future?, Chiba City, Japan, October 20-22, 1998, 1-13.

Collett, T.S., Bird, K.J. & Magoon, L.B. 1993. Subsurface temperatures and geothermal gradients on the North Slope of Alaska: Cold Regions Science and Technology 21, 275-293.

Collett, T.S. & Ginsburg, G.D. 1997. Gas Hydrates in the Messoyakha Gas Field of the West Siberian Basin - A Re-Examination of the Geologic Evidence: 7th ISOPE et al., International Offshore & Polar Conference Honolulu, Proceedings 1, 96-103.

Collett, T.S. & Ginsburg, G.D. 1998. Gas hydrate in the Messoyakha gas field of the West Siberian Basin - a re-examination of the geologic evidence: International Journal of Offshore and Polar Engineering 8/1, 22-29.

Collett, T.S., Godbole, S.P. & Economides, C.E. 1984. Quantification of in-situ gas hydrates with well logs: Proceedings, Annual Tech. Meeting Petroleumj Society CIM 35, 571-582.

Collett, T.S. & Ladd, John. 2000. Detection of gas hydrate with downhole logs and assessment of gas hydrate concentrations (saturations) and gas volumes on the Blake Ridge with electrical resistivity log data, In:

Paull, C.K., Matsumoto, R., Wallace, P.J., & Dillon, W.P. (eds). Proceedings of the Ocean Drilling Program, Scientific Results 164, 179-191.

Collett, T.S., Lee, M.W., Dallimore, S.R. & Agena, W.F. 1999. Seismic- and well-log-inferred gas hydrate accumulations on Richards Island. In: Dallimore, S.R., Uchida, T., & Collett, T.S. (eds). Scientific results from JAPEX/JNOC/GSC Mallik 2L-38 gas hydrate research well, Mackenzie Delta, Northwest Territories, Canada: Geological Survey of Canada Bulletin 544, 357-376.

Collett, T.S., Lewis, R.E., Dallimore, S.R., Lee, M.W., Mroz, T.H. & Uchida, T. 1999. Detailed evaluation of gas hydrate reservoir properties using JAPEX/JNOC/GSC Mallik 2L-38 gas hydrate research well downhole well-log displays. In: Dallimore, S.R., Uchida, T. & Collett, T.S. (eds). Scientific Results from JAPEX/JNOC/GSC Mallik 2L-38 Gas Hydrate Research Well, Mackenzie Delta, Northwest Territories, Canada. Geological Survey of Canada Bulletin 544, 295-312.

Committee Journal of International Gas Union. 1997. Development prospects for gas supply and demand worldwide 2000 - 2030: 20th World Gas Congress, Copenhagen.

Cook, J.G. & Laubitz, M.J. 1981. The thermal conductivity of two clathrate hydrates, Proceedings 17th Internation Thermal Conductivity Conference, Gaithersburg, MD.

Cook, R.A. & Davey, .J.F. 1990. Hydrocarbon exploration and potential: In Glasby, G.P. (Ed.) Antarctic sector of the Pacific, Elsevier Oceanography Series 51, 155-185.

Cooper, A.K., Barker, P.F. & Brancolini, G. (eds). 1995. Geology and seismic stratigraphy of the Antarctic margin: Geology and Seismic Stratigraphy of the Antarctic Margin, Antarctic Research Series 68, AGU, Washington, D.C. 301pp.

Cooper, A.K., Barrett, P.J., Hinz, K., Traube, V., Leitchenkov, G. & stagg, H.M.J. 1991. Cenozoic prograding sequences of the Antarctic continental margin: a record of glacio-eustatic and tectonic events: Marine Geology 102, 175-213.

Cooper, M.C., Selley, R.C. & Cartwright, J.A. 1998. Vertical gas migration mechanisms in the Central North Sea, studied with ultra high resolution digital 2D and 3D seismic data. In: Curzi, P.V. & Judd, A.G. (eds). Proceedings, 5th International Conference on gas in marine sediments, Bologna, 91-12 Sept 1998, 163-165.

Coren, F., Ceccone, G., Lodolo, E., Zitellini, N., Bonazzi, C. & Centonze, J. 1997. Morphology, seismic structure and tectonic development of the Powell Basin (Antarctica): Journal Geological Society London 154, 849-862.

Corey, A.T. 1994. Mechanics of immiscible fluids in porous media. Water resources Publications. 252pp.

Cox, J.L. 1983. Natural Gas Hydrates: Properties, Occurrence and Recovery, Butterworth Publishers, Boston, 1983.

Cragg, B. A., Parkes, R. J., Fry, J. C., Herbert, R. A., Wimpenny, J. W. T. and Getliff, J. M. 1990. Bacterial biomass and activity profiles within deep sediment layers. Proc. Ocean Drilling Program Scientific Results 112, 607-619.

Cragg, B. A., Parkes, R. J., Fry, J. C., Weightman, A. J., Rochelle, P. A., Maxwell, J. R., Kastner, M., Hovland, M., Whiticar, M. J. & Sample, J. C. 1995. The impact of fluid and gas venting on bacterial populations and processes in sediments from the Cascadia Margin accretionary system (sites 888-892) and the geochemical consequences. Proceedings of the Ocean Drilling Program, Scientific Results 146, 399-411.

Cragg, B.A., Parkes, R.J., Fry, J.C., Weightman, A.J., Rochelle, P.A. & Maxwell, J.R. 1996. Bacterial populations and processes in sediments containing gas hydrates (ODP Leg 146: Cascadia Margin). Earth and Planetary Science Letters 139, 497-507.

Cram, D.J. 1992. Molecular container compounds. Nature, 356, 29-36.

Cranston, R.E. 1996. Marine sediments as a source of atmospheric methane. Abstracts of Conference on Gas hydrates: Relevance to world margin stability and climatic change, September, 1996, 2.

Crill, P. M., Harriss, R. C. & Bartlett, K. B. 1991. Methane fluxes from Terrestrial Wetland Environments. In: Rogers, J. E. & Whitman, W. B. (eds). Microbial Production and consumption of Greenhouse Gases: Methane, Nitrogen Oxides and Halomethanes. American Society for Microbiology, Washington DC, 91-109.

Curray, J.R. 1994. Sediment volume and mass beneath the Bay of Bengal. Earth and Planetary Sciences Letters 125, 371-:383.

Curray, J.R. 1991. Possible greenshicst metamorphism at the base of a 22 km sedimentary section, Bay of Bengal. Geology 19, 1097-1100.

Dallimore, S.R. & Collett, T.S. 1995. Intrapermafrost gas hydrates from a deep core hole in the Mackenzie Delta, Northwest Territories, Canada. Geology 23/6, 527-530.

Dallimore, S.R. & Matthews, J.V. 1997. The Mackenzie Delta Borehole Project. Environmental Studies Research Funds Report 135, 1 CD ROM.

Dallimore, S.R., Uchida, T. & Collett, T.S. 1999. Summary. In: Dallimore, S.R., Uchida, T. & Collett, T.S. (eds). Scientific Results from JAPEX/JNOC/GSC Mallik 2L-38 Gas Hydrate Research Well, Mackenzie Delta, Northwest Territories, Canada. Geological Survey of Canada Bulletin, 544 (403pp), 1-10.

Darby, D.A., Naidu, A.S., Mowatt, T.C. & Jones, G. 1989. 24: Sediment composition and sedimentary processes in the Arctic Ocean. In: Herman, Y. The Arctic Seas. Climatology, Oceanography, Geology, and Biology. Van Nostrand Reinhold Co. New York, 657-720.

Davey, F.J. 1985. The Antarctic margin and its possible hydrocarbon potential: Tectonophysics 114, 443-470.

Davidson, D. W., Garg, S. K., Gough, S. R., Handa, Y. P., Ratcliffe, C. I., Ripmeester, J. A. & Tse, H. S. 1986. Laboratory analysis of a naturally occurring gas hydrate from sediment of the Gulf of Mexico. Geochim. Cosmochim. Acta 50, 619-623.

Davidson, D.W., El-Defrawy, M.K., Fuglem, M.O. & Judge, A.S. 1978. Natural gas hydrates in northern Canada, in Proceedings of the 3rd International Conference on Permafrost 1, National Research Council of Canada, 938-943.

Davidson, DW., Garg, S. K., Gough, S.R., Handa, Y.P., Ratcliffe, C.I., Ripmeester, J.A. & Tse, H.S. 1986. Laboratory analysis of a naturally occurring gas hydrate from sediment of the Gulf of Mexico. Geochim. Cosmochimica Acta 50, 619-623.

Davis, A.M., Huws, D.G. & Haynes, R. 1996. Geophysical ground-truthing experiments in Eckernfoerde Bay. Geo-Marine Letters 16, 160-166.

Davis, E.E., and Hyndman, R.D. 1989. Accretion and recent deformation of sediments along the northern Cascadia subduction zone: Geological Society America Bulletin 101, 1465-1480.

Davis, E.E., Hyndman, R.D. & Villinger, H. 1990. Rates of fluid expulsion across the Northern Cascadia accretionary prism: constraints from new heat flow and multichannel seismic reflection data. Journal of Geophysical Research 95, 8869-8889.

Davis, E.E., Villinger, H., MacDonald, R.D., Meldrum, R.D. & Grigel, J. 1997. A robust rapid-response probe for measuring bottom-hole temperatures in deep-ocean boreholes. Marine Geophysical Researches 19, 267-281.

Davy, H. 1811. On a combination of oxymuriatic gas and oxygen gas. Philosophical Transactions of the Royal Society 101, 155.

Dawson, A., Long, D. & Smith, D.E. 1988. The Storegga Slides: evidence from eastern Scotland for a possible tsunami. Marine Geology 82, 271-276.

de Boer, R.B., Houbolt, J.J.H.C. & Lagrand, J. 1985. Formation of gas hydrates in a permeable medium. Geologie en Mijnbouw 64, 245-249.

de Graaf, W., Wellsbury, P., Parkes, R.J. & Cappenberg, T.E. 1996. Comparison of acetate turnover in methanogenic and sulfate-reducing sediments by radio- and stable-isotope-labeling and specific inhibitors: evidence for isotopic exchange. Applied and Environmental Microbiology 62, 772-777.

de Roo, J.L., Peters, C.J., Lichtenthaler, R.N. & Diepen, G.A.N. 1983. Occurrence of methane hydrate in saturated and unsaturated solutions of sodium chloride and water in dependence of temperature and pressure. American Institute of Chemical Engineers Journal 29, 641-657.

de Wit, M.J. 1977. The evolution of the Scotia Arc as a key to the reconstruction of the southwestern Gondwanaland: Tectonophysics 37, 53-81.

Dehler, S. & Clowes, R.M. 1992. Integrated geophysical modelling of terranes and other structural features along the western Canadian margin: Canadian Journal Earth Sciences 29, 1492-1508.

DeMartin, B., Waite, W., Ruppel, C., Pinkston, J., Stern, L. & Kirby, S. 1999. Laboratory thermal conductivity measurements of methane hydrate and hydrate-sediment mixtures under simulated *in situ* conditions. EOS 80/17,. S337.

Dewhurst, D.N., Yang, Y. & Aplin, A.C. 1999. Permeability and fluid flow in natural mudstones. In: Aplin, A.C., Fleet, A.J. & Macquaker, J.H.S. (eds). Muds and Mudstones: Physical and Fluid Flow Properties. Geological Society London Special Publication 158, 23-43.

Dickens, G.R., Castillo, M.M. & Walker, J.C.G. 1997. A blast of gas in the latest Paleocene: simulating first-order effects of massive dissociation of oceanic methane hydrate. Geology 25, 259-262.

Dickens, G.R., O'Neil, J.R., Rea, D.K. & Owen, R.M. 1995. Dissociation of oceanic methane hydrate as a cause of the carbon isotope excursion at the end of the Paleocene: Paleoceanography 10, 965-971.

Dickens, G.R., Paull, C.K., Wallace, P. & ODP Leg 164 Scientific Party, 1997. Direct measurement of in situ methane quantities in a large gas-hydrate reservoir: Nature 385, 426-428.

Dickens, G.R. & Quinby-Hunt, M.S. 1994. Methane hydrate stability in seawater. Geophysical Research Letters 21, 2115-2118.

Dickens, G.R. & Quinby-Hunt, M.S. 1997. Methane hydrate stability in pore water: A simple theoretical approach for geophysical applications. Journal of Geophysical Research 102, 773-783.

Dillon, W.P., Austin, J.A., Jr., Scanlon, K.M., Edgar, N.T. & Parson, L.M. 1992. Accretionary margin of north-western Hispaniola: morphology, structure and development of part of the northern Caribbean plate boundary, Marine and Petroleum Geology 9/1, 70-88.

Dillon, W.P., Danforth, W.W., Huthchinson, D.R., Drury, R.M., Taylor, M.H. & Booth, J.S. 1998. Evidence for faulting related to dissociation of gas hydrate and release of methane off the southeastern Untied States. In: Henrtiet, J.-P. & Mienert, J., (eds). Gas Hydrates: Relevance to World Margin Stability and Climate Change. Geological Society London, Special Publication 137, 293-302.

Dillon, W.P. & Drury, R.M. 1996. Seismic reflection profiles on the Blake Outer Ridge near drillsites 994, 995, and 997, Ocean Drilling Program, Initial Reports 164, 47-53.

Dillon, W.P., Lee, M.Y. & Coleman, D.F. 1994. Identification of marine hydrates in situ and their distribution off the Atlantic coast of the United States, in Sloan, E.D., Jr., Happel, John, and Hnatow, M.A. (eds). International Conference on Natural Gas Hydrates, Annals of the New York Academy of Sciences 715, 364-380.

Dillon, W.P., Lee, M.W., Felhaber, K. & Coleman, D.F. 1993. Gas hydrates on the Atlantic continental margin of the united Staes — controls on concentration, (in) Howell, D.G. (ed). The Fuiture of Energy Gaes, U.S. Geological Survey Professional Paper 1570, 313-330.

Dillon, W.P., Fehlhaber, K., Coleman, D.F., Lee, M.W. & Hutchinson, D.R. 1995. Maps showing gas-hydrate distribution off the east coast of the United States: US Geological Survey Miscellaneous Field Studies Map (Scale 1:1,000,000), MF-2268, 2 sheets.

Dillon, W.P. & Max, M.D. 1999. Seismic reflections identify finite differences in gas hydrate resources, (November 1999) Offshore 59/11, 115-116, 148.

Dillon, W.P., Nealon, J.W., Taylor, M.H., Lee, M.W., Drury, R.M. & Anton, C.H. Seafloor collapse and methane venting associated with gas hydrate on the Blake Ridge — causes and implications to seafloor stability. In: Paull, C.K & Dillon, W.P. (eds). Natural Gas Hydrates: Occurrence, Distribution, and Dynamics. AGU monograph. 29pp., 15 figs. (in press).

Dillon, W.P. & Paull, C.K. 1983. Marine gas hydrate, II. Geophysical evidence, in: Cox, J.S. (ed). Natural Gas Hydrate: Properties, Occurrences, and Recovery, Butterworth Publishing, London, England, 73-90.

Dillon, W.P. & Popenoe, P. 1988. The Blake Plateau and Carolina Trough, In: Sheridan R.E. & Grow, J.A. (eds). The Atlantic Continental Margin, U.S. Geology of North America I-2, 291-328.

Dillon, W.P., Popenoe, P., Grow, J.A., Klitgord, K.D., Swift, B.A., Paull, C.K. & Cashman, K.V. 1983. Growth faulting and diapirism: Their relationship and control in the Carolina Trough, eastern North America, In: Watkins, J.S. & Drake, C.L. (eds). Studies in Continental Margin Geology, American Association of Petroleum Geologists, Memoir 34, 21-46.

Dixon, J. & Dietrich, J.R. 1990. Canadian Beaufort Sea and adjacent land areas, in Grantz, A., Johnson, L. & Sweeney, J.F. (eds). The Geology of North America, The Arctic Ocean region. Geological Society of America, Boulder, Colorado, L, 239-255.

Dobrynin, V.M., Korotajev, Yu. P. & Plyuschev, D.V. 1981. Gas hydrate--A possible energy resource, in Meyer, R.F., and Olson, J.C. (eds). Long-Term Energy Resources, Pitman Publishing, Boston, Massachusetts, 727-729.

DoE. 1987. Deep Source Gas Technology Status Report. U.S. Department of Energy Report DOE/METC-87/0250, 18pp.

DOE. 1987. Gas Hydrates Technology Status Report. U.S. Department of Energy, Office of Fossil Energy, Morgantown, West Virginia. DOE/METC-87/0246 (DOE)87001027), 25pp.

Domack, E.W., Jull, A. J.T., Anderson, J.B. & Linick, T.W. & Williams, C.R. 1989. Application of tandem accelerator mass-spectrometer dating to Pleistocene-Holocene sediments of the East Antarctic continental shelf: Quaternary Research 31, 277-287.

Domack, E.W., Jull, A. J.T., Anderson, J.B. & Linick, T.W. 1991. Mid-Holocene glacial recession from the Wilkes land continental shelf, East Antarctica: In Thomson, M.R.A., Crame J.A. & Thomson J. W. (eds). Geological Evolution of Antarctica, Cambridge, England, Cambridge University Press, 693-698.

Domenico, S.N. 1976. Effect of brine-gas mixture on velocity in an unconsolidated sand reservoir. Geophysics 41, 882-894.

Domenico, S.N. 1977. Elastic properties of unconsolidated porous sand reservoirs, Geophysics 42, 1339-1368.

Donaldson, K.H., Istok, J.D., Humphrey, M.D., O'Reilly, K.T., Hawleka, C.A. & Mohr, D.H. 1997. Development and testing of a kinetic model for oxygen transport in porous media in the presence of trapped gas. Ground Water 35, 270-279.

Doré, A., Lundin, E.R., Birkeland, O., Eliassen, P.E. & Jensen, L.N. 1997. The NE Atlantic margin: implications of late Mesozoic and Cenozoic events for hydrocarbon prospectivity, Petroleum Geoscience, 3, 117-131.

Doveton, J.H. 1986. Log Analysis of Subsurface Geology: Concepts and Computer Methods, Wiley, New York.

Dow, W.C., 1978. Petroleum source beds on continental slopes and rises. American Association Petroleum Geologists Bulletin 62. 1548-1606.

Downey, M.W. 1984. Evaluating seals for hydrocarbon accumulations. American Association Petroleum Geologists Bulletin 68, 1752-1763.

Driscoll, N.W., Weissel, J.K. & Goff, J.A. Potential for large-scale submarine slope failure and tsunami generation along the U.S. mid-Atlantic coast. Geology 28, 407-410.

Druffel, E.R.M., Williams, P.M., Bauer, J.E. & Ertel, J.R. 1992. Cycling of dissolved and particulate organic matter in the open ocean. Journal Geophysical Research 97, 15639-15659.

Duan, Z., Moller, N, Greenberg, J. & Weare, J.H. 1992. The prediction of methane solubility in natural waters to high ionic strength from to 250C and from 0 to 1600 bar. Geochmica et Cosmochimica Acta 56, 1451-1460; comment by Carroll, J.J., ibid. 56, 4301-4302.

Duan, Z., Moller, N. & Weare, J.H. 1992. An equation of state for the CH_4-CO_2-H_2O system: I. pure systems from 0 to 1000°C and 0 to 8000 bar; Geochimica et Cosmochimica Acta, 56, 2605-2617.

Dunbar, R.B., Anderson, J.B., Domack, E.W. & Jacobs, S.S. 1985. Oceanographic influences on sedimentation along the Antarctic continental shelf: American Geophysical Union. Antarctic Research Series 43, 291-312.

Durham, W.B. Kirby, S.H. & Stern, L.A. 1993. Flow of ices in the ammonia-water system. Journal of Geophysical Research 98/B10, 17,667-17,682.

Durham, W.B., Heard, H.C. & Kirby, S.H. 1983. Experimental deformation of polycrystalline H_2O ice at high pressure and low temperature: preliminary results. Journal Geophysical Research 88, B377-B392.

Durham, W.B., Mirkovich, V.V. & Heard, H.C. 1987. Thermal diffusivity of igneous rocks at elevated pressure and temperature. Journal of Geophysical Research 92, 11,615-11,634.

Dvorkin, J. & Nur, A. 1993. Rock physics for characterization of gas hydrates, *in* Howell, D.G. (ed). The Future of Energy Gases: U.S. Geological Survey Professional Paper 1570, 293-298.

Dvorkin, J. & Nur, A. 1996. Elasticity of high-porosity sandstones: theory for two North Sea data sets, Geophysic, 61, 1363-1370.

Dvorkin, J., Berryman, J. & Nur, A. 1999. Elastic moduli of cemented sphere packs, Mechanics of Materials 31, 461-469.

Dvorkin, J., Nur, A. & Yin, H. 1994. Effective properties of cemented granular material, Mechanics of Materials, 18, 351-366.

Dvorkin, J., Prasad, M., Sakai, A. & Lavoie, D. 1999. Elasticity of marine sediments, Geophysical Research Letters 26, 1781-1784.

Eidvin, T., Jansen, E. & Riis, F. 1993. Chronology of Tertiary fan deposits off the western Barents Sea: Implications for the uplift and erosion history of the Barents shelf. Marine Geology 112, 109-131.

Ecker, C., Dvorkin, J. & Nur, A. 1998. Estimating the amount of hydrate and free gas from surface seismic, SEG-98 Annual Meeting, New Orleans, Extended Abstracts, 566-569.

Ecker, C., Dvorkin, J. & Nur, A. 2000. Estimating the amount of hydrate and free gas from surface seismic. Geophysics 65, 565-573.

Edwards, R.N. 1997. On the resource evaluation of marine gas hydrate deposits using the seafloor transient electrical dipole-dipole method: Geophysics 62, 63-74.

EEZ-Scan 87. 1991. Atlas of the U.S. Exclusive Economic Zone, Atlantic continental margin, U.S. Geological Survey, Miscellaneous Investigations Series I-2054, 174pp.

Egeberg, P.K. & Barth, T. 1998. Contribution of dissolved organic species to the carbon and energy budgets of hydrate bearing deep sea sediments (Ocean Drilling Program Site 997 Blake Ridge). Chemical Geology 149, 25-35.

Eiken, O. & Austegard, A. 1987. The Tertiary orogenic belt of West-Spitsbergen; seismic expressions of the offshore sedimentary basins, Norsk Geologisk Tidsskrift 67, 383-394.

Eiken, O. & Hinz, K. 1993. Contourites in the Fram Strait. In: Contourites and bottom currents, D.A.V. Stow & J.C. Faugeresm, J.C. (eds). Sedimentary Geology 82, 15-32.

Eittreim, S.L. 1994. Transition from continental to oceanic crust on the Wilkes-Adelie margin of Antarctica: Journal Geophysical Research 99, 24,189-24,205.

Eittreim, S.L., Cooper, A.K. & Wannesson, J. 1995. Seismic stratigraphic evidence of ice-sheet advances on the Wilkes Land margin of Antarctica: Sedimentary Geology 96, 131-156.

Eittreim, S.L. & Smith, G. 1987. Seismic sequences and their distribution on the Wilkes Land margin: In Eittreim, S.L. & Hampton, M.A. (Eds.), The Antarctic Continental Margin, Geology and Geophysics of Offshore Wilkes Land, Circum Pacific Council for Energy and Mineral Resources, Earth Science Series, Houston, Texas 5A, 15-44.

Eldholm, O., Faleide, J.I. & Myhre, A.M. 1987. Continent-ocean transition at the western Barents Sea/Svalbard continental margin. Geology 15, 1118-1122.

Eldholm, O., Sundvor, E., Vogt, P.R., Hjelstuen, B.O., Crane, K., Nilsen, A.K. & Gladczenko, T.P. 1999. SW Barents Sea continental margin heat flow and Håkon Mosby Mud Volcano, *Geo-Marine Letters, 19,* 29-37.

Eldholm, O. & Talwani, M. 1977. Sediment distribution and structural framework of the Barents Sea. Geological Society of America 88, 1000-1014.

Eldholm, O., Thiede, J. & Taylor, E. 1989. Evolution of the Vøring volcanic margin. In: Eldholm, O. Thiede, J. & Taylor, E. et al. (eds). Proceedings of the Ocean Drilling Program, Scientific Results 104, 1033-1065.

Elliot, D.H. 1988. Antarctica: is there any oil and natural gas? Oceanus 31, 32-38.

Ellis, D.V. 1987. Well Logging for Earth Scientists. Elsevier, NY, 532pp.

Elverhøi, A., Hooke, R.L.B. & Solheim, A. 1998. Late Cenozoic erosion and sediment yield from the Svalbard-Barents Sea region: Implications for understanding erosion of glacierized basins. Quaternary Science Reviews, 17, 209-241.

Elvert, M., Suess, E. & Whiticar, M.J. 1999. Anaerobic methane oxidation associated with marine gas hydrates: superlight C-isotopes from saturated and unsaturated C-20 and C-25 irregular isoprenoids. Naturwissenschaften 86, 295-300.

Elvert, M., Suess, E.E., Greinert, J. & Whiticar, M.J. Archaea mediating anaerobic methane oxidation in deep-sea sediments at cold seeps of the eastern Aleutian subduction zone. Ogranic Geochemistry, (in press).

Embly, R.W. 1980. The role of mass transport in the distribution and character of deep-ocean sediments with special reference to the North Atlantic. Marine Geology 38, 28-50.

Englezos, P. & Hall, S. 1994. Phase equilibrium data on carbon dioxide hydrate in the presence of electrolytes, water soluble polymers and montmorillonite. Canadian Journal Chemical Engineering 72, 887-893.

Evans, D., King, E.L., Kenyon, N.H., Brett, C. & Wallis, D. 1996. Evidence for long-term instability in the Storegga Slide region off western Norway. Marine Geology, 130(3/4), 281-292.

Evans, R.L., Law, L.K., St. Louis, B., Cheesman, S. & Sananikone, K. 1999. The shallow porosity structure of the Eel Shelf, northern California: results of a towed electromagnetic survey. Marine Geology 154, 211-226.

Ewing, J.I. & Hollister, C.H. 1972. Regional aspects of deep sea drilling in the western North Atlantic, in Hollister, C.H., Ewing, J.I., et al., Initial Reports of the Deep Sea Drilling Project 11: U.S. Government Printing Office, 951-973.

Faleide, J.I., S.T. Gudlaugsson, O. Eldholm, A. Myhre, H.R. 1991. Jackson, Deep seismic transects across the sheared western Barents Sea-Svalbard continental margin, Tectonophysics 189, 73-89.

Faleide, J.I., Vågnes, E. & Gudlaugsson, S.T. 1993. Late Mesozoic-Cenozoic evolution of the western Barents Sea in a regional rift-shear tectonic setting, Marine Petroleum Geolog, 10, 186-214.

Faleide, J.I., Solheim, S., Fiedler, A., Hjelstuen, B.O., Andersen, E.S. & Vanneste, K. 1996. Late Cenozoic evolution of the western Barents Sea-Svalbard continental margin, Global and Planetary Change 12, 53-74.

Faraday, M. 1823. On fluid chlorine. Philosophical Transactions of the Royal Society 22A, 160-189.

Field, M.E. & Jennings, A.E. 1987. Seafloor gas seeps triggered by a northern California Earth Quake Marine Geology 77, 39-51.

Fink, C.R. & Spence, G.D. 1999. Hydrate distribution off Vancouver Island from multi-frequency single channel seismic reflection data: Journal Geophysical Research 104, 2909-2922.

Finley, P. & Krason, J. 1986. Geological evolution and analysis of confirmed or suspected gas hydrate localities. Volume 9. Formation and stability of gas hydrates of the Middle America Trench. U.S. Department of Energy Report DOE/MC21181 (DE88001007), 234pp.

Fisher, A.T. & Becker, K. 1993. A guide to ODP tools for downhole measurements. Ocean Drilling Program Technical Note 10, 148pp..

Fisher, A.T., Von Herzen, R.P., Blum, P. & Wang, K. 1999. Evidence for recent warming of shallow slope bottom water, offshore New Jersey. EOS 80, 172-173.

Fisher, C.R. 1990. Chemoautotrophic and methanotrophic symbioses in marine invertebrates, Reviews in Aquatic Sciences, 2 (3/4), 399-436.

Floodgate, G.D. & Judd, A.G. 1992 The origins of shallow gas, Continental Shelf Research 12, 1145-1156.

Floria, R.C. 1993. Geology and hydrocarbon prospects of Mahanadi Basin, India. Proceedings, Second Seminar on Petrographic Basins, India 1, 355-369.

Fofonoff, N.P. & Millard, R.C. 1982. Algorithms for computation of fundamental properties of seawater. UNESCO Technical Papers in Marine Science 44, 25-28.

Fogel, M.L., Sprague, E.K., Gize, A.P. & Frey, R.W. 1989. Diagenesis of organic matter in Georgia salt marshes. Estuarine Coastal Shelf Science 28, 211-230.

Franklin, L.J. 1980. In-situ hydrates-a potential gas source: Petroleum Engineer International, November, 112-122.

Franklin, L.J. 1981. Hydrates in Arctic Islands, in Bowsher, A.L., Proceedings of a Workshop on Clathrates (gas Hydrates) in the National Petroleum Reserve in Alaska, July 16-17, 1979, Menlo Park, California: U.S. Geological Survey Open-File Report 81-1298, 18-21.

Franks, F. & Reid, D.S. (eds). 1973. Chapter 5. Water: A Comprehensive Treatise 2, New York. Plenum, 524 pp.

Franzmann, P. D., Liu, Y., Balkwill, D. L., Aldrich, H. C., Conway de Macario, E. and Boone, D. R. 1997. Methanobacterium frigidum sp. Nov., a psychrophilic, H2-using methanogen from Ace Lake, Antarctica. International Journal Syst. Bacteriology 47, 1068-1072.

Fredlund, D.G. & Rahardjo, H. 1993. Soil mechanics for unsaturated soils. John Wiley Interscience, New York, 517pp.

Fry, B. 1986. Sources of carbon and sulfur nutrition for consumers in three meromictic lakes of New York State. Limnology Oceanography 31, 79-88.

Fuhrman, J.A., McCallum, K., Davis, A.A. 1992. Novel major archaebacterial group from marine plankton. Nature 356, 148-149.

Furuberg, L., Maloy, K.J. & Feder, J. 1996. Intermittent behaviour in slow drainage. Physical Reviews E53, 966-977.

Gabrielsen, R.H. & Færseth, R.B. 1988. Cretaceous and Tertiary reactivation of master fault zones of the Barents Sea, Norwegian Polar Institute Report 46, 93-97.

GAIL. 1996. National Gas Hydrate Programme-Expert Committee Report submitted to Secretary (Petroleum), Government of India, 36pp.

Gambeta, L.A. & Maldonado, A.P.R. 1990. Geophysical investigations in the Bransfield strait and in the Bellinghsausen Sea-Antarctica. In St.John, B. (ed). Antarctica as an Exploration Frontier - Hydrocarbon Potential, Geology and Hazards. American Association Petroleum Geologists Studies in Geology, Tulsa 31, 127-141.

Gammon, P.H., Kiefte, H. Clouter, M.J. & Denner, W.W. 1983. Elastic constants of artificial and natural ice samples by Brillouin Spectroscopy, Journal Glaciology 29, 433-460.

Ganguly, N., Spence, G.D., Chapman, N.R. & Hyndman, R.D. 2000. Heat flow variations from bottom simulating reflectors on the Cascadia margin. Marine Geology 164, 53-68.

Gassmann, F. 1951. Elasticity of porous media: Uber die elastizitat poroser medien: Vierteljahrsschrift der Naturforschenden Gesselschaft 96, 1-23.

Gautier, D.L., Dolton, G.L., Takahashi, K.I. & Varnes, K.L. 1995. National assessment of United States oil and gas resources on CD-ROM: U.S. Geological Survey Digital Data Series 30.

Gearing, P., Plucker, F.E. & Parker, P.L. 1977. Organic carbon stable isotope ratios of continental margin sediments. Marine Chemistry 5, 251-266.

Geoscience. April 1998. Gas hydrates: Resources? Hazards? Origins?. Keele, U.K., 24pp.

Gettrust, G., Wood, W., Lindwall, D., Chapman, R., Walia, R., Hannay, D., Louden, K., MacDonald, R., Spence, G. & Hyndman, R.D. 1999. New seismic study of deep sea gas hydrates results in greatly improved resolution. EOS 80, 439-440.

Ginnings, D.C. & Corruccini, R.J. 1947. An improved ice calorimeter - the determination of its calibration factor and the density of ice at 0°C, Journal of Research of the National Bureau of Standards, 38, 583-591, 1947. in: Petrenko, V.F. & Whitworth, R.W. (eds). Physics of Ice. Oxford University Press, New York, 1999).

Ginsburg, G.D., Guseinov, R.A., Dadashev, A.A., Ivanova, G.A., Kazantsev, S.A., Soloviev, V.A., Telepnev, Ye.V., Askery-Nasirov, R.E., Yesikov, A.D., Mal'tseva, V.I., Mashirov, Yu.G. & Shabayeva, I.Yu. 1992. Gas hydrate in the southern Caspian Sea: Izvestiya Akademii Nauk Serya Geologisheskaya 7, 5-20.

Ginsburg, G.D., Guseynov, R.A., Dadashev, A.A., Telepnev, E.V., Askeri-Nasirov, P.Ye., Yesikov, A.A., Mal'tseva, V.I., Mashiroav, Yu. G. & Shabayeva, I.Yu. 1992. Gas hydrates of the southern Caspian. International Geology Review 43, 765-782.

Ginsburg, G.D., Milkov, A.V., Soloviev, V.A., Egorov, A.V., Cherkashev, G.A., Vogt, P.R., Crane, K., Lorenson, T.D. & Khutorskoy, M.D. 1999. Gas Hydrate Accumulation at the Hakon Mosby Mud Volcano. Geo-Marine Letters 19, 57-67.

Ginsburg, G.D. & Soloviev, V.A. 1994. Mud volcano gas hydrates in the Caspian Sea. Bull. Geological Society Denmark 41, 95-100.

Ginsburg, G.D. & Soloviev, V.A. 1995. Submarine gas hydrate estimation: Theoretical and empirical approaches. Offshore Technology Conference, OTC Paper 7693. 1, 513-518.

Ginsburg, G.D., & V.A. Soloviev, 1997. Methane migration within the submarine gas-hydrate stability zone under deep-water conditions. Marine Geology 137, 49-57.

Ginsburg, G.D. & Soloviev, V.A. 1998. Submarine Gas Hydrates, Translated from Russian, Norma Publishers, St. Petersburg, Russia, 216 pp.

Ginsburg, G.D., Soloviev, V.A., Cranston, R.E., Lorenson, T.D. & Kvenvolden, K.A. 1993. Gas hydrate from the continental slope, offshore from Sakhalin Island, Okhotsk Sea: Geo-Marine Letters 13, 41-48.

Godbole, S.P. & Ehlig-Economides, C. 1985. Natural Gas Hydrates in Alaska: Quantification and Economic Evaluation: SPE California Regional Meeting, SPE 13593.

Goldberg, D. 1997. The role of downhole measurements in marine geology and geophysics, Review of Geophyics 35/3, 315-342.

Goldfinger, C., Torres, M.E. & Trehu, A. 1999. Possible strike-slip faults source for hydrate ridge methane vents, Cascadia margin: American Association Petroleum Geologists Bulletin 83, 688-689.

Gornitz, V. & Fung, I. 1994. Potential distribution of methane hydrates in the world's ocean: Global Biogeochemical Cycles 8, 335-347.

Grace, J.D. & Hart, G.F. 1986. Giant gas fields of northern West Siberia: The American Association of Petroleum Geologists Bulletin 70/7, 830-852.

Granli, J.R.,Arntsen, B., Sollid, A. &. Hilde, E. 1999. Imaging through gas-filled sediments using marine shear-wave data. Geophysics 64, 668-677.

Grantz, A. & May, S.D. 1982. Rifting history and structural development of the continental margin north of Alaska. In: Watkins, J.S. & Drake, C.L. (eds). Studies in Continental Margin Geology. American Association of Petroleum Geologists Memoir 34, 77-100.

Grantz, A., Boucher, G.W. & Whitney, O.T. 1976. Possible solid gas hydrate and natural gas deposits beneath the continental slope of the Beaufort Sea: U.S. Geological Survey Circular 733, 17pp.

Grantz, A., Hart, P.E. & Kvenvolden, K.A. 1989. Seismic reflection character, distribution, estimated volume and stability of gas hydrate deposits beneath the Arctic Ocean north of Alaska. American Geophysical Union. EOS 70/43, 1152.

Grantz, A., Moore, T.E. & Roeske, S.M. 1989. Transect of Alaska from the Gulf of Alaska to the Arctic Ocean, 28th international geological congress; (abss), Resumes 28, 1579-1580. In: Eldholm, O., Thiede, J., Taylor, E., et al. (eds). Proceedings of the Ocean Drilling Program, Scientific Results 104, 319-326.

Grauls, D., J.-P. Blanché & J.-L. Poudré. 1999. Hydrate sealing efficiency from seismic AVO and hydromechanical approaches. Paper presented at second JNOC workshop on Gas Hydrates Technologies, Tokyo, Japan, Aug, 1999.

Grevemeyer, I., A. Rosenberger, A. & Villinger, H. 2000. Natural gas hydrates on the continental slope off Pakistan: constraints from seismic techniques. Geophysical Journal International 140, 295-310.

Grunow, A.M., Dalziel, I.W.D., Harrison, T.M. & Heizler, M.T. 1992. Structural geology and geochronology of subduction complexes along the margin of Gondwanaland: New data from the Antarctic Peninsula and southernmost Andes: Geological Society America Bulletin 104, 1497-1514.

Gudmundsson, J. S., Parlaktuna, M. & Khokhar, A. A. 1994. Storing natural gas as frozen hydrate. SPE Production and Facilities. 69-73.

Gudlaugsson, S.T., Faleide, J.I., Johansen, S.E. & Breivik, A.J. 1998. Late Paleozoic structural development of the south-western Barents Sea. Marine and Petroleum Geology 15, 73-102.

Guerin, G., Goldberg, D. &. Meltser, A. 1999. Characterization of in situ elastic properties of gas hydrate-bearing sediments on the Blake Ridge, Journal Geophysical Research 104, 17,781-17,795.

Ha'aretz. 2000. Fifteen years of natural gas found off Ashkelon. By Amiram Cohen Ha'aretz Energy Correspondent Thursday, March 9, 2000.

Haimila, N.E., Kirschner, C.E., Nassichuk, W.W., Ulmishek, G. & Procter, R.M. 1990. Sedimentary basins and petroleum resource potential of the Arctic Ocean region, in: Grantz, A., Johnson, L. & Sweeney, J.F. (eds). The Geology of North America, The Arctic Ocean region: Geological Society America, Boulder, Colorado, L, 503-537.

Hammerschmidt, E.G. 1934. Formation of gas hydrates in natural gas transmission lines. Industrial and Engineering Chemistry 26, 851.

Hammond, R.D. & Gaither, J.R. 1983. Anomalous seismic character -- Bering Sea shelf. Geophysics 48, 590-605.

Hampton, M.A., Lee, H. J. & Locat, J. 1996. Submarine landslides, Reviews of Geophysics 34, 33-59.

Handa, Y.P. 1986. Journal of Chemical Thermodynamics 18, 891.

Handa, Y.P. 1990. Effect of hydrostatic pressure and salinity on the stability of gas hydrates. Journal Physical Chemistry 94, 2652-2657.

Handa, Y.P. & Stupin, D. 1992. Thermodynamic properties and dissociation characteristics of methane and propane hydrates in 70-A°-radius silica gel pores. Journal of Physical Chemistry 96, 8599-8603.

Handa, Y.P. 1986. Compositions, enthalpies of dissociation, and heat capacities in the range 85 to 270 K for clathrate hydrates of methane, ethane, and propane, and enthalpy of dissociation of isobutane hydrate, as determined by a heat-flow calorimeter, Journal Chemical Thermodynamics 18, 915-921.

Handa, Y.P. 1990. Effect of hydrostatic pressure and salinity on the stability of gas hydrates. Journal Physical Chemistry 94, 2652-2657.

Handa, Y.P., Hawkins, R.E. & Murray, J.J. 1984. Calibration and testing of a Tian-Calvet heat-flow calorimeter: Enthalpies of fusion and heat capacities for ice and tetrahydrofuran hydrate in the range 85 to 270 K. Journal Chemical Thermodynamics 16, 623-632.

Haq, B.U. 1984. Paleoceanography - A synoptic overview of 200 million years of ocean history. In: Haq, B.U. & Milliman, J.D. (eds). Marine Geology and Oceanography of Arabian Sea and Coastal Pakistan. Van Nostrand Reinhold, New York, 201-231.

Haq, B.U. 1993. Deep sea response to eustatic change and the significance of gas hydrates for continental margin stratigraphy: Special Publication International Association Sedimentologists 18, 93-106.

Haq, B.U., Hardenbol, J. & Vail, P.R. 1987. Chronology of fluctuating sea level since the Triassic: Science 235, 1156-1167.

Haq, B.U. 1998a. Natural gas hydrates: searching for the long-term climatic and slope-stability records. In: Henriet, J-P. & Mienert, J. (eds). Gas hydrates: Relevance to world margin stability and climate change. Geolological Society London, Special Publications 137, 303-318.

Haq, B.U. 1998b. Gas hydrates: Greenhouse nightmare? Energy panacea or pipe dream?: GSA Today, Geological Society of America, 8/11, 1-6.

Harrington, J.F. & Horseman, S.T. 1999. Gas transport properties of clays and mudrocks. In Aplin, A.C., Fleet, A.J. & Macquaker, J.H.S. eds. Muds and Mudstones: Physical and Fluid Flow Properties. Geological Society London Special Publication 158, 107-124.

Harrison, W.E. & Curiale, J.A., 1982, Gas hydrates in sediments of holes 497 and 498A, Deep Sea Drilling Project Leg 67, in Aubouin, J., von Huene, R., et al., Initial Reports of the Deep Sea Drilling Project 67. U.S. Government Printing Office, 591-594.

Harvey, L.D.D. & Huang, Z. 1995. Evaluation of potential impact of methane clathrate destabilization on future global warming: Journal of Geophysical Research 100, 2905-2926.

Hashin, Z. & Shtrikman, S. 1963. A variational approach to the elastic behavior of muliphase materials, Journal Mechanical Physics Solids 11, 127-140.

Hasiotis, T., Papatheodorou, G. & Kastanos, N. 1996. A pockmark field in the Patras Gulf (Greece) and its activation during the 14/7/93 seismic event. Marine Geology 130/3-4, 333-344.

Hawkings, T.J. & Hatelid, W.G. 1975. The regional setting of the Taglu Field, in Canada's Continental Margins and Offshore Petroleum Exploration: Canadian Society of Petroleum Geologist Press, 633-647.

Hayes, D.E. 1978. Geophysical Atlas of the East and Southeast Asian Seas. Office of Naval Research. Lamont Dougherty Geophysical Observatory.

Heard, H.C., Durham, W.B., Boro, C. & Kirby, S.H. 1990. A triaxial deformation apparatus for service at $77 \leq T \leq 273$ K. In: Duba, A.G., et al. (eds). The Brittle-Ductile Transition in Rocks, Geophysical Monograph 56. American Geophysical Union, Washington, D. C., 225-228.

Hearn, N., Hooton, R.D. & Mills, R.H. 1994. Pore structure and permeability . In: Kleiger, P. & Lamondm J.F. (eds). Significance of tests and properties of Concrete and concret-making materials. ASTM STP 169C, .240-261.

Hedberg, H. D. 1974. Role of Methane generation to undercompacted shales, shale diapirs and mud volcanoes. Kleiger, P. & Lamondm J.F. (eds). 58, 661-673.

Heggland, R. 1997. Detection of gas migration from a deep source by the use of exploration 3-D seismic data. Marine Geology 137, 41-47.

Helgerud, M.B., Dvorkin, J., Nur, A., Sakai, A. & Collett, T. 1999a. Elastic-wave velocity in marine sediments with gas hydrates: Effective medium modeling, Geophysical Research Letters 26, 2021-2024.

Helgerud, M.B., Waite, W.F., Nur, A., Pinkston, J., Stern, L., Kirby, S. & Durham, B. 1999b. Laboratory measurement of compressional and shear wave speeds through methane hydrate (abs). EOS Supplement 80, T51A-03.

Henriet, J.-P., & Mienert, J. (eds). 1998. In: Gas Hydrates: Relevance to World Margin Stability and Climate Change. Geological Society London, Special Publication 137.

Henry, P. Thomas, M. & Clennell, M.B. 1999. Formation of natural gas hydrates in marine sediments: 2. Thermodynamic calculations of stability conditions in porous marine sediments. Journal of Geophysical Research, 104, 23005-23020.

Herron, E.M. & Tucholke, B.E. 1976. Seafloor magnetic patterns and basementstructure in the southwestern Pacific: In Hollister CD, Craddock C, et al., Initial Reports Deep Sea Drilling Program 35, Washington, U.S. Government Printing Office, 263-278.

Hesse, R. & Harrison, W.E. 1981. Gas hydrates causing pore water freshening and oxygen isotope fractionation in deep-water sedimentary sections of terrigenous continental margins. Earth and Planetary Science Letters 55, 453-462.

Hill, R. 1952. The elastic behavior of crystalline aggregate, Proceedings Physical Society London A65, 349-354.

Hindle, A.D. 1997. Petroleum migration pathways and charge concentration: A three-dimensional model. American Association Petroleum Geologists Bulletin 81, 1451-1481.

Hinrichs, K.U., Hayes, J.M., Sylva, S.P., Brewer, P.G. & DeLong, E.F. 1999. Methane-consuming archaebacteria in marine sediments. Nature 398, 802-805.

Hitchon, B. 1974. Occurrence of natural gas hydrates in sedimentary basins, in Kaplan E.R., ed., Natural Gases in Marine Sediments: Plenum, New York, 195-225.

Hitchon, B., Underschultz, J.R., Bachu, S. & Sauveplane, C.M. 1990, Hydrogeology, geopressures and hydrocarbon occurrences, Beaufort-Mackenzie basin: Bulletin Canadian Petroleum Geology 38, 215-235.

Hjelstuen, B.O., Eldholm, O., Faleide, J.I. & Vogt, P.R. 1992. Regional setting of Håkon Mosby Mud Volcano, SW Barents Sea, Geo-Marine Letters, 19, 22-28,1999.Hyndman, R.D., & G.D. Spence. A seismic study of methane hydrate marine bottom simulating reflectors. Journal of Geophysical Research 97 (B5), 6683-6698.

Ho, C.K. & Webb, S.W. 1998. Capillary barrier performance in heterogeneous porous media. Water Resources Research 34, 603-609.

Hobro, J.W. 1999. Three-dimensional tomographic inversion of combined reflection and refraction travel-time data: University of Cambridge, Ph.D. thesis.

Hobro, J.W., Singh, S.& Minshull, T. 1998. Tomographic seismic studies of the methane hydrate stability zone in the Cascadia margin. In: Henriet, J.P. & Mienert, J. (eds). Gas Hydrates: Relevance to World Margin Stabilihy and Climate Change. Geological Society London Special Publlication 137, 133-140.

Hoehler, T.M., Alperin, M.J., Albert, D.B. & Martens, C.S. 1994. Field and laboratory studies of methane oxidation in an anoxic marine sediment: Evidence for a methanogen-sulfate reducer consortium. Global Biogeochemical Cycles 8, 451-463.

Holbrook, W.S., Hoskins, H., Wood, W.T., Stephen, R.A., Lizarralde, D. & Leg 164 Science Party, 1996. Methane hydrate and free gas on the Blake Ridge from vertical seismic profiling. Science 273, 1840-1843.

Holder, G.D., Malone, R.D., Lawson, W.F. 1987. Effects of gas composition and geothermal properties on the thickness and depth of natural-gas-hydrate zone: Journal Petroleum Technology, September, 1147-1152.

Hollister, C.D. & Ewing, J.I. 1972. Sites 102, 103, 104 Blake Bahama Outer Ridge (northern end). Initial Reports of the Deep Sea Dnlling Project 11, 135-218.

Holtz, R.D. & Kovacs, W.D. 1981. An introduction to geotechnical engineering: Prentice-Hall, Inc., Inglewood Cliffs, New Jersey, 733pp.

Honza, E. (ed). 1978. Geological investigation of the Okhotsk and Japan Seas off Hokkaido. Cruise Report 11, Geological Survey Japan, 72pp.

Houghton, J.T., Meira Filho, L.G., Callander, B.A., Harris, N., Kattenberg, A. & Maskell, K. 1995. Climate Change 1995, Intergovernmental Panel on Climate Change, Cambridge University (in press).

Hovland, M. 1992. Pockmarks and gas charged sediments in the eastern Skageerak. Continental Shelf Research 12, 1111-1119

Hovland, M. 1998. Are there commercial deposits of methane hydrates in ocean sediments? In JNOC-TRC report "Methane Hydrates-Resources in the Near Future?" 17-21. JNOC, Japan.

Hovland, M. & Curzi, P.V. 1989. Gas seepage and assumed mud diapirism in the Italian central Adriatic Sea. Marine & Petroleum Geology 6, 161-169.

Hovland, M., Gallagher, J.W., Clennell, M./B. & Lekvam, K. 1997. Gas hydrate and free gas volumes in marine sediments: Example from the Niger Delta front. Marine and Petroleum Geology 14, 245-255.

Hovland M. & Judd, A.G. 1988. Seabed Pock-marks and Seepages, Impact on Geology. Graham and Trotman Ltd, London, 293pp.

Hovland, M., Judd, A.G. & Söderburg, P. 1994. Gas accumulation in and migration through low permeability, fine-grained sediments. Proceedings Hedberg research Conference, Near Surface Expression of Hydrocarbon Migration, Vancouver, B.C., Canada, 24-28 April 1994; American Association Petroleum Geologists, 1-4.

Hovland, M., Løseth, H., Bjørkum, P.A., Wensaas, L. & Arntsen, B. 1999. Seismic detection of shallow high pressure zones. Offshore December. 99, 94-96.

Hovland, M., Lysne, D. & Whiticar, M.J. 1995. Gas hydrate and sediment gas composition, ODP Hole 892A, offshore Oregon, USA. In: Carson, B. Westboork, G.K. and Musgrave, R.J. eds. Proceedings ODP, Scientific Results, 146, College Station TX (Ocean Drilling Program), 151-161.

Howitt, F. 1971. Permafrost geology at Prudhoe Bay, Alaska: World Petroleum, 42/8, 28-38.

Huang, C.-Y. 1993. Bathymetric ridges and troughs in the active arc-continent collision region off southeastern Taiwan. reply and discussions. Journal Geological Society China 36, 91-109.

Hubbard, H.M. 1991. The real cost of energy. Scientific American 264/4, 36-42.

Huber, M. & Sloan, L.C. 1999. Warm climate transitions: A general circulation modeling study of the Late Paleocene Thermal Maximum: Journal of Geophysical Research 104/D14, 16, 633-16,655.

Huber, R., Kurr, M., Jannasch, H.W. & Stetter, K.O. 1989. A novel group of abyssal methanogenic archaebacteria (Methanopyrus) growing at 110°C. Nature 424, 833-834.

Hugget, Q.J., Somers, M.L., Cooper, A.K. & Stubbs, A.R. 1992 Interference fringes on GLORIA sisescan sonar images from the Bering Sea and their implications: Marine Geophysical Researches 14, 47-63.

Hullar, M.A.J., Fry, B., Peterson, B.J. & Wright, R.T. 1996. Microbial utilization of estuarine dissolved organic carbon: a stable isotope tracer approach tested by mass balance. Applied Environmental Microbiology 62, 2489-2493

Hunt, J.M. 1996 Petroleum Geochemistry and Geology, 2nd Ed., W.H. Freeman, San Francisco, California.

Hunt, R.E. 1984. Geotechnical Engineering Investigation Manual: McGraw-Hill Book Company, New York, 983pp.

Hyndman, R. D. & Davis, E.E. 1992. A mechanism for the formation of methane hydrate and seafloor bottom-simulating-reflectors by vertical fluid expulsion. Journal Geophysical Research 97, 7025-7041.

Hyndman, R.D., Foucher, J.P., Yamano, M. & Fisher, A. 1992. Scientific Team of Ocean Drilling Program Leg 131, Deep sea bottom-simulating-reflectors:calibration of the base of the hydrate stability field as used for heat flow estimates, Earth and Planetary Science Letters 109, 289-301.

Hyndman, R.D., Moore, G.F. & Moran, K. 1993. Velocity, porosity, and pore-fluid loss from the Nankai subduction zone accretionary prism: I.A. Hill, et al. (eds). Proceedings Ocean Drill. Program Scientific Results 131. College Station, TX, 211-220.

Hyndman, R.D.& Spence, G.D. 1992. A seismic study of methane hydrate marine bottom-simulating reflectors, Journal Geophysical Research 97, 6683-6698.

Hyndman, R.D., Spence, G.D., Yuan, T. & Davis, E.E. 1994. Regional geophysics and structural framework of the Vancouver Island margin accretionary prism: Ocean Drill. Program Inititial Reports 146, 399-419.

Hyndman, R.D. & Wang, K., 1993. Thermal constraints on the zone of major thrust earthquake failure: the Cascadia subduction zone: Journal Geophysical Research 98, 2039 – 2060.

Hyndman, R.D., Yorath, C.J., Clowes, R.M. & Davis, E.E. 1990. The northern Cascadia subduction zone at Vancouver Island: seismic structure and tectonic history. Canadian Journal Earth Sciences 27, 313-329.

Hyndman, R. D., Yuan, T. & Moran, K. 1999. The concentration of deep sea gas hydrates from downhole electrical resistivity logs and laboratory data, Earth and Planetary. Science Letters 172, 167-177.

Ichikawa, Y., Nonaka, A., Urano, T & Yasuhara, Y. 1998. Development of mud cooling system installed onboard of drill ship. in: JNOC-TRC, proceedings of the International Symposium on Methane Hydrates, resources in the near Future?, Chiba City, Japan, October 20-22, 1998, 161-172.

Impey, M.D., Grindrod, P., Takase, H. & Worgan, K.J. 1997. A capillary network model for gas migration in low permeability media. SIAM J. Appl. Math. 57, 357-608.

Ingram, G.M. & Urai, J.L. 1999. Top-seal leakage through faults and fractures: the role of mudrock properties. . In: Aplin, A.C., Fleet, A.J. & Macquaker, J.H.S. (eds). Muds and Mudstones: Physical and Fluid Flow Properties. Geological Society London Special Publication 158, 125-135.

Ishihara, T., Tanahashi, T., Sato, M., Okuda, Y. 1996. Preliminary report of geophysical and geological surveys of the west Wilkes Land margin. Proc. NIPR Symp. Antarctic Geoscience 9, 91-108.

Islam, M.R. 1994. A new recovery technique for gas production from Alaskan gas hydrates. Journal of Petroleum Science and Engineering 11, 267-281.

Ivanhoe, L.F. 1980. Antarctica-operating conditions and petroleum prospects: Oil & Gas Journ. 78, 212-220.

Ivanov, V.L. 1985. The geological prerequisites for petroleum prediction in Antarctica: International Geological Reviews 27, 757-769.

Iversen, N. & Jørgensen, B. B. 1985. Anaerobic methane oxidation rates at the sulfate methane transition in marine sediments from the Kattegat and Skaggerak (Denmark). Limnology and Oceanography, 30, 944-955.

Iversen, N. & Jorgensen, B.B. 1993. Diffusion coefficients of sulfate and methane in marine sediments influence of porosity. Geochim. Cosmochim. Acta 57, 571-578.

Jackson, H.R., Forsyth, D.A., Hall, J.K. & Overton, Q. 1991. Chapter 10. Seismic reflection and refraction. In: Grantz, A., Johnson, L. & Sweeney, J.F. (eds). The Arctic Ocean region. Geological Society of America, The Geology of North America, V.L, 153-170. Also Plate 5.

Jaeger, J.C. 1956. Conduction of heat in an infinite region bounded internally by a circular cylinder of a perfect conductor. Australian Journal of Physics 9, 167-179.

Jaeger, J.C. 1958. The measurement of thermal conductivity and diffusivity with cylindrical probes. EOS 39, 708-710.

Jaffe, A.M. & Manning, R.A. 2000. The Shocks of a World of Cheap Oil. Foreign Affairs 79/1, 16-29.

Jansa, L.F. & MacQueen, J.M. 1978. Stratigraphy and hydrocarbon potential of the central North Atlantic Basin. Geoscience Canada 5, 176-183.

Jansen, E., Befring, S., Bugge, T. Eidvin, T., Holtedahl, H. & Sejrup, H.P. 1987. Large submarine slides on the Norwegian continental margin: sediments, transport and timing, Marine Geology 78, 77-107.

Jeanthon, C., Reysenbach, A. L., Lharidon, S., Gambacorta, A., Pace, N. R., Glenat, P. & Prieur, D. 1995. Thermotoga subterranea sp-nov, a new thermophilic bacterium isolated from a continental oil-reservoir. Archives Of Microbiology 164, 91-97.

JNOC. 2000. News release by JNOC on January 20, 2000. Official news release JNOC's URL <http://www.jnoc. go.jp/what.html> in Japanese. Additional information: <http://www.jhd. go.jp/cue/TUHO/1999/suiroe /47_tempe.html>. Personal (rather than official) English translation by Ko-Ichi Nakamura of Geological Survey of Japan communicated originally to W.P. Dillon of the U.S. Geological Survey.

JNOC. October 20-22. 1998. Proceedings of the International Symposium on Methane Hydrates: Resources in Near Future?, JNOC-TRC, Chiba City, Japan, 399pp.

JNOC-TRC. 1998. Proceedings of the International Symposium on Methane Hydrates, resources in the near Future?, Chiba City, Japan, October 20-22, 1998, 399pp.

Jones, M.E. 1994. Mechanical principles of sediment deformation. Ch.2 In: Maltman, A.J. (ed). Geological Deformation of Sediments. Chapman and Hall, London, 362pp.

Jones, R.D. & Morita, R.Y. 1983. Methane oxidation by *Nitrosococcus oceanus* and *Nitrosomonas europaea*. Applied and Environmental Microbiology 45, 401-410.

Jones, R.E., Beeman, R.E. & Suflita, J.M. 1989. Anaerobic metabolic processes in the deep terrestrial subsurface. Geomicrobiology J. 7: 117-130.

Jørgensen, B.B. 1982. Mineralisation of organic matter in the sea bed - the role of sulphate reduction. Nature 296, 643-645.

Jouzel, J., et al. 1993. Extending the Vostock ice-core record of paleoclimate to the penultimate glacial period: Nature 364, 407-412.

Judd. A.G. & Hovland, M. 1992. The evidence of shallow gas in marine sediments. Continental Shelf Research 12, 1081-1096.

Judd, A.G. & Sim, R. 1998. Shallow gas migration mechanisms in deep water sediments. In: Ardus, D.A., Hobbs, R., Horsnell, M., Jardine, R., Long, D. & Sommerville, J. (eds). Offshore Site Investigation and Foundation Behaviour: New Frontiers. Society of Underwater Technology, London, 163-174.

Judge, A.S. & Majorowicz, J.A. 1992. Geothermal conditions for gas hydrate stability in the Beaufort-Mackenzie area: the global change aspect: Global and Planetary Change, 98/2-3, 251-263.

Judge, A.S., Smith, S.L. & Majorowicz, J. 1994. The current distribution and thermal stability of natural gas hydrates in the Canadian Polar Regions, in Proceedings Fourth International Offshore and Polar Engineering Conference, Osaka, Japan, 307-313.

Judge, A.S., Taylor, A.E., Burgess, M. & Allen, V.S. 1981. Canadian geothermal data collection--northern wells 1978-80: Earth Physics Branch, EMR Canada, Geothermal Series 12.

Kaplan, I.R. & Laberg, J.S. (eds). 1974. Natural Gases in Marine Sediments. New York: Plenum, 324pp.

Karisiddaiah, S.M.M. Veerayya, K.H. Vora, & Wagle, B.G. 1993. Gas-charged sediments on the inner continental shelf off western India. Indian Marine Geology 110, 143-152.

Karl, D. M. 1995. The Microbiology of Deep-Sea Hydrothermal Vents.CRC Press Inc., Boca Raton, Florida

Kastner, M., Kvenvolden, K.A. & Lorenson,T.D. 1998. Chemistry, isotopic composition, and origin of a methane-hydrogen sulfide hydrate at the Cascadia subduction zone: Earth and Planetary Science Letters 156, 173-183.

Kastner, M., Kvenvolden, K.A., Whiticar, M.J., Camerlenghi, A. & Lorenson, T.D. 1995. Relation between pore fluid chemistry and gas hydrates associated with bottom-simulating reflectors at the Cascadia Margin, sites 889 & 892. Proceedings of the Ocean Drilling Program, Scientic Results 146/1, 175-187.

Katz, M.E., Pak, D.K., Dickens, G.R. & Miller, K.G. 1999. Carbon input during the latest Paleocene thermal maximum. Science 286, 1531-1533.

Katzman, R., Holbrook, W.S. & Paull, C.K. 1994. Combined vertical-incidence and wide-angle seismic study of a gas hydrate zone, Blake Ridge. Journal Geophysical Research 99, 17,975-17,995.

Kavoun, M. & Vinnikovskaya, O. 1994. Seismic stratigraphy and tectonics of the northwestern Weddell Sea (Antarctica) inferred from marine geophysical surveys: Tectonophysics 240, 299-341.

Kayen, R.E. & Lee, H.J. 1991. Pleistocene slope instability of gas hydrate-ladden sediment of the Beaufort Sea margin, Marine Geotech. 10, 125-141.

Kayen, R.E. & Lee, H.J. 1993. Slope stability in regions of sea-floor gas hydrate: Beaufort Sea Continental Slope, In: W.C. Schwab, Lee, H.L. & Twichell, D.C. (eds). Submarine Landslides: Selective Studies in the U.S. Exclusive Zone, U.S. Geological Survey Bulletin 2002, 97-103.

Kelley, C.A., Coffin, R.B. & Cifuentes, L.A. 1998. Stable isotope evidence for alternate carbon sources in the Gulf of Mexico. Limnology Oceanography 43, 1962-1969.

Kennett, J.P., Cannariato, K.G., Hendy, I.L. & Behl, R. J. 2000. Carbon isotopic evidence for methane hydrate instability during Quaternary interstadials: Science 288,:128-133.

Kennett, J.P. & Frackler-Adams, B.N. 2000. Relationship of clathrate instability to sediment deformation in the upper Neogene of California, Geology 28, 215-218.

Kennett, J.P., Stott, L.D. 1991. Abrupt deep sea warming, paleoceanographic changes and benthic extinctions at the end of Palaeocene: Nature 353, 319-322.

Kennicutt II, M.C., Brooks, J.M. & Cox, H.B. 1993. The origin and distribution of gas hydrates in marine sediments. In: Organic Ceochemistry, Engel, M.H. & S. A. Macko, A. (eds). Plenum, New York, NY, 535-544.

Kenyon, N.H., 1987. GLORIA study of the Indus Fan. Institute of Oceanographic Sciences RRS Charles Darwin Cruise 20, Cruise Report 198, 17pp.

Kim, Y., Kim, H.-S., Larter, R.D., Camerlenghi, A., Gambetta, L.A.P. & Rudowski, S. 1995. Tectonic deformation in the upper crust and sediments at the South Shetland trench: In Cooper, A.K., Barker, P.F. & Brancolini, G. (Eds.), Geology and Seismic Stratigraphy of the Antarctic Margin, Antarctic Research Series 68, AGU, Washington, D.C., 157-166.

Kimura, G., Silver, E., Blum, P. & ODP Leg 170 Shipboard Party. 1997. Proceedings of the Ocean Drilling Program, Initial Reports 170, 458pp.

King, E.C. & Barker, P.F. 1988. The margins of the South Orkney microcontinents: Journal Geological Society London 145, 317-331.

King, G. M. 1991. Measurement of acetate concentrations in marine pore water by using an enzymatic approach. Applid Environmental Microbiology 57, 3476-3481.

Klepeis, K.A. & Lawver, L.A. 1996. Tectonics of the Antarctic-Scotia plate boundary near Elephant and Clarence Islands, West Antarctica: Journal Geophysical Researches 101, 20,211-20,231.

Klusman, R.W. 1993. Soil gas and related methods for natural resource exploration. Wiley. 492pp.

Koch, P.J., Zachos, J. & Gingerrich, P. 1992. Correlation between isotope records in marine and continental carbon reservoirs near the Paleocene-Eocene boundary: Nature 358, 319-322.

Kolla, V. & Coumes, F. 1987. Morphology, internal structure, seismic stratigraphy, and sedimentation of Indus Fan. American Association of Petroleum Geologists Bulletin 73/6, 650-677.

Kopf, A., Robertson, A.H.F., Clennell, M.B. & Flecker, R. 1998. Mechanisms of mud extrusion on the Mediterranean Ridge accretionary prism. Geo-marine Letters 18, 97-114.

Korenaga, J., Holbrook, W.S. Singh, S.C. & Minshull, T.A. 1997. Natural gas hydrates on the southeast US margin: Constraints from full waveform and traveltime inversions of wide-angle seismic data. Journal Geophysical Researches 102, 15,345-15,365.

Kotelnikova, S. & Pedersen, K. 1997. Evidence for methanogenic Archaea and homoacetogenic Bacteria in deep granitic rock aquifers. FEMS Microbiology Reviews 20, 339-349.

Kotkoskie, T.S., Al-Ubaidi, B., Wildeman, T.R. & Sloan, E.D. 1990. Inhibition of gas hydrates in water-based drilling muds, Proceedings of the Society of Petroleum Engineers Annual Technical Conference and Exhibition 4, 359-368.

Krason, J. & Ciesnik, M. 1985. Geological evolution and analysis of confirmed or suspected gas hydrate localities, Volume 5--Gas hydrates in the Russian literature: Report for U.S. Department of Energy, Office of Fossil Energy, Morgantown Energy Technology Center, Morgantown, West Virginia, DE-AC21-84MC21181, 164pp.

Krason, J. & Ciesnik, M.S. 1987. Geological evolution and analysis of confirmed or suspected gas hydrate localities. V. 10. Basin analysis, formation and stability of gas hydrates of the aleutian Trench and the Bering Sea. U.S. Department of Energy, DOE/MC/21181-1950 (DE88001008), 152pp.

Krause, F.K. 2000. Clathrate hydrate frost heave structures in a carbonate mud-mound: Meiklejohn Peak, Nevada, USA, 2000 Am Asoc. Petroleum Geologists Annual Convention 1.9, A81.

Kremlev, A.N. & Ginsburg, G.D. 1989. The first results of the search for submarine gas hydrates in the Black Sea (the 21st expedition of the RV Yevpatoriya): Geologia i Geophysica 4, 110-111.

Kulm, L.D. & Suess, E. 1990. Relationship between carbonate deposits and fluid venting: Oregon accretionary prism: Journal Geophysical Research 95, 8899-8915.

Kuramoto S., Joshima, M. & Okamura, Y. 1998. The relationship between methane gas hydrates and submarine active faults. (in Japanese) Abstracts, the 104th Annual Meeting of the Geological Society of Japan, 242.

Kuster, G.T. & Toksöz, M.N. 1974. Velocity and attenuation of seismic waves in two-phase media, 1, Theoretical formulation, Geophysics 39, 587-606.

Kuuskraa, V.A. 1985. Feasibility of Methane Recovery from Gas Hydrates: for presentation at the U.S. Dept of Energy Peer Review of Unconventional Gas Recovery Gas Hydrates Program.

Kuuskraa, V.A. 1998. Hydrates contain vast store of world gas resources. Oil and Gas Journal, May 11, 90-95.

Kuuskraa, V.A. & Hammershaimb, E.C. 1984. The energy ballance of hydrate recovery. In: 1984 International gas research conference, 47-58.

Kuuskraa, V.A., Hammershaimb, E.C., Holder, G.D.& Sloan, E.D., 1983, Handbook of gas hydrate properties and occurrence. U.S. Department of Energy, DOE/MC/19239-1546. U.S. Government Printing Office, 234pp.

Kvenvolden, K.A. 1988. Methane hydrate-A major reservoir of carbon in the shallow geosphere? Chemical Geology 71, 41-51.

Kvenvolden, K.A. 1988. Methane hydrates and global climate. Biogeochemical Cycles 2, 221-229.

Kvenvolden, K.A. 1993. Gas hydrates - Geological perceptive and global change: Reviews of Geophysics 31, 173-187.

Kvenvolden, K.A. 1993. A primer on gas hydrate. In: Howell, D.G., et al. (eds). The Future of Energy Gases. U.S. Geological Survey Professional Paper 1570, 279-291.

Kvenvolden, K.A. 1993. Gas hydrate as a potential energy resource -- a review of their methane content, in Howell, D.G., ed., The Future of Energy Gases: U.S. Geological Survey Professional Paper 1570, 555-561.

Kvenvolden, K.A. 1995. A review of the geochemistry of methane in natural gas hydrate: Organic Geochemistry 23, 997-1008.

Kvenvolden, K.A. 1998. A primer on the geological occurrence of gas hydrates. In: Henriet, J-P. & Mienert, J. (eds). Gas hydrates: Relevance to world margin stability and climate change. Geological Society London, Special Publication 137, 9-30.

Kvenvolden, K.A. (ed). 1998. Panel Discussion. Methane Hydrate as a resourc in the near future. Proceedings JNOC-TRC meeting on 'Methane hydrates: Resources in the near future?'. Chiba City, Japan, October 20-22, special suppliment, 29pp.

Kvenvolden, K.A. 1999. Potential effects of gas hydrate on human welfare: Proceedings of the National Academy of Sciences (USA) 96, 3420-3426.

Kvenvolden, K.A & Barnard, L.A. 1983. Gas hydrates of the Blake Outer Ridge, Site 533, Deep Sea Drilling Project Leg 76. In: Sheridan, R.E., Gradstein, F., et al. (eds). Initial Reports of the Deep Sea Drilling Project 76: U.S. Government Printing Office, 353-365.

Kvenvolden, K.A. & Barnard, L.A. 1983. Hydrate of natural gas in continental margins, In: Watkins, J.S. & Drake, C.L., (eds). Studies in Continental Margin Geology: American Association of Petroleum Geologists Memoir 34, 631-640.

Kvenvolden, K.A. & Claypool, G.E. 1988. Gas hydrate in oceanic sediment: U.S. Geological Survey Open-File Report 88-216, 50pp.

Kvenvolden, K.A., Claypool, G.E., Threlkeld, C.N. & Sloan, E.D. 1984. Geochemistry of a naturally occurring massive marine gas hydrate: Organic geochemistry 6, 703-713.

Kvenvolden, K.A., Ginsburg, G.D. & Soloviev, V.A. 1993. Worldwide distribution of subaquatic gas hydrates: Geo-Marine Letters 13, 32-40.

Kvenvolden, K.A., Golan-Bac, M. & Rapp, J.B. 1987. Hydrocarbon geochemistry of sediments offshore from Antarctica: Wilkes Land continental margin: In Eittreim, S.L. & Hampton, M.A. (eds). The Antarctic continental margin: Geology and Geophysics of offshore Wilkes Land. Circum Pacific Council for Energy and Mineral Resources, Earth Science Series, Houston, Texas 5A, 205-213.

Kvenvolden, K.A., Golan-Bac, M., McDonald, T.J., Pflaum, R.C. & Brooks, J.M. 1989. 15. Hydrocarbon gases in sediment of the Voring Plateau, Norwegian Sea. In: Eldholm, O., Thiede, J., Taylor, E., et al Proceeding of the Ocean Drilling Program, Scientific Results 104, 319-326.

Kvenvolden, K.A. &d Kastner, M. 1990. Gas hydrates of the Peruvian outer continental margin. In: Suess, I., von Huene, R., et al., (eds). Proceedings of the Ocean Drilling Program, Scientific Results, College Station, Texas 112, 517-526.

Kvenvolden, K.A. & McDonald, T.J. 1982. Gas hydrates in the Middle America Trench, DSDP Leg 84. EOS 63, 101.

Kvenvolden, K.A. & McDonald, T.J. 1985. Gas hydrate of the Middle America Trench, Deep Sea Drilling Project Leg 84, In: von Huene, R., Aubouin, J., et al. (eds). Initial Reports Deep Sea Drilling Project, Washington, D.C., U.S. Government Printing Office 84, 667- 682.

Kvenvolden, K.A. & McMenamin, M.A. 1980. Hydrates of natural gas: A review of their geologic occurrence. U.S. Geological Survey Circular 825, 11pp.

Kvenvolden, K.A. & von Huene, R. 1985. Natural gas generation in sediments of the convergent margin of the eastern aleution trench area In: Howel, D.G. (ed). Tectonostratigraphic Terranes of the Circum-Pacific Region. Circum-Pacific Council for Energy and Mineral Resources Earth Science Series 1, 31-49.

Laberg, J.S. & Andreassen., K. 1996. Gas hydrate and free gas indications within the Cenozoic succession of the Bjornoya Basin, western Barents Sea. Marine and Petroleum Geology 13, 921 -940.

Lachenbruch, A.H. 1968. Rapid estimation of the topographic disturbance to superficial thermal gradients: Reviews of Geophysics 6, 365 – 400.

Lai, M. C. & Gunsalus, R. P. 1992. Glycine betaine and potassium-ion are the major compatible solutes in the extremely halophilic methanogen Methanohalophilus strain Z7302. Journal Bacteriology 174, 4.

Lamontagne, R.A., Swinnenton, J.W., Linnebon, V.J & Smith, W.O. 1973. Methane concentrations in various marine environments. Journal Geophysical Research 78, 5317-5324.

Lance, S., Henry, P., Le Pichon, X., Lallement, S., Chamley, H., Rostek, F., Faugeres, J.-C., Gonthier, E. & Olu, K. 1998. Submersible study of mud volcanoes seaward of the Barbados accretionary wedge: sedimentology, structure and rheology. Marine Geology 145, 255-292.

Lancelot, Y. & Ewing, J.I. 1972. Correlation of natural gas zonation and carbonate diagenesis in Tertiary sediments from the north-west Atlantic, in Hollister, C.D., Ewing, J.I., et al., Initial Reports of the Deep Sea Drilling Project 11: U.S. Government Printing Office, 791-799.

Lance, S., Henry, P., Le Pichon, X., Lallement, S., Chamley, H., Rostek, F., Faugeres, J.-C., Gonthier, E. & Olu, K. 1998. Submersible study of mud volcanoes seaward of the Barbados accretionary wedge: sedimentology, structure and rheology. Marine Geology 145, 255-292.

Lancelot, Y. & Ewing, J.I. 1972. Correlation of natural gas zonation and carbonate diagenesis in Tertiary sediments from the north-west Atlantic, in Hollister, C.D., Ewing, J.I., et al., Initial Reports of the Deep Sea Drilling Project 11: U.S. Government Printing Office, 791-799.

Landvik, J.Y., Mangerud, J. & Salvigsen, O. 1988. Glacial history and permafrost in the Svalbard area: Proceedings of the Fifth International Conference on Permafrost, Trodheim, Norway, August 2-5, 1988: Tapir Publishers, Trondheim, Norway, 194-198.

LaPlanca, S. & Post, B. 1960. Thermal expansion of ice, Acta Crystalographica 13, 503-505.

Larter, R.D. & Barker, P.F. 1989. Seismic stratigraphy of the Antarctic Peninsula Pacific margin: a record Pliocene-Pleistocene ice volume and paleo-climate: Geology 17, 731-734.

Larter, R.D. & Barker, P.F. 1991a. Effects of ridge crest-trench interactionon Antarctic-Phoenix spreading: forces on a young subducting plate: Journal Geophysial Research 96, 19,583-19,607

Larter, R.D. & Barker, P.F. 1991b. Neogene interaction of tectonic and glacialprocesses at the Pacific margin of the Antarctic Peninsula: International Association Sedimentologists Special Publication 12, 165-186

Lashof, D.A. & Ahuja, D.R. 1990. Relative contribution of greenhouse gas emissions to global warming. Nature 344, 529-531.

Lee, M.W. & Collett,T.S. 1999. Amount of gas hydrate estimated form compressional- and shear-wave velocities at the JAPEX/JNOC/GSC Mallik 2L-38 gas hydrate research well. Iin: Dallimore, S.R., Uchida, T. & Collett, T.S. (eds.). Scientific Results from JAPEX/JNOC/GSC Mallik 2L-38 Gas Hydrate Research Well, Macenzie Delta, Northwest Territories, Canada. Geological Society Canada Bulletin 554, 313-322.

Lee, M.W. & Dillon, W.P. 2000. Amplitude blanking related to gas hydrate concentration. Marine Geophysical Researches. (in press).

Lee, M.W., Hutchinson, D.R., Agena, W.F., Dillon, W.P., Miller, J.J. & Swift, B.A. 1994. Seismic character of gas hydrates on the southeastern U.S. continental margin, Marine Geophysical Researches 16, 163-184.

Lee, M.W., Hutchinson, D.R., Collett, T.S. & Dillon, W.P. 1996. Seismic velocities for hydrate-bearing sediments using weighted equation. Journal Geophysical Researches 101, 20,347-20,358.

Lee, M.W., Hutchinson, D.R., Dillon, W.P., Miller, J.J., Agena, W.F. & Swift, B.A. 1993. Method of estimating the amount of in-situ gas hydrates in deep marine sediments. Marine and Petroleum Geology 10, 493-506.

Lee, R.W., Thuesen, E.V. & Childress. J.J. 1992a. Ammonium and free amino acids as nitrogen sources for the chemoautotrophic symbiosis *Solemya reidi* Bernard (Bivalvia: Protobranchia). Journal Experimental Marine Biology Ecology 158, 75-91.

Lee, R.W., Thuesen, E.V., Childress, J.J. & Fisher., C.R. 1992b. Ammonium and free amino acid uptake by a deep-sea mussel (*Bathymodiolus* sp., undescribed) containing methanotrophic bacterial symbionts. Marine Biology 113, 99-106.

Lein A., Vogt, P.R., Crane, K., Egorov, A. & Ivanov, M. 1999. Chemical and isotopic evidence for the nature of the fluid in CH_4-containing sediments of the Håkon Mosby Mud Volcano. Geo-Marine Letters 19, 76–83.

Levitus, S., Antonov, J.I., Boyer, T.P. & Stephens, C. 2000. Warming of the world ocean. Science, 287, 2225-2229.

Lewan, M.D. & Fisher, J.B. 1994. Organic acids from petroleum source rocks. In: Pittman, E. D. & Lewan, M. D. (eds). Organic acids in Geological Processes. Srpinger-Verlag, New York, 70-114.

Lewellen, R.I. 1973. The occurrence and characteristics of nearshore permafrost, northern Alaska: Proceedings of the Second International Conference on Permafrost, Takutsk, USSR, July 13-28, 1973: National Academy of Sciences, Washington D.C., 131-135.

Li, X. & Yortsos, Y.C. 1995. Theory of multiple bubble growth in porous media by solute diffusion. Chemical Engineering Science 50, 1247-1271.

Lindsey, J.F., Holliday, D.W. & Hulbert, A.G. 1991. Sequence stratigraphy and the evolution of the Ganges-Brahmaputra Delta Complex. American Association of Petroleum Geologists Bulletin, 1233-1254.

Liu, C.-S., Lundberg, N., Reed, D.L. & Huang, Y.-L. 1993. Morphological and seismic characteristics of the Kaoping Submarine Canyon. Marine Geology 111, 93-108.

Lodolo, E., Camerlenghi, A. & Brancolini, G. 1993. A bottom simulating reflector on the South Shetland Margin, Antarctic Peninsula: Antarctic Science 5, 207-210.

Lodolo, E., Tinivella, U., Pellis, G. & the R/V OGS-Explora Party 1998. Seismic investigation of the Bottom Simulating Reflectors on the South Shetland Margin: Terra Antartica Reports 2, 71-74.

Lonsdale, M.J. 1990. The relationship between silica diagenesis, methane, and seismic reflections on the South Orkney microcontinent: In Barker, P.F., Kennett, J.P. et al. (eds). Proceedings of the Ocean Drilling Program, Scientific Research 113, College Station, Texas, 27-36.

Losh, S. 1998. Oil migration in a major growth fault: structural analysis of Pathfinder core, South Eugene Island Block 330, Offshore Loiuisiana. American Association Petroleum Geologists Bulletin 82, 1694-1710.

Lovley, D.R., Chapelle, F.H. & Woodward, J.C. 1994. Use of dissolved H_2 concentrations to determine the distribution of microbially catalyzed redox reactions in anoxic ground water. Environmental Science Technology 28, 1005-1210.

Løvø, V., Elverhoi, A., Antonsen, P., Solheim, A., Butenko, G., Gregersen, O. & Liestol, O. 1990. Submarine permafrost and gas hydrates in the northern Barents Sea. Norsk Polarinstitutt Rapportserie 56, 171pp.

Lowrie, A. & Max, M.D. 1999. The extraordinary promise and challenge of gas hydrates. World Oil, September 49, 53-55.

Lowry, M.L. & Miller, C.T. 1995. Pore-scale modeling of nonwetting-phase residual in porous media. Water Resources Research 31, 455-473.

Lu, R.S., Lee, C.-S. & Kuo, S.-Y. 1977. An isopach map for the offshore area of Taiwan and Luzon. Acta Oceanographica Taiwanica 7, 1-9.

MacDonald, G.J. 1990. Role of methane clathrates in past and future climates. Climatic Change 16, 247-281.

MacDonald, G.T. 1990. The future of methane as an energy resource: Annual Reviews of Energy 15, 53-83.

MacDonald, I.R., Boland, G.S., Baker, J.S., Brooks, J.M., Kennicutt II, M.C. & Bidigare, R.R. 1989. Gulf of Mexico chemosynthetic communities. Marine Biology 101, 235-247.

MacDonald, I.R., Buthman, D.B., Sager, W.W., Peccini, M.B. & Guinasso, N.L. Pulsed flow of oil from a mud volcano. Geology (in review).

MacDonald, I.R., Guinasso, N.L., Ackleson, S.G., Amos, J.F., Duckworth, R. Sassen, R & Brooks, M.J. 1993. Natural oil slicks in the Gulf of Mexico visible from space. Journal Geophysical Research. 98-C9, 16351-16364.

MacDonald, I.R., Guinasso, NL. Jr., Sassen, R., Brooks, L., Lee, J.M. & Scott, K.T. 1994. Gas Hydrate that breaches the sea floor on the continental slope of the Gulf of Mexico. Geology 22, 699-704.

Mackay, J.R. 1972. Offshore permafrost and ground ice, southern Beaufort Sea, Canada: Canadian Journal of Earth Sciences 9/11, 1550-1561.

MacKay, M.E., Jarrard, R.D., Westbrook, G.K., Hyndman, R.D. & the Shipboard Scientific Party of ODP Leg 146. 1994. Origin of bottom simulating reflectors: Geophysical evidence from the Cascadia accretionary prism: Geology 22, 459-462.

MacKay, M.E., Moore, G.F., Cochrane, G.R., Moore, J.C. & Kulm, L.D. 1992. Landward vergence and oblique structural trends in the Oregon accretionary prism: implications and effect on fluid flow: Earth Planetary Science Letters 109, 477-491.

Majorowicz, J. A., Jessop, A.M. & Judge, A.S. 1995. Geothermal Regime, in Dixon, J. (ed). Geological Atlas of the Beaufort-Mackenzie Area. Geological Survey of Canada Miscellaneous Report 59.

Majorowicz, J.A. & Osadetz, K.G. 1999. Basic geological and geophysical controls bearing on gas hydrate distribution and volumes in Canada: Geological Survey of Canada Open-File Report.

Majorowicz, J.A., Jones, F.W. & Judge, A.S. 1990. Deep subpermafrost thermal regime in the Mackenzie Delta basin, northern Canada--analysis from petroleum bottom-hole temperature data: Geophysics 55, 362-371.

Makogon, Y.F. 1981. Hydrate of natural gas: Pennwell Publishing Company, Tulsa, Oklahoma, 237pp.

Makogon, Yu., F. 1997. Hydrates of Hydrocarbons, PennWell, Tulsa, Oklahoma. ISBN: 0878147187,

Makogon, Y.F. 1988. Natural gas hydrates--the state of study in the USSR and perspectives for its use: Proceedings of the Third Chemical Congress of North America, Toronto, Canada, June 5-10, 1988, 18pp.

Makogon, Yu. F., Trebin, F.A., Trofimuk, A.A., Tsarev, V.P. & Cherskiy, N.V. 1972. Detection of a pool of natural gas in a solid (hydrated gas) state (in English): Doklady Akademii Nauk SSSR 196, 203-206; Doklady-Earth Science Section 196, 197-200.

Makogon, Yu. F., Trofimuk, A.A., Tsarev, V.P. & Cherskiy, N.V. 1973. Possible origin of natural gas hydrates at floors of seas and oceans (in English): Academiya Nauk SSSR, Sibirskoe Otdeleniye Geologiya I Geofizika, 4, 3-6; International Geology Review 16 (1974), 553-556.

Makogon, Yu. F., Tsarev, V.I. & Cherskiy, V.V. 1972. Formation of large natural gas fields in zones of permanently low temperatures (in English): Doklady Akademii Nauk SSSR 205, 700-703; Doklady-Earth Science Section 205 (1973), 215-218.

Maldonado, A., Larter, R.D. & Aldaya, F. 1994. Forearc tectonic evolution of the South Shetland Margin, Antarctic Peninsula: Tectonics 13, 1345-1370.

Maldonado, A., Zitellini, N., Leitchenkov, G., Balanya, J.C., Coren, F., Galindo-Zaldivar, J., Lodolo, E., Jabaloy, A., Zanolla, C., Rodriguez-Fernandez, J. & Vinnikovskaya, O. 1998. Small ocean basin development along the Scotia-Antarctica plate boundary and in the northern Weddell Sea: Tectonophysics 296, 371-402.

Mandl, G. 1988. Mechanics of tectonic faulting. Elsevier, Amsterdam, 407pp.

Markl, R.G., Bryan, G.M. & Ewing, J.L. 1970. Structure of the Blake Bahama Outer Ridge: Journal of Geophysical Research 75, 4539-4555.

Martens, C.S., Chanton, J.P. & Paull, C.K. 1991. Biogenic methane from abyssal brine seeps at the base of the Florida escarpment.Geology 19, 851-854.

Martens, C.S. & Klump, J.V. 1980. Biogeochemical cycling in an organic-rich coastal marine basin, 1. Methane sediment-water exchange processes. Geochim. Cosmochim. Acta 44, 471-490.

Martin, J.B., Kastner, M., Henry, Le Pichon & Lallement. 1996. Chemical and isoptopic eveidence for sources of fluids in a mud volcano field seaward of the Barbados accretionary wedge. Journal Geophysical Researches 101, 20325-20345.

Massana, R., Murray, A.E., Preston, C.M. & DeLong, E.F. 1997. Vertical distribution and phylogenetic characterization of marine palnktonic Archaea in the Santa Barbara Channel. Applied Environmental Microbiology 63, 500-556.

Masson D.G., Bett, B.J. & Birch, K.G. 1997 Atlantic margin environmental survey. Sea Technology 38/10, 52-59.

Masters, C.D., Root, D.H. & Attanasi, E.D. 1991. Resource constraints in petroleum production potential: Science 253, 146-152.

Mathews, M.A. & von Huene, R. 1985. Site 570 methane hydrate zone In Orlofsky, S. (ed). DSDP Initerum Reports (U.S. Govt. Printing Office) 84, 773-790.

Mathews, M.A., 1986. Logging characteristics of methane hydrate. The Log Analyst 27, 26-63.

Matsumoto, R. 1995. Feasibility of methane hydrate under the sea as a natural gas resource (in Japanese). Journal Japanese Association Petroleum Technology 60, 147-156.

Matsumoto, R., Lu, H., Hiroki, Y., Waseda, A., Baba, K., Yagi, M. & Fujii, T. 1998. Gas hydrate drilling in the Nankai Trough offshore Cape Omaezaki. central Japan. Abstracts, Gas Hydrates Symposimn, Geoscience 98 (Keele, UK), 13.

Matsumoto, R., Watanabe, Y., Satoh, M., Okada, H., Hiroki, Y., Kawasaki, M. & ODP Leg 64 Shipboard Scientific Party. 1996. Distribution and occurrence of marine gas hydrates (in Japanese with English abstract). Preliminary results of ODP Leg 164: Blake Ridge Drilling-. Journal Geological Society Japan 102, 932-944.

Matthews, M.D. 1996. Migration - a view from the top. In: Schumacher, D. & Abrams, M.A. (eds) Hydrocarbon migration and its near-surface expression, American Association of Petroleum Geologists Memoir 66, 139-155.

Mavko, G., Mukerji, T. & Dvorkin, J. 1998. The rock rhysics handbook: Tools for seismic analysis in porous media, Cambridge University Press, New York, 329pp.

Max, M.D. 1990. Natural gas clathrates: geoacoustic inducement of low bottom loss. Marine Technology Society Conference, Washington DC, April 1990, 578-583.

Max, M.D. 1990. Gas hydrate and acoustically laminated sediments: probable environmental cause of anomalously low acoustic-interaction bottom loss in deep ocean sediments. Naval Research Laboratory Report 9235, 68pp.

Max, M.D. 1999. Oceanic methane hydrate: New fuel source? EEZ Technology 4, 25-28.

Max, M.D. & Chandra, K., 1998 The dynamic oceanic hydrate system: Production constraints and strategies: Proceedings Offshore Technology Conference, Houston, 4-7 May, 1998. OTC Paper 8684, 217-226.

Max, M.D. & Clifford, S. 2000. The state, potential distribution, and biological implications of methane in the Martian crust. Journal of Geophysical Research-Planets 105/E2, 4165-4171.

Max, M.D. & Dillon, W.P. 1998. Oceanic methane hydrate: Character of the Blake Ridge hydrate stability zone and potential for methane extraction. Journal of Petroleum Geology 21, 343-358.

Max, M.D. & Dillon, W.P. 1999. Oceanic methane hydrate: The character of the Blake Ridge hydrate stability zone and the potential for methane extraction: Author's correction. Journal of Petroleum Geology 22, 227-228.

Max, M.D. & Dillon, W.P. 2000. Methane hydrate (clathrate): Natural gas hydrate: a frozen asset?. Chemistry & Industry (U.K.), 1, January, 16-18.

Max, M.D., Dillon,, W.P. & Drury, R.M. M.D. 1996. Resource implications of hydrate development for the United States (abs): Some hydrate-gas volume estimates for the US East Coast. Marine Studies Group and Mineral Deposits Studies Group of the Geological Society of London, 29-30 October, 1996, 6.

Max, M.D., Dillon, W.P. & Malone, R.D. 1991. Proceedings: Report on National Workshop on Gas Hydrates April 23 and 24, 1991, U.S. Geological Survey, Reston, VA, U.S. Department of Energy DOE/METC - 91/6/24, DE 91016654, 38pp.

Max, M.D., Dillon, W.P., Nishimura, C. & Hurdle, B.G. 1999. Sea-floor methane blow-out and global firestorm at the K-T boundary. Geo-Marine Letters 18/4, 285-291.

Max, M.D., John, V.T. & Pellenbarg, R.E. 1999. Methane hydrate fuel storage for all-electric ships: an opportunity for technological innovation. Anals of the New York Academy of Sciences, Special publication of papers from the Third International Conference on gas hydrates (in press).

Max, M.D. & Lowrie, A. 1993. Natural gas hydrates: Arctic and Nordic Sea potential. In: Vorren, T.O., Bergsager, E., Dahl-Stamnes, Ø.A., Holter, E., Johansen, B., Lie, E. & Lund, T.B. Arctic Geology and Petroleum Potential, Proceedings of the Norwegian Petroleum Society Conference, 15-17 August 1990, Tromsø, Norway. Norwegian Petroleum Society (NPF), Special Publication 2 Elsevier, Amsterdam, 27-53.

Max, M.D.& Lowrie, A. 1996. Methane hydrate: A frontier for exploration of new gas resources. Journal Petroleum Geology 19, 41-56.

Max, M.D. & Lowrie, A. 1997. Oceanic methane hydrate development: Reservoir character and extraction. Offshore Technology Conference, 235-240.

Max, M.D. & Lowrie, A. 1996. Oceanic methane hydrates: a frontier gas resource. Journal Petrolroleum Geology 19, 41-56.

Max, M.D. & Lowrie, A., 1997. Oceanic methane hydrate development: Reservoir character and extraction. Offshore Technology Conference 5-8 May, Houston Texas, Procedings, 235-240.

Max, M.D. & Miles, P.R. 1999. Marine survey for gas hydrate. Proceedings American Association Petroleum Geologists Annual General Meeting, San Antonio, April 1999, A89-90.

Max, M.D. & Miles, P.R. 1999. Marine Survey for Gas Hydrate: Offshore Technology Conference Houston, Texas. OTC paper 10727, 12pp.

Max, M.D. & Ohta, Y. 1988. Did major fractures in continental crust control orientation of the Knipovich Ridge-Lena Trough segment of the plate margin? Polar Research 6, 85-93.

Max, M.D. & Pellenbarg, R.E. 1999. Desalination through Methane Hydrate. U.S. Patent 5,873,262, issued February 23, 1999.

McCollom, T.M. & Shock, E.L. 1997. Geochemical constraints on chemolithoautotrophic metabolism by microorganisms in seafloor hydrothermal systems. Geochimica et Cosmochimica Acta. 61, 4375-4391.

McDonnell, S.L., Max, M.D., Cherkis, N.Z. & Czarnecki, M.F. 1998. Likelihood of methane hydrate deposits along the northern margin of the South China Sea and northwestern Philippine Sea. In: EOS 79, F462.

McDonnell, S.L., Max, M.D., Cherkis, N.Z. & Czarnecki, M.F. Tectono-sedimentary Controls on the Likelihood of Gas Hydrate Occurrence near Taiwan. Marine Geology (in press).

McGinnis J.P. & Hayes D.E. 1995. The roles of down-slope and along-slope depositional processes: southern Antarctic Peninsula margin: In Cooper, A.K., Barker, P.F. & Brancolini, G. (eds). Geology and Seismic Stratigraphy of the Antarctic Margin, Antarctic Research Series 68, AGU, Washington, D.C. 141-156.

McIver, R.D. 1974. Hydrocarbon gas (methane) in canned Deep Sea Drilling Project core samples, In: Kaplan, I.R., (ed). Natural Gases in Marine Sediments: Plenum, New York, 63-69.

McIver, R.D. 1975. Hydrocarbon gases in canned core samples from Leg 28 sites 271, 272, & 273, Ross Sea. In: Initial Reports of the Deep Sea Drilling Project, 28, U.S. Government Printing Office, Washington, D.C., 815-817.

McIver, R.D. 1981. Gas hydrate. In: Meyer, R.F. & Olson, J.C. (eds). Long-Term Energy Resources. Pitman Publishing, Boston, Massachusetts, 713-726.

Mckenzie, D. 1978. Some remarks on the development of sedimentary basins, Earth and Planetary Science Letters 40, 25-32.

Meier-Augenstein, W. 1995. On line recording of 13C/12C ratios and mass spectra in one gas chromatographic analysis. High Resolution Chromatography 18, 28-32.

Melnikov, V. & Nesterov, A. 1996. Modelling of gas hydrates formation in porous media. Second International Conference on Natural Gas Hydrates, 541-548.

Meyer, R.F. 1981. Speculations on oil and gas resources in small fields and unconventional deposits, in: Meyer, R.F. & Olson, J.C. (eds). Long-Term Energy Resources. Pitman Publishing, Boston, Massachusetts, 49-72.

Mi, Y. 1998. Seafloor sediment coring and multichannel seismic studies of gas hydrate, offshore Vancouver Island. M.Sc. thesis, University of Victoria, Victoria, Canada.

Mienert, J. & Bryn, P. 1997. Gas hydrate drilling conducted on the European Margin. EOS 78, 567, 571.

Mienert, J. & Posewang, J. 1999. Evidence of shallow- and deepwater gas hydrate destabilizations in North Atlantic polar continental margin sediments. Geo-Marine Letters 19, 143- 149.

Mienert, J., Posewang, J. & Baumann, M. 1998. Gas hydrates along the northeastern Atlantic margin: possible hydrate-bound margin instabilities and possible release of methane. In: Henriet, J.P. & Mienert, J. (eds). Gas Hydrates: Relevance to World Margin Stabilihy and Climate Change. Geological Society London Special Publication 137, 275-291.

Mienert, J, Posewang J. & Lukas D. Changes in the hydrate stability zone on the Norwegian Margin and its consequence for methane and carbon releases into the oceanosphere. SFB-313. Synthese. In: Schäfer P., Ritzrau W, Schlüter M, & Thiede. (eds). The northern North Atlantic: A changing environment . Springer Verlag, in press.

Miles, P.R. 1995. Potential distribution of methane hydrate beneath the European continental margins. Geophysical Research Letters, 22/23, 3179-3182.

Miles, P.R. & Max, M.D. 1999. Mapping Natural Gas Hydrates with Tuned Detection Tool: Offshore Magazine, August, 138.

Miles, P.R., Schaming, M., Casas, A., Sachpazi, M. & Marchetti, A. 1997 Capturing a European legacy: EOS 78, 582.

Miller, R.D. 1980. Freezing phenomena in soils, in D. Hillel, D (ed). Introduction to Soil Physics. Academic. San Diego, CA, 254-299.

Minshull, T.A., Singh, S.C. &, Westbrook, G.K. 1994. Seismic velocity structure at a gas hydrate reflector, offshore western Colombia, from full waveform inversion. Journal Geophysical Researches 99, 4715-4734.

Mitchell, B. & Tinker, J. 1980. Antarctica and its resources. An Earthscan Publication. International Institute for Environment & Development, London, 162pp.

Mitchell, R., MacDonald, I.R. & Kvenvolden, K. 1999. Estimates of total hydrocarbon seepage into the Gulf of Mexico based on satellite remote sensing images. EOS Supplement. 80/49, OS242

Miyairi, M., Akihisa, K., Uchida, T., Collett, T.S. & Dallimore, S.R. 1999. Well-log interpretation of gas-hydrate-bearing formations in the Mallik 2L-38 gas hydrate research well, *in* Dallimore, S.R., Uchida, T., & Collett, T.S., eds, Scientific Results from JAPEX/JNOC/GSC Mallik 2L-38 Gas Hydrate Research Well, Mackenzie Delta, Northwest Territories, Canada, Geological Survey of Canada Bulletin 544, 281-293.

Molochushkin, E.N. 1978. The effect of thermal abrasion on the temperature of the permafrost in the coastal zone of the Laptev Sea: Proceedings of the Second International Conference on Permafrost, Takutsk, USSR, July 13-28, 1973: National Academy of Sciences, Washington D.C., 90-93.

Moudrakovski, I.L., Ratcliffe, C.I., McLaurin, G.E., Simard, B. & Ripmeester, J.A. 1999. Hydrate layers on ice particles and superheated ice. a H NMR microimaging study. Journal Physical Chemistry A103, 4969-4971.

Mountain, G.S. & Tucholke, B.E. 1985. Mesozoic and Cenozoic geology of the U.S. Atlantic continental slope and rise, In: Poag, C.W. (ed). Geologic Evolution of the United States Atlantic Margin, Van Nostrand Reinhold, New York, 293-341.

Murphy, W.F. III. 1982. Effects of Microstructure and Pore Fluids on the Acoustic Properties of Granular Sedimentary Materials. Ph.D. Dissertation, Stanford University.

Myhre, A.M. & Eldholm, O. 1988. The western Svalbard margin (74 °N-80 °N), Marine and Petroleum Geology 5, 134-156.

Naini, B.R. & Talwani, M. 1982. Structural framework and the evolutionary history of the continental margin of western India. American Association of Petroleum Geologists Memoir 34, 167-191.

Nakamura, A., Kuramoto, S., Matsumoto, T. & Kimura, M. 1997. Acoustic sedimentary structure and geological structure in the Japan Basin (in Japanese with English abstract). JAMSTEC Journal of Deep Sea Research. Japan Marine Science and Technology Center 13, 615-657.

Nassichuk, W.W. 1983. Petroleum potential in Arctic North America and Greenland: Cold Regions Science and Technology 7, 51-88.

Nassichuk, W.W. 1987. Forty years of northern non-renewable natural resource development: Arctic, 40/4, 274-284.

National Petroleum Council. 1992. The potential for natural gas in the United States: The National Petroleum Council, Publishers, v. I and II, 520pp.

Nazina, T.N., Ivanova, A.E., Borzenkov, I.A., Belyaev, SS. & Ivanov, M.V. 1995. Occurrence and geochemical activity of microorganisms in high-temperature, water-flooded oil fields of Kazakhstan and Western Siberia. Geomicrobiology Journal 13, 181-192.

Neglia, S. 1979. Migration of fluids in sedimentary basins. 198063, 573-597.

NGRI. 1998. Analysis of single channel seismic data from the continental margins of India for exploration of gas hydrates (NGRI Tech. Report No. NGRI-98LITHOS-221), 53pp.

NGRI. May 1996. Gas Hydrate Exploration along the Continental Margins of India - Evaluation of Available Geophysical and Geological Data (NGRI Technical Report 96). LITHOS 193, 199pp.

Nilsen, R.K., Torsvik, T. & Lien, T. 1996. Desulfotomaculum thermocisternum sp nov, a sulfate reducer isolated from a hot North Sea oil reservoir. International Journal of Systematic Bacteriology 46, 397-402.

NIO. June 1999. Bottom Simulating Reflections on the Western Continental Margin of India (From in-house seismic reflection data), Document NIO/CON-3/99, 42pp.

NIO. November 1997. Gas Hydrate Resource Map of India (NIO/SP-25/97), 56pp.

Nisbet, E.G. 1990. The end of ice age: Canadian Journal of Earth Sciences 27, 148-157.

Nisbet, E.G. & Piper, D.J.W. 1998. Giant submarine landslides, Nature 392, 329-330.

Norris, R.D. & Rohl, U. 1999. Carbon cycling and chronology of climate warming during the Paleocene/Eocene transition: Nature 401, 775-778.

Nottvedt, A., Livbjerg, F., Midboe, P.S. & Rasmussen, E. 1992. Hydrocarbon potential of the central Spitsbergen Basin, in Vorren, T.O. et al. (eds). Arctic Geology and Petroleum Potential: Norwegian Petroleum Society Special Publication 2, 333-361.

NPC. (National Petroleum Council) 1999. Natural Gas - Meeting the Challenges of the Nation's Growing Natural Gas Demand, Draft released via http://www.npc.org. US Department of Energy 1998. Annual Energy Outlook, With Projections to 2020.

Nunn, J.A. 1996. Buoyancy-driven propagation of isolated fluid-filled fractures: Implications for fluid transport in Gulf of Mexico geopressured sediments: Journal Geophysical Research 101, 2963-2970.

Nur, A., Mavko, G., Dvorkin, J. & Galmudi, D. 1998. Critical porosity: A key to relating physical properties to porosity in rocks, The Leading Edge 17, 357-362.

Offshore. 2000. SINTEF turning problems of gas hydrates to advantage: Offshore, April, 194.

Okuda, Y. 1996. Research on gas hydrates for resource assesment in relation to national drilling programme in Japan. In: Proceedings of the 2nd International Coference on natural gas hydtrates, 633-639.

Oremland, R.S. 1988. Biogeochemistry of methanogenic bacteria. in: Zehnder, A.J.B. (ed). Biology of Anaerobic Microorganisms. John Wiley, New York, 641-705.

Osterkamp, T.E. & Fei, T. 1993. Potential occurrence of permafrost and gas hydrates in the continental shelf near Lonely, Alaska: Proceedings of the Sixth International Conference on Permafrost, Beijing, China, July 5-9, 1993: National Academy of Sciences, Washington D.C., 500-505.

Ostrander, W.J. 1984. Plane-wave reflection coefficients for gas sands at non-normal angles of incidence. Geophysics 49, 1637-1648.

Oswin, J. & Dunn, J. 1988. Frequency, power and depth performance of ClassIV flextensional transducers, in: Decarpigny, H. (ed). Power sonic and ultrasonic transducers design. Springer Verlag, 121-133.

Pakulski, J.D., Benner, R. Amon, R., Eadie, B. & Whitledge, T. 1995. Community metabolism and nutrient cycling in the Mississippi River plume: evidence for intense nitrification at intermediate salinities. Marine Ecological Progress Series 117, 207-218.

Panda, P.K. 1994. Geothermal maps of India and their significance in resources assessments. Petroleum Asia Journal 17, 202-210.

Pandit, B.I. & King, M.S. 1982. Elastic wave propagation in propane gas hydrates. In: French, H.M. (ed). Fourth Canadian Permafrost Conference, 335-342.

Pankhurst, R.J. 1990. The Paleozoic and Andean magmatic arcs of West Antarctica and southern South America: In: Kay, S.M. & Rapela, C.W. (eds.), Plutonism from Antarctica to Alaska. Geological Society America Special Paper 241, 1 7

Park, A., Dewers, T. & Ortoleva, P. 1990. Cellular and oscillatory self-induced methane migration. Earth Science Reviews 29, 249-265.

Parkes, R.J., Cragg, B.A., Bale, S.J., Getliff, J.M., Goodman, K., Rochelle, P.A., Fry, J.C., Weightman, A.J. & Harvey, S.M. 1994. Deep bacterial biosphere in Pacific Ocean sediments. Naturc 371, 410-413.

Parkes, R.J., Cragg, B.A. & Wellsbury, P. 2000. Recent studies on bacterial populations and processes in marine sediments: a review. Hydrogeological Reviews, (in press).

Pauling, L. & Marsh, R.E. 1952. The crystal structure of chlorine hydrate. Proceedings National Association of Science, 38, 112-118.

Paull, C.K. 1997. Drilling for gas hydrates: Ocean Drilling Program Leg 164. In: Proceedings of the Offshore Technology Conference, Houston, Texas, 5-8 May, 1997, OTC Paper 8294, 1-8.

Paull, C.K., Beulow, W.J., Ussler, W., III, & Borowski, W.S. 1996. Increased continental-margin slumping frequency during sea-level lowstands above gas hydrate- bearing sediments, Geology 24, 143-146.

Paull, C.K., Borowski, W.S., Rodriguez, N.M. & ODP Leg 164 Shipboard Scientific Party. 1998. Marine gas hydrate inventory: preliminary results of ODP Leg 164 and implications for gas venting and slumping associated with the Blake Ridge gas hydrate field. In: Henriet, J.-P. & Mienert, J., (eds). Gas Hydrates: Relevance to World Margin Stability and Climate Change: Geological Society London Special Publication 137, 153-160.

Paull, C.K., Chanton, J.P., Neumann, A.C., Coston, J.A. & Martens, C.S. 1992. Indicators of methane-derived carbonates and chemosynthetic organic carbon deposits. Examples from the Florida Escarpment. Palaios 7, 361-375.

Paull, C.K. & Dillon, W.P. 1981. The appearance and distribution of the gas hydrate reflector off the southeastern United States. U.S. Geological Survey Open File Report 80 - 88, 24pp.

Paull, C.K. & Dillon, W.P. 1981. Appearance and distribution of the gas hydrate reflection in the Blake Ridge region, offshore southeastern United states, U.S. Geological Survey Miscellaneous Field Studies Map MF-1252.

Paull, C.K., Martens, C.S., Chanton, J.P., Neumann, A.C., Coston, A.C., Jull, T.J.A. & Toolin, L.J. 1989. Old carbon in living organisms and young $CaCO_3$ cements from abyssal brine seeps. Nature 342, 166-168.

Paull, C.K., R. Matsumoto, P. Wallace, and et. al. 1996. Proceedings of the Ocean Drilling Program, Initial Reports 164, Ocean Drilling Program, College Station, TX, 623pp.

Paull, C.K. & Matsumoto, R. 2000. Leg 164 Overview. In: Proceedings ODP, Scientific Results 164. Paull, C.K., Matsumoto, R., Wallace, P.J. & Dillon, W.P. (eds). 3-10, Ocean Drilling Program, College Station, TX.

Paull, C.K., Ussler, III, W. & Borowski, W.S. 1994. Sources of biogenic methane to form marine gas hydrates- In situ production or upward migration?: Annals of the New York Academy of Sciences 715, 392-409.

Paull, C.K., Ussler, W. III, Dillon, W.P. 1991. Is the extent of glaciation limiter by marine gas-hydrates? Geophysical Research Letters. 18, 432-434.

PCAST. 1997. Exectutive Summary in: Federal Energy Research and Development for the Challenges of the Twenty-First Century. The Presidents Committee of Advisors on Science and Technology. Gibbons, J.H. (Chair and ed). The White House, U.S. Government. (November) (T: 202-456-6100, F: 202-456-6026). (Internet: <http://www.whitehouse.gov/ WH/EOP /OSTP/html/OSTP_ HOME .html>), 33pp.

Pearson, C., Murphy, J. & Hermes, R. 1986. Acoustic and resistivity measurements on rock samples containing tetrahydrofuran hydrates: Laboratory analogues to natural gas hydrate deposits, Journal Geophysical Researches 91, 14132-14138.

Pearson, C.F., Halleck, P.M., McGuire, P.L., Hermes, R. & Mathews, M. 1983. Natural gas hydrate deposits: A review of in situ properties, Journal Physical Chemistry 87, 4180-4185.

Pecher, I.A., Booth, J.S., Winters, W.J., Mason, D.H., Relle, M.K. & Dillon, W.P. 1999. Gas hydrate distribution in sands – results from seismic laboratory studies: AGU Spring Meeting supplement to Eos, April 27, S338.

Pecher, I.A., Holbrook, W.S., Lizarralde, D., Stephen, R.A., Hoskins, H., Hutchinson, D.R. & Wood,W.T. 1997b. Shear waves through methane hydrate-bearing sediments - results from a wide-angle experiment during ODP Leg 164, EOS 78, F340.

Pecher, I., Holbrook, W.S., Stephen, R.A., Hoskins, H., Lizarralde, D., Hutchinson, D.R. & Wood,W.T. 1997a. Offset vertical seismic profiling for marine gas hydrate exploration - is it a suitable technique? First results from ODP Leg 164, Proc. 29th Offshore Technology Conference, 193-200.

Pecher, I.A., Minshull, T.A., Singh, S.C. & von Huene, R. 1996a. Velocity structure of a bottom simulating reflector offshore Peru: Results from full waveform inversion, Earth Planetarey Science Letters 139, 459-469.

Pecher, I., Ranero, C.R., von Heune, R., Minshull, T. & Singh, S.C. 1998. The nature and distribution of bottom simulating reflectors at the Costa Rica convergent margin. Geophysical Journal International 133, 219-229.

Pecher, I.A., von Huene, R., Ranero, C., Kukowski, N., Minshull, T.A. & Singh, S.C. 1996b. Formation mechanisms of free gas beneath the hydrate stability zone at convergent margins - geophysical evidence from bottom simulating reflectors at the Peruvian and Pacific Costa Rican margins, Proceedings International conference on natural gas hydrates, 593-600.

Pedersen, T., Wangen, M. & Johansen, H. 1996. Flow along fractures in sedimentary basins. In: Jamtveit, B. & Yardley, B.W.D. (eds). Fluid Flow and Transport in Rocks. Chapman & Hall, 203-213-223.

Pelz, O., Cifuentes,, L.S., Kelley, C.A., Trust, B.A. & Coffin, R.B.. 1998. Tracing the assimilation of organic compounds using _^{13}C analysis of unique amino acids in the bacterial cell wall. FEMS Microbiology Ecology 25, 229-240.

Pelz, O., Hesse, C., Tesar, M., Coffin, R.B. & Abraham, W.-R. 1997. Development of methods to measure carbon isotopic ratios of bacterial biomarkers in the environment. Isotopes Environmental Health Studies 33, 131-144.

Perry, R.K., Fleming, H.S., Cherkis, N.Z. & Feden, R. 1980. Bathymetry of the Norwegian, Greenland and Western Barents Sea. Geological Society America Map and Chart Series MC-21.

Perry, R.K., Fleming, H.S., Weber, J.R., Kristoffersen, Y., Hall, J.K., Grantz, A., Johnson, G.L., Cherkis, N.Z. & Larsen, B. 1986. Bathymetry of the Arctic Ocean (1:4,704,075 at Lat. 78 N). Naval Research Laboratory - Acoustics Division, Acoustic Media Characterization Branch. Geological Society of America Map and Chart Series, MC-56.

Peterson, B., Fry, B., Hullar, M., Saupe, S. & Wright. R. 1994. The distribution and stable carbon isotopic composition of dissolved organic carbon in estuaries. Estuaries 18, 111-121.

Peterson, B.J. & Fry, B. 1987. Stable isotopes in ecosystem studies. Annual Review of Ecological Systems 18, 293-320.

Petit, J.R., Jouzel, J., Raynaud, D. et al. 1999. Climate and atmospheric history of the past 420,000 years from the Vostock ice core, Antarctica: Nature 399, 429-436.

Pewe, T.L. 1983. The periglacial environment in North America during Wisconsin Time, in Porther, S.C. (ed). The Late Pleistocene: University of Minnesota Press, Minneapolis, Minnesota, 157-189.

Pflaum, R.C., Brooks, J.M., Cox, H.B., Kennicutt, M.C. & Sheu, D-D. 1986. Molecular and isotopic analysis of core gases and gas hydrates, Deep Sea Drilling Project Leg 96. In: Bouma, A.H., Coleman, J.M., Meyer, A.W., et al. (eds). Initial Reports of the Deep Sea Drilling Project 96: U.S. Government Printing Office, 781-784.

Pimenov, N., Gebruk, A., Moskalev L. & Vogt, P.R. Microbial processes of cabon cycle as the base of the food chain of Håkon Mosby mud volcano benthic community. Geo-Marine Letters 19, 89-96.

Plass-Dülmer, C., Koppmann, R.K., Ratte M. & Rudolph, J. 1995. Light non methane hydrocarbons in seawater. Global Biogeochemical Cycles 9, 79-100.

Popenoe, P. & Dillon,W.P. 1996. Characteristics of the continental slope and rise off North Carolina from GLORIA and seismic-reflection data: The interaction of downslope and contour current processes, Chapt. 4, In: Gardner, J.V., Field, M.E. & Twichell, D.C. (eds). Geology of the United States' Seafloor: The View from GLORIA, Cambridge University Press, Cambridge, U.K., 59-79.

Popenoe, P., Schmuck, E.A. & Dillon, W.P. 1993. The Cape Fear Landslide: Slope failure associated with diapirism and gas hydrate decomposition. In: Submarine Landslides: Selective Studies in the U. S. Exclusive Economic Zone. U.S. Geological Survey Bulletin, 2002, 40-53.

Poropkari, A.L., Babu, C.P.P. & Mascarenhas, A. 1993. New evidences for enhanced preservation of organic carbon in contact with oxygen minimum zone on the western continental margin of India. Journal Marine Geology 111, 7-13.

Posewang, J. 1997. Nachweis von Gashydraten und freiem Gas in den Sedimenten des nordwesteuropäischen Kontinentalabhanges mit hochauflösenden reflexionsseismischen Methoden und HF-OBS-Daten, pp. 137, Berichte aus dem Sonderforschungsbereich 313, 68, Phd thesis, Universität Kiel, Germany.

Posewang, J. & Mienert, J. 1999. High-resolution seismic studies of gas hydrates west of Svalbard. Geo-Marine Letters 19, 150-156

Posewang, J. & Mienert, J. 1999a. The enigma of double BSRs: Indicators for changes in the hydrate stability field?, Geo-Marine Letters 19, 157-163.

Potential Gas Committee. 1981. Potential Supply of Natural Gas in the United States. Colorado School of Mines, Golden, Colorado, 119 pp.

Premuzic, E.T. 1980. Organic carbon and nitrogen in the surface sediments of world oceans and seas: distribution and relationship to bottom topography. Brookhaven National Laboratory. Report BNL 51084 for the U.S. Department of Energy, 117pp.

Prensky, S. 1995. A review of gas hydrates and formation evaluation of hydrate-bearing reservoirs, Transactions SPWLA 36th Annual Symposium, paper GGG, Paris, France.

Prior, D.B. & Coleman, J.M. 1984 Submarine slope instability, In: Brunsden, D. & Prior, D.B. (eds). Slope Instability, John Wiley & Sons Ltd., 419-455.

Procter, R.M., Taylor, G.C. & Wade, J.A. 1984. Oil and natural gas resources of Canada 1983: Geological Survey of Canada Paper 83-31, 59pp.

Rao, Y.H. 1999. C-Program for the calculation of gas hydrate stability zone thickness. Computers & Geosciences 25, 705-707.

Rastogi, A., Deka, B, Budhiraja, I.L. & Agarwal, G.C. 1999. Possibility of large deposites of gas hydrates in deeper waters of India. (Copyright 1999, Taylor & Francis) Marine Georesources and Geotechnology 17, 49-63.

Ratcliffe, E.H. 1960. The thermal conductivities of ocean sediments. Journal Geophysical Research 65, 1535-1541.

Rau, G.H., Takahashi, T. & DesMarais, D.J. 1989. Latitudinal variation in plankton _^{13}C: implilcation for CO_2 and productivity in past oceans. Nature 341, 516-518.

Rau, G. H., Takahashi, T., DesMarais, D.J. & Sullivan, C.W. 1991. Particulate organic matter _^{13}C variations across the Drake Passage. Journal Geophysical Research 96, 131-135.

Ravot, G., Magot, M., Fardeau, M.L., Patel, B., Prensier, G., Egan, A., Garcia, J.L. & Ollivier, B. 1995. Thermotoga elfii sp-nov, a novel thermophilic bacterium from an african oil-producing well. International Journal of Systematic Bacteriology 45, 308-314.

Rebesco, M., Camerlenghi, A. & Zanolla, C. 1998. Bathymetry and morphogenesis of the continental margin west of the Antarctic Peninsula: Terra Antartica 5, 715-725.

Rebesco, M., Larter, R.D., Barker, P.F., Camerlenghi, A. & Vanneste, L.E. 1994. The history of sedimentation on the continental rise west of the Antarctic Peninsula: Terra Antarctica 1, 277-279.

Rebesco, M., Larter, R.D., Barker, P.F., Camerlenghi, A. & Vanneste, L.E. 1997a History of Sedimentation on the Continental Rise West of the Antarctic Peninsula: In Cooper, A.K. & Barker, P.F.(eds). Geology and Seismic Stratigraphy of the Antarctic Margin, Part 2, Antarctic Research Series 71, AGU, Washington, D.C., 29-49.

Rebesco, M., Larter, R.D., Camerlenghi, A. & Barker, P.F. 1996. Giant Sediment Drifts on the Continental Rise West of the Antarctic Peninsula: Geo-Marine Letters 16, 65-75.

Rees, G.N., Grassia, G.S., Sheehy, A.J., Dwivedi, P.P. & Patel, B. 1995. Desulfacinum infernum gen-nov, sp-nov, a thermophilic sulfate- reducing bacterium from a petroleum reservoir. International Journal of Systematic Bacteriology 45, 85-89.

Rempel, A.W. & Buffett, B. 1997. Formation and accumulation of gas hydrate in porous media. Journal Geophysical Research 102, 10,151-10,164.

Rempel, A.W. & B.A. Buffett, Mathematical models of gas hydrate accumulation. In: Henriet, J.-P. & Mienert, J. (eds). Geological Society London Special Publication 137, 63-74, 1998.

Reuss, A. 1929. Berechnung der fliessgrenzev von mischkristallen auf grund der plastizitätsbedingung für einkristalle, Zeitschrift für Angewandte Mathematik und Mechanik 9, 49-58.

Revil, A., Cathles, L.M. III, Shosa, J.D., Pezard, P.A. & de Larouziere, F.D. 1998. Capillary sealing in sedimentary basins: a clear field example. Geophysical Research Letters 25, 389-392.

Riis. F., Vollset, J. & Sand, M. 1986. Tectonic development of the western margin of the Barents Sea and adjacent areas. In: M.T. Halbouty (ed.), Future Petroleum Provinces of the World, American Association of Petroleum Geologists Memoir 40, 661-676.

Ritger, S., Carson B. & Suess, E. 1987. Methane-derived authigenic carbonates formed by subduction-induced pore-water expulsion along the Oregon/Washington margin: Geological Society America Bulletin. 98, 147-156.

Rivkina, E., Gilichinsky, D., Wagener, S., Tiedje, J. & McGrath, J. 1998. Biogeochemical activity of anaerobic microorganisms from buried permafrost sediments. Geomicrobiology Journal, 187-193.

Roberts, S.J., Nunn, J.A., Cathles, L. & Cipriani, F.-D. 1996. Expulsion of abnormally-pressured fluids along faults. Journal Geophysical Research 101, B12, 28231-28252.

Rodriquez, N.M. & Paull, C.K. 2000 Data Report: ^{14}C dating of sediments of the uppermost Cape Fear slide plain: Constraints on the timing of this massive submarine landslide. In: Paull, C.K., Matsumoto, R., Wallace, P.J. & Dillon, W.P., (eds). Proceedings ODP, Scientific Results, 164, College Station TX (Ocean Drilling Program), 325-327.

Romankevich, E.A. 1984. Geochemistry of organic matter in the ocean. Springer-Verlag, Berlin, New York, 334pp.

Rønnevik, H. & Jacobsen, H.-P. 1984. Structural highs and basins in the western Barents Sea. In: A.M. Spencer (ed). Petroleum Geology of the North European Margin. Graham and Trotman, London, 19-32.

Ross, R.G. & Anderson, P. 1982. Effect of guest molecule size on the thermal conductivity and heat capacity of clathrate hydrates. Canadian Journal of Chemistry 60, 881.

Ross, R.G., Andersson, P. & Backstrom, G. 1981. Unusual PT dependence of thermal conductivity for a clathrate hydrate, Nature 290, 322-323.

Rothwell, R.G., Thomson, J. & Kohler, G. 1998. Low sea-level emplacement of a very large Late Pleistocene 'Megaturbidite' in the western Mediterranean Sea. Nature 392, 377-380.

Rowe, M.M. & Gettrust, J.F. 1993. Faulted structure of the bottom simulating reflector on the Blake Ridge, western North Atlantic. Geology 21, 833-836.

Rowe, M.M. & Gettrust, J.F. 1993. Fine structure of methane hydrate bearing sediments on the Blake Outer Ridge as determined from deep-tow multichannel seismic data. Journal Geophysical Research 98, 463-473.

Ruppel, C. 1997. Anomalously cold temperatures observed at the base of the gas hydrate stability zone on the U.S. Atlantic passive margin, Geology 25/8, 699-702.

Ruppel, C. & Dickens, G. 1999. Climate change and methane gas hydrate reservoirs: quantitative constraints on methane release during the Late Paleocene thermal maximum. EOS 80, S333-334.

Ruppel, C. & M. Kinoshita. 2000. Fluid, methane, and energy flux in an active margin gas hydrate province, offshore Costa Rica. Earth and Planetary Science Letters (in press).

Ruppel, C., Von Herzen, R.P. & Bonneville, A. 1995. Heat flux through an old (~175 Ma) passive margin: offshore southeastern USA. Journal of Geophysical Research 100, 20,037-20,058.

Sættem, J. Bugge, T., Fanavoll, S., Goll, R.M., Mørk, A., Mørk, M.B.E., Smelror, M. & Verdenius, J.G. 1994. Cenozoic margin development and erosion of the Barents Sea: Core evidence from the southwest of Bjørnøya, Marine Geology 118, 257-281.

Sachs, W. & Meyn, V. 1995. Pressure and temperature dependence of the surface tension in the system natural gas/water: Principles of investigation and the first precise experimental data for pure methane/water at 25°C up to 46.8 MPa. Colloids and Surfaces A94, 291-301.

Sackett, W.M. 1991. A history of the _^{13}C composition of oceanic plankton. Marine Chemistry 34, 153-156.

Sahimi, M. 1994. Applications of percolation theory. Taylor & Francis, London 258pp.

Sahimi, M. 1995. Flow and transport in porous media and fractured rock. VCH publishers, 312pp.

Sakai, A. 1998. Seismic studies related to gas hydrates in the Nankai Trough. Proceedings of the International Symposium on Mathane Hydrates, Resources in the near future?, In: JNOC-TRC, proceedings of the International Symposium on Methane Hydrates, resources in the near Future?, Chiba City, Japan, October 20-22, 1998, 61.

Sakai, A. 1999. Is gas hydrate disseminated in pore or cemented on grains? An analysis for the Mackenzie Delta VSP survey. EOS, AGU Spring Meeting.

Sakai, A. 1999. Velocity analysis of vertical seismic profile (VSP) survey at JAPEX/JNOC/GSC Mallik 2L-38 gas hydrate research well, and related problems for estimating gas hydrate concentration, Geological Survey of Canada Bulletin 544, 323-340.

Salvador, A. 1987. Late Triassic-Jurassic paleogeopraphy and origin of Gulf of Mexio basin. Bulletin American Association of Petroleum Geologists. 71, 419-451.

Saricks, C.L. 1989. Environmental considerations of a fuel-flexible transportation system. Ppresented at the 1988 Air Pollution Control Association Specialy Conference on Environmental Challenges in Energy Utilization Cleveland, Ohio, 12-13 Octorer, 1988. Department of Energy Technical Information Center Document DE89 003659.

Sassen, R. 1980. Biodegradation of crude oil and mineral deposition in a shallow Gulf Coast salt dome. Organic Geochemistry 2:153-1266.

Sassen, R. & MacDonald, I.R.. 1994. Evidence of structure H hydrate, Gulf of Mexico continental slope: Organic Geochemistry 22, 1029-1032.

Sassen, R. & Macdonald, I.R. 1997. Hydrocarbons of experimental and natural gas hydrates, Gulf of Mexico continental slope. Organic Geochemistry 26, 289-293.

Sassen, R., MacDonald, I.R., Requego, A.G., Guinasso Jr., N.L., Kennicutt II, M.C., Sweet, S.T. & Brooks, J.M. 1994. Organic geochemistry of sediments from chemosynthetic communities, Gulf of Mexico slope. Geo-Marine Letters 14, 110-119.

Sassen, R., Roberts, H.H., Aahron, P., Larkins, J., Chinn, E.W. & Carney, R. 1993. Chemosythetic bacterial mats at cold hydrocarbon seeps, Gulf of Mexico continental slope. Organic Geochemistry 20, 77-89.

Satoh, M., Maekawa T. & Okuda Y. 1996. Estimation of amount of methane and resources of Natural gas hydrates in the world and around Japan. (in Japanese with English abstract). Journal Geological Society Japan 102, 959-971.

Saunders, D.F., Burston, K.R. & Thompson, C.K. 1999. Model for hydrocarbon microseepage and related near-surface alterations. Gas Hydrates 83, 170-185.

Scharma, G.D., Kamath, V.A., Godbole, S.P., Patil, S.L., Paranjpe, S.G., Mutalik, P.N. & Nadem, N. 1987. Development of Alaskan gas hydrate resources. Annual Report, U. S. Department of Energy DOE/MC/61114-2608 (DE88010270), 233pp.

Schiff, S. Aravena, L.R., Trumbore, S.E. & Dillon, P.J 1990. Dissolved organic carbon cycling in forested watersheds: A carbon isotope approach. Water Resources Research 26, 2949-2957.

Schlee, J. S., Dillon, W. P., Popenoe, Peter, Robb, J.M. & O'Leary, D.W. 1992. GLORIA mosaic of the deep sea floor off the Atlantic coast of the United States, U.S. Geological Survey, Miscellaneous Field Studies Map MF-2211.

Schlenk, W., Jr. 1951. Organic occlusion compounds. Fortschung Chemica Forsch., 2, 92 145,

Schlomer, S. & Krooss, B.M. 1997. Experimental characterization of the hydrocarbon sealing efficiency of cap rocks. Mar. & Petrol. Geol. 14, 565-580.

Schlumberger, 1989. Log interpretation principles/applications, Houston, TX (Schlumberger Educ. Services).

Schmuck, E.A. & Paull, C.K. 1993. Evidence for gas accumulation associated with diapirism and gas hydrates at the head of the Cape Fear slide. Geo-Marine Letters 13, 145-152.

Scholl, D.W. & Creager, J.S. 1973. Geologic synthesis of Leg 19 (DSDP) results; far north Pacific, and Aleutian Ridge, and Bering Sea. In: Creager, J.S. & Scholl, D.W. (eds). Initial Reports DSDP, U.S. Printing Office, Washington, D.C., 897-913

Scholl, D.W. & Hart, P.E. 1993. Velocity and amplitude structures on seismic-reflection profiles--possible massive gas-hydrate deposits and underlying gas accumulations in the Bering Sea Basin. In: Howell, D.G. (ed). The Future of Energy Gases. U.S. Geological Survey Professional Paper 1570, 331-351.

Scholwalter, T. 1979. Mechanics of secondary hydrocarbon migration and entrapment, American Association of Petroleum Geologists Bulletin 63, 723-760.

Schonheit, P., Kristjensson, J.K. & Thauer, R.K. 1982. Kinetic mechanism for the ability of sulfate-reducers to out-compete methanogensfor acetate. Archives for Microbiology 132, 285-288.

Schultheiss, P.J. & Gunn, D.E. 1985. The permeability and consolidation of deep-sea sediments. Insitutute of Oceanographic Sciences, Report 201, 94pp.

Schumacher, D. & Abrams, M.A. 1996. Hydrocarbon migration and its near surface expression. American Association Petroleum Geologists Memoir 66 Tulsa OK.

Schwab, H.L., Lee, & Twichell, D.C. Submarine Landslides: Selective Studies in the U.S. Exclusive Zone, U.S. Geological Survey Bulletin 2002.

Scott, A. R., Kaiser, W. R. & Ayers, J., W.B. 1994. Thermogenic and secondary biogenic gases, San Juan Basin, Colorado and New Mexico - Implications for coalbed gas productivity. American Association Petroleum Geologists Bulletin 78/8, 1186-1209.

Scranton, M.I. 1988. Temporal variations in the methane content of the Cariaco Trench. Deep Sea Research 35, 1511-1523.

Secor, D.T. 1965. Role of fluid pressure in jointing. American Journal Science 253, 633-646.

Sehgal, S.V., Ganerjee, & Chandra, K. 1991. Gas hydrate- A review. ONGC Bulletin 28, 179-202.

Sergiyenko, S.I. & Maydak, V.I. 1982. Formation conditions for heavy oil pools and possible hydrate formation in the northeastern part of the Timan-Pechora province: Geologiya Nefti i Gaza 8, 33-35.

Seuss, E., von Huene, R., et al., 1988. Procedings ODP Interim Reports 112. College Station, TX (Ocean Drilling Program).

Severinghaus, J.P. & Brook, E.J. 1999. Abrupt climate change at the end of the last glacial period inferred from trapped air in polar ice: Science 286, 930-934.

Sharp, M., Parkes, J., Cragg, B., Fairchild, I.J., Lamb, H. & Tranter, M. 1999. Widespread bacterial populations at glacier beds and their relationship to rock weathering and carbon cycling. Geology 27, 107 - 110.

Sheriff, R.E. & Geldart, L.P. 1995. Exploration Seismology, Cambridge University Press, Cambridge, U.K., 592pp.

Shipboard Scientific Party. 1985. Site 570 (Leg 84). In: von Huene, R., et al. (eds). Proceedings, Deep Sea Drilling Project, Initial Reports, Washington D.C., U.S. Government Printing Office, 67, 283-336.

Shipboard Scientific Party. 1986. Sites 614-624 (Leg 96). In: Bouma, A.H., et al. (eds). Proceedings, Deep Sea Drilling Project, Initial Reports, Washington D.C., U.S. Government Printing Office, 96, 3-424.

Shipboard Scientific Party. 1990. Site 796 (Leg 127). In: Tamake, K., et al. (eds). Proceedings, Ocean Drilling Program, Initial Reports, College Station, Texas, 127, 247-322.

Shipboard Scientific Party. 1991. Site 808 (Leg 128). In: Taira, A., et al., Proceedings, Ocean Drilling Program, Initial Reports, College Station, Texas, 131, 71-269.

Shipboard Scientific Party. 1994. Sites 892 and 889 (Leg 146). In: Westbrook, G.K., et al. (eds). Proceedings, Ocean Drilling Program, Initial Reports, College Station, Texas, 146, 301-396.

Shipboard Scientific Party. 1996. Sites 994, 995, and 997 (Leg 164). In: Paull, C.K., et al. (eds). Proceedings, Ocean Drilling Program, Initial Reports, College Station, Texas, 164, 99-623.

Shipboard Scientific Party, 1997. Site 1041. In: Kimura, G., Silver, E., Blum, P., et al. (eds). Proceedings of the Ocean Drilling Program, Initial Reports, College Station, Texas 170, 153-188.

Shipboard Scientific Party 1999. Leg 178 Summary: Antarctic glacial history and sea-level change. In: Barker, P.F., Camerlenghi, A., Acton, G.D., et al., Proceedings ODP, Initial Tepts., 178: College Station, TX (OCean Drilling Program), 1-58.

Shipley, T.H. & Didyk, B.M. 1982. Occurrence of methane hydrates offshore southern Mexico, in Watkins, J.S., Moore, J.C., et al., Initial Reports of the Deep Sea Drilling Project 66. U.S. Government Printing Office, 547-555.

Shipley, T.H., Houston, M.H., Buffler, R.T., Shaub, F.J., McMillen, K.J., Ladd, J.W. & Worzel, J.L. 1979. Seismic reflection evidence for widespread occurrence of possible gas-hydrate horizons on continental slopes and rises, American Association Petroleum Geologists Bulletin 63, 2204-2213.

Shipley, T., Stoffa, P. & Dean, D. 1990. Underthrust sediments, fluid migration paths, and mud volcanoes associated with the accretionary wedge off Costa Rica: Middle America Trench. Journal Geophysical Research 95/B6, 8743-8752, .

Shosa, J. & Cathles, L. 1996. Capillary exit pressure as a basin sealing mechanism. (abstract), AAPG Annual Convention, San Diego, Program with Abstracts. Tulsa, Oklahoma, A129.

Shpakov, V.P., Tse, J.S., Tulk, C.A., Kvamme, B. & Belosludov, V.R. 1998. Elastic moduli calculation and instability in structure I methane clathrate hydrate, Chemistry Physics Letters 282, 107-114.

Sieburth J.M. 1993. C1 bacteria in the watercolumn of Chesapeake Bay, USA. I Distribution of sub-populations of O2-tolerant, obligately anaerobic, methanylotrophic methanogens that occur in microniches reduced by the bacterial consortium. Marine Ecology Progress Series 95, 67-80.

Silfer, J.A., Engel, M.H., Macko, S.A. & Jumeau, E.J. 1991. Stable carbon isotope analysis of amino acid enantiomers by conventional isotope ratio mass spectrometry and combined gas chromatography/isotope ratio mass spectrometry. Analytical Chemistry 63, 370-374.

Sills, G.C., Wheeler, S.J., Thomas, S.D. & Gardner, T.N. 1991. Behaviour of offshore soils containing gas bubbles. Geotechnique 41, 227-241.

Sim, R. & Judd, A.G. 1995. Simple gas migration modelling. In: Wever, T.F. (ed). Proceedings Workshop on Modelling Methane-rich Sediments of Eckernförde Bay, 26-30 June 1995, Kiel, Germany. FWG Report 22, 35-38.

Sinclair, J.L. & Ghiorse, W.C. 1989. Distribution of aerobic bacteria, protozoa, algae and fungi in deep subsurface sediments. Geomicrobiology Journal, 15-31.

Singh, S.C. & Minshull, T.A. 1994. Velocity structure of a gas hydrate reflector at Ocean Drilling Program site 889 from global seismic waveform inversion: Journal Geophysical Research 99, 24,221-24,233.

Singh, S.C., Minshull, I.A. & Spence. G.D. 1993. Velocity structure of a gas hydrate reflector. Science 260, 204-207.

Skogseid, J. & Eldholm, O. 1987. Early Cenocoic crust at the Norwegian continental margin and conjugate Jan Mayen Ridge. Journal of Geophysical Research 92(B11), 11,471-11,491.

Skogseid, J. & Eldholm, O. 1989. Vøring Plateau continental margin: seismic interpretation, stratigraphy, and vertical movements. In: Eldholm, O., Thiede, J., Taylor, E., et al. (eds). Proceedings of the Ocean Drilling Program, Scientific Results 104, 993-1030.

Skogseid, J., Planke, S., Faleide, J.I., Pedersen, T. Eldholm, O. & Neverdal, F. NE Atlantic continental rifting and volcanic margin formation, Dynamics of the Norwegian Margin, Geological Society of London Special Publication, 167, (in press).

Sloan, E.D. 1990. Clathrate hydrates of natural gases, Marcel Dekker, New York, 641pp.

Sloan, E.D. 1998. Clathrate hydrate of natural gases, Second Edition: Marcel Dekker Inc., Publishers, New York, New York, 705pp.

Sloan, E.D. 1998. Gas Hydrates: Review of Physical/Chemical Properties. Energy & Fuels 12, 191-196.

Smith, J.H.S. & Wennekers, J.H.N. 1977. Geology and hydrocarbon discoveries of Canadian Arctic Islands. American Association of Petroleum Geologists Bulletin, 61/1, 1-26.

Smith, S.L. & Judge, A.S. 1993. Gas hydrate database for Canadian arctic and selected east coast wells: Geological Survey of Canada Open File Report 2746, 7pp.

Smith, S.L. & Judge, A.S. 1995. Estimates of methane hydrate volumes in the Beaufort-Mackenzie region, Northwest Territories: Geological Survey of Canada, Current Research 1995-B, 81-88.

Soloviev, V.A. & Ginsburg, G.D. 1997. Water segregation in the course of gas hydrate formation and accumualtion in submarine gas-seepage fields. Marine Geology 137, 59-68.

Soter, S. 1999. Macroscopic seismic anomalies and submarine pockmarks in the Corinth-Patras rift, Greece. Tectonophysics 308, 275-290.

Spence, G.D., Minshull, T.A. & Fink, C. 1995. Seismic structure of a marine methane hydrate reflector off Vancouver Island: ODP Scientific Results 146, 163-174.

St .John, B. 1984. Antarctica-Geology and hydrocarbon potential: In Halbouty, M.T. (ed). Future petroleum provinces of the World, Tulsa, OK, American Association Petroleum Geologists Memoir 40, 55-100.

Stein, C.A. & Stein, S. 1993. Constraints on Pacific Midplate Swells from Global Depth-Age and Heat Flow - Age Models, in: The Mesozaic Pacific: Geology, Tectonics and Volcanism, Geophysical Monograph 77, 53-76.

Stern, L.A., Circone, S., Kirby, S.H., Pinkston, J.C. & Durham, W.B. 1998c. Dissociation of methane hydrate at 0.1 MPa, and short term preservation by rapid depressurization. EOS 79/45, 462.

Stern, L.A., Hogenboom, D.L., Durham W.B., Kirby S.H. & Chou I-M. 1998b. Optical cell evidence for superheated ice under gas-hydrate-forming conditions, Journal Physical Chemistry B 102/15, 2627-2632.

Stern, L.A., Kirby, S.H. & Durham, W.B. 1996, Peculiarities of methane clathrate hydrate formation and solid-state deformation, including possible superheating of water ice, Science, 273, 5283, 1843-1848.

Stern, L.A., Kirby, S.H. & Durham, W.B. 1998a. Polycrystalline methane hydrate: synthesis from superheated ice, and low-temperature mechanical properties, Energy & Fuels, 12/2, 201-211.

Stetter, K. O. 1996. Hyperthermophilic procaryotes. FEMS Microbiology Reviews 18, 149-158.

Stevens, T.O. & Mckinley, J.P. 1995. Lithoautotrophic microbial ecosystems in deep basalt aquifers. Science 270, 450-454.

Stoll, R.D. & Bryan, G.M. 1979. Physical properties of sediments containing gas hydrates. Journal Geophysical Research 84, 15,101-15,116.

Stoll, R.D., Ewing, J.I. & Bryan, G.M. 1971. Anomalous wave velocities in sediments containing gas hydrates: Journal of Geophysical Research 76, 2090-2094.

Suess, E., Bohrmann, G., Greinert, J. & Lausch, E. 1999. Flammable ice: Scientific American 281/5, 76-83.

Suess, E., Torres, M.E., Bohrmann, G., Collier, R.W., Grientner, J., Linke, P., Rehter, G., Trehu, A., Wallmann, K., Winckler, G. & Zulegger, E. 1999b, Gas hydrate destabilization: enhanced dewatering, benthic material turnover, and large methane plumes at the Cascadia convergent margin: Earth Planetary Science Letters 170, 1-15.

Summerhayes, C.P., Bornhold, B.D. & Embley, R.W. 1979. Surficial slides and slumps on the continental slope and rise of South West Africa: A reconnaissance study, Marine Geology 31, 265-277.

Taira, A., Hill, I., Firth, J.V. & Shipboard Scientific Party. 1991. Proceedings ODP, Initerum Reports 131, Ocean Drilling Program, College Station, TX.

Tamaki, K., Pisciotto, K., Allan, J. & Shipboard Scientific Party. 1990. Proceedings ODP Initerum Reports 127, Ocean Drilling Program, College Station, TX.

Tanahashi, M., Eittreim, S. & Wannesson, J. 1994. Seismic stratigraphic sequences of the Wilkes Land margin: Terra Antartica 1, 391-393.

Tanahashi, M., Saki, T. Oikawa, N. & Sato, S. 1987. An interpretation of the multichannel seismic reflection profiles across the continental margin of the Dumont d'Urville Sea, off Wilkes Land, East Antarctica. In: Eittreim, S.L. & Hampton, M.A. (eds). The Antarctic continental margin: Geology and Geophysics of offshore Wilkes Land. CPCEMR Earth Science Series 5A, Houston, Texas, 1-14.

Taylor, A.E. 1988. A constraint to the Wisconsinan glacial history, Canadian Arctic Archipelago: Journal of Quaternary Science 3/1, 15-18.

Taylor, A.E., Burgess, M., Judge, A.S. & Allen, V.S. 1982. Canadian geothermal data collection-northern wells, 1981: Earth Physics Branch, EMR Canada, Geothermal Series 13.

Taylor, A.E., Dallimore, S.R. & Judge, A.S. 1996. Late Quaternary history of the Mackenzie-Beaufort region, Arctic Canada, from modeling of permafrost temperatures: Canadian Journal of Earth Sciences 33, 62-71.

Taylor, M.H., Dillon, W.P. & Pecher, I.A. 2000. Trapping and migration of methane associated with the gas hydrate stability zone at the Blake Ridge Diapir: new insights from seismic data, Marine Geology 164, 79-89.

Taylor, B., Huchon, P., Klaus, A., et al. 1999. Proceedings ODP Init Reports 180 [CD-ROM]. Ocean Drilling Program, Texas A&M University, College Station, TX 77854-9547, USA.

Taylor, A.E., Judge, A.S. & Allen, V. 1988. The automatic well temperature measuring system installed at Cape Allison C-47, offshore well, Arctic Islands of Canada, 2. data retrieval and analysis of the offshore regime: Journal of Canadian Petroleum Technology.

Theide, J., Altenbach, A., Bleil, U., Mudie, P., Pfirman, S., Sundvor, E. et al. 1990. Properties and history of the central eastern Arctic sea floor. Polar Record 26, 1-6.

Thiede, J., Clark, D.L. & Herman, Y. 1990. Late Mesozoic and Cenozoic paleoceanography of the northern polar oceans, in: The Arctic Ocean region,. Grantz, A, Johnson, L & Sweeney, J.F. (eds). The geology of North America. Geological Society America, 427-458.

Thiel, V., Peckmann, .J., Seifert, R., Wehrung, Pl, Reitner, J. & Michaelis, W. 1999. Highly isotopically depleted isoprenoids: molecular markers for ancinet methane venting. Geochimica et Cosmochimica Acta 63 (23/24), 3959-3966.

Thomas, E. 1998. Biogeography of Late Paleocene benthic foraminiferal extinction. In: Aubrey, M.-P. et al. (eds). Late Paleocene-Early Eocene Biotic and Climatic Events in the Marine and Terrestrial record. Columbia University Press, 214-243.

Thorpe R.B., Pyle, J.A. & Nisbet. E.G., 1998. What does the ice-core record imply concerning the maximum climatic impact of possible gas hydrate release at Termination 1A? In: Henriet, J-P. & Mienert, J. (eds). Gas hydrates: Relevance to world margin stability and climate change. Geological Society London Special Publication 137, 319-326.

Tinivella U., Lodolo E., Camerlenghi A. & Boehm, G. 1998. Seismic tomography study of a bottom simulating reflector off the South Shetland Islands (Antarctica). In: Henriet, J.-P. & Mienert, J. (eds): Gas hydrates: Relevance to world margin stability and climate change. Geological Society, London Special Publication 137, 141-151.

Tinivella, U & Accainao, F. 2000. Compressional velocity structure and Poissons's ratio in marine sediments with gas hydrate and free gas by inversion of reflected and refracted seismic data (South Shetland Islands, Antarctica). Marine Geology 164, 13-27.

Tohidi, B., Danesh, A. & Todd, A.C. 1995. Modelling single and mixed electrolyte solutions and its applications to gas hydrates, Chemical Engineering Research Design 73, 464-472.

Tohidi, B., Danesh, A. & Todd, A.C. 1997. On the mechanism of gas hydrate formation in subsea sediments. (abs) American Chemical Society 213, 37-39.

Tomlinson J.S., Pudsey C.J., Livermore R.A., Larter R.D. & Barker P.F. 1992. Long- Range Side Scan Sonar (GLORIA) Survey of the AntarcticPeninsula Pacific Margin: In Yoshida, Y., Kaminuma, K. & Shiraishi K., (eds.), Recent Progress in Antarctic Earth Science, Terra Sci Publ., Tokyo, 423-429.

Tono, S. 1998. Japanese strategy for the R&D of mehtane hydrate: resources evaluation and operational point of view. Abstracts, Gas Hydrates Symposimn, Geoscience 98 (Keele, U.K), 10.

Tono, S. 1998. Preface. In: JNOC-TRC, proceedings of the International Symposium on Methane Hydrates, resources in the near Future?, Chiba City, Japan, October 20-22, 1998.

Trehu, A., Torres, M.E., Moore, G.F., Suess, E. & Bohrmann, G. 1999. Temporal and temporal evolution of a gas hydrate-bearing accretionary ridge on the Oregon continental margin: Geology 27, 939-942.

Trehu, A.M., Lin, G., Maxwell, E., and Goldfinger, C. 1995. A seismic reflection profile across the Cascadia subduction zone offshore central Oregon: New constraints on methane distribution and crustal structure: Journal Geophysical Research 100, 15101-15116.

Trofimuk, A.A., Cherskiy, N.V. & Tsarev V.P. 1973. Accumulation of natural gases in zones of hydrate formation in the hydrosphere: Doklady Akademii Nauk SSSR 212, 931-934.

Trofimuk, A.A., Cherskiy, N.V. & Tsarev V.P. 1975. The biogenic methane resources of the oceans: Doklady Academii Nauk SSSR 225, 936-943.

Trofimuk, A.A., Cherskiy, N.V., & Tsarev, V.P. 1977. The role of continental glaciation and hydrate formation on petroleum occurrences, in Meyer, R.F., ed., Future Supply of Nature-Made Petroleum and Gas: Pergamon Press, New York, New York, 919-926.

Trofimuk, A.A., Cherskiy, N.V. & Tsarev V.P. 1979. The gas-hydrate sources of hydrocarbons: Priroda. 1, 18-27.

Tryon, M.D., Brown, K.M., Torres, M.E., Trehu, A.M., McManus, J. & Collier, R.W. 1999. Measurements of transience and flow cycling near episodic methane gas vents, Hydrate Ridge, Cascadia. Geology 27, 1075-1078.

Tse, J.S. & White, M.A. 1988. Origin of glassy crystalline behavior in the thermal properties of clathrate hydrates: A thermal conductivity study of tetrahydrofuran hydrate, Journal Physical Chemistry 92, 5006-5011.

Tsuji, Y, Furutani, A., Matsuura, S. & Kanamori, K. 1998. Exploratory Surveys for Evaluation of Methane Hydrates in the Nankai Trough Area, Offshore Central Japan. In: JNOC-TRC, proceedings of the International Symposium on Methane Hydrates, resources in the near Future?, Chiba City, Japan, October 20-22, 1998, 15-26.

Tucholke, B.E., Bryan, G.M. & Ewing, J.I. 1977. Gas-hydrate horizons detected in seismic-profiler data from the western North Atlantic: American Association of Petroleum Geologists Bulletin 61, 698-707.

Turoff, M. & Hiltz, S.R. 1995. Computer-Based Delphi Processes, in: Adler, M. & E. Ziglio, E. (eds). Gazing into the Oracle: The Delphi method and its application to social policy and public health. Jessica Kingsley Publishers,London, 56-88pp.

Twichell, D.C. & Cooper, A.K. 2000. Relation between seafloor failures and gas hydrates in the Gulf of Mexico: a regional comparison: Program, AAPG Annual Meeting, New Orleans, LA, April 16-19, A150.

Tzirita, A. 1992. In situ detection of natural gas hydrates using electrical and thermal properties. Offshore Technology Research Center. 220pp.

Vågnes, E. 1997. Uplift at thermo-mechanically coupled ocean-continent transforms: Modeled at the Senja Fracture Zone, southwestern, Barents Sea. Geo-Marine Letters, 17, 100-109.

Vasil'ev, V.G., Makogon, Yu. F., Trebin, F.A., Trofimuk, A.A. & Cherskiy, N.V. 1970. The property of natural gases to occur in the Earth crust in a solid state and to form gas hydrate deposits: Otkrytiya v SSSR 1968-1969, 15-17.

Veevers, J.J. 1987. The conjugate continental margins of Antarctica and Australia: In: Eittreim, S.L. & Hampton, M.A. (eds). The Antarctic continental margin: geology and geophysics of offshore Wilkes Land. Circum-Pac. Council for Energy and Natural Research. Earth Science Series 5A, Houston, Texas, 45-73.

Vogt, P.R. Magnetic anomalies and crustal magnetization. In: Vogt, P.R. & Tucholke, B.E. (eds). The Geology of North America, M, The Western North Atlantic Region. Geological Society of America, 229-256.

Vogt P.R., et al. 1997. Haakon Mosby mud volcano provides unusual example of venting. EOS 78/48, 556-557.

Vogt, P.R., Crane, K., Sundvor, E., Max, M.D., & Pfirman, S.L. 1994. Methane-generated (?) pockmarks on young, thickly sedimented oceanic crust in the Arctic: Vestnesa Ridge, Fram Strait, Geology 22, 255-258.

Vogt, P.R., Gardner, J. & Crane, K. 1999. The Norwegian-Barents-Svalbard (NBS) continental margin: Introducing a natural laboratory of mass wasting, hydrates, and ascent of sediment, pore water, and methane. Geo-Marine Letters 19, 2–21.

Vogt, P., Gardner, J., Crane, K., Sundvor, E., Bowles, F. & Cherkashev, G. 1999. Ground-Truthing 11-12 kHz sidescan sonar imagery in the Norwegian-Greenland Sea- Part I: Pockmarks on the Vestnesa Ridge and Storegga Slide margin, Geo-Marine Letters 19, 97-110.

Vogt, P.R. & Sundvor, E. 1996. Heat flow highs on the Norwegian-Barents-Svalbard continental slope; deep crustal fractures, dewatering, or "memory in the mud?", Geophysical Research Letters 23, 3571-3574.

Von Herzen, R. & Maxwell, A.E. 1959. The measurement of thermal cunductivity of deep-sea sediments using the needle-probe method. Journal Geophysical Research 64, 1557-1563.

Von Herzen, R. & Maxwell, A.E. 1959. The measurement of thermal cunductivity of deep-sea sediments using the needle-probe method, Journal Geophysical Ressearch 64, 1557-1563.

von Huene, R. & Pecher, I. 1999. Vertical tectonics and the origins of BSRs along the Peru margin: Earth Planetary Science Letters 166, 47-55.

von Huene, R., Aubouin, J., et al. 1985. Initerum Reports DSDP, 84, Washington, D.C. U.S. Govt. Printing Office.

von Huene, R., Pecher, I.A. & Gutscher, M.-A. 1996. Development of the accretionary prism along Peru and material flux after subduction of Nazca Ridge. Tectonics 15, 19-33.

von Reden, K.F., McNichol, A.P., Peden, J.C., Elder, K.L., Gagnon, A.R. & R.J. Schneider, RJ. 1997. AMS measurements of the ^{14}C distribution in the Pacific Ocean. Nuclear Instruments and Methods B 123, 438-442.

Vorobyova, E., Soina, V., Gorlenko, M., Minkovskaya, N., Zalinova, N., Mamukelashvili, A., Gilichinsky, D., Rivkina, E. & Vishnivetskaya, T. 1997. The deep cold biosphere: facts and hypothesis. FEMS Microbiology Reviews 20, 277-290.

Vorren, T.O., Bergsager, E., Dahl-Stamnes, Ø.A., Holter, E., Hohansen, B., Lie, E. & Lund, T.B. (eds). 1993 Arctic Geology and Petroleum Potential. Norwegian Petroleum Society Special Publication 2, 751pp.

Vorren, T.O., Laberg, J.S., Blaume, F., Dowdeswell, J.A., Kenyon, N.H., Mienert, J., Rumohr, J. & Werner, F. 1998. The Norwegian-Greenland Sea continental margins: morphology and late Quaternary sedimentary processes and environment, Quaternary Science Reviews 17 (1-3), 273-302.

Wagner, G., Frette, V., Birovljev, A., Jossang, T., Meakin, P & Feder, J. 1996. Fractal structures in secondary migration. In: Jamtveit, B. & Yardley, B.W.D. (eds). Fluid Flow and Transport in Rocks. Chapman & Hall, 203-212.

Waite, W., et al. 2000. Personal Communication to Dr. J. Dvorkin, 30 April, 2000.

Waite, W.F., Helgerud, M.B., Nur, A., Pinkston, J., Stern, L.A., Kirby, S.H. &. Durham, W.B. 1998. First measurements of P- and S-wave speed on pure methane gas hydrate, EOS79, F463.

Waite, W.F., Helgerud, M.B., Nur, A., Pinkston, J.C., Stern, L. A., Kirby, S.H. & Durham, W.B. 2000. Laboratory measurements of compressional and shear wave speeds through pure methane hydrate, Annals of the New York Academy of Sciences, 3rd International Conference on Gas Hydrates, (in press) 8 pp.

Wakishima, R., Imazato, M., Nara, M., Aumann, J.T. & Hyland, C. 1998. The development of a pressure and temperature core sampler (PTCS) for the recovery of in-situ methane hydrates. in: JNOC-TRC, proceedings of the International Symposium on Methane Hydrates, resources in the near Future?, Chiba City, Japan, October 20-22, 1998, 107-120.

Walia, R. & Hannay, D. 1999. Source and receiver geometry corrections for deep towed multichannel seismic data: Geophysical Research Letters 26, 1993-1996.

Walia, R., Y. Mi, Hyndman, R. D. & Sakai, A. 1999. Vertical seismic profile (VSP) in the JAPEX/JNOC/GSC Mallik 2L-38 gas hydrate research well, In: Scientific Results from JAPEX/JNOC/GSC Mallik 2L-38 Gas Hydrate Research Well, Mackenzie Delta, Northwest Territories, Canada, Dallimore, S.R., Uchida, T. & Collett; T.S. (eds). Geological Survey of Canada, Bulletin 544, 341-355.

Wang, K., Davis, E.E. & van der Kamp, G. 1998. Theory for the effects of free gas in subsea formations in tidal pore pressure variations and seafloor displacements. Journal Geophysical Research 103, 12339-12353.

Wang, Z. & Nur, A. 1992. Seismic and acoustic velocities in reservoir rocks, II, SEG Geophysics reprint series, SEG, Tulsa, 457pp.

Wannesson, J., Pelras, M., Petitperrin, B., Perret, M. & Segoufin, J. 1985. A geophysical survey of the Adelie margin, East Antarctica: Mar. and Petrol. Geology 2, 192-201.

Ward, B.B. 1987. Kinetic studies on ammonia and methane oxidation by *Nitrosococcus oceanus*. Archives Microbiology 147, 126-133.

Ward, B.B. & Kilpatrick, K.A. 1990. Relationship between substrate concentration and oxidation of ammonium and methane in a stratified water column. Continental Shelf Research 10, 1193-1208.

Waseda, A. 1998. Organic carbon content, bacterial methanogenesis, and accumulation of gas hydrates in marine sediments, Geochemical Journal 32, 143-157.

Watts, N.L. 1987. Theoretical aspects of cap rocks and hydrocarbon seals for single- and two-phase hydrocarbon columns. Marine & Petroleum Geology 4, 274-307.

Weaver, J.S. & Stewart, J.M. 1982. In situ hydrates under the Beaufort shelf, in French, M.H. ed., Proceedings of the Fourth Canadian Permafrost Conference, 1981: National Research Council of Canada, The Roger J.E. Brown Memorial Volume, 312-319.

Webster, M. 1994. Webster's Dictionary, Boston. Houghton Mifflin Co., 1536pp.

Wellsbury, P., Goodman, K., Cragg, B.A. & Parkes, R.J. 2000. The geomicrobiology of deep marine sediments from Blake Ridge containing methane hydrate (Sites 994, 995 and 997). Proceedings of the Ocean Drilling Program, Scientific Results 164, 379-391.

Wellsbury, P., Herbert, R.A. & Parkes, R.J. 1996. Bacterial activity and production in near-surface estuarine and freshwater sediments. FEMS Microbiology Ecology 19, 203-214.

Wellsbury, P. & Parkes, R.J. 1995. Acetate bioavailability and turnover in an estuarine sediment. FEMS Microbiology Ecology 17, 85-94.

Westbrook, G.K., Carson, B., Musgrave, R.J., et al., 1994. Procedings Ocean Drilling Program, Initerum Reports: 146/1, College Station, TX. 399-419.

Wetzel, A. 1993. The transfer of river load to deep-sea fans: a quantitative approach. American Association of Petroleum Geologists Bulletin 10, 1679-1692.

Wheeler, S.J. Sham, W.K. & Thomas, S.D. 1990. Gas pressure in unsaturated offshore soils. Canadian Geotechnical Journal 27, 79-89.

White, D.C., Davis, W.M., Nickels, J.S., King, J.D. & Bobbie, R.J. 1979. Determination of the sedimentary microbial biomass by extractable lipid phosphate. Oecologia 40, 51-62.

Whiticar, M.J. 1999. Carbon and hydrogen isotope systematics of bacterial formation and oxidation of methane. Chemical Geology 161, 291-314.

Whiticar, M.J. & Faber, E. 1989. 16. Molecular and stable isotope composition of headspace and total hydrocarbon gases at ODP Leg 104, Sites 642, 643, and 644, Vøring Plateau, Norwegian Sea. In: Eldholm, O., Thiede, J., Taylor, E., et al. (eds). Proceeding of the Ocean Drilling Program, Scientific Results 104, 327-334.

Whiticar, M.J., Faber, E. & Schoell, M. 1986. Biogenic methane formation in marine and freshwater environments: CO2 reduction vs acetate fermentation - isotope evidence. Geochim. Cosmochim Acta 50, 693-709.

Whiticar, M. Hovland, M., Kastner, M. & Sample, J.C. 1995. Organic chemistry of gases, fluids, and hydrates at the Cascadia accretionary margin: *i:n* Carson, B., Westbrook, G.K., Musgrave, R.J. & Suess, E. (eds). Proceedings Ocean Drill. Program, Scientific Results 146, 385-397.

Whitman, W.B., Coleman, D.C. & Wiebe, W.J. 1998. Prokaryotes: The unseen majority. Proceedings of the U.S. National Academy of Sciences 95, 6578-6583.

Williams, P.M. & Druffel, E.R.M. 1987. Radiocarbon in dissolved organic matter in the central Nort Pacific Ocean. Nature 330, 246-248.

Willoughby, E. & Edwards, R.N. 2000. Shear velocities in Cascadia from seafloor compliance measurements: Geophysical Research Letters (in press).

Winters W.J., Pecher, I.A., Mason, D.H., Booth, J.S. & Dillon, W.P. 2000. Physical properties of sediment containing natural and laboratory-formed gas hydrate: Program, American Association Petroleum Geologists Annual Meeting, New Orleans, LA, April 16-19, A159.

Winters W.J., Pecher, I.A., Booth, J.S., Mason, D.H., Relle, M.K. & Dillon, W.P. 1999. Properties of samples containing natural gas hydrate from the JAPEX/JNOC/GSC Mallik 2L-38 gas hydrate research well, determined using Gas Hydrate And Sediment Test Laboratory Instrument (GHASTLI), *in* Dallimore, S.R., Uchida, T., & Collett, T.S., eds, Scientific Results from JAPEX/JNOC/GSC Mallik 2L-38 Gas Hydrate Research Well, Mackenzie Delta, Northwest Territories, Canada, Geological Survey of Canada Bulletin 544, 241-250.

Wood, W.T., Holbrook, W.S. & Hoskins, H. 2000. In situ measurements of P-wave attenuation in the methane hydrate and gas-bearing sediments of the Blake Ridge, In: Proceedings Ocean Drilling Program Scientific Results, College Station, Texas.

Wood, W.T. & Ruppel, C. 2000. Seismic and thermal investigations of the Blake Ridge gas hydrate area: a synthesis, in: Paull, C.K., Matusumoto, R., Wallace, P.J. & Dillon, W.P. (eds). Proceedings of the Ocean Drilling Program, Scientific Results 164, 253-264.

Wright, J.A. & Louden, K.E. 1989. CRC hand book of Seaffloor heat flow, CRC Press Inc., Florida.

wvusd. 2000. <http://dbhs.wvusd.k12.ca .us/Chem-History/Faraday-Chlorine-1823.html>.

Xu, W. & Ruppel, C. 1999. Predicting the occurrence, distribution, and evolution of methane gas hydrate in porous marine sediments. Journal of Geophysical Research 104, 5081-5096.

Yakushev, V.S. & Collett, T.S. 1992. Gas hydrates in Arctic regions: risk to drilling and production. Proceedings, Second Int. Offshore and Polar Eng. Conf., San Francisco, June 14-19, 669-673.

Yamano, M., Foucher, J.-P., Kinoshita, M., Fisher, A., Hyndman, R.D. & ODP Leg131 Shipboard Scientific Party. 1992. Heat flow and fluid flow regime in the western Nankai accretionary prism. Earth Planetary Science Letters 109, 451-462.

Yamano, M., Honda, S. & Uyeda, S. 1984 Nankai Trough; a hot trench? Marine Geophysical Research 6, 187-203.

Yamano, M., Uyeda, S., Aoki, Y. & Shipley, T.H. 1982. Estimates of heat flow derived from gas hydrates. Geology 10, 339-343.

Yamazaki, A. 1997. MITI's plan of R&D for technology of methane hydrate development as domestic gas resources. Production, Treatment and Underground Storage of Natural Gas, International Gas Union, 20th World Gas Conference, 355-370.

Yassir, N.A. & Bell, J.S. 1994. Relationships between pore pressures, stresses and present-day geodynamics in the Scotian Shelf, Offshore Eastern Canada. American Association Petroleum Geologists Bulletin 78, 1863-1880.

Yefremova, A.G. & Gritchina, N.D. 1981. Gas hydrates in sediments beneath seas and the problem of their exploitation: Geologiya Nefti i Gaza 2, 32-35.

Yefremova, A.G. & Zhizhchenko, B.P. 1974. Occurrence of crystal hydrates of gas in sediments of modern marine basins: Doklady Akademii Nauk SSSR 214, 1179-1181.

Yuan, J. & Edwards, R.N. 2000. Towed seafloor electromagnetics and assessment of gas hydrate deposits, Geophysical Research Letters, (in press).

Yun, J.W., Orange, D.L. & Field, M.E. 1999. Subsurface gas offshore northern Californoia and its link to submarine geomorphology. Marine Geology 154, 357-368.

Yuan, T., Hahar, K.S., Roop Chand, R.D., Hyndamn, G.D. & Chapman, N.R. 1998. Marine gas hydrates: Seismic observations of bottom-simulating reflectors off the west coast of Canada and the east coast of India. Geohorizons, March, 1998, 5-15.

Yuan, T., Hyndman, R.D., Spence, G.D. & Desmons, B. 1996. Seismic velocity increase and deep-sea gas hydrate concentration above a bottom-simulating reflector on the northern Cascadia continental slope, Journal Geophysical. Research 101 13,655-13,671.

Yuan, T., Spence, G.D., Hyndman, R.D., Minshull, T.A. & Singh, S.C. 1999. Seismic velocity studies of a gas hydrate bottom-simulating reflector on the northern Cascadia continental margin; amplitude modelling and full waveform modelling: Journal Geophysical Research 104, 1179-1191.

Zakrewski, M. & Handa, Y.P. 1993. Thermodynamic properties of ice and of tetrahydrofuran hydrate in confined geometries. Journal of Chemical Thermodynamics 25, 631-637.

Zatsepina, O. & Buffett, B.A. 1997. Phase equilibrium of gas hydrate: implications for the formation of hydrate in the deep sea-floor. Geophysical Research Letters 24, 1567-1570.

Zatsepina, O.Y. & Buffett, B.A. 1998. Thermodynamic conditions for the stability of gas hydrate in the seafloor. Journal Geophysical Research 103, 24127-24139.

Zelles, L. Bai, Q.Y., Beck, T. & Beese, F. 1992. Signature of fatty acids in phospholipids and lipopolysaccharides as indicators of microbial biomass and community structure in agricultural soils. Soil Biology and Biochemistry. 24:317-323.

Zengler, K., Richnow, H. H., RosselloMora, R., Michaelis, W. & Widdel, F. 1999. Methane formation from long-chain alkanes by anaerobic microorganisms. Nature 401, 266-269.

Zhang, W, Durham, W.B., Stern, L.A. & Kirby, S.H. 1999. Experimental deformation of methane hydrates: new results. EOS 80/17, S337.

Zhang, W. et al. 1999: EOS, American Geophysical Union Spring Meeting Suppliment 80, S337.

Zimmerman, R.W. & King, M.S. 1986. The effect of the extent of freezing on seismic velocities in unconsolidated permafrost, Geophysics 51, 1285-1290.

Zobell, C.E. 1963. Organic geochemistry of sulfur. In: Organic Geochemistry. Breger, I.A. (ed). Monograph 16, Earth Sciences Series. 543-528, Macmillin, New York.

Zoeppritz, K., Über Reflexion und Durchgang seismischer Wellen durch Unstetigkeitsflächen über Erdbebenwellen VII B, Nachr. der königl. Ges. der Wiss. zu Göttingen, K1, 57-84, 1919.

Zuehlsdorff, L., Spiess, V., Huebscher, C., Villinger, H. & Rosenberger, A. 2000. BSR occurrence, near surface reflectivity anomalies and small scale tectonism imaged in a multi-frequency seismic data set from the Cascadia accretionary prism. Geology Rundschau 88, 655-667.

Zwart, G., Moore, J.C. & Cochrane, G.R. 1996. Variations in temperature gradients identify active faults in the Oregon accretionary prism. Earth Planetary Science Letters 139, 485-495.

LIST OF CONTRIBUTING AUTHORS

Square brackets are: [telephone / fax].
E-mail addresses are indicated by: <***@***>.
<http: = website>

Dr. Karin Andreassen
Department of Geology
University of Tromsø
Dramsveien 201
N-9037 Tromsø, Norway
[77 64 44 20 / 47-77 64 56 00]
<karina@ibg.uit.no>

Dr. Christian Berndt
Department of Geology
University of Tromsø
Dramsveien 201
N-9037 Tromsø, Norway
[47-77 64 64 79 / 47-77 64 56 00]
<Christian.Berndt@bg.uit.na>

Ing. Klaas J. Bil
Shell International Exploration and Production
Research and Technical Services
P.O. Box 60
2280 AB Rijswijk
The Netherlands
[31703112852 / 31703113366]
< k.j.bil@siep.shell.com>

Dr. Peter G. Brewer
Monterey Bay Aquarium Research Institute
P.O. Box 628
7700 Sandholt Road
Moss Landing CA 95039-9644
[408--775-1706 / (408)-775-1620]
< brpe@mbari.org>
Home Page: http://www.mbari.org

Dr. Angelo Camerlenghi
Istituto Nazionale di Oceanografia e di Geofisica Sperimentale - OGS
P.O. Box 2011 I-34016 Trieste Italy
Borgo Grotta Gigante 42/c I-34010 Sgonico (TS) Italy
[39-040-2140253: 39-040-21401 (main desk) / 39-040-327307] telex: 460329 OGS I
<acamerlenghi@ogs.trieste.it>
<http://www.ogs.trieste.it/GDL/INTE>

Dr. Jeffrey P. Chanton
Department of Oceanography
Florida State University
Tallahassee FL 32306
[850-644-7493 / 850-644-2581]
<jchanton@mailer.fsu.edu>

Dr.N. Ross Chapman
School of Earth and Ocean Sciences
University of Victoria
Victoria, B.C., Canada V6T 1W5
[250 472-4340 /]
<chapman@uvic.ca>

Dr. Michael Czarnecki
Naval Research Laboratory
Washington DC 20375-5350
[202-767-0522 / 202-767-0167]

Dr. Susan Circone
Earthquake Hazards Team
U. S. Geological Survey
345 Middlefield Rd, MS/ 977
Menlo Park, CA 94025
[650-329-5674 / 650-329-5163]
<scircone@usgs.gov>

Dr. M. B. Clennell
Centro de Pesquisa em Geofísica e Geologia
Universidade Federal da Bahia
Rua Caetano Moura, 123,
Campus Universitário de Ondina
[55-71-2370408/332 94 33 / 55-71-2473004]
<clennell@cpgg.ufba.br>
web: www.cpgg.ufba.br

Dr. R. B.Coffin
Richard B. Coffin, Ph.D.
Code 6115, Chemistry Division
Naval Research Laboratory
4555 Overlook Ave., SW
Washington, DC 20375
[202-767-0065 / 202-404-8515]
<rcoffin@ccsalpha3.nrl.navy.mil>

Dr. Timothy S. Collett
U.S. Geological Survey
Denver Federal Center
Box 25046, MS-939
Denver, CO 80225
[303.236.5731 / 303.236.0459]
<tcollett@usgs.gov>

Dr. Scott R. Dallimore,
Geological Survey of Canada
9860 W. Saanich Rd.
Sidney, B.C., V8L 4B2
[. / .]
<sdallimo@NRCan.gc.ca>

Dr. William P. Dillon
Coastal and Marine Geology Program
U.S. Geological Survey
Quissett Campus
Woods Hole, MA 02543
[508-457-2224: 508-548-8700 / 508-457-2310]
<bdillon@usgs.gov>

Dr. William B. Durham
U.C. Lawrence Livermore National Laboratory
P.O. Box 808
Livermore, CA 94550
[925-422-7046 / 925-423-1057]
<durham1@llnl.gov>

Dr. Jack Dvorkin
Geophysics Department
Stanford University
Stanford, CA 94305-2215
[650-725-9296 / 650-725-7344]
<jack@pangea.Stanford.edu>

Dr. R. Nigel Edwards
Department of Physics
University of Toronto
Toronto, Ontario, Canada M5S 1A7
[410-978-2267 /]
<edwards@geophy.physics.utoronto.ca>

Dr. Dave S. Goldberg
Director, Borehole Research
Lamont-Doherty Earth Observatory
of Columbia University
Palisades, New York 10964, U.S.A.
[914-365-8674 / 914-365-3182]
<goldberg@ldeo.columbia.edu>
<http://www.ldeo.columbia.edu/BRG/>

Dr. Kenneth S. Grabowski
[2]Naval Research Laboratory
Code 6370
4555 Overlook Ave, S.W.
Washington DC 20375
Phone: 202-767-5738, Fax: 202-767-5301
email: grabowski@nrl.navy.mil

Dr. Bilal U. Haq
Director of Marine Geology and Geophysics
National Science Foundation
Division of Ocean Sciences
4201 Wilson Blvd.
Arlington VA 22230
703-306-1586
<bhaq@nsf.gov>

Michael B. Helgerud
Geophysics Department
Stanford University
Mitchell Bldg., RM B63
Stanford, CA 94305-2215
[650-723-7910 (650-723-0166) / 650-725-7344]
< helgerud@pangea.stanford.edu>
<http://pangea.stanford.edu/~helgerud/index.html>

Dr. W. Steven Holbrook
Associate Professor
Dept. of Geology and Geophysics
University of Wyoming
Laramie, WY 82071
[307-766-2427 / 307-766-6679]
<steveh@hekla.gg.uwyo.edu>
<steveh@uwyo.edu>
<http://azure.whoi.edu/~steveh/myinfo.html>

Dr. Martin Hovland
Statoil,
Stavanger
N4001 Norway.
mhovland@statoil.no

Dr. Roy D. Hyndman
Pacific Geoscience Centre
Geological Survey of Canada
Sidney, B.C., Canada V8L 4B2
[(250-363-6428 /]
<hyndman@hyndman@pgc.nrcan.gc.ca>
<http://www.pgc.nrcan.gc.ca/>

Dr. Alan Judd,
Centre for Marine and Atmospheric Sciences
School of Sciences
University of Sunderland
Benedict Building, St. Georges Way
Sunderland, SR2 7BW, U.K.
[+44-191 515 2729 / +44-191 515 2741]
<alan.judd@sunderland.ac.uk>

Dr. Stephen H. Kirby
Earthquake Hazards Team
U. S. Geological Survey
345 Middlefield Rd, MS/ 977
Menlo Park, CA 94025
[650-329-4847 / 650) 329-5163]
<skirby@usgs.gov>

Dr. Keith A. Kvenvolden
USGS, MS 999
345 Middlefield Rd.
Menlo Park, CA 94025
[650-329-4196 /]
<kk@octopus.wr.usgs.gov>

Emanuele Lodolo
Istituto Nazionale di Oceanografia e di
Geofisica Sperimentale - OGS
Osservatorio Geofisico Sperimentale
P.O. Box 2011 I-34016 Trieste Italy
Borgo Grotta Gigante 42/c I-34010 Sgonico
(TS) Italy
[339-040-21401 / 39-040-327307]

David H. Mason
384 Woods Hole Road
Woods Hole, MA 02543 USA
[508-457-2358,/ 508-457-2310]

Dr. Michael D. Max
Marine Desalination Systems (Research)
Suite 302, 1121 Connecticut Ave. NW
Washington DC 20036
[202-466-7337 / 202-955-5879]
[202-333-7420 / 202-333-7420]
<xeres@erols.com>

Sheila L. McDonnell
Naval Research Laboratory
Washington DC 20375-5350
[202-767-0522 / 202-767-0167]
<sheila@hp8c.nrl.navy.mil>

Prof. Dr. Jüergen Mienert
Department of Geology
University of Tromsø
Dramsveien 201
N-9037 Tromsø, Norway
[47-77 64 44 46 / 47-77 64 56 00]
<Juergen.Mienert@ibg.uit.no>
<http://www.ibg.uit.no/geologi/>

Dr. Peter R. Miles
Room 786/11
Southampton Oceanography Centre
Empress Dock
SOUTHAMPTON SO14 3ZH, United
Kingdom
[44-(0)1703-596560 / 44-(0)1703-596554]
<P.Miles@soc.soton.ac.uk>

Dr. Amos Nur
Geophysics Department
Stanford University
Stanford, CA 94305
[650-723-9526 / 650-723-1188
nur@pangea.stanford.edu

Prof. R. John Parkes
University of Bristol
Department of Earth Sciences
Wills Memorial Building
Queens Road
Bristol BS8 1RJ
U.K.
[44-0117-954-5427) / 0117-925-3385]
<J.Parkes@bris.ac.uk>

Dr. Charles Paull,
Senior Scientist
Monterey Bay Aquarium Research Institute
7700 Sandholdt Road
Moss Landing, CA 95039-9644
[831-775-1886 / 831-775-1620]
<paull@mbari.org>
<http://www.mbari.org/paull>

Dr. Ingo A. Pecher
Univ. of Texas Inst. for Geophysics
4412 Spicewood Springs Rd., Bldg. 600
Austin, TX 78759-8500, USA
[512-232-3203 / 512-471-8844]
<ingo@ig.utexas.edu>

Dr. Robert E. Pellenbarg
Chemistry Division
Naval Research Laboratory
4555 Overlook Ave., SW
Washington, DC 20375
[202-767-2479 / 202-404-8515]

Dr. Edward T. Peltzer, III
Monterey Bay Aquarium Research Institute
7700 Sandholdt Road
Moss Landing, CA 95039-9644
[831-775-1851 / 831-775-1620]
<etp3@mbari.org>
<etp3@cirrus.shore.mbari.org>
<http://www.mbari.org/~etp3>

Michael Riedel
School of Earth and Ocean Sciences
University of Victoria
Victoria, B.C., Canada V6T 1W5
[(250- 721-6193 /]
<mriedel@geosun1.seos.uvic.ca>

Dr. Carolyn D. Ruppel
Assistant Professor of Geophysic
School of Earth and Atmospheric Sciences
Georgia Tech
Atlanta, GA 30332-034,0 U.S.A.
[404 894-0231 /404 894-5638]
<cdr@piedmont.eas.gatech.edu>
<http://hydrate.eas.gatech.edu/cdr/ruppel.html

Dr. George D. Spence
School of Earth and Ocean Sci.
University of Victoria
Victoria, B.C., V8W 2Y2 CANADA
[250-721-6187 /]
< gspence@uvic.ca>
<http://geosun1.seos.uvic.ca>

Dr. Laura A. Stern
U.S.G.S.
345 Middlefield Road, MS-977
Menlo Park, CA 94025
[650-329-4811 / 650-329-5163]
<lstern@usgs.gov>

Dr. W. Ussler
Monterey Bay Aquarium Research Institute
7700 Sandholdt Road
Moss Landing, CA 95039-9644
[831-775-1879 / 831-775-1620]
<methane@mbari.org

Dr. Peter Wellsbury
University of Bristol
Department of Geology
Wills Memorial Building
Queens Road
Bristol BS8 1RJ
U.K.
[0117-928-9000 / 0117-925-3385]
<Peter.Wellsbury@bristol.ac.uk>

Dr. William F. Waite
U.S. Geological Survey
345 Middlefield Rd., MS/ 977
Menlo Park, CA 94025
[650-329-4843 / 650/329-5163]
<wwaite@usgs.gov>

William Winters
U.S. Geological Survey
Woods Hole, MA 02543
[508-457-2358 /]
<wwinters@usgs.gov>

Ms. Jian Yuan
Department of Physics
University of Toronto
Toronto, Ontario, Canada M5S 1A7
[416-978-5177 /]
<jian@geophy.physics.utoronto.ca>

Coastal Systems and Continental Margins

1. B.U. Haq (ed.): *Sequence Stratigraphy and Depositional Response to Eustatic, Tectonic and Climatic Forcing*. 1995 ISBN 0-7923-3780-8
2. J.D. Milliman and B.U. Haq (eds.): *Sea-Level Rise and Coastal Subsidence*. Causes, Consequences, and Strategies. 1996. ISBN 0-7923-3933-9
3. B.U. Haq, S.M. Haq, G. Kullenberg and J.H. Stel (eds.): *Coastal Zone Management Imperative for Maritime Developing Nations*. 1997 ISBN 0-7923-4765-X
4. D.R. Green and S.D. King (eds.): *Coastal and Marine GeoInformation Systems: Applying the Technology to the Environment*. 2001 ISBN 0-7923-5686-1
5. M.D. Max (ed.): *Natural Gas Hydrate in Oceanic and Permafrost Environments*. 2000 ISBN 0-7923-6606-9

KLUWER ACADEMIC PUBLISHERS – DORDRECHT / BOSTON / LONDON